Digital Control Systems

Digital Control Systems

Theoretical Problems and Simulation Tools

Anastasia Veloni
Nikolaos I. Miridakis

CRC Press is an imprint of the
Taylor & Francis Group, an **informa** business

CRC Press
Taylor & Francis Group
6000 Broken Sound Parkway NW, Suite 300
Boca Raton, FL 33487-2742

Printed on acid-free paper

International Standard Book Number-13: 978-1-138-03920-9 (Hardback)

Visit the Taylor & Francis Web site at
http://www.taylorandfrancis.com

and the CRC Press Web site at
http://www.crcpress.com

Professor Anastasia Veloni dedicates this book to the

honor of her parents, Nikolaos and Eystathia.

Dr. Nikolaos I. Miridakis dedicates this book to his daughter, Ioanna,

his wife, Melpomeni, and his parents, Ioannis and Panayota.

Contents

Preface

Automatic control systems represent one of the most important fields of applied science and engineering. This is due to the fact that automation is today typically related to the development of all forms of technology.

A significant breakthrough in the area of control systems is the transformation of signals such that they can be connected to computing devices. A prerequisite, and more so, a current trend, is the fact that control systems are based on systems' computational gain, which enables them to work faster, more efficiently, and with greater reliability.

This book is solely dedicated to digital control systems and aims to provide the reader with both theoretical and applied scientific knowledge regarding the broader field of digital control systems.

A particular emphasis is given to the analysis of the fundamental theory, which is further simplified to numerous case studies with respect to digital control systems. In addition, at the end of each chapter, there is a detailed handbook, which contains the mathematical formulas required to solve certain exercises.

Each chapter is accompanied by a rich collection of solved exercises using a corresponding analytical methodology, whereby deeper understanding and consolidation of digital control systems is made simple and comprehensible.

In Chapter 1, some introductory concepts are analyzed in order to familiarize the reader with the required terminology on the subject.

Chapter 2 presents the z-transform, and its relation to the Laplace transform is also explained. Additionally, all the calculation methodologies of the inverse z-transform are analyzed here in order to study the behavior of a digital control system in the time domain.

Chapter 3 deals with the transfer function of sampled data systems, and its derivation is explicitly defined and explained. Moreover, Mason's formula is presented, and the method in which it can be used with the aid of the signal flow diagram to obtain the pulse transfer function.

Chapter 4 analyzes all the discretization methods of continuous time system transfer function, such as the method of z-transform, the exponential method, the differential backward method, the differential forward method, the method of the bilinear transform (with and without frequency change), and the method of pole-zero matching.

Chapter 5 analyzes the description of discrete-time systems in the state space. All the applicable forms are analyzed, such as direct form, canonical form, controllable canonical form, observable canonical form, and Jordan canonical form. The concepts of controllability and observability are thoroughly explained, and the discretization of continuous-time systems is analyzed into the state space, by using the integration and differentiation methods.

Chapter 6 analyzes the stability of discrete-time control systems. Certain related criteria are also analyzed, such as the unit circle, Routh (with Möbius transform) and Jury. The method of root locus is also explained as well as the Nyquist and Bode stability criteria.

In Chapter 7, the steady-state errors (i.e., position, velocity, and acceleration) are analyzed. Furthermore, the time and harmonic responses of discrete-time control systems are also studied.

Chapter 8 explains how digital control systems are designed to meet the given requirements or specifications. The indirect and direct design methods are analyzed here. The digital PID controller, the deadbeat digital controller, and the digital LEAD/LAG filters are also analyzed.

Finally, the simulation tools used throughout this textbook will be presented and briefly described in Chapter 9.

We hope that the readers will benefit from this manuscript, and enjoy their reading at the same time.

Authors

Professor Anastasia Veloni is a lecturer at the Piraeus University of Applied Sciences, Department of Computer Engineering, Greece. She has extensive teaching experience in a variety of courses in the automatic control area and is author/coauthor of four textbooks, and her research interests lie in the areas of signal processing and automatic control.

Dr. Nikolaos I. Miridakis is with the Piraeus University of Applied Sciences, Department of Computer Engineering, Greece, where he is currently an adjunct lecturer and a research associate. Since 2012, he has been with the Department of Informatics, University of Piraeus, where he is a senior research associate. His main research interests include wireless communications, and more specifically interference analysis and management in wireless communications, admission control, multicarrier communications, MIMO systems, statistical signal processing, diversity reception, fading channels, and cooperative communications.

1

Introduction

1.1 Introduction

One of the most important theoretical and practical aspects of automatic control is the systems' interconnection through a *feedback* process and generally the concept of feedback such that the system input will always depend on its corresponding output.

A remarkable technological breakthrough in these systems is their transformation such that they can be connected to computing units (computerization). Currently, control systems are based on computers' power gain, which enables them to work faster, more efficiently, and with greater reliability.

Nowadays, the control of a system performance is mainly carried out with the aid of microcontrollers or microprocessors. Several procedures, described by analog systems, can be controlled using digital systems. For the signal processing purpose, the utilization of filters is widespread; the latter stands for certain devices that either allow the components of a desired signal to pass, by rejecting the unwanted ones (i.e., band-pass filters) or correct the distorted signal components (i.e., equalizers). Also, satisfactory compensators are used in order to modify the dynamic response of control systems so as to maximize the response of the integrated system.

The increasing flexibility of state-of-the-art digital processors drives the above procedures to rely mainly on digital systems or digital filters. However, the interconnection of analog and digital systems requires an appropriate configuration so that both analog and digital signals can be processed. The switching between analog and digital signals is obtained by the sampling of signals and the so-called zero-order preservation.

By using the term "system," we explicitly define a physical part of the natural world where we assume that it comprises a set of components which simultaneously operate in a prescribed manner so as to achieve a certain goal. A system communicates with the environment through signals. Control systems can be classified, according to the type of signals that are being processed, in continuous-time or *analog*, in discrete-time or *digital*, and mixed or *hybrid*.

The continuous-time control systems, also called continuous data-flow systems or analog systems, include elements that only produce or process signals in continuous time. The discrete-time control systems, also called *discrete data-flow control systems* or *sample-data systems* or *digital systems,* include components which produce or process discrete signals to one or more parts of the system. Systems which process or produce discrete signals in some parts and continuous signals to the others are called *mixed* or *hybrid.*

By definition, computers are digital systems. Hence, all input data streams are in a digital form, that is, digital signals. Yet, various signals in the natural world are continuous time (e.g., position and temperature). Thus, they must be transformed to digital signals prior to being processed by computer systems. To convert an analog signal to a discrete one, the fundamental *sampling* mode-of-operation is required.

1.2 Description of Analog and Digital Control Systems

When speaking of automatic control we basically refer to closed-loop systems. The latter systems include steps, such as the physical system we wish to control; feedback and gain levels; and of course the controller itself, which is selected depending on the form of the underlying physical system. In an era where almost everything works under the guidance and supervision of the computer, the realization of the automatic control problem with discrete (or digital) manner is quite important.

It is known that the analog automatic control is very crucial, not only because it is used in many scientific fields, but also because many digital controllers implement analog control algorithms.

The basic form of the operating diagram (block diagram) of an analog closed-loop control system with a single input and a single output is shown in Figures 1.1 and 1.2 (where the PID controller's structure is presented in detail).

The *process* (or procedure or controlled system—plant) is the system or subsystem, which is controlled by the closed-loop system.

The *reference input* is an external signal applied to the closed-loop system, whereas it stands as a trigger for a specific behavior of the controlled system. Usually, it represents the ideal or desired output of the system.

The *controller* (or *regulator*) is the element which generates the control signal, which is input signal to the controlled system. The implementation of the analog regulator requires a system which usually consists of several components and devices.

The *feedback signal* is a related function of the controlled output. It is algebraically added to the reference signal in order to provide the error signal, which activates the control system.

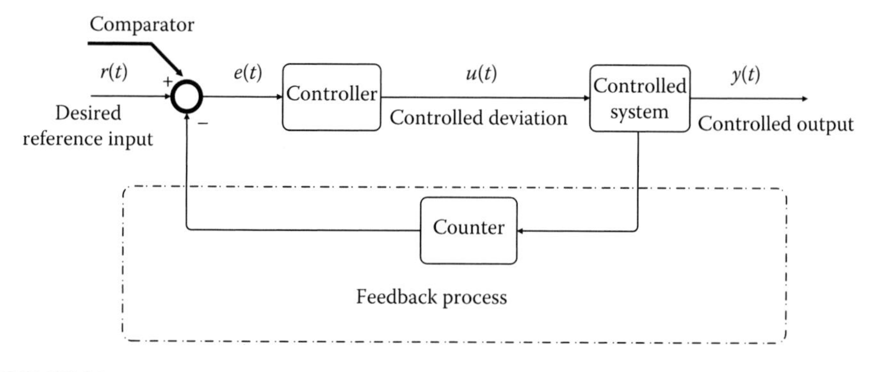

FIGURE 1.1
Analog automatic control system.

Analog controllers and, generally, continuous-time systems are described by the Laplace transform or by differential equations and are correspondingly processed to produce continuous-time signals. Digital controllers and, in general, discrete-time systems are described by the *Z-transform* or by *differential equations*. The problem of developing digital controllers is substantially related to the construction of a PC program.

In the case we wish to use a computer for the automatic control of a process, then the analog controller should be replaced with a digital one and the calculation of the error signal and the dynamic response of the analog can be implemented via the digital one, as shown in Figures 1.3 and 1.4.

The resulting system is a mixed *sampled-data system*, which includes both continuous- and discrete-time signals.

As shown in Figure 1.3, the control system is a hybrid system that exhibits both continuous and discrete dynamic behavior, where its main digital part is the controller, part of which is the computer.

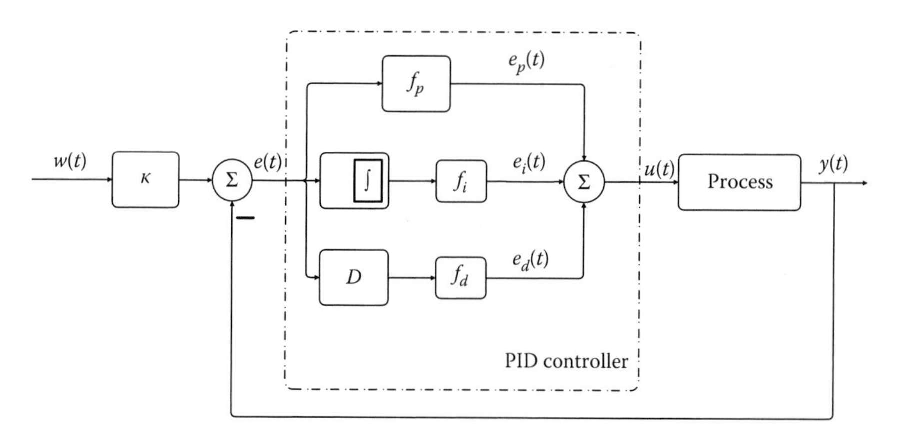

FIGURE 1.2
Analog automatic control system with a PID controller.

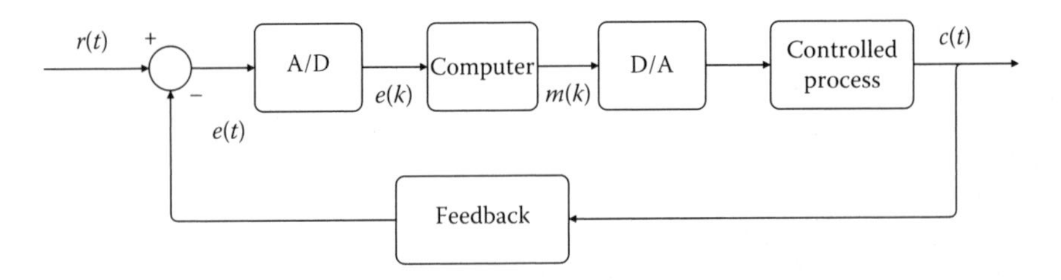

FIGURE 1.3
Digital control system.

The *controller* computes the error signal (or the difference) and generates an appropriate input signal to the controlled process in order to provide the desired output signal. Using the controller, a transfer function is implemented, the design of which can be made via a suitable algorithm that is programmed into the PC or by using special equipment. The computer may be a digital filter or a microprocessor, depending on the complexity and size of the system.

By appropriately adjusting the controller's parameters, we are able to intervene in the dynamics of the closed-loop system such that the system meets its requirements and specifications.

In summary, there are *two fundamental differences* between the analog and digital control systems.

- First, the digital system computes output samples and not the continuous signal.
- Second, the digital controller is described by difference equations; so the differential equations which represent the analog controller must be converted to difference equations. Thereby, the signals received by a digital controller are given in a discrete form, while the signals which enter the controlled system are continuous. Therefore, the continuous signals of the controlled output and the reference input should be sampled so as to provide the appropriate input of a discrete controller and then the signal, generated by the controller, must be converted to being continuous, to produce an input for the controlled system.

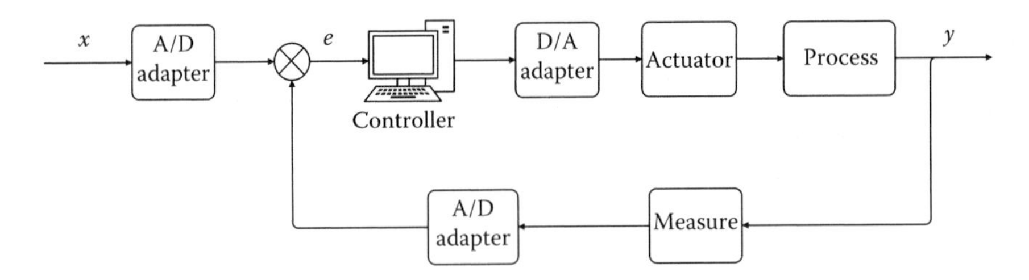

FIGURE 1.4
Computer controlled system (CCS).

To ensure communication compatibility between the controlled continuous-time system and discrete controller it is necessary to connect them through elements, which achieve the conversion of digital signals to analog and vice versa. These elements are the interface between the system and the computer and are called, respectively, *digital to analog converter (DAC)* and *analog to digital converter (ADC)*. The A/D converter essentially operates as a sampler. The D/A converter is able to transform a signal from the discrete-time to the continuous-time domain and is implemented with the aid of a restraint system.

The sampling process and the process of converting analog signals into discrete ones, and vice versa, plays quite an important role for the controller's performance.

Sampling process: Sampling a continuous-time signal replaces the original continuous signal from a sequence of values to discrete-time points. The sampling is used when a control system includes a digital controller, since sampling and quantization is necessary to input data in such a controller. Moreover, a sampling procedure appears whenever necessary control measurements are generated in an intermittent manner. For instance, in a radar tracking system, where the radar antenna scans a certain area around azimuth and elevation, corresponding information is taken once for each rotation/cycle of the antenna. Thus, the radar scan mode produces sampled data. In another example, a sampling procedure is necessary every time a controller or large-scale computer is shared at several time points in order to save cost. Then, a control signal is sent to each module only periodically and the signal is therefore converted to a sampled-data signal.

The sampling or discretization procedure is usually followed by a *quantization* process. During the quantization process, the sampled analog amplitude is replaced by a corresponding digital amplitude (i.e., binary number). Afterward, the digital signal is processed by the computer. The computer's output is sampled and fed to a hold circuit. The hold circuit output is a continuous-time signal and is fed to the actuator.

1.3 Advantages of Digital Control Systems and Applications

Computer-assisted control produces significant advantages as compared to conventional typical analog regulators. Some of these *advantages* are

- Great flexibility in modifying the controller's features. Indeed, these features can easily be altered by modifying the program. In contrast, in analog control systems, changing the controller characteristics is usually a nontrivial task and quite expensive because a substitution of elements and devices is required.

- Data processing is easy. Complex calculations can be performed easily and quickly. Analog controllers do not have this capability.
- They present better technical behavior in comparison to the analog control systems in terms of reliability and sensitivity to disturbances.
- They have an improved stability, a lower physical weight and, in many cases, a lower implementation cost.

Nevertheless, computer-assisted control presents some *disadvantages* compared to analog control systems such as

- The errors introduced during both the sampling process of continuous systems and the quantization of discrete-time signals.
- The difficulty in the digital control system design, particularly if the process is complicated. In the digital control of a complex process, the designer must have a good knowledge of the process to be controlled and should be able to obtain its corresponding mathematical model. This mathematical model can be obtained in the form of differential equations or difference equations or in some other form. The designer must be familiar with the measurement technology related to the process output and other variables involved in the process. Also, the designer must have a relatively good knowledge of computer systems and modern control theory. In this regard, a good knowledge of simulation techniques is useful.

The most challenging part in the design of control systems is the accurate modeling of the physical unit or process. When designing a digital controller, it is necessary to recognize the fact that the mathematical model of a unit or process, in many cases, represents only an approximation of the relevant physics. Exceptions are the cases when modeling electromechanical and hydromechanical systems, since they can be accurately and efficiently modeled. For example, the modeling of a robotic-arm system can be achieved with quite high accuracy.

The computer-assisted control technique has been applied in a variety of systems and processes such as industrial control, telecommunication systems, wireless/wired networks, nuclear and chemical reactors, terrestrial—maritime and air transport systems, weapons systems, far-distant system control, robotics, space applications, biotechnology, medicine, biology, etc.

2

z-Transform

2.1 Introduction

Analog systems are designed and analyzed with the use of *Laplace transforms*. On the other hand, discrete-time systems are analyzed using a similar technique called *z-transform*.

The basic lines of reasoning are the same for both scenarios: After determining the impulse response of the system, the response of any other input signal can be extracted by simple arithmetic operations. The behavior and the stability of the system can be predicted from the zeros and poles of the transfer function.

As Laplace transform converts the differential equations into algebraic terms with respect to s, *z-transform converts the difference equations into algebraic terms with respect to z*. Both transformations are matching a complex quantity to the points of a region of the complex plane.

It should be noted that the z-plane (i.e., the domain of z-transform) is organized in a *polar* form, while the s-plane (i.e., the domain of Laplace transform) is in a *Cartesian* form.

2.2 From Laplace Transform to z-Transform

The z-transform greatly facilitates the study and design of nonlinear time-varying discrete-time systems, because it transforms the difference equation that describes the system into an algebraic equation.

In Figure 2.1, the procedure followed by using z-transform, where there are three steps to resolve the differential equation (D.E.) and the direct solution of the given D.E. via higher mathematics, which is much more laborious, are given.

To show that z- and Laplace transforms are two parallel techniques, the Laplace transform, which is already known, will be used, and capitalizing on it, the mathematical expression of z-transform will be developed.

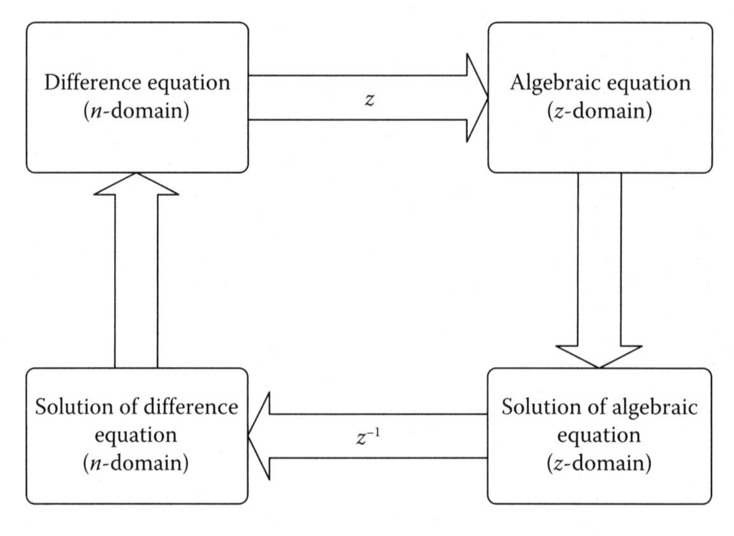

FIGURE 2.1
Solution of D.E. using z-transform.

The Laplace transform is an integral-based transform that connects the temporal representation of the signal—that is $x(t)$—with its representation in the s domain, namely, $X(s)$.

The Laplace transform is defined as $X(s) = \int_{t=-\infty}^{+\infty} x(t) \cdot e^{-st} dt$, where s is a complex number.

By substituting $s = \sigma + j\omega$ one is able to reach an alternative form of $X(s)$ function, which is $X(\sigma,\omega) = \int_{t=-\infty}^{+\infty} x(t) \cdot e^{-\sigma t} \cdot e^{-j\omega t} dt$. Inserting in the latter expression $e^{-j\omega t} = \cos(\omega t) - j\sin(\omega t)$ we have that

$$
\begin{aligned}
X(\sigma,\omega) &= \int_{t=-\infty}^{+\infty} x(t) \cdot e^{-\sigma t} \cdot [\cos(\omega t) - j\sin(\omega t)] dt \Rightarrow \\
X(\sigma,\omega) &= \int_{t=-\infty}^{+\infty} x(t) \cdot e^{-\sigma t} \cos(\omega t) dt - j \int_{t=-\infty}^{+\infty} x(t) \cdot e^{-\sigma t} \sin(\omega t) dt
\end{aligned}
\tag{2.1}
$$

Based on Equation 2.1, $x(t)$ signal is analyzed in sine and cosine waves, whose width exponentially varies in the time domain according to the relation $e^{-\sigma t}$. Each point of the complex s-plane is determined by the real and imaginary part, that is, the parameters σ and ω. At every point of s-plane, one can calculate the complex quantity $X(\sigma, \omega)$.

The real part of $X(\sigma, \omega)$ arises by multiplying the $x(t)$ signal with a cosine waveform having a frequency ω whose amplitude decreases exponentially with a rate σ and is then integrated for all time instances.

Thus,

$$\Re e\{X(\sigma,\omega)\} \sim \int_{t=-\infty}^{+\infty} x(t)\cdot e^{-\sigma t}\cdot \cos(\omega t)dt \tag{2.2}$$

The imaginary part is obtained by using a similar manner, that is, by multiplying the $x(t)$ signal with a sine waveform of frequency ω whose amplitude decreases exponentially with a rate σ. Hence,

$$\Im m\{X(\sigma,\omega)\} \sim \int_{t=-\infty}^{+\infty} x(t)\cdot e^{-\sigma t}\cdot \sin(\omega t)dt \tag{2.3}$$

Based on the above representation of the Laplace transform, one can formulate the z-transform, that is, the corresponding transformation relation for discrete signals in *three steps*.

Step 1: The first step is the most obvious: Change the signal from continuous to discrete, that is, $x(t) \rightarrow x[n]$ and of course the integral should be replaced with a sum. Thereby: $\int_{t=-\infty}^{+\infty} \rightarrow \sum_{n=-\infty}^{+\infty}$ so

$$X(\sigma,\omega) = \sum_{n=-\infty}^{+\infty} x[n]\cdot e^{-\sigma n}\cdot e^{-j\omega n} \tag{2.4}$$

Despite the fact that $x[n]$ signal is a discrete one, $X(\sigma, \omega)$ is continuous since σ and ω variables can take continuous values.

In the case of the Laplace transform, one could go through any point (σ, ω) (not quite any point; the integral will not converge for points not belonging in the region of convergence) of the complex plane and define the real and imaginary part of $X(\sigma, \omega)$ by integrating over time, as previously explained.

If the case of z-transform, one may again go through up to any point of the complex plane, but replace integration with summation.

Step 2: In the second step, polar coordinates are introduced to represent the exponential $e^{-\sigma n}$.

The exponential signal $y[n] = e^{-\sigma n}$ can be written as: $y[n] = r^{-n}$, where, apparently, the substitution $e^{\sigma} = r$ has taken place, holding that $\sigma = \ln r$.

It is noteworthy that:

In the form $y[n] = e^{-\sigma n}$, $y[n]$ increases with time when $\sigma < 0$.

In the form $y[n] = r^{-n}$, $y[n]$ increases with time when $r < 1$.

In the form $y[n] = e^{-\sigma n}$, $y[n]$ decreases with time when $\sigma > 0$.

In the form $y[n] = r^{-n}$, $y[n]$ decreases with time when $r > 1$.

In the form $y[n] = e^{-\sigma n}$, $y[n]$ remains unchanged when $\sigma = 0$.

In the form $y[n] = r^{-n}$, $y[n]$ remains unchanged when $r = 1$.

Hence,

$$X(r,\omega)=\sum_{-\infty}^{\infty}x[n]r^{-n}\cdot e^{-j\omega n} \tag{2.5}$$

Step 3: The substitution $z=r\cdot e^{j\omega}$ is performed, therefore the standard form of z-transform arises.

$$X(z)=\sum_{-\infty}^{\infty}x[n]\cdot z^{-n} \tag{2.6}$$

The z-transform (Equation 2.6) is a valuable tool for analyzing discrete linear time-invariant (LTI) systems. It provides capabilities for

- Efficient calculation for the response of a LTI system (the convolution in the discrete-time domain) $y(n) = x(n)\star h(n)$ is computed as a product in the z-transform domain: $Y(z) = X(z)H(z)$, so $y(n) = \mathrm{IZT}(Y(z))$.
- Stability analysis of a LTI system (via calculation of the region of convergence).
- Description of LTI with regard to its behavior in the frequency domain (low pass filter, band pass filter, etc.).

2.2.1 Comparison of *s*- and *z*-Planes in the Region of Convergence

The main differences between s- and z-planes are presented in Figure 2.2.

The points of s-plane are described by two parameters: σ parameter corresponds to the real axis that determines the exponential rate of reduction,

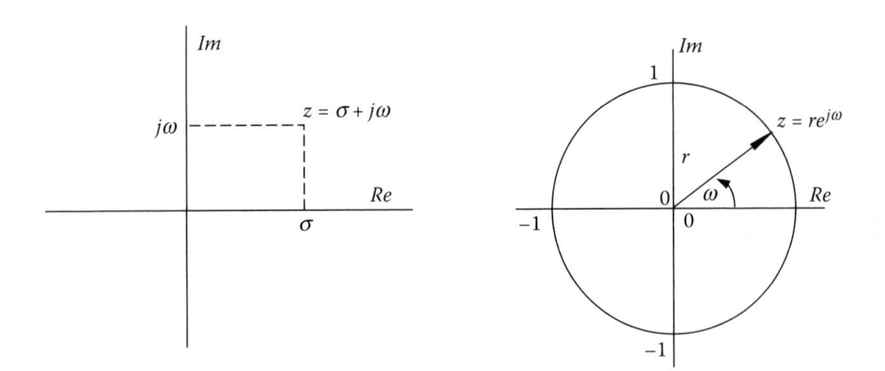

FIGURE 2.2
s-plane is orthogonal, while z-plane is polar.

while ω parameter corresponds to the imaginary axis, which determines the oscillation frequency. Both parameters are arranged in a rectangular array on the s-plane.

The resulting geometry arises from the fact that for each s number, its position is determined by the relation:

$$s = \sigma + j\omega.$$

z-plane is *polar-based*. Each complex number (z) is determined by its distance (r) from the origin, corresponding to the exponential rate of reduction (recall the expression $X(r, \omega)$), and ω parameter corresponds to the angle of r, across the positive horizontal semi-axis. The polar arrangement of z arises from the relation $z = r \cdot e^{j\omega}$ or $z = re^{j\omega} = r(\cos \omega + j \sin \omega)$.

As a consequence of the above differences, vertical lines in s-plane become circles in z-plane. This is due to the fact that $\sigma = \ln r$, which has previously been implemented. For instance, the imaginary axis of s-plane, that is, the line formed by setting $\sigma = 0$, will reflect to a circle of radius $r = 1$ into the z-plane. Indeed, the relation $\sigma = 0$ corresponds to $0 = \ln r$ in z-plane and, hence, $r = 1$.

Lines parallel to the imaginary axis located in the left half-plane ($\sigma < 0$) correspond to concentric circles which are located within the unit circle in the z-plane. A similar behavior is realized in lines located in the right s-half-plane.

For example, a causal continuous-time system is *stable* if its poles are in the left s-half-plane. Correspondingly, a causal discrete-time system is stable if its poles are located inside the unit circle.

2.3 z-Transform Properties

The properties of the z-transform have a very similar meaning to the properties of the Laplace transform. The most important properties of the z-transform, commonly used for problem solving, are presented below.

2.3.1 Time Shift

If $X(z)$ denotes the z-transform of $x[n]$ function, then the corresponding transform of $x[n-N]$ is given by $z^{-N}X(z)$. The time shift operation adds or subtracts the axes' origin or infinity from the region of convergence of $X(z)$.

If...	Then...
$x[n] \leftrightarrow X(z)$	$x[n-N] \leftrightarrow z^{-N}X(z)$

2.3.2 Linearity

Let $x[n]$ be a function, which arises from the linear combination of two functions $x_1[n]$ and $x_2[n]$ with regions of convergence Π_1 and Π_2, respectively. The region of convergence of $x[n]$ includes the intersection of $\Pi_1 \cap \Pi_2$.

If...	Then...
$x[n] \leftrightarrow X(z)$	$ax_1[n]+\beta x_2[n] \overset{ZT}{\leftrightarrow} aX_1(z)+\beta X_2(z)$

2.3.3 Time Reverse

If the z-transform of the $x[n]$ function is $X(z)$, with region of convergence Π, then the transform of $x[-n]$ is $X(1/z)$ with region of convergence $1/\Pi$.

If...	Then...
$x[n] \leftrightarrow X(z)$	$x[-n] \overset{ZT}{\leftrightarrow} X\left(\frac{1}{z}\right)$

2.3.4 Convolution

Let two functions $x_1[n]$ and $x_2[n]$ with corresponding z-transforms and regions of convergence $x_1[n] \leftrightarrow X_1(z)$ where $z \in \Pi_1$ and $x_2[n] \leftrightarrow X_2(z)$ where $z \in \Pi_2$. The transform for the signals' convolution $x_1[n]$ and $x_2[n]$ is $x_1[n] * x_2[n] \leftrightarrow X_1(z)X_2(z) = X(z)$, where the region of convergence of $X(z)$ is identical or includes the intersection of regions of convergence of $X_1(z)$ and $X_2(z)$.

If...	Then...
$x[n] \leftrightarrow X(z)$	$x_1[n] * x_2[n] \leftrightarrow X_1(z)X_2(z)$

2.3.5 Differentiation in *z*-Domain

Let $X(z)$ be the transform of $x[n]$ function with region of convergence Π. Then, $nx[n] \leftrightarrow -z(dX(z)/dz)$ with the same region of convergence.

If...	Then...
$x[n] \leftrightarrow X(z)$	$nx[n] \leftrightarrow -z(dX(z)/dz)$

2.3.6 Initial and Final Value Theorem

If $x[n] = 0$ and $n < 0$, then $x[0] = \lim_{z\to\infty} X(z)$ (I.V.T.)

For causal stable systems, it holds that:

$$x[\infty] = \lim_{n\to\infty} x[n] = \lim_{z\to 1} (z-1)\,X(z) \quad \text{(F.V.T.)}$$

If...	Then...
$x[n] \leftrightarrow X(z)$	$x[0] = \lim_{z \to \infty} X(z)$
$x[n] \leftrightarrow X(z)$	$x[\infty] = \lim_{n \to \infty} x[n] = \lim_{z \to 1} (z-1)X(z)$

2.4 Inverse z-Transform

The implementation of z-transform results in the transportation from the discrete-time domain to z-domain. The opposite procedure is implemented with the aid of the inverse z-transform.

The inverse z-transform is defined by

$$x[n] = Z^{-1}[X(z)] = \frac{1}{2\pi j} \oint_c X(z) z^{n-1} dz \tag{2.7}$$

where c is a closed contour within the region of convergence of $F(z)$ which includes the intersection of real and imaginary axes of z-complex plane.

Due to the fact that the calculation of the involved integral is quite cumbersome, usually the calculation is made in the form of tables, which provide the timing functions of basic complex functions. In general, these tables cover only some cases, thus some other methods can be used for calculating the inverse z-transform.

There are *three methods* for calculating the inverse transform of a function $X(z)$

1. Method of *power series expansion*
2. Method of *partial fraction expansion*
3. Method of *complex integration* (via the residue theorem)

2.4.1 Division Method

Using this method, *certain samples* of the inverse z-transform are calculated; a corresponding analytical expression is not provided.

Dividing the numerator by the denominator of the $X(z)$ function, $X(z)$ takes a series formation in terms of z.

2.4.2 Method of Partial Fraction Expansion

Partial fractions expansion is particularly useful method for the analysis and design of systems, because the impact of any characteristic root or eigenvalue becomes straightforward.

The facilitation of partial fraction expansion occurs for $X(z)/z$, and not for $X(z)$.

We distinguish *three cases* of partial fraction expansion for $X(z)/z$, according to the form of its poles.

- *Case of distinct real poles*: In this case, $X(z)/z$ is expanded in a fractional sum series as follows:

$$\frac{X(z)}{z} = \frac{B(z)}{(z-p_1)\cdots(z-p_n)} = \frac{c_1}{(z-p_1)} + \cdots + \frac{c_n}{(z-p_n)} \tag{2.8}$$

- The c_i coefficients are computed using the *Heaviside formula* (Heaviside formula—Oct. 1931) *for distinct poles*, hence,

$$c_i = \lim_{z \to p_i} \left\{ (z-p_i) \frac{B(z)}{(z-p_1)(z-p_2)...(z-p_n)} \right\} \tag{2.9}$$

- *Case of nondistinct real poles (poles with multiplication factor n—multiple real poles)*: In this case, $X(z)/z$ is expanded in a fractional sum series as follows:

$$\frac{X(z)}{z} = \frac{B(z)}{(z-p_1)^n \cdots (z-p_n)} = \frac{c_{11}}{(z-p_1)} + \frac{c_{12}}{(z-p_1)^2} + \cdots$$

$$\frac{c_{1n}}{(z-p_1)^n} + \frac{c_2}{(z-p_2)} \cdots + \frac{c_n}{(z-p_n)} \tag{2.10}$$

- The c_i coefficients are computed using the Heaviside formula *for multiple poles*, hence,

$$c_{ij} = \frac{1}{(n-j)!} \lim_{z \to p_i} \left\{ \frac{d^{(n-j)}}{dz^{(n-j)}} (z-p_i)^n \frac{X(z)}{z} \right\} \tag{2.11}$$

- The remaining coefficients are computed via Equation 2.9.
- *Case of complex roots*: In this case, the coefficient of the numerator for one of the complex roots is computed via Equation 2.9 or 2.11. Therefore, the coefficient in the numerator of the term which has the conjugate root of the former at the denominator becomes its corresponding conjugate.

2.4.3 Method of Complex Integration

This method is quite general and is used when one or more partial fractions of the expanded $F(z)$ are not included into the lookup tables of z-transform.

This method relies on the definition formula of the *inverse z-transform*.

The utilization of Equation 2.7 requires the use of the *residue theorem*, which is given by

$$\oint F(z)z^{n-1}dz = 2\pi j \sum residues[F(z)z^{n-1}] \tag{2.12}$$

In the latter expression, also known as **Cauchy's formula**, Σ stands for the sum of residues for the poles of $F(z)$, which includes the c curve.

Combining the above expressions, we have that

$$f[n] = \sum residues[F(z)z^{n-1}] \tag{2.13}$$

- If there is a simple first-order pole of $F(z)\cdot z^{n-1}$ (i.e., $z = \alpha$), then its residual is given by

$$F(z)z^{n-1}(z-\alpha)\Big|_{z=\alpha} \tag{2.14}$$

- If there is an m-order pole of $F(z)z^{n-1}$, then its residual is given by

$$\frac{1}{(m-1)!}\frac{d^{(m-1)}}{dz^{(m-1)}}F(z)z^{n-1}(z-\alpha)^m\Bigg|_{z=\alpha} \tag{2.15}$$

2.5 Formula Tables

Tables 2.1 and 2.2.

2.6 Solved Exercises

EXERCISE 2.1

Compute the z-transform of unit-step function $u[n] = \begin{cases} 1 & \text{when} \quad n \geq 0 \\ 0 & \text{when} \quad n < 0 \end{cases}$.

Solution

Based on the definition of z-transform, we have that

TABLE 2.1

z-Transform for Elementary Functions

	$x[n]$	$X(z)$	Π.Σ.
1	$\delta[n]$	1	All z
2	$u[n]$	$\frac{z}{z-1}$	$\lvert z\rvert > 1$
3	$-u[-n-1]$	$\frac{z}{z-1}$	$\lvert z\rvert < 1$
4	$\delta[n-m]$	z^{-m}	All z except 0 $(m > 0)$ or ∞ $(m < 0)$
5	$a^n u[n]$	$\frac{z}{z-a}$	$\lvert z\rvert > \lvert a\rvert$
6	$-a^n u[-n-1]$	$\frac{z}{z-a}$	$\lvert z\rvert < \lvert a\rvert$
7	$na^n u[n]$	$\frac{az}{(z-a)^2}$	$\lvert z\rvert > \lvert a\rvert$
8	$-na^n u[-n-1]$	$\frac{az}{(z-a)^2}$	$\lvert z\rvert < \lvert a\rvert$
9	$(n+1)a^n u[n]$	$\left[\frac{z}{z-a}\right]^2$	$\lvert z\rvert > \lvert a\rvert$
10	$(\cos\Omega n)u[n]$	$\frac{z^2(\cos\Omega)z}{z^2-(2\cos\Omega)z+1}$	$\lvert z\rvert > 1$
11	$(\sin\Omega n)u[n]$	$\frac{(\sin\Omega)z}{z^2-(2\cos\Omega)z+1}$	$\lvert z\rvert > 1$
12	$(r^n \cos\Omega n)u[n]$	$\frac{z^2-(r\cos\Omega)z}{z^2-(2r\cos\Omega)z+r^2}$	$\lvert z\rvert > r$
13	$(r^n \sin\Omega n)u[n]$	$\frac{(r\sin\Omega)z}{z^2-(2r\cos\Omega)z+r^2}$	$\lvert z\rvert < r$

$$X(z) = \sum_{-\infty}^{+\infty} u[n]\cdot z^{-n} = \sum_{n=0}^{+\infty} 1\cdot z^{-n} = \sum_{n=0}^{+\infty} (z^{-1})^n = \frac{1}{1-z^{-1}} = \frac{z}{z-1} \tag{2.1.1}$$

Unit-step function affects the limits of summation.

The sequence $\sum_{n=0}^{+\infty}(z^{-1})^n$ is the infinite sum series of a decreasing geometric progression with first term 1 and ratio z^{-1}.

Consequently, it must hold that $|z| < 1$ or $|z| > 1$

The condition $|z| > 1$ defines the *region of convergence* of the transform, that is, the set of values of the complex z-plane for which the sum of z-transform converges.

TABLE 2.2

z-Transform Properties

	Property	Function	Transform	R.C.
		$x[n]$	$X(z)$	R
		$x_1[n]$	$X_1(z)$	R_1
		$x_2[n]$	$X_2(z)$	R_2
1	*Linearity*	$ax_1[n]+\beta x_2[n]$	$aX_1(z)+\beta X_2(z)$	$R'\supset R_1\cap R_2$
2	Time shift	$x[n-N]$	$z^{-N}X(z)$	$R'\supset R$
3	Multiplication with z_0^n	$z_0^n x[n]$	$X\left(\frac{z}{z_0}\right)$	$R'=\lvert z_0\rvert R$
4	Multiplication with $e^{j\Omega n}$	$e^{j\Omega n}x[n]$	$X(e^{-j\Omega}z)$	$R'=R$
5	*Time reverse*	$x[-n]$	$X\left(\frac{1}{z}\right)$	$R'=\frac{1}{R}$
6	Multiplication with n	$nx[n]$	$-z\frac{dX(z)}{dz}$	$R'=R$
7	Sum	$\sum_{k=-\infty}^{n} x[n]$	$\frac{1}{1-z^{-1}}X(z)$	$R'\supset R\cap\{\lvert z\rvert>1\}$
8	Convolution	$x_1[n]*x_2[n]$	$X_1(z)\cdot X_2(z)$	$R'\supset R_1\cap R_2$
9	Initial value	If $x[n]=0$ when $n<0$, then $x[0]=\lim_{z\to\infty}X(z)$		
10	Final value	For stable systems $\lim_{n\to\infty}x[n]=\lim_{z\to 1}(z-1)X(z)$		

EXERCISE 2.2

Compute the z-transform of the function $x[n]=a^n\cdot u[n]$.

Solution

Based on the definition of z-transform, we have that $X(z)=\sum_{n=-\infty}^{\infty}a^n u[n]\cdot z^{-n}=\sum_{n=0}^{\infty}(a\cdot z^{-1})^n$, unit-step function restricts the sum bounds from $n=0$ to infinity.

The term $(a\cdot z^{-1})^n$ is the general expression of a geometric progression with $a_1=1$ and $\omega=\alpha\cdot z^{-1}$.

In order for $\sum_{n=0}^{\infty}(az^{-1})^n$ to converge, the geometric progression should be decreasing, that is, $\lvert a\cdot z^{-1}\rvert<1$.

With this constraint at hand

$$\sum_{n=-\infty}^{\infty}(a\cdot z^{-1})^{-n}=\frac{1}{1-az^{-1}}=\frac{1}{1-\frac{a}{z}}=\frac{z}{z-a} \tag{2.2.1}$$

Hence,

$$x[n] = a^n \cdot u[n] \leftrightarrow X(z) = \frac{z}{z-a} \tag{2.2.2}$$

The above transform applies only when

$$|a| \cdot |z^{-1}| < 1 \quad \text{or} \quad |a| < |z| \quad \text{or} \quad |z| > |a|$$

The condition $|z| > |a|$ defines the *region of convergence* of the transform, that is, the set of values of the complex z-plane for which the z-transform converges.

EXERCISE 2.3

Compute the z-transform of the functions $x_1[n] = \delta[n-2]$ and $x_2[n] = \delta[n+2]$.

Solution

The functions $x_1[n]$ and $x_2[n]$ are two time-shifted impulse functions.

The function $x_1[n] = \delta[n-2]$ is equal to zero except the case when $n = 2$.

The z-transform is $X_1(z) = \sum_{n=-\infty}^{n=+\infty} x_1[n]z^{-n} = z^{-2}$ and produces a double pole at $z = 0$. The region of convergence is the entire complex plane except the point (0,0), yet it includes infinity.

Correspondingly, for the function $x_2[n]$, it will be expressed as $X_2(z) = \sum_{n=-\infty}^{n=+\infty} x_2[n]z^{-n} = z^2$ with a pole reaching infinity. The region of convergence is the entire complex plane including (0,0), yet excepting infinity.

EXERCISE 2.4

Compute the z-transform and the region of convergence for the functions $x[n] = (5/6)^n u[n]$ and $y[n] = (5/6)^{n+5} u[n+5]$.

Solution

The function $x[n]$ takes nonzero values only when $n \geq 0$, hence, it is a causal function.

The function $y[n]$ arises by shifting $x[n]$ for 5 time units to the left $y[n] = x[n+5]$, therefore, there are some negative values of the n variable for which $y[n] \neq 0$. Thereby, $y[n]$ is a noncausal function.

From the transform of (5), given in the transformation table, it holds that $X(z) = (z/(z-5/6))$ with a region of convergence $|z| < (5/6)$.

The transform of $y[n]$ can be easily computed with the aid of the time shift property (property 2—Table 2.2). The function is given as $Y(z) = z^{-5}(z/z-5/6) = (z^{-4}/z-5/6)$.

The poles and regions of convergence for $X(z)$ and $Y(z)$ functions are presented in the following table.

$x[n] = \left(\frac{5}{6}\right)^n u[n]$

$y[n] = \left(\frac{5}{6}\right)^{n+5} u[n+5]$

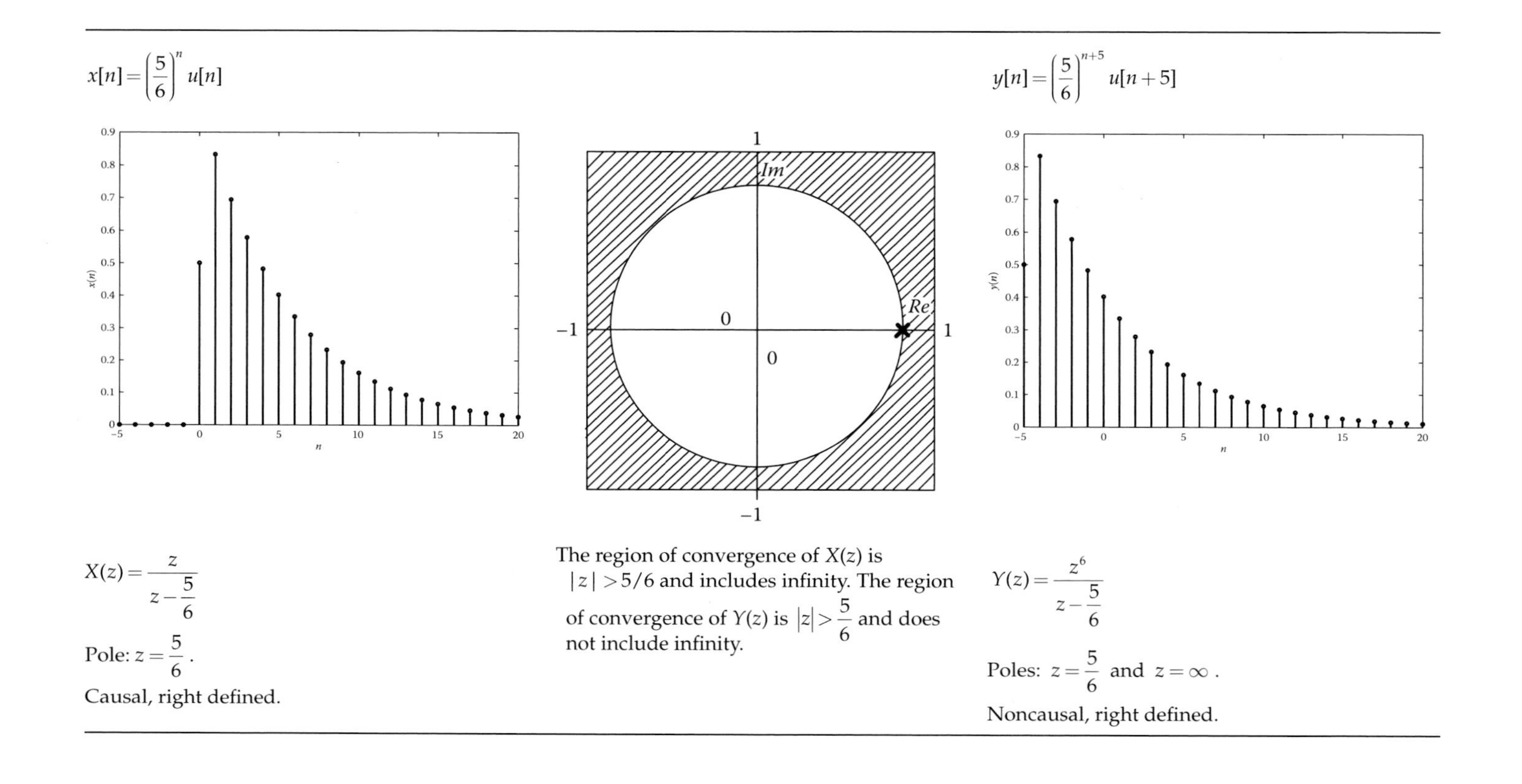

$X(z) = \frac{z}{z - \frac{5}{6}}$

Pole: $z = \frac{5}{6}$.

Causal, right defined.

The region of convergence of $X(z)$ is $|z| > 5/6$ and includes infinity. The region of convergence of $Y(z)$ is $|z| > \frac{5}{6}$ and does not include infinity.

$Y(z) = \frac{z^6}{z - \frac{5}{6}}$

Poles: $z = \frac{5}{6}$ and $z = \infty$.

Noncausal, right defined.

EXERCISE 2.5

Compute the z-transform and the region of convergence for the functions $x[n] = -(5/4)^n u[-n-1]$ and $y[n] = -(5/4)^{n-3}u[-n+2]$.

Solution

The function $y[n]$ arises from the time shift of $x[n]$ for 3 time units to the right, $y[n] = x[n-3]$, hence, as shown in the figure, there are some positive values of the variable n for which $y[n] \neq 0$. The function $y[n]$ is noncausal.

From the transform of (5), given in the transform lookup table, it holds that $X(z) = (z/(z-5/4))$ with region of convergence $|z| < 5/4$.

The transform of $y[n]$ can be easily computed with the aid of the time shift property (property 2—Table 2.2). The function is given as

$$Y(z) = \frac{z^{-2}}{z-(5/4)} = \frac{1}{z^2(z-5/4)} \tag{2.5.1}$$

The poles and the regions of convergence of $X(z)$ and $Y(z)$ are presented in the table (page 21).

Computation of z-transform of the function $y[n] = -(5/4)^{n-3}u[-n+2]$ without using the time shift property.

Based on the definition of the z-transform, we get

$$Y(z) = \sum_{n=-\infty}^{n=+\infty} y[n]z^{-n} = -\sum_{n=-\infty}^{n=+\infty}\left(\frac{5}{4}\right)^{n-3} u[-n+2]z^{-n} = -\sum_{n=-\infty}^{n=2}\left(\frac{5}{4}\right)^{n-3} z^{-n}$$
$$= -\left(\frac{5}{4}\right)^{-3}\sum_{n=-\infty}^{n=2}\left(\frac{5}{4}\right)^{n} z^{-n} = -\left(\frac{5}{4}\right)^{-3}\sum_{n=-\infty}^{n=2}\left(\frac{5}{4}z^{-1}\right)^{n}$$

or

$$Y(z) = \sum_{n=-\infty}^{n=+\infty} y[n]z^{-n} = -\left(\frac{5}{4}\right)^{-3}\sum_{n=-\infty}^{n=2}\left(\frac{5}{4}z^{-1}\right)^{n} \tag{2.5.2}$$

The sum $\sum_{n=-\infty}^{n=2}(5/4z^{-1})^n$ is written as $\sum_{n=-\infty}^{n=2}(5/4z^{-1})^n$

$$= \cdots + \left(\frac{5}{4}z^{-1}\right)^{-k} + \cdots \left(\frac{5}{4}z^{-1}\right)^{-2} + \left(\frac{5}{4}z^{-1}\right)^{-1} + \left(\frac{5}{4}z^{-1}\right)^{0} + \left(\frac{5}{4}z^{-1}\right)^{1} + \left(\frac{5}{4}z^{-1}\right)^{2} \tag{2.5.3}$$

An elegant way to build in the relation given by the sum of infinite terms of a decreasing geometric progression is to set $(5/4z^{-1}) = \Lambda$.

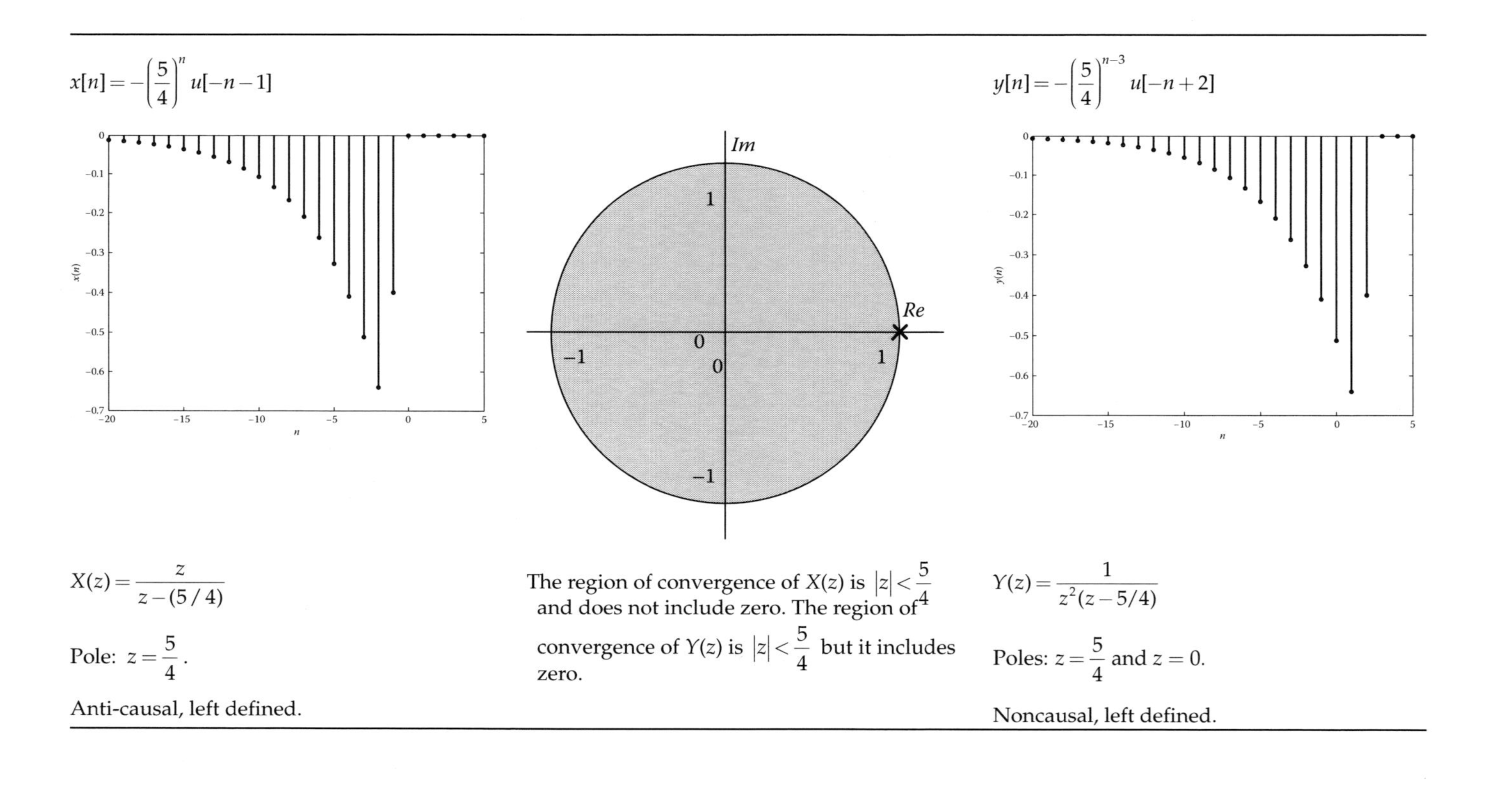

$x[n] = -\left(\frac{5}{4}\right)^{n} u[-n-1]$

$X(z) = \frac{z}{z-(5/4)}$

Pole: $z = \frac{5}{4}$.

Anti-causal, left defined.

The region of convergence of $X(z)$ is $|z| < \frac{5}{4}$ and does not include zero. The region of convergence of $Y(z)$ is $|z| < \frac{5}{4}$ but it includes zero.

$y[n] = -\left(\frac{5}{4}\right)^{n-3} u[-n+2]$

$Y(z) = \frac{1}{z^2(z-5/4)}$

Poles: $z = \frac{5}{4}$ and $z = 0$.

Noncausal, left defined.

Then,

$$\sum_{n=-\infty}^{n=2}\left(\frac{5}{4}z^{-1}\right)^n = \cdots + \Lambda^k + \cdots \Lambda^2 + \Lambda^1 + \Lambda^0 + \Lambda^{-1} + \Lambda^{-2}$$

or

$$\sum_{n=-2}^{n=\infty}\Lambda^n = \cdots + \Lambda^k + \cdots \Lambda^2 + \Lambda^1 + \Lambda^0 + \Lambda^{-1} + \Lambda^{-2}.$$

Re-expressing the sum series in an ascending order with respect to the exponent n, it holds that $\sum_{n=-2}^{n=\infty}\Lambda^n = \Lambda^{-2} + \Lambda^{-1} + \Lambda^0 + \Lambda^1 + \Lambda^2 + \cdots + \Lambda^k + \cdots$, which is the infinite sum series of a geometric progression with first term $\Lambda^{-2} = ((5/4)z^{-1})^{-2}$ and ratio $\Lambda = (4/5)z$.

In order to become a decreasing geometric progression, it should hold that $|\Lambda| < 1$ or $|(4/5)z| < 1$ or $|z| < 5/4$.

Substituting into $a_1/(1-\Lambda)$, which provides the infinite sum series of a decreasing geometric progression, it stems that

$$\frac{a_1}{1-\Lambda} = \frac{\left(\frac{5}{4}z^{-1}\right)^2}{1-\frac{4}{5}z} = -\frac{\left(\frac{5}{4}z^{-1}\right)^2}{\frac{4}{5}z-1} = -\frac{5}{4}\frac{\left(\frac{5}{4}z^{-1}\right)^2}{z-\frac{5}{4}} = -\left(\frac{5}{4}\right)^3\frac{z^{-2}}{z-\frac{5}{4}} \tag{2.5.4}$$

Substituting in Equation 2.5.2, $Y(z)$ is computed as

$$Y(z) = \sum_{n=-\infty}^{n=+\infty} y[n]z^{-n} = -\left(\frac{5}{4}\right)^{-3}\sum_{n=-\infty}^{n=2}\left(\frac{5}{4}z^{-1}\right)^n = \frac{z^{-2}}{z-\frac{5}{4}} = \frac{1}{z^2\left(z-\frac{5}{4}\right)} \tag{2.5.5}$$

EXERCISE 2.6

Compute the z-transform and the region of convergence for the function $x[n] = (1/3)^n u[n] + 2^n u[-n-1]$.

Solution

The given function can be considered as the sum of $x_1[n]$ and $x_2[n]$ where $x_1[n] = (1/3)^n u[n]$ and $x_2[n] = 2^n u[-n-1]$.

The transform of $x_1[n]$ directly results from the transform (5) of Table 2.1 which is $x_1[n] = (1/3)^n u[n] \leftrightarrow z/((z-(1/3))) = X_1(z)$ with region of convergence $|z| > 1/3$.

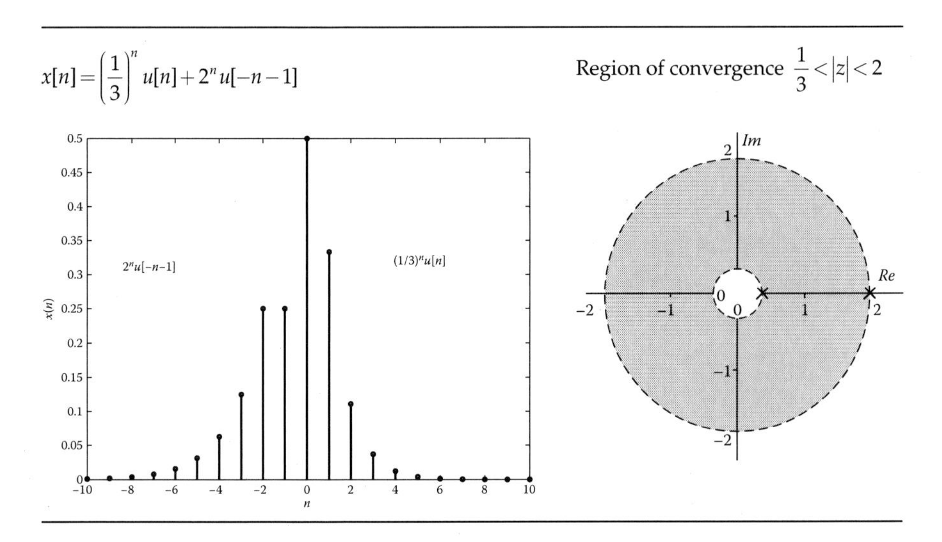

Based on transform (6), we have $-2^n u[-n-1] \leftrightarrow z/(z-2)$.

From the property of linearity, if $y[n] \leftrightarrow Y(z)$ then $ay[n] \leftrightarrow aY(z)$ hence $x_2[n] = 2^n u[-n-1] \leftrightarrow -z/(z-2)$ with region of convergence $|z| < 2$.

Consequently,

$$X(z)=\frac{z}{z-\frac{1}{3}}+\frac{z}{z-2}$$

$$=\frac{z^2-2z-z^2+\frac{1}{3}z}{\left(z-\frac{1}{3}\right)(z-2)}=\frac{-\frac{5}{3}z}{\left(z-\frac{1}{3}\right)(z-2)}=-5\frac{z}{3z^2-7z+2}$$

The poles of $X(z)$ are $z_1 = 2.$ and $z_2 = 1/3$. Since $x[n]$ is bilateral, the region of convergence will be the ring between the two poles, that is, it will be defined by the relation $1/3 < |z| < 2$, which is reflected at the intersection of the regions of convergence of $X_1(z)$ and $X_2(z)$.

EXERCISE 2.7

Compute the z-transform for the function $x[n] = n \cdot a^n u[n]$.

Solution

According to the transform (5), $x[n] = a^n u[n] \leftrightarrow (z/(z-a)) = X(z)$. The derivative of $X(z)$, with respect to z, is expressed as

$$\frac{d}{dz}\left(\frac{z}{z-a}\right)=\frac{(z)'(z-a)-(z-a)'\cdot z}{(z-a)^2}=\frac{1\cdot(z-a)-1\cdot z}{(z-a)^2}=\frac{-a}{(z-a)^2} \qquad (2.7.1)$$

Based on the property $n \cdot x[n] \to -z(dX(z)/dz)$, it holds that

$$\mathbf{Z}\{n \cdot a^n u[n]\} = \frac{az}{(z-a)^2} \tag{2.7.2}$$

EXERCISE 2.8

Compute the z-transform of the function: $x[n] = (1/3)^n u[-n]$.

Solution

Based on the transform (6), we have $-(1/3)^n u[-n-1] \leftrightarrow (z/(z-(1/3)))$, when

$$|z| < \frac{1}{3} \tag{2.8.1}$$

The function is time shifted by one unit, that is, n is replaced with $n-1$ so

$$\begin{aligned} &-\left(\frac{1}{3}\right)^{n-1} u[-(n-1)-1] \leftrightarrow \frac{z}{z-\frac{1}{3}} z^{-1} \quad \text{or} \\ &-\left(\frac{1}{3}\right)^{-1}\left(\frac{1}{3}\right)^{n} u[-n] \leftrightarrow \frac{1}{z-\frac{1}{3}} \dot{\eta} \left(\frac{1}{3}\right)^{n} u[-n] \leftrightarrow -\frac{1}{3}\frac{1}{z-\frac{1}{3}} \end{aligned} \tag{2.8.2}$$

EXERCISE 2.9

Compute the z-transform of the function:

$$x[n] = n\left(\frac{1}{2}\right)^n u[n-2]$$

Solution

According to the transform (5), we have $(1/2)^n u[n] \leftrightarrow (z/z-1/2)$, when $|z| < 1/3$.

Based on the time shift property we have

$$\begin{aligned} &\left(\frac{1}{2}\right)^{n-2} u[n-2] \leftrightarrow z^{-2}\frac{z}{z-\dfrac{1}{2}} \Rightarrow \\ &\left(\frac{1}{2}\right)^{2}\left(\frac{1}{2}\right)^{n-2} u[n-2] \leftrightarrow \left(\frac{1}{2}\right)^{2} \frac{z^{-1}}{z-\frac{1}{2}} \Rightarrow \left(\frac{1}{2}\right)^{n} u[n-2] \leftrightarrow \left(\frac{1}{4}\right)\frac{z^{-1}}{z-\frac{1}{2}} \end{aligned}$$

Capitalizing on the differentiation property

$$n\left(\frac{1}{2}\right)^{n-2} u[n-2] \leftrightarrow -z\frac{d}{dz}\left[\frac{1}{4}\frac{z^{-1}}{z-\frac{1}{2}}\right] = -\frac{z}{4}\frac{-z^{-2}(z-\frac{1}{2})-z^{-1}}{(z-\frac{1}{2})^2}$$

or

$$n\left(\frac{1}{2}\right)^{n-2} u[n-2] \leftrightarrow -\frac{z}{4}\cdot\frac{-z^{-1}+\frac{1}{2}z^{-2}-z^{-1}}{(z-\frac{1}{2})^2}$$

where after some manipulations:

$$n\left(\frac{1}{2}\right)^{n-2} u[n-2] \leftrightarrow \frac{z^{-2}}{2}\cdot\frac{1-\frac{1}{4}z^{-1}}{(1-\frac{1}{4}z^{-1})^2}$$

EXERCISE 2.10

Define the initial and final value of the system's impulse when

$$Y(z) = \frac{2z^2}{(z-1)(z-\alpha)(z-\beta)}, \quad |\alpha|,|\beta|<1$$

Solution

From the initial value theorem

$$y[0] = \lim_{z\to\infty} Y(z) = \lim_{z\to\infty}\frac{2z^2}{(z-1)(z-\alpha)(z-\beta)} = 0 \tag{2.10.1}$$

From the final value theorem

$$y[\infty] = \lim_{z\to 1}(z-1)\,Y(z) = \lim_{z\to 1}\frac{2z^2}{(z-\alpha)(z-\beta)} = \frac{2}{(1-\alpha)(1-\beta)} \tag{2.10.2}$$

EXERCISE 2.11

Find the first four coefficients of the function $x[n]$ when:

$$X_1(z) = \frac{4z^{-1}}{z^{-2}-2z^{-1}+2}$$

,

Solution

a. For $X_1(z)$:

$$X_1(z) = \frac{4z^{-1} \cdot z^2}{(z^{-2} - 2z^{-1} + 2) \cdot z^2} = \frac{4z}{2z^2 - 2z + 1} \tag{2.11.1}$$

By the polynomial division of numerator and denominator of $X_1(z)$, it holds that

$$\begin{array}{l|l}
4z & 2z^2 - 2z + 1 \\ \cline{2-2}
-4z + 4 - 2z^{-1} & 2z^{-1} + 2z^{-2} + z^{-3} + \cdots \\ \hline
\quad 4 - 2z^{-1} & \\
\quad -4 + 4z^{-1} - 2z^{-2} & \\ \hline
\qquad 2z^{-1} + 2z^{-2} & \\
-2z^{-1} - 2z^{-2} - 2z^{-3} & \\ \hline
-2z^{-3} & \\ \hline
\end{array}$$

$$X_1(z) = 2z^{-1} + 2z^{-2} + z^{-3} + \cdots$$

Therefore, we get

$$x_1[n] = \{\underset{\uparrow}{0}, 2, 2, 1, \ldots\} \tag{2.11.2}$$

b. For $X_2(z)$, we provide another method for a polynomial division

$$\begin{array}{r}
z^{-1} + 0z^{-2} - 3z^{-3} - 4z^{-4} \\ \hline
z^3 + 2z + 4 \,\big)\, z^2 - 1 \qquad\qquad\qquad\qquad \\
\underline{z^2 + 2 + 4z^{-1}} \qquad\qquad\qquad\quad \\
-3 - 4z^{-1} \qquad\qquad\qquad\quad \\
\underline{-3 \qquad - 6z^{-2} - 12z^{-3}} \qquad \\
-4z^{-1} + 6z^{-2} + 12z^{-3} \qquad \\
\underline{-4z^{-1} \qquad - 8z^{-3} - 16z^{-4}} \\
6z^{-2} + 20z^{-3} + 16z^{-4}
\end{array}$$

Therefore, we get

$$
\begin{aligned}
&X_2(z)=z^{-1}+0z^{-2}-3z^{-3}-4z^{-4}+\cdots\\
&\Rightarrow x_2[n]=0\delta[n]+1\delta[n-1]+0\delta[n-2]-3\delta[n-3]+\cdots
\end{aligned}
\tag{2.11.3}
$$

EXERCISE 2.12

Compute the inverse z-transform of the function

$$F(z)=\frac{z^{-1}}{(1-\alpha z^{-1})^2(1-\beta z^{-1})}=\frac{z^2}{(z-\alpha)^2(z-\beta)}$$

Solution

It holds that

$$f[n]=\sum residues\left[\frac{z^2}{(z-\alpha)^2(z-\beta)}z^{n-1}\right] \tag{2.12.1}$$

The residue of the simple pole $z=\beta$ is

$$\left.\frac{z^2z^{n-1}}{(z-\alpha)^2}\right|_{z=\beta}=\left.\frac{z^{n+1}}{(z-\alpha)^2}\right|_{z=\beta}=\frac{\beta^{n+1}}{(\beta-\alpha)^2} \tag{2.12.2}$$

The residue of the double pole $z=\alpha$ is

$$\left.\frac{d}{dz}\left(\frac{z^2}{z-\beta}z^{n-1}\right)\right|_{z=\alpha}=\left.\frac{[n+1]z^n(z-\beta)-z^{n+1}}{(z-\beta)^2}\right|_{z=\alpha}=\alpha^n\frac{[n+1](\alpha-\beta)-\alpha}{(\alpha-\beta)^2} \tag{2.12.3}$$

Hence,

$$f[n]=\frac{\alpha^n[n(\alpha-\beta)-\beta]+\beta^{n+1}}{(\alpha-\beta)^2} \tag{2.12.4}$$

EXERCISE 2.13

Compute the inverse z-transform of the function: $F(z)=z^{-2}+z^{-1}+1/0.2z^{-2}+0.9z^{-1}+1$ using (a) the partial fraction expansion and (b) complex integration.

Solution

a. Computation of $f(k)$ using the method of partial fraction expansion.

$$F(z)=\frac{z^{-2}+z^{-1}+1}{0.2z^{-2}+0.9z^{-1}+1}=\frac{z^2+z+1}{(z+0.4)(z+0.5)} \tag{2.13.1}$$

$$\frac{F(z)}{z}=\frac{z^2+z+1}{z(z+0.4)(z+0.5)}=\frac{k_1}{z}+\frac{k_2}{z+0.4}+\frac{k_3}{z+0.5} \tag{2.13.2}$$

Calculation of k_i

$$k_1=\lim_{z\to 0}\frac{F(z)}{z}z=\frac{1}{0.2}=5$$

$$k_2=\lim_{z\to -0.4}\frac{F(z)}{z}(z+0.4)=-19 \tag{2.13.3}$$

$$k_3=\lim_{z\to -0.5}\frac{F(z)}{z}(z+0.5)=15$$

Hence,

$$F(z)=5-\frac{19z}{z+0.4}+\frac{15z}{z+0.5} \tag{2.13.4}$$

So, $f(k)$ becomes

$$f(k)=Z^{-1}[F(z)]=(5\delta(k)-19(-0.4)^k+15(-0.5)^k)u(k) \tag{2.13.5}$$

b. Computation of $f(k)$ using the method of complex integration.

$$f(k)=Z^{-1}\left[F(z)\right]=\sum residuesF(z)z^{k-1}\Rightarrow$$

$$f(k)=\sum residues\frac{(z^2+z+1)z^{k-1}}{(z+0.4)(z+0.5)} \tag{2.13.6}$$

For $k = 0$, we have the poles: $z = 0, z = -0.4, z = -0.5$

$$residue\ z=0 \rightarrow \left.\frac{z^2+z+1}{(z+0.4)(z+0.5)}\right|_{z=0} = \frac{1}{0.2} = 5, k=0 \quad (2.13.7)$$

$$residue\ z=-0.4 \rightarrow \left.\frac{(z^2+z+1)z^{k-1}}{z+0.5}\right|_{z=0.4} = 7.6(0.4)^{k-1} = -19(0.4)^k \quad (2.13.8)$$

$$residue\ z=-0.5 \rightarrow \left.\frac{(z^2+z+1)z^{k-1}}{z+0.4}\right|_{z=-0.5} = -7.5(-0.5)^{k-1} = 15(-0.5)^k$$

(2.13.9)

Thereby:

$$f(k) = (5\delta(k) + 19(-0.4)^k + 15(-0.5)^k)u(k) \quad (2.13.10)$$

EXERCISE 2.14

A discrete-time system has a transfer function:

$$H(z) = \frac{4z^2}{z^2 - \frac{1}{4}}$$

a. Calculate the output $y(n)$, if the input is the unit-function $x(n) = n$.
b. Verify the above result, by deriving the first four output values with the aid of the infinite division method.

Solution

a. Calculation of $y(n)$.

$$y(n) = z^{-1}\{Y(z)\} = z^{-1}\{H(z)X(z)\} \Rightarrow$$

$$y(n) = z^{-1}\left\{\frac{4z^2}{z^2 - \dfrac{1}{4}} \cdot \frac{z}{(z-1)^2}\right\} \quad (2.14.1)$$

$$\frac{Y(z)}{z} = \frac{4z^2}{(z-\frac{1}{2})\cdot(z+\frac{1}{2})\cdot(z-1)^2}$$
$$= \frac{k_1}{(z-\frac{1}{2})} + \frac{k_2}{(z+\frac{1}{2})} + \frac{k_{31}}{(z-1)} + \frac{k_{32}}{(z-1)^2} \quad (2.14.2)$$

Calculation of k_i

$$k_1 = \lim_{z \to 1/2} \frac{Y(z)}{z}\left(z - \frac{1}{2}\right) = 4$$

$$k_2 = \lim_{z \to -1/2} \frac{Y(z)}{z}\left(z + \frac{1}{2}\right) = -\frac{4}{9}$$

$$k_{31} = \lim_{z \to 1} \frac{d}{dz}\left(\frac{Y(z)}{z}(z-1)^2\right) = -\frac{32}{9} \tag{2.14.3}$$

$$k_{32} = \lim_{z \to 1} \frac{Y(z)}{z}(z-1)^2 = \frac{16}{3}$$

Thus,

$$Y(z) = 4\frac{z}{z - \frac{1}{2}} - \frac{4}{9}\frac{z}{z + \frac{1}{2}} - \frac{32}{9}\frac{z}{z-1} + \frac{16}{3}\frac{z}{(z-1)^2} \tag{2.14.4}$$

By using the inverse z-transform, the output $y(n)$ of the system is computed.

$$y(n) = \left(4\left(\frac{1}{4}\right)^n - \frac{4}{9}\left(-\frac{1}{2}\right)^n - \frac{32}{9} + \frac{16}{3}n\right)u(n) \tag{2.14.5}$$

b. Verification with the infinite division method

$$Y(z) = \frac{4z^3}{\left(z^2 - \frac{1}{4}\right)\cdot(z-1)^2} = \frac{4z^3}{z^4 - 2z^3 + \frac{3}{4}z^2 + \frac{1}{2}z - \frac{1}{4}} \tag{2.14.6}$$

$4z^3$	$z^4 - 2z^3 + \frac{3}{4}z^2 + \frac{1}{2}z - \frac{1}{4}$
$-4z^3 + 8z^2 - 3z^{-2} + z^{-1}$	$4z^{-1} + 8z^{-2} + 13z^{-3} + \cdots$
$8z^2 - 3z^{-2} + z^{-1}$	
$-8z^2 + 16z^{-6} + 4z^{-1} + 2z^{-2}$	$\Downarrow$
$13z^{-8} - 3z^{-1} + 2z - 2$	$y(0) = 0, y(1) = 4$
$-13z + 26 - \frac{39}{4}z^{-1} - \frac{13}{2}z^{-2} + \frac{13}{4}z^{-3}$	$y(2) = 8, y(3) = 13$

From the expression (2.14.5), for $n = 0, 1, 2, 3$, exactly the same values can be obtained.

(2.14.5)$\Rightarrow$ $n=0 \rightarrow y(0)=0$

$n=1 \rightarrow y(1)=4$

$n=2 \rightarrow y(2)=8$

$n=3 \rightarrow y(3)=13$

EXERCISE 2.15

Compute the inverse z-transform of

$$F_1(z) = \frac{10z}{(z-1)(z-0.5)}, \quad F_2(z) = \frac{2z^3+z}{(z-2)^2(z-1)}$$

$$F_3(z) = \frac{z^2+z+2}{(z-1)(z^2-z+1)}, \quad F_4(z) = \frac{z(z+1)}{(z-1)(z-.25)^2}$$

Solution

a.

$$\frac{F_1(z)}{z} = \frac{k_1}{z-1} + \frac{k_2}{z-0.5} = \frac{20}{z-1} + \frac{20}{z-0.5} \tag{2.15.1}$$

Thus,

$$\begin{aligned} F_1(z) &= 20\frac{z}{z-1} - 20\frac{z}{z-0.5} \Rightarrow \\ f_1(k) &= IZT[F_1(z)] = 20(1-0.5^k)u(k) \end{aligned} \tag{2.15.2}$$

b. *A-solution method*: Partial fraction expansion

$$\frac{F_2(z)}{z} = \frac{2z^2+1}{(z-2)^2(z-1)} = \frac{k_1}{z-1} + \frac{k_{21}}{z-2} + \frac{k_{22}}{(z-2)^2} \tag{2.15.3}$$

$$k_1 = \left.\frac{2z^2+1}{(z-2)^2(z-1)}(z-1)\right|_{z=1} = \frac{3}{1} = 3 \tag{2.15.4}$$

$$k_{22} = \lim_{z\to 2}\left[\frac{2z^2+1}{z-1}\right] = \frac{9}{1} = 9 \tag{2.15.5}$$

$$k_{21} = \lim_{z\to 2}\left[\frac{2z^2+1}{z-1}\right]' = \lim_{z\to 2}\left[\frac{4z(z-1)-(2z^2+1)}{(z-1)^2}\right] = \frac{8-(9)}{1} = -1 \tag{2.15.6}$$

Hence,

$$\frac{F_2(z)}{z} = \frac{3}{z-1} - \frac{1}{z-2} + \frac{9}{(z-2)^2} \tag{2.15.7}$$

or

$$F_2(z) = 3\frac{z}{z-1} - \frac{z}{z-2} + 9\frac{z}{(z-2)^2} \overset{IZT}{\Rightarrow}$$
$$f_2(k) = Z^{-1}[F_2(z)] = (3 - 2^k + 9k2^{k-1})u(k) \tag{2.15.8}$$

B-solution method: Complex integration

$$F_2(z) = \frac{2z^3 + z}{(z-2)^2(z-1)} = \frac{z(2z^2+1)}{(z-2)^2(z-1)} \Rightarrow$$
$$F_2(z)z^{k-1} = \frac{z^k(2z^2+1)}{(z-2)^2(z-1)} \tag{2.15.9}$$

Using the complex integration method, we get

$$residue\Big|_{z=1} \Rightarrow \lim_{z\to 1}\frac{z^k(2z^2+1)}{(z-2)^2} = \frac{3}{1} = 3 \tag{2.15.10}$$

$$residue\Big|_{z=2} \Rightarrow \lim_{z\to 2}\left[\frac{d}{dz}\left(\frac{z^k(2z^2+1)}{z-1}\right)\right] = \lim_{z\to 2}\left[\left(\frac{2z^{k+2}+z^k}{z-1}\right)'\right] =$$

$$\lim_{z\to 2}\left[\frac{\left(2(k+2)z^{k+1} + kz^{k-1}\right)(z-1) - (2z^{k+2}+z^k)}{(z-1)^2}\right] =$$

$$\frac{\left(2(k+2)2^{k+1} + k2^{k-1}\right) - \left(2\cdot 2^{k+2} + 2^k\right)}{1} =$$

$$2k2^{k+1} + 2^2 2^{k+1} + k2^{k-1} - 2\cdot 2^{k+2} - 2^k =$$

$$4k2^k + 8\cdot 2^k + \frac{k}{2}2^k - 8\cdot 2^k - 2^k = \left(4k + \frac{k}{2}\right)2^k - 2^k = \frac{9k}{2}2^k - 2^k$$

$$residue\Big|_{z=2} = 9k\cdot 2^{k-1} - 2^k \tag{2.15.11}$$

so

$$f_2(k) = (3 + 9k2^{k-1} - 2^k)u(k) \tag{2.15.12}$$

c.

$$\frac{F_3(z)}{z}=\frac{z^2+z+2}{z(z-1)(z^2-z+1)}=\frac{k_1}{z}+\frac{k_2}{z-1}+\frac{k_3}{z-e^{j\frac{\pi}{3}}}+\frac{k_4=\overline{k_3}}{z-e^{-j\frac{\pi}{3}}} \quad (2.15.13)$$

$$k_1=\lim_{z\to 0}\left(\frac{F_3(z)}{z}z\right)=\frac{2}{(-1)1}=-2 \quad (2.15.14)$$

$$k_2=\lim_{z\to 1}\left(\frac{F_3(z)}{z}(z-1)\right)=\frac{1+1+2}{(1-1+1)1}=4 \quad (2.15.15)$$

$$k_3=\lim_{z\to e^{j\frac{\pi}{3}}}\left(\frac{F_3(z)}{z}(z-e^{j\frac{\pi}{3}})\right)=\frac{e^{j\frac{2\pi}{3}}+e^{j\frac{\pi}{3}}+2}{e^{j\frac{\pi}{3}}\left(e^{j\frac{\pi}{3}}-1\right)\left(e^{j\frac{\pi}{3}}-e^{-j\frac{\pi}{3}}\right)}=$$

$$\frac{-0.5+j0.866+0.5+j0.866+2}{(0.5+j0.866-1)2j\sin\frac{\pi}{3}e^{j\frac{\pi}{3}}}=\frac{2+j1.732}{(-0.5+j0.866)2j0.866e^{j\frac{\pi}{3}}}=$$

$$\frac{2.645^{\angle 40.9^\circ}}{1^{\angle 60^\circ}1^{\angle 120^\circ}1^{\angle 90^\circ}1.732}=1.53^{\angle -229.1^\circ}$$

$$\overline{k_3}=k_4=1.53^{\angle 229.1^\circ} \quad (2.15.16)$$

Thus,

$$F_3(z)=-2+4\frac{z}{z-1}+1.53e^{-j229.1}\frac{z}{z-e^{j\frac{\pi}{3}}}+1.53e^{j229.1}\frac{z}{z-e^{-j\frac{\pi}{3}}} \quad (2.15.17)$$

$$f_3(k)=-2\delta(k)+4u(k)+1.53e^{-j229.1}(e^{j\frac{\pi}{3}})^k+1.53e^{j229.1}(e^{-j\frac{\pi}{3}})^k \Rightarrow$$

$$f_3(k)=-2\delta(k)+4u(k)+1.53(e^{-j229.1}e^{j\frac{k\pi}{3}}+e^{j229.1}e^{-j\frac{k\pi}{3}})=$$

$$-2\delta(k)+4u(k)+1.53(e^{j(k\frac{\pi}{3}-229.1)}+e^{-j(k\frac{\pi}{3}-229.1)})$$

$$f_3(k)=\left(-2\delta(k)+4u(k)+3.06\cos\left(\frac{k\pi}{3}-229.1\right)\right)u(k) \quad (2.15.18)$$

d.

$$\frac{F(z)}{z}=\frac{(z+1)}{(z-1)(z-0.25)^2}=\frac{A}{z-1}+\frac{B}{(z-0.25)^2}+\frac{C}{(z-0.25)} \quad (2.15.19)$$

Subsequently, another calculation method for A, B, and C is provided.

$$A = \left.\frac{(z+1)}{(z-0.25)^2}\right|_{z=1} = \frac{2}{(0.75)^2} = 3.56 \tag{2.15.20}$$

To derive B, we proceed to a multiplication with $(z-0.25)^2$

$$\frac{(z+1)}{(z-1)} = \frac{A}{z-1}(z-0.25)^2 + B + C(z-0.25)$$

Therefore, at $z = 0.25$, we have

$$B = \left.\frac{(z+1)}{(z-1)}\right|_{z=0.25} = \frac{1.25}{-0.75} = -\frac{5}{4}\frac{4}{3} = -\frac{5}{3} = -1.67 \tag{2.15.21}$$

Calculation of C

$$\frac{d}{dz}\left[\frac{(z+1)}{(z-1)}\right] = \frac{d}{dz}\left[\frac{A}{z-1}(z-0.25)^2\right] + C,$$

$$C = \frac{d}{dz}\left.\left[\frac{(z+1)}{(z-1)}\right]\right|_{z=.25} = \frac{1(z-1)-1(z+1)}{(z-1)^2} \Rightarrow$$

$$C = \left.\frac{-2}{(z-1)^2}\right|_{z=.25} = \frac{-2}{(-0.75)^2} = -3.56 \tag{2.15.22}$$

Thus,

$$F(z) = \frac{(z+1)}{(z-1)(z-0.25)^2} = \frac{3.56z}{z-1} + \frac{-1.67z}{(z-0.25)^2} + \frac{-3.56z}{(z-0.25)} \tag{2.15.23}$$

The second term can be written as

$$-1.67\frac{z}{(z-0.25)^2} = \frac{-1.67}{0.25}\frac{0.25z}{(z-0.25)^2} = -6.68\frac{0.25z}{(z-0.25)^2}$$

Thus,

$$f(k) = \left[3.56 - 6.68 \cdot k(0.25)^k - 3.56(0.25)^k\right]u(k) \tag{2.15.24}$$

EXERCISE 2.16

Compute the inverse z-transform of $X(z) = (z^2 + 6z)/(z^2 - 2z + 2)(z - 1)$

Solution

$$\frac{X(z)}{z} = \frac{k_1}{z-1} + \frac{k_2}{z-1+j} + \frac{k_3}{z-1-j} \tag{2.16.1}$$

The k_i coefficients are calculated by using Heaviside's formula.

$$k_1 = \lim_{z=1} \frac{z^2+6z}{(z^2-2z+2)(z-1)}(z-1) = 7 \tag{2.16.2}$$

$$\begin{aligned} k_3 &= \lim_{z=1+j} \frac{z^2+6z}{(z^2-2z+2)(z-1)}(z-1-j) = \frac{1+j+6}{(1+j-1)(1+j-1+j)} \\ &= \frac{7+j}{j \cdot 2j} = \frac{7+j}{2j^2} = -\frac{7+j}{2} = -\frac{7}{2} - \frac{j}{2} \end{aligned} \tag{2.16.3}$$

$$k_2 = -\frac{7}{2} + \frac{j}{2} = \overline{k_3} \tag{2.16.4}$$

Hence,

$$X(z) = 7\frac{z}{z-1} + \left(-\frac{7}{2} + j\frac{1}{2}\right)\frac{z}{z-1+j} + \left(-\frac{7}{2} - j\frac{1}{2}\right)\frac{z}{z-1-j} \tag{2.16.5}$$

$$(2.16.5) \overset{IZT}{\Rightarrow} x(k) = 7 + \left(-\frac{7}{2} + j\frac{1}{2}\right)\left(\sqrt{2}\right)^k e^{-j\frac{\pi}{4}k} + \left(-\frac{7}{2} - j\frac{1}{2}\right)\left(\sqrt{2}\right)^k e^{j\frac{\pi}{4}k} \Rightarrow$$

$$\begin{aligned} x(k) &= 7 - \frac{7}{2}\left(\sqrt{2}\right)^k e^{-j\frac{\pi}{4}k} - \frac{7}{2}\left(\sqrt{2}\right)^k e^{j\frac{\pi}{4}k} + \\ &\quad + j\frac{1}{2}\left(\sqrt{2}\right)^k e^{-j\frac{\pi}{4}k} - j\frac{1}{2}\left(\sqrt{2}\right)^k e^{j\frac{\pi}{4}k} \Rightarrow \end{aligned}$$

$$x(k) = 7 - \frac{7}{2}\left(\sqrt{2}\right)^k \left[e^{j\frac{\pi}{4}k} + e^{-j\frac{\pi}{4}k}\right] - j\frac{1}{2}\left(\sqrt{2}\right)^k \left[e^{j\frac{\pi}{4}k} - e^{-j\frac{\pi}{4}k}\right] \Rightarrow$$

$$x(k) = 7 - \frac{7}{2}\left(\sqrt{2}\right)^k 2\cos\frac{\pi}{4}k - j\frac{1}{2}\left(\sqrt{2}\right)^k 2j\sin\frac{\pi}{4}k \Rightarrow$$

$$x(k) = \left(7 - 7\left(\sqrt{2}\right)^k \cos\frac{\pi}{4}k + \left(\sqrt{2}\right)^k \sin\frac{\pi}{4}k\right)u(k) \tag{2.16.6}$$

EXERCISE 2.17

Compute the inverse z-transform of $X(z) = (z^3 + 1)/(z^3 - z^2 - z - 2)$

Solution

The denominator of the given function is expressed as

$$A(z) = z^3 - z^2 - z - 2 = (z-2)(z + 0.5 + j0.866)(z + 0.5 - j0.866) \quad (2.17.1)$$

So, $X(z)/z$ is analyzed into a partial fraction expansion, such that

$$\frac{X(z)}{z} = \frac{c_0}{z} + \frac{c_1}{z + 0.5 + j0.866} + \frac{\overline{c}_1}{z + 0.5 - j0.866} + \frac{c_3}{z-2} \quad (2.17.2)$$

The c_i coefficients are calculated by using Heaviside's formula.

$$\left.\begin{aligned} c_0 &= \left[\frac{X(z)}{z}(z)\right]_{z=0} = \frac{1}{-2} = -0.5 \\ c_1 &= \left[\frac{X(z)}{z}(z + 0.5 + j0.866)\right]_{z=-0.5-j0.866} = 0.429 + j0.0825 \\ c_3 &= \left[\frac{X(z)}{z}(z-2)\right]_{z=2} = 0.643 \end{aligned}\right\} \quad (2.17.3)$$

Replacing the corresponding values of Equation 2.17.3 , $X(z)$ is calculated and by using the inverse z-transform, $x[n]$ is obtained.

$$\begin{aligned} X(z) &= c_0 + \frac{c_1 z}{z + 0.5 + j0.866} + \frac{\overline{c}_1 z}{z + 0.5 - j0.866} + \frac{c_3 z}{z-2} \\ &= c_0 + \frac{c_1}{1 + 0.5 + j0.866z^{-1}} + \frac{\overline{c}_1}{1 + 0.5 - j0.866z^{-1}} + \frac{c_3}{1 - 2z^{-1}} \end{aligned} \quad (2.17.4)$$

$$x[n] = c_0\delta[n] + c_1(-0.5 - j0.866)^n u[n] + \overline{c}_1(-0.5 + j0.866)^n u[n] + c_3 2^n u[n] \quad (2.17.5)$$

Based on Equation 2.17.6 and substituting in Equation 2.17.5, we arrive at $x[n]$.

$$\left.\begin{aligned} |p_1| &= \sqrt{(0.5)^2 + (0.866)^2} = 1 \\ \angle p_1 &= \pi + \tan^{-1}\frac{0.866}{0.5} = \frac{4\pi}{3} \text{ rad} \\ |c_1| &= \sqrt{(0.429)^2 + (0.0825)^2} = 0.437 \\ \angle c_1 &= \tan^{-1}\frac{0.0825}{0.429} = 0.19 \text{ rad} \quad (10.89^\circ) \end{aligned}\right\} \quad (2.17.6)$$

$$x[n] = c_0\delta[n] + c_1(-0.5 - j0.866)^n u[n] + \bar{c}_1(-0.5 + j0.866)^n u[n] + c_3 2^n u[n]$$
$$= c_0\delta[n] + 2|c_1||p_1|\cos(\angle p_1 n + \angle c_1) + c_3(2)^n u[n]$$
$$= -0.5\delta[n] + 0.874\cos\left(\frac{4\pi}{3}n + 0.19\right) + 0.643(2)^n u[n] \tag{2.17.7}$$

The above result can be also be verified with the aid of MATLAB®, by implementing the following code:

```
num = [1 0 0 1];
den = [1 -1 -1 -2 0];
[r, p] = residue(num, den)
r =      p =

0.6429         2.0000
0.4286 - 0.825i        -0.5000 + 0.8660i
0.4286 + 0.825i      -0.5000 - 0.8660i
-0.5000      0

The first 20 output samples are calculated by:
num = [1 0 0 1];
den = [1 -1 -1 -2 0];
x = filter(num, den, [1 zeros(1,19)]);
```

EXERCISE 2.18

Compute the inverse z-transform of $F(z) = \dfrac{2z}{(z-2)(z-1)^2}$

Solution

a. Using the division method

$$F(z) = \frac{2z}{z^3 - 4z^2 + 5z - 2}$$

$$\underline{2z^{-2} + 8z^{-3} + 22z^{-4} + 52z^{-5} + 114z^{-6} + \cdots}$$
$$Z^3 - 4z^2 + 5z^{-2}|2z$$
$$\underline{2z^{-8} + 10z^{-1} - 4z^{-2}}$$
$$8 - 10z^{-1} + 04z^{-2}$$
$$\underline{8 - 32z^{-1} + 40z^{-2} - 16z^{-3}}$$
$$22z^{-1} - 36z^{-2} + 016z^{-3}$$
$$\underline{22z^{-1} - 88z^{-2} + 110z^{-3} - 44z^{-4}}$$
$$52z^{-2} - 094z^{-3} + 044z^{-4}$$
$$\underline{52z^{-2} - 208z^{-3} + 260z^{-4} - 104z^{-5}}$$
$$114z^{-3} - 216z^{-4} + 104z^{-5}$$

Hence,

$$F(z) = \sum_{n=0}^{\infty} f_n z^{-n} = 2z^{-2} + 8z^{-3} + 22z^{-4} + 52z^{-5} + 114z^{-6} + \cdots \tag{2.18.1}$$

The first 7 samples of $f(n)$ are

n	0	1	2	3	4	5	6	…
$f(n)$	0	0	2	8	22	52	114	…

b. Using the partial fraction expansion

$$F(z)=\frac{2z}{(z-2)(z-1)^2}=\frac{k_1 z}{z-2}+\frac{k_2 z}{z-1}+\frac{k_3 z}{(z-1)^2} \tag{2.18.2}$$

Another way to calculate k_1 is obtained by multiplying both parts of the equation with $(z–2)$, dividing with z, and setting $z\rightarrow 2$.

$$\frac{2z}{(z-1)^2}=k_1 z+\frac{k_2 z(z-2)}{z-1}+\frac{k_3 z(z-2)}{(z-1)^2} \tag{2.18.3}$$

$$\frac{2}{(z-1)^2}=k_1+\frac{k_2(z-2)}{z-1}+\frac{k_3(z-2)}{(z-1)^2} \tag{2.18.4}$$

$$\left.\frac{2}{(z-1)^2}\right|_{z=2}=k_1+\left.\frac{k_2(z-2)}{z-1}\right|_{z=2}+\left.\frac{k_3(z-2)}{(z-1)^2}\right|_{z=2} \tag{2.18.5}$$

Hence, $k_1 = 2$.

Similarly, in order to obtain k_3, we multiply both parts of the equation with $(z–1)^2$, divide with z and set $z \rightarrow 1$.

$$\frac{2}{(z-2)}=\frac{k_1(z-1)^2}{z-2}+k_2(z-1)+k_3 z \tag{2.18.6}$$

Thus, $k_3 = -2$

In order to calculate k_2, both parts of expression (2.18.6) are differentiated and we set $z\rightarrow 1$.

$$\frac{2}{(z-2)}=\frac{k_1(z-1)^2}{z-2}+k_2(z-1)+k_3 z \tag{2.18.7}$$

$$-\frac{2}{(z-2)^2}=k_1\left[\frac{2(z-1)}{z-2}-\frac{2(z-1)^2}{(z-2)^2}\right]+k_2 \tag{2.18.8}$$

Hence, $k_2 = -2$
Thus, $F(z)$ becomes:

$$F(z) = \frac{2z}{z-2} - \frac{2z}{z-1} - \frac{2z}{(z-1)^2} \tag{2.18.9}$$

EXERCISE 2.19

Compute the inverse z-transform of

a. $H(z) = (z-1)(z+0.8)/(z+0.5)(z+0.2)$,
b. $H(z) = (z^2-1)(z+0.8)/(z-0.5)^2(z+0.2)$

Solution

a. $H(z) = \dfrac{(z-1)(z+0.8)}{(z+0.5)(z+0.2)}$,

$$\frac{X(z)}{z} = \frac{C_1}{z} + \frac{C_2}{z-0.5} + \frac{C_3}{z+0.2}$$

$$C_1 = \frac{X(z)}{z} z\Big|_{z=0} = 8, \qquad C_2 = \frac{X(z)}{z}(z-0.5)\Big|_{z=0.5} = -1.857,$$

$$C_3 = \frac{X(z)}{z}(z+0.2)\Big|_{z=-0.2} = -5.143,$$

$$X(z) = 8 - \frac{1.857z}{z-0.5} - \frac{5.143z}{z+0.2}$$

$$x[n] = 8\delta[n] - 1.857(0.5)^n u[n] - 5.143(-0.2)^n u[n]$$

b. $H(z) = \dfrac{(z^2-1)(z+0.8)}{(z-0.5)^2(z+0.2)}$

$$\frac{X(z)}{z} = \frac{C_1}{z} + \frac{C_2}{z+0.2} + \frac{C_3}{z-0.5} + \frac{C_4}{(z-0.5)^2}$$

$$C_1 = \frac{X(z)}{z} z\Big|_{z=0} = -16, \qquad C_2 = \frac{X(z)}{z}(z+0.2)\Big|_{z=-0.2} = 5.88,$$

$$C_4 = \frac{X(z)}{z}(z-0.5)\Big|_{z=0.5} = -2.79,$$

$$C_3 = \frac{d}{dz}\left(\frac{(z^2-1)(z+0.8)}{z(z+0.2)}\right)\Bigg|_{z=0.5} = 11.12$$

$$X(z) = -16 - \frac{5.88z}{z+0.2} - \frac{11.12z}{z-0.5} - \frac{2.79z}{(z-0.5)^2}$$

$$x[n] = -16\delta[n] + 5.88(-0.2)^n u[n] + 11.12(0.5)^n u[n] - 2.79n(0.5)^n u[n].$$

EXERCISE 2.20

Solve the following difference equation, where the initial values are zero.

$$y[k+2] - 0.3y[k+1] + 0.02y[k] = (0.01)(0.3)^k u[k]$$

Solution

We apply the z-transform in the given equation

$$z^2Y(z) - 0.3zY(z) + 0.02Y(z) = 0.01\frac{z}{z-0.3} \tag{2.20.1}$$

$$(2.20.1) \Rightarrow Y(z)(z^2 - 0.3z + 0.02) = 0.01\frac{z}{z-0.3} \Rightarrow$$

$$Y(z) = \frac{0.01z}{(z-0.1)(z-0.2)(z-0.3)} \tag{2.20.2}$$

$$\frac{Y(z)}{z} = \frac{A}{(z-0.1)} + \frac{B}{(z-0.2)} + \frac{C}{(z-0.3)} \tag{2.20.3}$$

$$\left.\begin{aligned} A &= \left.\frac{0.01}{(z-0.2)(z-0.3)}\right|_{z=0.1} = \frac{0.01}{(-0.1)(-0.2)} = 0.5 \\ B &= \left.\frac{0.01}{(z-0.1)(z-0.3)}\right|_{z=0.2} = \frac{0.01}{(0.1)(-0.1)} = -1 \\ C &= \left.\frac{0.01}{(z-0.1)(z-0.2)}\right|_{z=0.3} = \frac{0.01}{(0.2)(0.1)} = 0.5 \end{aligned}\right\} \tag{2.20.4}$$

Thus,

$$y[k] = (0.5(0.1)^k - (0.2)^k + 0.5(0.3)^k)u[k] \tag{2.20.5}$$

We further elaborate on the solution. From the given difference equation, the first nonzero terms are

$k = 0$: $y[2] - 0.3y[1] + 0.02y[0] = (0.01)$ and $y[2] = 0.001$
$k = 1$ $y[3] - 0.3(0.01) = (0.01)(0.3)$ and $y[3] = 0.006$

From the analytical solution of expression (2.20.5), we have that

$$\begin{aligned} y[0] &= (0.5 - 1 + 0.5) = 0 \\ y[1] &= (0.5(0.1) - (0.2) + 0.5(0.3)) = (0.05 - 2 + 0.15) = 0 \\ y[2] &= (0.5(0.01) - (0.04) - 0.5(0.09)) \\ &= 0.005 - 0.04 + 0.045 = 0.001 \\ y[3] &= (0.5(0.1)^3 - (0.2)^3 + 0.5(0.3)^3) \\ &= 0.0005 - 0.008 + 0.0135 = 0.006 \end{aligned}$$

The derived values match the previously derived solution; so, the result is valid.

EXERCISE 2.21

Solve the second-order difference equation $y[k+2]+5y[k+1]+6y[k] = 5x[k+2]$ with initial conditions $y(0)=0$, $y(1)=2$.

Solution

We apply z-transform in the given equation, such that

$$Z[y(k+2)-5y(k+1)+6y(k)]=0 \Rightarrow$$
$$z^2Y(z)-z^2y(0)-zy(1)-5(z\cdot Y(z)-z\cdot y(0))+6Y(z)=0 \Rightarrow$$
$$z^2Y(z)-2z-5zY(z)+6Y(z)=0 \Rightarrow$$
$$Y(z)(z^2-5z+6)=2z \Rightarrow Y(z)=\frac{2z}{z^2-5z+6} \tag{2.21.1}$$

Then,

$$\frac{Y(z)}{z}=\frac{2}{z^2-5z+6}=\frac{2}{(z-2)(z-3)}=\frac{k_1}{z-2}+\frac{k_2}{z-3} \tag{2.21.2}$$

$$k_1=\lim_{z\to 2}\frac{Y(z)}{z}(z-2)=-2 \tag{2.21.3}$$

$$k_2=\lim_{z\to 3}\frac{Y(z)}{z}(z-3)=2 \tag{2.21.4}$$

Thus,

$$Y(z)=z\left[\frac{Y(z)}{z}\right]=-\frac{2z}{z-2}+\frac{2z}{z-3} \Rightarrow$$

$$y(k)=IZT\left[Y(z)\right]=(-2(2)^k+2(3)^k)u(k) \tag{2.21.5}$$

EXERCISE 2.22

Solve the second-order difference equation $y(k+2)-5y(k+1)+6y(k)=0$ with the initial conditions $y(0)=-12$, $y(1)=59$.

Solution

We apply z-transform in the given equation and we obtain $(Y(z)/z)$, such that

$$z^2Y(z) - z^2y(0) - zy(1) + 5[zY(z) - zy(0)] + 6Y(z) = 5\times\frac{z}{z-1}$$

$$z^2Y(z) - z^2\times(-12) - z\times(59) + 5zY(z) - 5z\times(-12) + 6Y(z) = 5\times\frac{z}{z-1}$$

$$(z^2 + 5z + 6)Y(z) = 12z^2 + z + 5\times\frac{z}{z-1}$$

$$\begin{aligned} Y(z) &= \frac{12z^2 + z}{z^2 + 5z + 6} + \frac{1}{z^2 + 5z + 6}\times\frac{5z}{z-1} \\ \frac{Y(z)}{z} &= \frac{(12z+1)(z-1)+5}{(z+2)(z+3)(z-1)} = \frac{K_1}{z+2} + \frac{K_2}{z+3} + \frac{K_3}{z-1} \end{aligned} \tag{2.22.1}$$

We calculate the corresponding numerators as

$$\begin{aligned} K_1 &= \frac{(12z+1)(z-1)+5}{(z+2)(z+3)(z-1)}\times(z+2)\bigg|_{z=-2} = \frac{64}{3} \\ K_2 &= \frac{(12z+1)(z-1)+5}{(z+2)(z+3)(z-1)}\times(z+3)\bigg|_{z=-3} = \frac{-135}{4} \\ K_3 &= \frac{(12z+1)(z-1)+5}{(z+2)(z+3)(z-1)}\times(z-1)\bigg|_{z=1} = \frac{5}{12} \end{aligned} \tag{2.22.2}$$

Substituting the values of K_i into $Y(z)/z$, we multiply with z, hence,

$$Y(z) = \frac{64/3z}{z+2} + \frac{135/4z}{z+3} + \frac{5/12z}{z-1} \tag{2.22.3}$$

By using the inverse z-transform, we get

$$y(k) = \left[\frac{64}{3}(-2)^k - \frac{135}{4}(-3)^k + \frac{5}{12}\right]u(k) \tag{2.22.4}$$

EXERCISE 2.23

Calculate the unit-step response of a system with the difference equation

$$y[n] + 1.5y[n-1] + 0.5y[n-2] = x[n] - x[n-1], \quad y[-1] = 2,\ y[-2] = 1$$

Solution

We apply z-transform in the given equation taking into consideration the initial conditions.

$$
\begin{aligned}
Y(z) &= \frac{-(a_1y[-1]+a_2y[-2])z^2 - a_2y[-1]z}{z^2+a_1z+a_2} + \frac{b_0z^2+b_1z}{z^2+a_1z+a_2}X(z) \\
&= \frac{-((1.5)(2)+(0.5)(1))z^2-(0.5)(2)z}{z^2+1.5z+0.5} + \frac{z^2-z}{z^2+1.5z+0.5}\left(\frac{z}{z-1}\right) \\
&= \frac{-3.5z^2-z}{z^2+1.5z+0.5} + \frac{z^2}{z^2+1.5z+0.5} \qquad (2.23.1)
\end{aligned}
$$

Equation (2.23.1) can be further simplified to

$$
Y(z) = \frac{-2.5z^2-z}{z^2+1.5z+0.5} = \frac{0.5z}{z+0.5} - \frac{3z}{z+1} \qquad (2.23.2)
$$

The inverse z-transform gives

$$
y[n] = 0.5(-0.5)^n - 3(-1)^n, \quad n = 0, 1, 2, \ldots \qquad (2.23.3)
$$

The corresponding MATLAB code, which calculates and depicts the system's response is

```
num = [1 -1 0];
den = [1 1.5.5];
n = 0:20;
x = ones(1, length(n));
zi = [-1.5*2-0.5*1, -0.5*2];
y = filter(num, den, x, zi);
stem(y,'Linewidth',3)
```

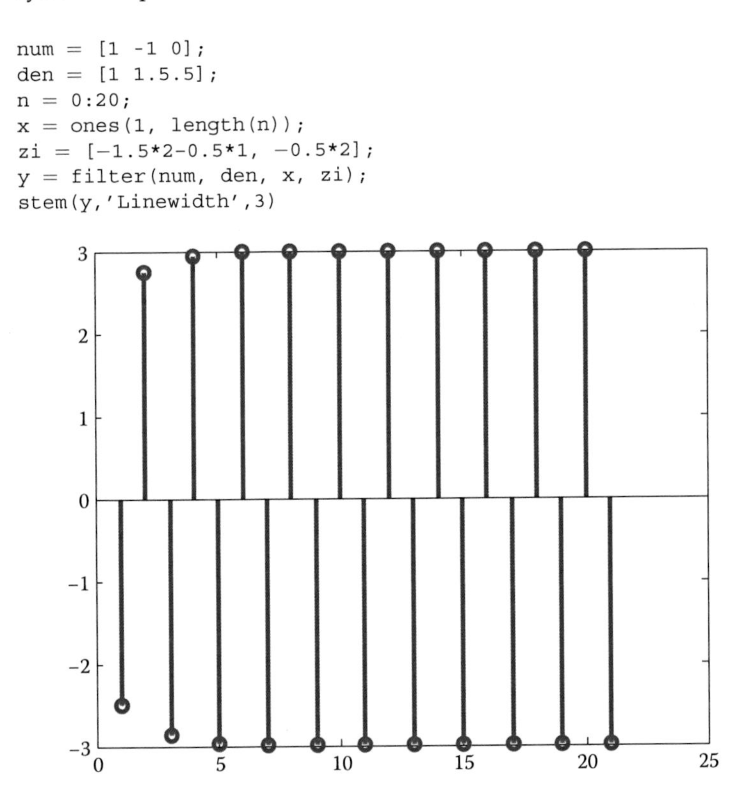

EXERCISE 2.24

Compute the z-transform of function $x(n) = [3\ 5\ 4\ 3]$, $0 \le n \le 3$ using MATLAB.

Solution

A—solution method

```
syms z
x0=3; x1=5; x2=4; x3=3;
Xz=x0*(z^0)+x1*(z^-1) +x2*(z^-2)+x3*(z^-3)
pretty(X)
```

$$X(z) = 3 + \frac{5}{z} + \frac{4}{z^2} + \frac{3}{z^3}$$

B—solution method

```
syms z
x=[3 5 4 3];
n=[0 1 2 3];
X=sum(x.*(z.^-n))
pretty(X)
```

$$X(z) = 3 + \frac{5}{z} + \frac{4}{z^2} + \frac{3}{z^3}$$

EXERCISE 2.25

Compute the z-transform of $f(n) = 2^n$ using MATLAB.

Solution

```
syms n z
f = 2^n;
ztrans(f)
simplify(ans)
```

$$F(z) = z/(z-2)$$

To verify the above result, we calculate the inverse z-transform of $F(z) = z/(z - 2)$;

```
syms n z
F = z/(z - 2);
iztrans(F)
ans = 2^n
```

EXERCISE 2.26

Compute the z-transform of: $\delta(n)$, $u(n)$, $n \cdot u(n)$, $a^n u(n)$, $na^n u(n)$, $\cos(\omega_0 n)u(n)$, $\sin(\omega_0 n)u(n)$, $a^n\cos(\omega_0 n)u(n)$, $a^n\sin(\omega_0 n)u(n)$ using MATLAB.

Solution

In the following table, the main transform relations are provided, and their accuracy is verified with the aid of *ztrans* and *iztrans* functions.

It is noteworthy that the ztrans function calculates the single-sided transform Ztrans $(n \geq 0)$" so the unit-step function can be omitted.

Discrete-Time Domain	z-Domain	Commands	Result
$x(n)$	$X(z)$	syms n z a w	
$\delta(n)$	1	f=dirac(n); ztrans(f,z)	ans = dirac(0) % $\delta(0)=1$
$u(n)$	$\frac{z}{(z-1)}$	f=heaviside(n) ztrans(f,z)	ans = z/(z−1)
$n \cdot u(n)$	$\frac{z}{(z-1)^2}$	ztrans(n,z)	ans = z/(z−1)^2
$a^n u(n)$	$\frac{z}{z-a}$	F=z/(z−a); f=iztrans(F,n)	f = a^n
$n\,a^n u(n)$	$\frac{a\,z}{(z-a)^2}$	f=n*a^n; ztrans(f,z)	ans = z*a/(−z+a)^2
$\cos(\omega_0 n)u(n)$	$\frac{z^2 - z\cos(\omega_0)}{z^2 - 2z\cos(\omega_0)+1}$	f=cos(w*n) ztrans(f,z)	ans = (−z+cos(w))*z/ (−z^2+2*z*cos(w)−1)
$\sin(\omega_0 n)u(n)$	$\frac{z\sin(\omega_0)}{z^2 - 2z\cos(\omega_0)+1}$	f=sin(a*n); ztrans(f,z)	ans = z*sin(a)/ (z^2−2*z*cos(a)+1)
$a^n \cos(\omega_0 n)u(n)$	$\frac{z^2 - a\,z\cos(\omega_0)}{z^2 - 2a\,z\cos(\omega_0)+a^2}$	f=(a^n)*cos(a*n) ztrans(f,z) simplify(ans)	ans = −(−z+cos(a)*a)*z/ (z^2−2*z* cos(a)*a+a^2)
$a^n \sin(\omega_0 n)u(n)$	$\frac{az\sin(\omega_0)}{z^2 - 2az\cos(\omega_0)+a^2}$	f=(a^n)*sin(a*n); ztrans(f,z) simplify(ans)	ans = z*sin(a)*a/(z^2−2*z* cos(a)*a+a^2)

EXERCISE 2.27

Rewrite the following function in a partial fraction expansion formulation.

$$X(z) = \frac{z^2 + 3z + 1}{z^3 + 5z^2 + 2z - 8}$$

Solution

```
% Calculation of denominator's roots
A=[1 5 2 -8];
riz=roots(A);
% Calculation of numerators of the partial fractions
```

```
syms z
X=(z^2+3*z+1)/(z^3+5*z^2+2*z-8);
c1=limit((z-riz(1))*X,z,riz(1))
c2=limit((z-riz(2))*X,z,riz(2))
c3=limit((z-riz(3))*X,z,riz(3))
```

Hence, we get

$$X(z)=\frac{z^2+3z+1}{z^3+5z^2+2z-8}=\frac{1/2}{z+4}+\frac{1/6}{z+2}+\frac{1/3}{z-1}$$

EXERCISE 2.28

Rewrite the following function in a partial fraction expansion formulation.

$$X(z)=\frac{z^2+3z+1}{z^3-3z+2}$$

Solution

A—solution method

```
% Calculation of denominator's roots
>> A=[1 0 -3 2];
>> riz=roots(A)
riz = -2.0000  1.0000  1.0000
```

Observe the existence of a double root at point 1.0000

$X(z)$ can be expressed as

$$X(z)=\frac{c_1}{z-\lambda_1}+\frac{c_2}{(z-\lambda_1)^2}+\cdots+\frac{c_r}{(z-\lambda_1)^r}+\frac{c_{r+1}}{z-\lambda_{r+i}}+\cdots+\frac{c_n}{z-\lambda_n}.$$

The coefficients $c_1 \ldots c_n$ can be calculated as

$$\begin{aligned} c_i &= \lim_{z\to\lambda_i}\frac{1}{(r-i)!}\frac{d^{r-1}((s-\lambda_i)^r X(z))}{dz^{r-1}}, && i=1,\ldots,r \\ c_i &= \lim_{z\to\lambda_i}(z-\lambda_i)X(z), && i=r+1,\ldots,n \end{aligned}.$$

```
% Calculation of c1
syms z
X=(z^2+3*z+1)/(z^3-3*z+2);
c1=limit((z-riz(1))*X,z,riz(1))
% Calculation of c2 (i=1) - 2 common roots (r=2)
r=2
% definition of (z-λi)^r X(z)
f=((z-1)^r)*X;
% Definition d^(r-1)((z - λi)^r X(z))/dz^(r-1)
par=diff(f,z,r-1);
```

```
% Calculation 1/(r − i)
fact=1/factorial(r-1);
% Calculation c2
c2=limit(fact*par,z,1)
% Calculation c3 (i=2)
par=diff(f,z,r-2);
fact=1/factorial(r-2);
limit(fact*par,z,1)
```

Therefore,

$$X(z) = \frac{z^2+3z+1}{z^3-3z+2} = \frac{-1/9}{z+2} + \frac{10/9}{z-1} + \frac{5/3}{(z-1)^2}$$

B—solution method

$X(z)$ can be converted to partial fraction expansion via the *residue* command.

```
% Define the coefficients of numerator and denominator
num=[ 1 3 1];
den=[ 1 0 -3 2]
% Use of residue command
[R,P,K]=residue(num,den)
```

$X(z)$ can now be expressed as a fractional expansion $X(z) = \frac{-1/9}{z+2} + \frac{10/9}{z-1} + \frac{5/3}{(z-1)^2}$, which is the same as in the previous result.

EXERCISE 2.29

Expand the following function into partial fractions $X(z) = 3z^3 + 8z^2 + 4/z^2 + 5z + 4$.

Solution

```
n= [ 3 8 0 4]
d=[ 1 5 4];
[R,P,K]=residue(n,d);
```

$X(z)$ can be written as

$$X(z) = \frac{20}{z+4} + \frac{3}{z+1} + 3z - 7$$

Let us verify the above result using an alternative definition of the residue command.

```
R=[ 20 3];
P=[-4 -1];
K=[ 3 -7];
[B,A]=residue(R,P,K);
```

Then, $X(z) = B(z)/A(z) = 3z^3 + 8z^2 + 4/z^2 + 5z - 4$

EXERCISE 2.30

Solve the following difference equation $y(n) + 0.5y(n-1) + 2y(n-2) = 0.9^n$

Solution

The general definition of a difference equation is provided by

$$y(n) = \sum_{k=0}^{q} b_k x(n-k) + \sum_{k=1}^{p} a_k y(n-k), \quad a_i,\, i = 1,\ldots,n \quad \text{constants}$$

The following procedure is applied:
Taking z-transform at both parts of the given equation

$$Z\{y(n) + 0.5y(n-1) + 2y(n-2)\} = Z\{0.9^n\}$$

Due to the linearity, we get

$$Z\{y(n)\} + 0.5Z\{y(n-1)\} + 2Z\{y(n-2)\} = \frac{10z}{10z-9}$$

Calculate the z-transforms of the following

$$Y(z) + 0.5z^{-1}Y(z) + 2z^{-2}Y(z) = \frac{z}{z-0.9}$$

Solve the resultant expression in terms of $Y(z)$:

$$Y(s) = \frac{z}{(z-0.9)(1+0.5z^{-1}+2z^{-2})} = \frac{z^3}{(z-0.9)(z^2+0.5z+2)}$$

Calculate the inverse z-transform of $Y(z)$, i.e., calculate $y(n)$, which is the desired solution.

The corresponding MATLAB code is given as follows:

```
syms t s Y
X=ztrans(0.9^n,z) % z - transform for the 2nd part of the
difference equation
Y1=z^(-1)*Y; % Define Z{y(n − 1) } as Y1
Y2=z^(-2)*Y; % Define Z{y(n − 2) } as Y2
```

Next, we take X into the left part of the difference equation and we define a variable G, which is equal to the entire left part (a polynomial of Y).

```
G=Y + 0.5*Y1 + 2*Y2-X; % Mainly, G includes variables Y and z
SOL=solve(G,Y); % Solution of Y using the solve command
```

Hence, we get

$$\frac{z^3}{(z-0.9)(z^2+0.5z+2)}$$

```
y=iztrans(SOL,n); % the inverse z-transform is applied and
the desired result is obtained.
```

EXERCISE 2.31

a. Solve the following difference equation, using the z-transform: $y(n) - y(n-1) = u(n)$.
b. Plot the results for the range $0 \leq n \leq 50$.

Solution

a.
```
syms n z Y
x=heaviside(n);
X=ztrans(x,z);
Y1=z^(-1)*Y;
G=Y-Y1-X;
SOL=solve(G,Y);
y=iztrans(SOL,n)
```

The solution for $y(n)$ is y = 1 + n

b.
```
n1=0:50;
yn=subs(y,n,n1);
stem(n1,yn);
legend('LISI y(n)');
```

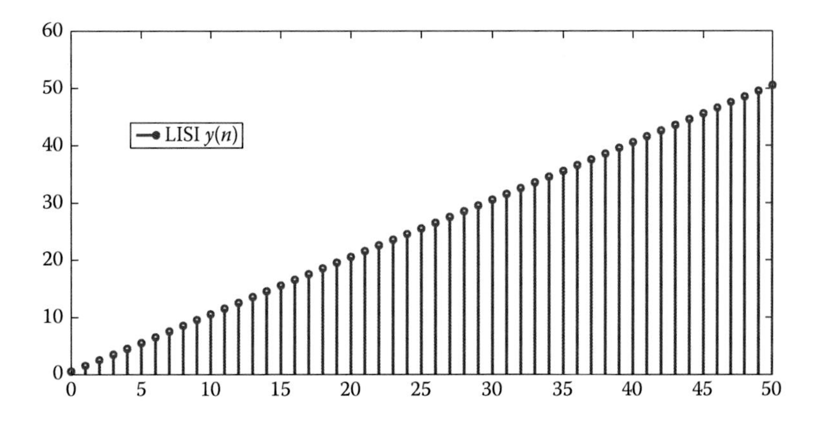

EXERCISE 2.32

a. Solve the following difference equations using the z-transform

$$y(n)-y(n-1)=x(n)+x(n-1) \quad \text{and}$$
$$y(n)+1.5y(n-1)+0.5y(n-2)=x(n)+x(n-1)$$

where $x(n)=0.8^n$.

b. Plot the results for the range $0 \le n \le 20$.

c. Verify the accuracy of the solution by replacing the result in the given difference equation.

Solution

1. $y(n)-y(n-1)=x(n)+x(n-1)$, $x(n)=0.8^n$

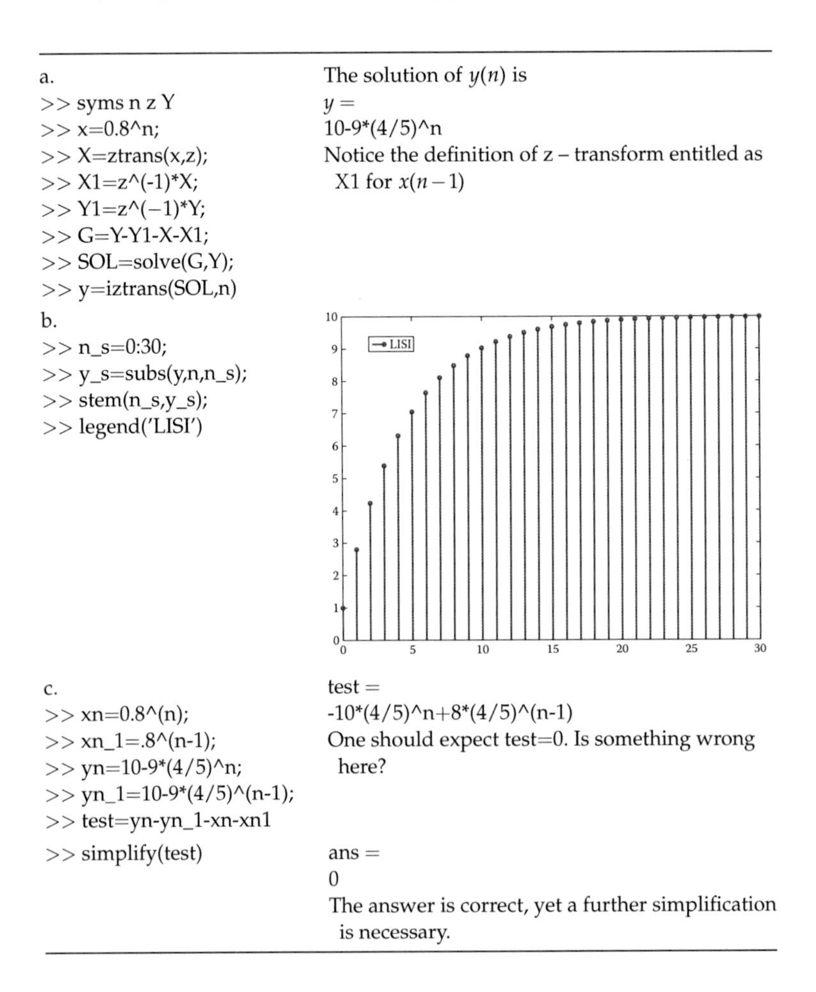

a.

```
>> syms n z Y
>> x=0.8^n;
>> X=ztrans(x,z);
>> X1=z^(-1)*X;
>> Y1=z^(−1)*Y;
>> G=Y-Y1-X-X1;
>> SOL=solve(G,Y);
>> y=iztrans(SOL,n)
```

The solution of $y(n)$ is

y =

10-9*(4/5)^n

Notice the definition of z – transform entitled as X1 for $x(n-1)$

b.

```
>> n_s=0:30;
>> y_s=subs(y,n,n_s);
>> stem(n_s,y_s);
>> legend('LISI')
```


c.

```
>> xn=0.8^(n);
>> xn_1=.8^(n-1);
>> yn=10-9*(4/5)^n;
>> yn_1=10-9*(4/5)^(n-1);
>> test=yn-yn_1-xn-xn1
```

test =

-10*(4/5)^n+8*(4/5)^(n-1)

One should expect test=0. Is something wrong here?

```
>> simplify(test)
```

ans =

0

The answer is correct, yet a further simplification is necessary.

2. $y(n)+1.5y(n-1)+0.5y(n-2)=x(n)+x(n-1)$, $x(n)=0.8^n$

a.

```
>> syms n z Y
>> x=0.8^n;
>> X=ztrans(x,z);
>> X1=z^(-1)*X;
>> Y1=z^(-1)*Y;
>> Y2=z^(-2)*Y;
>> G=Y+1.5*Y1+0.5*Y2-X-X1;
>> SOL=solve(G,Y);
>> y=iztrans(SOL,n)
```

The solution of $y(n)$ is

```
y=
5/13*(−1/2)^n+8/13*(4/5)^n
```

b.

```
>> n_s=0:20;
>> y_s=subs(y,n,n_s);
>> stem(n_s,y_s);
>> legend('LISI y(n)')
```

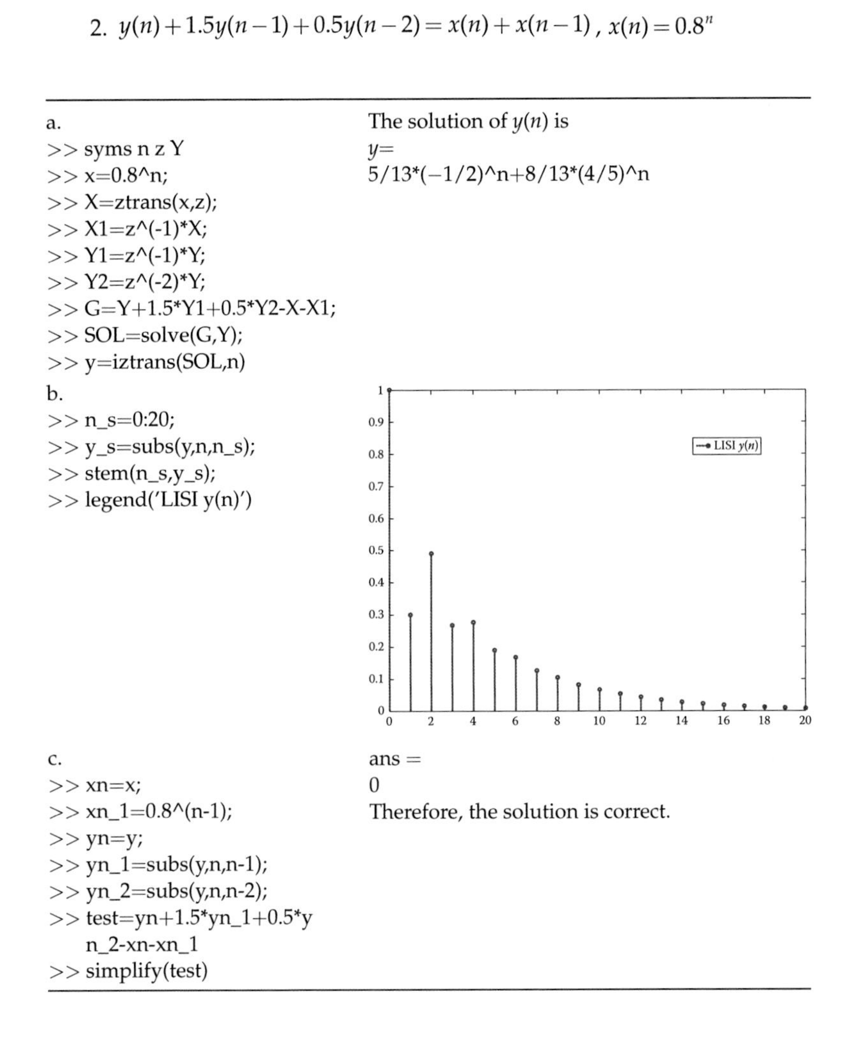

c.

```
>> xn=x;
>> xn_1=0.8^(n-1);
>> yn=y;
>> yn_1=subs(y,n,n-1);
>> yn_2=subs(y,n,n-2);
>> test=yn+1.5*yn_1+0.5*y
   n_2-xn-xn_1
>> simplify(test)
```

```
ans =
0
```

Therefore, the solution is correct.

EXERCISE 2.33

Solve the following difference equation: $y(k+2)+0.8y(k+1)+0.1y(k) = u(k)$ using MATLAB.

Solution

We will use the function **`recur.m`** (function y = recur(a,b,n,x,x0,y0);), where the solution of a difference equation of the following form is provided

$$y[n]+a1*y[n-1]+a2*y[n-2]\cdots+an*y[n-N]=$$
$$b0*x[n]+b1*x[n-1]+\cdots+bm*x[n-M]$$

The corresponding MATLAB code is

```
% a = [a1 a2 ... aN], b = [b0 b1 ... bM]
a = [0.8 0.1]; b = [0 0 1];
x0 = [0 0]; y0 = [0 0];
k =0:5;
x = ones(1,6);
y = recur(a,b,k,x,x0,y0)
stem(k,y);xlabel('k');ylabel('y(k)');title('Stem plot of
system step response')
```

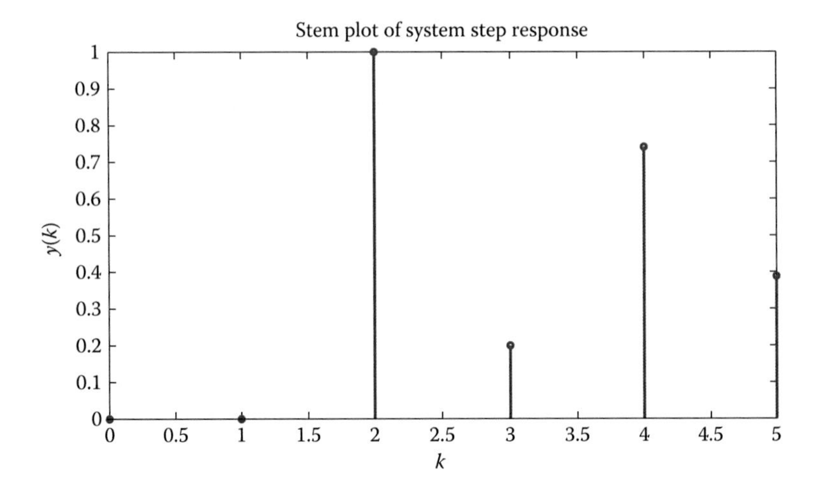

The function recur.m can be retrieved from the following web-address: https://www.mathworks.com/matlabcentral/fileexchange/2148-fundamentals-of-signals-and-systems-using-the-web-and-matlab/content/recur.m

3

Transfer Function

3.1 Introduction

The transfer function is defined as the ratio of the z-transform of the output for a linear invariant system to the z-transform of its input, when the initial conditions are zero and corresponds to a relation which describes the dynamics of the system under consideration.

Consider the system of Figure 3.1. Its transfer function is given by

$$G(z) = \frac{Y(z)}{X(z)} = 0 \tag{3.1}$$

The transfer function represents the z-transform of the impulse response.

3.2 Open-Loop Sampled-Data Control System

Consider the system of Figure 3.2, which corresponds to an open-loop sampled-data system.

In this case, the output ($Y(z)$) of the given system is

$$Y(z) = Z[G_0(s)G_p(s)]G_c(z)X(z) = G(z)G_c(z)X(z) \tag{3.2}$$

where

$G_p(s)$ is the transfer function of the system under control,

$G_c(z)$ is the transfer function of the digital controller,

$G(z)$ is the transfer function of the analog system in discrete time, and

$G_0(s)$ is the transfer function of the system with which $G_p(s)$ is discretized.

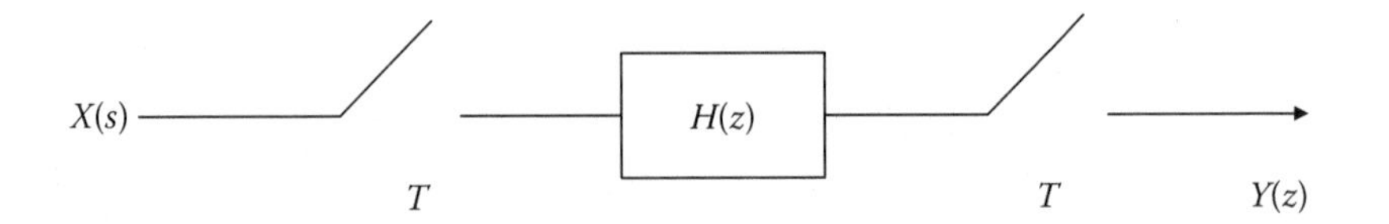

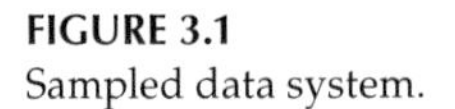
FIGURE 3.1
Sampled data system.

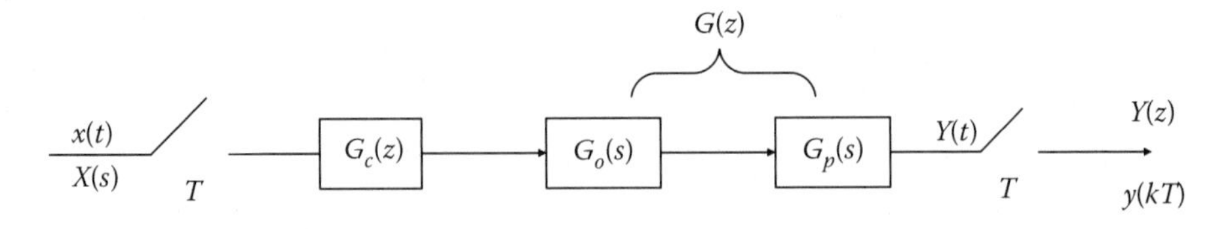

FIGURE 3.2
Open-loop sampled-data system.

3.3 Closed-Loop Sampled-Data Control System

Consider the system of Figure 3.3, which corresponds to a closed-loop sampled-data control system.

In this case, the output ($Y(z)$) of the given system is

$$Y(z) = \frac{G_c(z)G(z)}{1 + G_c(z)Z[H(s)G_0(s)G_p(s)]} \tag{3.3}$$

The expression (3.3) deduced from the above block diagram and will subsequently be explained.

According to Figure 3.3 we have that

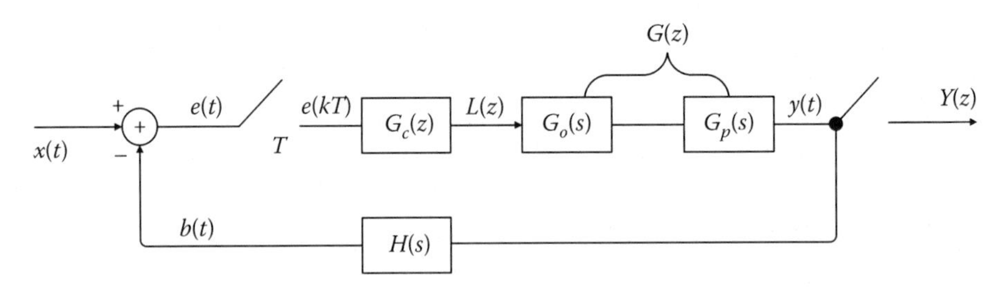

FIGURE 3.3
Closed-loop sampled-data system.

$Y(z) = G(z)G_c(z)E(z) \quad (1)$
$E(z) = X(z) - B(z) \quad (2)$
$B(z) = Z[H(s)G_0(s)G_p(s)]L(z) \quad (3)$
$L(z) = G_c(z)E(z) \quad (4)$
$(3),(4) \Rightarrow B(z) = Z[H(s)G_0(s)G_p(s)]G_c(z)E(z) \quad (5)$
$(2),(5) \Rightarrow E(z) = X(z) - Z[H(s)G_0(s)G_p(s)]G_c(z)E(z) \ (6)$

Thus

$$E(z) = \frac{X(z)}{1 + G_c(z)Z[H(s)G_0(s)G_p(s)]} \quad (7)$$

$$(1),(7) \Rightarrow Y(z) = \frac{G_c(z)G(z)}{1 + G_c(z)Z\left[H(s)G_0(s)G_1(s)\right]} \quad (8)$$

The transfer function of the closed-loop system is

$$G_{cl}(z) = \frac{Y(z)}{X(z)} = \frac{G_c(z)G(z)}{1 + G_c(z)Z\left[H(s)G_0(s)G_p(s)\right]} \tag{3.4}$$

For $H(s) = 1$ it holds that

$$G_{cl}(z) = \frac{G_c(z)G(z)}{1 + G_c(z)G(z)} \tag{3.5}$$

It should be noted that the expression (3.3), which provides the system's output in z-domain, can be dynamically changed according to the form of the sampled data system, correspondingly.

In the Formula Tables of this chapter, several block diagrams for digital control systems are presented, including their corresponding transfer function in terms of z.

3.4 Signal Flow Graphs

Signal flow graphs (S.F.Gs), similar to block diagrams, provide an overview of the system and represent an alternative representation of the relationship among the variables of the system. S.F.G theory was developed by *S. J. Mason* (July 1953) and is implemented in any system without the need to simplify the functional diagram, which is a particularly laborious process for complex diagrams. A flow graph consists of nodes, branches, and loops.

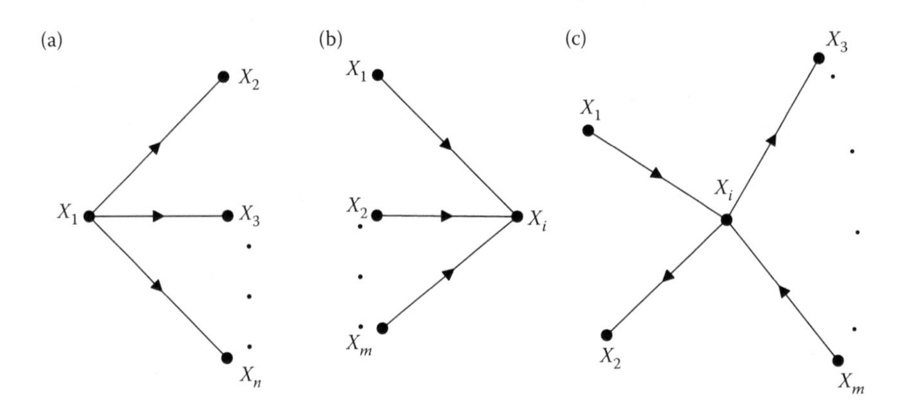

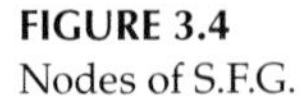
FIGURE 3.4
Nodes of S.F.G.

Each *node* denotes a certain variable (signal) and corresponds to one of the following categories:

a. *Source or input node*: The node from which one or more branches start and in which no branch ends (Figure 3.4a).
b. *Sink node*: The node which receives one or more branches and from which no branch starts (Figure 3.4b).
c. *Mixed node*: The node which has incoming and outcoming branches (Figure 3.4c).

In addition, each branch connects two nodes and can be described by two features, namely, the direction and gain. The *direction* stands for the signal's direction from one node to another (Figure 3.5), while its *gain* is the a factor or the transfer function, which connects the variables x_1 and x_2. The direction and gain are related to each other as $x_2 = ax_1$.

Path is a branch sequence of the same direction (e.g., x_1, x_2, x_3, x_4 in Figure 3.6).

Forward path is the end-to-end path between the input and output node (e.g., x_1, x_2, x_3, x_4 in Figure 3.6).

Loop represents the closed path which starts and ends at the same node (e.g., x_2, x_3, x_2 in Figure 3.6). Two loops, within S.F.G., are called *nontouching* loops in the case when they do not contain any common node to each other.

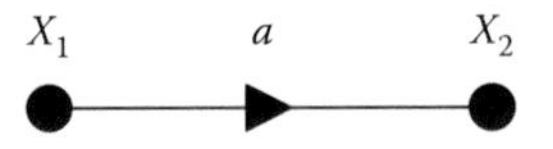

FIGURE 3.5
Branch of S.F.G.

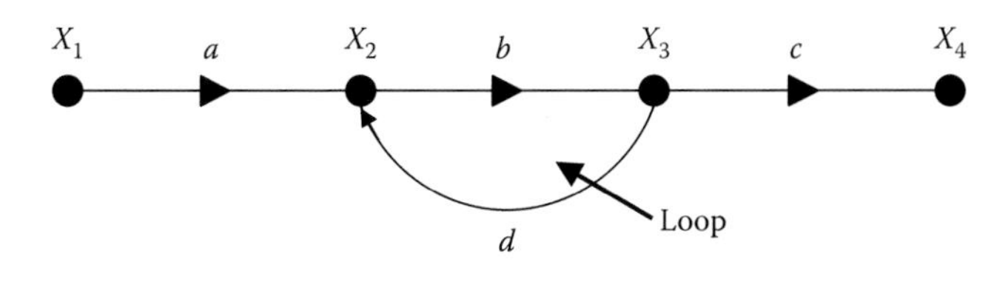

FIGURE 3.6
Loop of S.F.G.

3.5 Mason's Formula

Mason's formula (Mason's gain formula—1953) provides the relation between the input and output of a given system via S.F.G., directly, without successive simplifications and is expressed as

$$G(s) = \frac{\sum_{n=1}^{k} T_n \Delta_n}{\Delta} \tag{3.6}$$

where

T_n = the gain of the n-th direct path which connects the input and output.

Δ = The determinant of the graph, which is given by

$$\Delta = 1 - \sum L_1 + \sum L_2 - \sum L_3 + \cdots \tag{3.7}$$

with

$$\begin{cases} L_1 & \text{is the loop gain} \\ \Sigma L_1 & \text{is the sum of each loop gain at the signal flow graph} \end{cases}$$

$$\begin{cases} L_2 & \text{is the gain product of two nontouching loops} \\ \Sigma L_2 & \text{is the sum of all the per two nontouching loop gains} \end{cases}$$

and so on.

Δ_n = The subdeterminant of T_n, which is obtained from (3.7), without taking into account the branches that are adjacent to the nth forward path.

In the case when we deal with a complex block diagram and we need to calculate the control ratio O/I (output/input), then we convert the block diagram to S.F.G. and apply the general output–input gain equation.

The exact procedure followed using Mason's formula in digital control systems is

1. Design S.F.G. directly from the block diagram of the sampled data system.
2. Observing the flow graph, one can notice certain discontinuities due to the involvement of the samplers; thus, we are able to redesign S.F.G. removing the latter discontinuities, with the aid of the mathematical model. Apparently, the newly derived S.F.G. is equivalent to the previous one.
3. Calculate the transfer function by using Mason's formula at the modified S.F.G.

3.6 Difference Equations

Difference equations correspond to discrete-time systems, while differential equations correspond to continuous-time systems. In the difference equation: $y[n] + y[n-2] = x[n]$, $x[n]$ denotes the system's input, which is a known signal, while $y[n]$ is the corresponding output, which represents the unknown parameter. The question posed by the above difference equation is as follows: Determine a signal $y[n]$, which gives a known signal $x[n]$, when is added to itself, yet shifted by 2 units to the right, namely, $y[n-2]$.

The general form of an N-degree difference equation is

$$\sum_{k=0}^{N} b_k y[n-k] = \sum_{m=0}^{M} \alpha_m x[n-m] \tag{3.8}$$

and to solve it, $(N+M)$ initial conditions of $y[-1], y[-2], \ldots, y[-N]$ and $x[-1], x[-2], \ldots, x[-M]$ should be known.

NOTE: In the solved exercises presented below, the concept of sampled signal will be used. The sample signal extracted from the sampling of its analog counterpart is mathematically expressed as (Figure 3.7):

$$y^*(t) = x(t)\cdot\delta_T(t) = x(t)\cdot\sum_{k=0}^{\infty}\delta(t-kT) = \sum_{k=0}^{\infty} x(kT)\delta(t-kT) \tag{3.9}$$

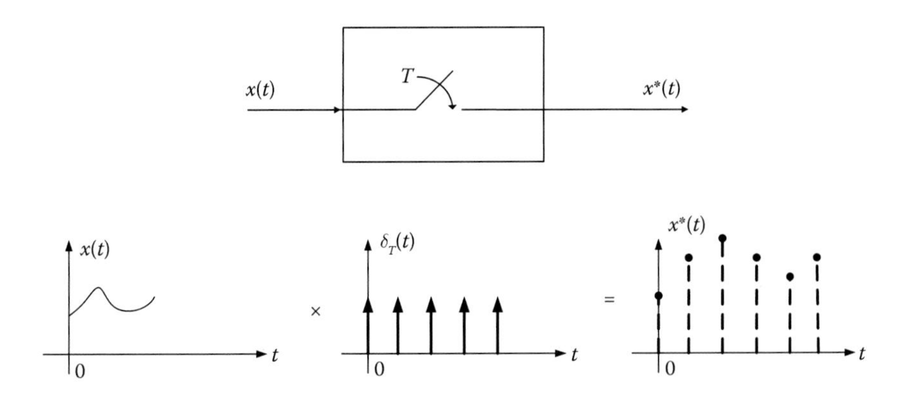

FIGURE 3.7
Sampled signal.

Since: $\delta_T(t) = \sum_{k=0}^{\infty} \delta(t - kT) = \sum_{k=-\infty}^{+\infty} C_n e^{jk\omega_s t} = \frac{1}{T} \sum_{k=-\infty}^{+\infty} e^{jk\omega_s t} \leftarrow \omega_s = 2\pi/T$

$$\left(C_n = \frac{1}{T} \int_{-T/2}^{T/2} \delta_T(t) e^{jk\omega_s t} dt = \frac{1}{T} \int_{0_-}^{0_+} \delta_T(t) e^{jk\omega_s t} dt = \frac{1}{T} \int_{0_-}^{0_+} \delta_T(t) dt = \frac{1}{T} \right)$$

We have

$$x^*(t) = x(t) \cdot \frac{1}{T} \sum_{k=-\infty}^{+\infty} e^{jk\omega_s t} \xrightarrow[\text{transformation}]{\text{Laplace}} X^*(s) = \frac{1}{T} \sum_{k=-\infty}^{+\infty} X(s + jk\omega_s) \tag{3.10}$$

$$\left(\text{where: } X(s) = L[x(t)] = \int_{0}^{\infty} x(t) e^{-st} dt \right)$$

$$\text{The frequency spectrum of } x^*(t): \quad X^*(j\omega) = \frac{1}{T} \sum_{k=-\infty}^{+\infty} X[j(\omega + k\omega_s)] \tag{3.11}$$

The frequency spectrum of $x(t)$ and $x^*(t)$ are presented in Figure 3.8a and b, respectively.

It holds that

$$Z[X^*(s)] = X(z) \tag{3.12}$$

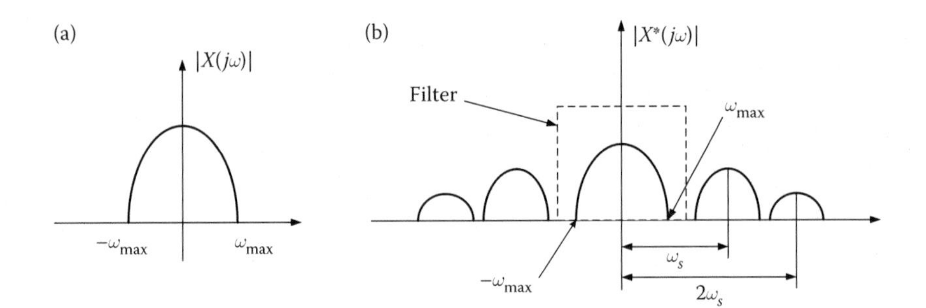

FIGURE 3.8
Frequency spectrum of analog and sampled signal.

3.7 Formula Tables

Tables 3.1 throguh 3.3.

TABLE 3.1
Transfer Function of Closed-Loop Sampled-Data Systems

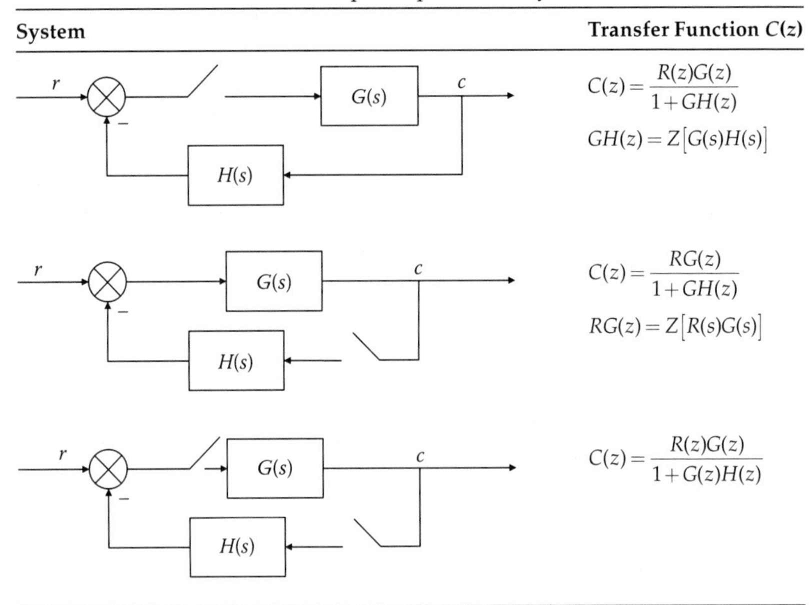

System	Transfer Function $C(z)$
(diagram 1)	$C(z)=\dfrac{R(z)G(z)}{1+GH(z)}$ $GH(z)=Z[G(s)H(s)]$
(diagram 2)	$C(z)=\dfrac{RG(z)}{1+GH(z)}$ $RG(z)=Z[R(s)G(s)]$
(diagram 3)	$C(z)=\dfrac{R(z)G(z)}{1+G(z)H(z)}$

TABLE 3.2

Transfer Function of Closed-Loop Sampled-Data Systems

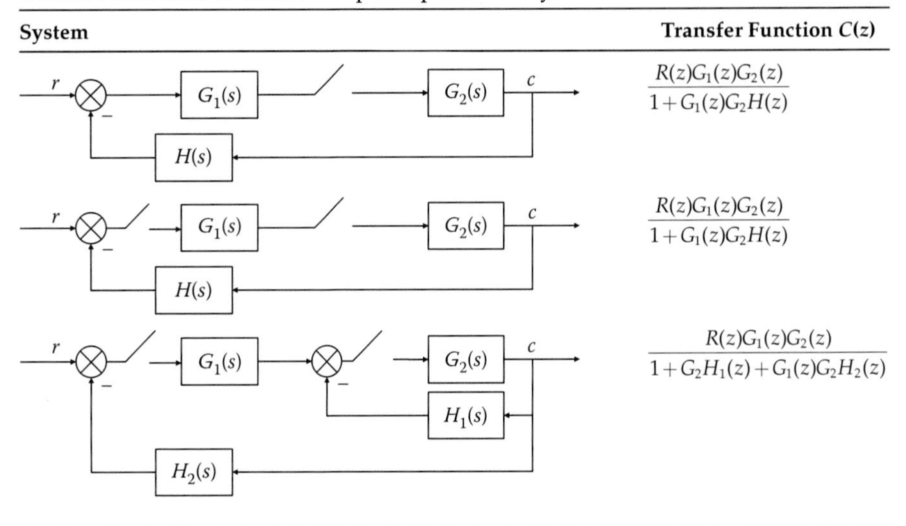

System	Transfer Function $C(z)$
	$\dfrac{R(z)G_1(z)G_2(z)}{1+G_1(z)G_2H(z)}$
	$\dfrac{R(z)G_1(z)G_2(z)}{1+G_1(z)G_2H(z)}$
	$\dfrac{R(z)G_1(z)G_2(z)}{1+G_2H_1(z)+G_1(z)G_2H_2(z)}$

TABLE 3.3

Transfer Function of Closed-Loop Sampled-Data Systems

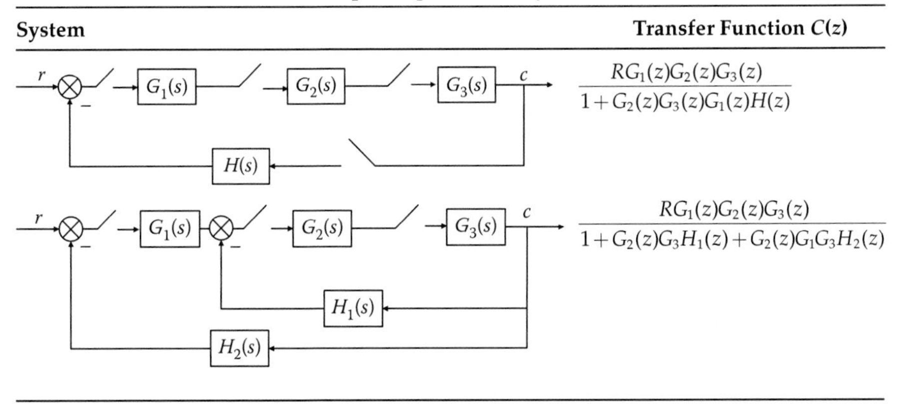

System	Transfer Function $C(z)$
	$\dfrac{RG_1(z)G_2(z)G_3(z)}{1+G_2(z)G_3(z)G_1(z)H(z)}$
	$\dfrac{RG_1(z)G_2(z)G_3(z)}{1+G_2(z)G_3H_1(z)+G_2(z)G_1G_3H_2(z)}$

3.8 Solved Exercises

EXERCISE 3.1

Find the difference equation for the system of the following scheme.

Let: $K = 10$, $T = 0.5$ s and $r(t) = u(t)$.

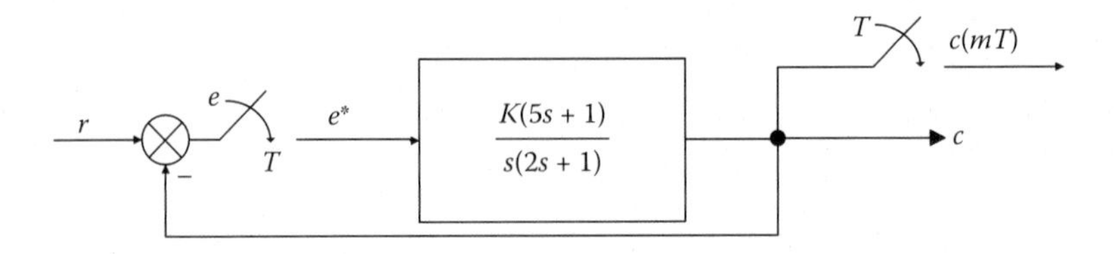

Solution

From the above scheme, we have that

$$\frac{C(s)}{e^*(s)} = \frac{K(5s+1)}{s(2s+1)} \overset{ILT}{\Rightarrow} 2\frac{d^2c(t)}{dt^2} + \frac{dc(t)}{dt} = 5K\frac{de^*(t)}{dt} + Ke^*(t) \quad (3.1.1)$$

The expression (3.1.1) takes the form of expression (3.1.3), after discretization, which is the difference equation of the entire system.

$$\begin{aligned} &2\frac{d^2c(t)}{dt^2} \approx 2\frac{c(k)-2c(k-1)+c(k-2)}{T^2}; \quad \frac{dc(t)}{dt} \approx \frac{c(k)-c(k-1)}{T} \\ &5K\frac{de^*(t)}{dt} \approx 5K\frac{e^*(k)-e^*(k-1)}{T}; \quad Ke^*(t) = Ke^*(k) \end{aligned} \quad (3.1.2)$$

Thus, the difference equation becomes

$$\frac{2+T}{T^2}c(k) - \frac{4+T}{T^2}c(k-1) + \frac{2}{T^2}c(k-2) = \frac{5K+KT}{T}e^*(k) + \frac{5K}{T}e^*(k-1) \Rightarrow$$

$$c(k) - \frac{4+T}{2+T}c(k-1) + \frac{2}{2+T}c(k-2) = \frac{5KT+KT^2}{2+T}e^*(k) + \frac{5KT}{2+T}e^*(k-1) \Rightarrow$$

$$c(k) = \frac{5KT+KT^2}{2+T}e^*(k) + \frac{5KT}{2+T}e^*(k-1) + \frac{4+T}{2+T}c(k-1) - \frac{2}{2+T}c(k-2) \quad (3.1.3)$$

For $K = 10$, $T = 0.5$ s, we get

$$(3.1.3) \Rightarrow c(k) = 11e^*(k) + 10e^*(k-1) + 1.8c(k-1) - 0.8c(k-2) \quad (3.1.4)$$

It holds that

$$e^*(k) = r(k) - c(k) = 1 - c(k) \quad (3.1.5)$$

So

$$c(k) = 1.75 - \frac{41}{60}c(k-1) - \frac{4}{60}c(k-2) \tag{3.1.6}$$

If $c(0) = 0$, then substituting in expression (3.1.6) and solving it successively, it stems that

$$c(1) = 1.75 - \frac{41}{60}c(0) - \frac{4}{60}c(-1) = 1.75$$

$$c(2) = 1.75 - \frac{41}{60}c(1) - \frac{4}{60}c(0) \approx 0.554$$

$$c(3) = 1.75 - \frac{41}{60}c(2) - \frac{4}{60}c(1) \approx 1.255$$

$$c(m) = 1.75 - \frac{41}{60}c(m-1) - \frac{4}{60}c(m-2) \tag{3.1.7}$$

EXERCISE 3.2

Find the transfer function of the following discrete systems:

a. $y(k) + 0.5y(k-1) = 2x(k)$
b. $y(k) + 2y(k-1) - y(k-2) = 2x(k) - x(k-1) + 2x(k-2)$
c. $y(kT) + 3y(kT-T) + 4y(kT-2T) + 5y(kT-3T)$
   $= r(kT) - 3r(kT-T) + 2r(kT-2T)$

Solution

To derive the corresponding transfer functions, it suffices to apply z-transform onto the difference equations, by assuming zero initial conditions.

a. $y(k) + 0.5y(k-1) = 2x(k)$ (3.2.1)

$$Y(z)(1+0.5z^{-1}) = 2X(z) \;\Rightarrow\; H(z) = \frac{2}{1+0.5z^{-1}} = \frac{2z}{z+0.5}$$

b. $y(k) + 2y(k-1) - y(n-2) = 2x(n) - x(n-1) + 2x(k-2).$ (3.2.2)

$$Y(z)(1+2z^{-1} - z^{-2}) = X(z)(2 - z^{-1} + 2z^{-2})$$

$$\Rightarrow H(z) = \frac{2 - z^{-1} + 2z^{-2}}{1+2z^{-1} - z^{-2}} = \frac{2z^2 - z + 2}{z^2 + 2z - 1}$$

c. $Y(z) + 3Y(z)z^{-1} + 4Y(z)z^{-2} + 5Y(z)z^{-3} = R(z) - 3R(z)z^{-1} + 2R(z)z^{-2}$

$$\Rightarrow G(z) = \frac{Y(z)}{R(z)} = \frac{1-3z^{-1}+2z^{-2}}{1+3z^{-1}+4z^{-2}+5z^{-3}} = \frac{z^3 - 3z^2 + 2z}{z^3+3z^2+4z+5} \tag{3.2.3}$$

EXERCISE 3.3

Derive the difference equations of the systems with transfer functions given as

$$G(z) = \frac{z^4 + 3z^3 + 2z^2 + z + 1}{z^4 + 4z^3 + 5z^2 + 3z + 2} \quad \text{and} \quad H(z) = \frac{z}{z^2 - 1.7z + 0.72}$$

Solution

a.

$$G(z) = \frac{Y(z)}{R(z)} = \frac{1 + 3z^{-1} + 2z^{-2} + z^{-3} + z^{-4}}{1 + 4z^{-1} + 5z^{-2} + 3z^{-3} + 2z^{-4}} \tag{3.3.1}$$

So, we have

$$\begin{aligned} &Y(z)(1 + 4z^{-1} + 5z^{-2} + 3z^{-3} + 2z^{-4}) \\ &= R(z)(1 + 3z^{-1} + 2z^{-2} + z^{-3} + z^{-4}) \end{aligned} \tag{3.3.2}$$

Applying the inverse z-transform in (3.3.2), it yields the desired result.

$$\begin{aligned} &y(kT) + 4y(kT - T) + 5y(kT - 2T) + 3y(kT - 3T) + 2y(kT - 4T) \\ &= r(kT) + 3r(kT - T) + 2r(kT - 2T) + r(kT - 3T) + r(kT - 4T) \end{aligned} \tag{3.3.3}$$

b.

$$H(z) = \frac{z}{z^2 - 1.7z + 0.72} = z^{-1} \frac{1}{1 - 1.7z^{-1} + 0.72z^{-2}} \tag{3.3.4}$$

It holds that $Y(z) = H(z)U(z)$. Hence

$$(1 - 1.7z^{-1} + 0.72z^{-2})Y(z) = z^{-1}U(z) \tag{3.3.5}$$

We are transferred from the z-domain to the time domain; thereby, the requested difference equation is presented as

$$y_k = 1.7y_{k-1} - 0.72y_{k-2} + u_{k-1} \tag{3.3.6}$$

EXERCISE 3.4

Derive the transfer function of the following discrete systems and show that they are different to each other.

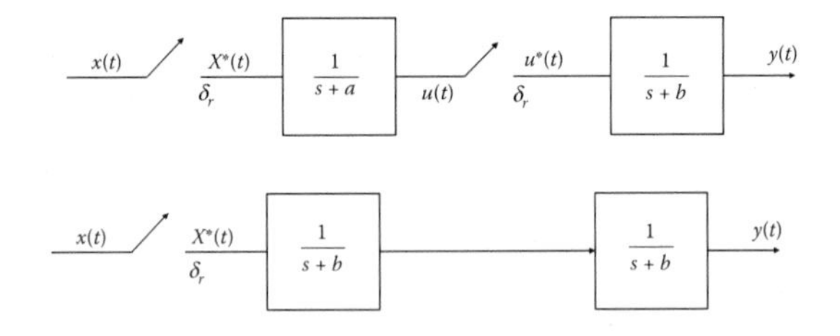

Solution

For the first system, observe that the transfer function is given by

$$\frac{Y(z)}{X(z)} = G(z)H(z) = \frac{z}{z-e^{-aT}} \cdot \frac{z}{z-e^{-bT}} \tag{3.4.1}$$

For the second system, the transfer function is given by

$$\begin{aligned}\frac{Y(z)}{X(z)} &= ZT\left[\frac{1}{(s+a)(s+b)}\right] = ZT\left[\frac{1}{b-a}\left(\frac{1}{s+a} - \frac{1}{s+b}\right)\right] \\ &= \frac{1}{b-a}\left(\frac{z}{z-e^{-aT}} - \frac{z}{z-e^{-bT}}\right) = \frac{1}{b-a}\left[\frac{(e^{-aT}-e^{-bT})z^{-1}}{(1-e^{-aT}z^{-1})(1-e^{-bT}z^{-1})}\right]\end{aligned} \tag{3.4.2}$$

These systems are not equivalent since their corresponding transfer functions are different.

EXERCISE 3.5

Derive the output function of the following scheme, into the z-domain, when the input function is $z(t) = \mu(t)$.

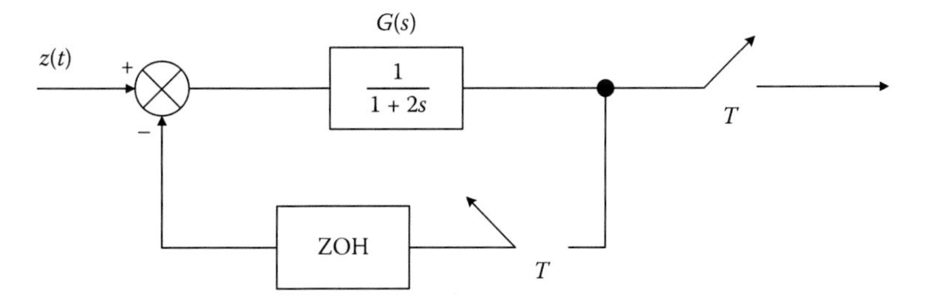

Solution

The expression of the output function in z-domain is

$$C(z) = \frac{z[R(s)G(s)]}{1 + z[G_{zoh}(s)G(s)]} \tag{3.5.1}$$

where

$$z[R(s)G(s)] = RG(z) = z\left[\frac{k}{s(1+2s)}\right] = \frac{k(1-e^{-0,5T})z}{(z-1)(z-e^{-0,5T})} \tag{3.5.2}$$

and

$$G_{zoh}G(z) = z\left[\frac{1-e^{-Ts}}{s}\frac{k}{1+2s}\right] = \frac{k(1-e^{-0,5T})}{z-e^{-0,5T}} \tag{3.5.3}$$

Substituting, we get

$$G(z)=\frac{k(1-e^{-0,5T})z}{(z-1)(z-e^{-0,5T}+k(1-e^{-0,5T}))}\tag{3.5.4}$$

For $k=1$ and $T=0.314$ s, we have that

$$C(z)=\frac{0.145z}{(z-1)(z-0.709)}\tag{3.5.5}$$

EXERCISE 3.6

Are the following systems equivalent?

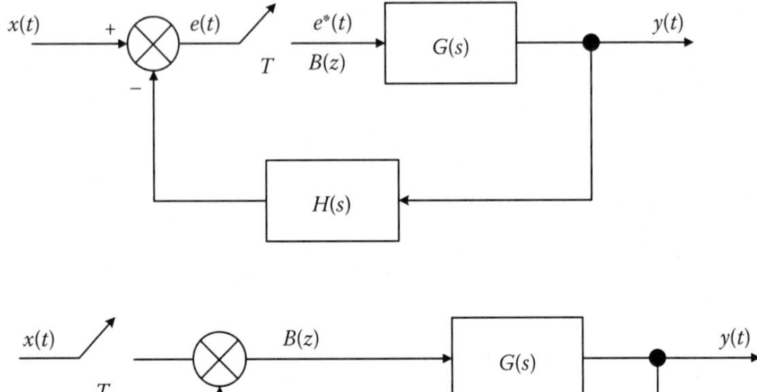

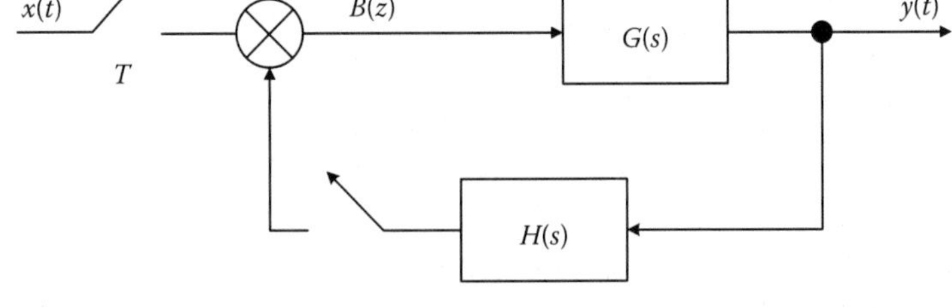

Solution

The most usual case is to use a sampler in the error channel $e(t)$, as the first scheme shows. Since the system is linear, the sampler can be moved from the error signal to the system's input and to the output of feedback unit.

We have:

$$Y(z)=G(z)E(z)\tag{3.6.1}$$

where

$$E(z)=X(z)-z[G(s)H(s)]E(z)\tag{3.6.2}$$

Hence

$$E(z)=\frac{X(z)}{1+z[G(s)H(s)]}\tag{3.6.3}$$

(3.6.1),(3.6.3):

$$\Rightarrow \frac{Y(z)}{X(z)} = \frac{G(z)}{1 + z[G(s)H(s)]} \tag{3.6.4}$$

EXERCISE 3.7

Find the expression that provides the output $Y(z)$ as a function of the input and the included system parameters, as shown in the following scheme.

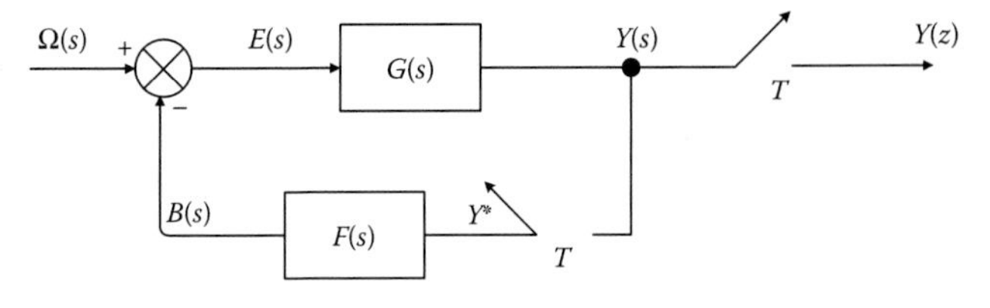

Solution

We compute $Y(z)$ from the system's model equations. Note that the output of each sampler is marked with a superscript (*), for example, Y^* corresponds to the postsampled $Y(s)$ used for the feedback channel.

From the above block diagram, we have that

$$E(s) = \Omega(s) - B(s) \tag{3.7.1}$$

$$B(s) = F(s)Y^*(s) \tag{3.7.2}$$

$$Y(s) = G(s)E(s) \overset{(1)}{=} G(s)(\Omega(s) - B(s)) \tag{3.7.3}$$

$$(3.7.2),(3.7.3) \Rightarrow Y(s) = G(s)\Omega(s) - G(s)F(s)Y^*(s)) \tag{3.7.4}$$

By observing the expression (3.7.4), it is clear that both $Y(s)$ and $Y^*(s)$ are present; thus, to derive $Y^*(s)$ as a common factor, we apply a starred Laplace transform as follows.

$$\begin{aligned}
&(3.7.4)\overset{*}{\Rightarrow}(Y(s))^* = [G(s)\Omega(s) - G(s)F(s)Y^*(s))]^* \Rightarrow \\
&Y^*(s) = [G(s)\Omega(s)]^* - [G(s)F(s)Y^*(s))]^* = G\Omega^*(s) - Y^*(s)[G(s)F(s)]^* \\
&\qquad = G\Omega^*(s) - Y^*(s)GF^*(s) \Rightarrow \\
&Y^*(s) + Y^*(s)GF^*(s) = G\Omega^*(s) \Rightarrow Y^*(s)(1 + GF^*(s)) = G\Omega^*(s) \Rightarrow
\end{aligned}$$

$$Y^*(s) = \frac{G\Omega^*(s)}{1 + GF^*(s)} \tag{3.7.5}$$

Using z-transform in (3.7.5), we get

$$Y(z) = \frac{G\Omega(z)}{1 + GF(z)} \tag{3.7.6}$$

where $G\Omega(z) = z[G(s)\Omega(s)]$ and $GF(z) = z[G(s)F(s)]$

EXERCISE 3.8

Find the expression that provides the output $Y(z)$ as a function of the input and the included system parameters, as shown in the following scheme:

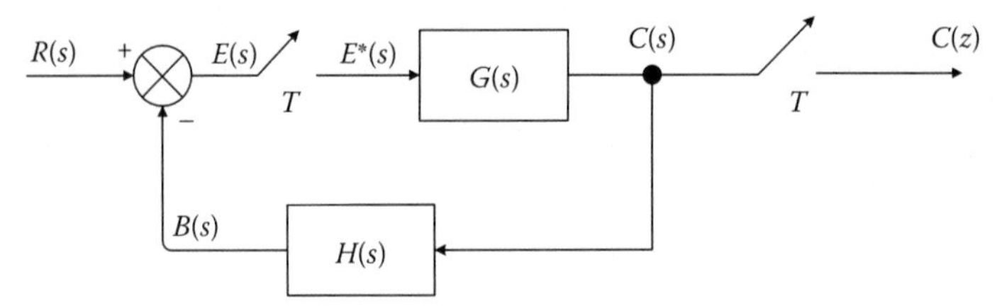

Solution

We compute $Y(z)$ from the system's model equations.

From the above block diagram, we have that

$$E(s) = R(s) - B(s) \tag{3.8.1}$$

$$B(s) = H(s)C(s) \tag{3.8.2}$$

$$C(s) = G(s)E^*(s) \tag{3.8.3}$$

$$(3.8.2) \Rightarrow B(s) = H(s)G(s)E^*(s) \tag{3.8.4}$$

$$(3.8.1),(3.8.4) \Rightarrow E(s) = R(s) - H(s)G(s)E^*(s) \tag{3.8.5}$$

By observing the expression (3.8.5), it is clear that both $E(s)$ and $E^*(s)$ are present; thus, to derive $E^*(s)$ as a common factor, we apply a starred Laplace transform as follows:

$$\begin{aligned}&(3.8.5)\overset{*}{\Rightarrow} E^*(s) = R^*(s) - HG^*(s)E^*(s) \Rightarrow\\ &E^*(s) + HG^*(s)E^*(s) = R^*(s) \Rightarrow\\ &E^*(s)\left[1 + HG^*(s)\right] = R^*(s) \Rightarrow\end{aligned}$$

$$E^*(s) = \frac{R^*(s)}{1 + HG^*(s)} \tag{3.8.6}$$

$$(3.8.3),(3.8.5) \Rightarrow C(s) = G(s)E^*(s) \overset{*}{\Rightarrow} C^*(s) = G^*(s)E^*(s) \Rightarrow E^*(s) = \frac{C^*(s)}{G^*(s)} \tag{3.8.7}$$

$$(3.8.6),(3.8.7) \Rightarrow \frac{C^*(s)}{G^*(s)} = \frac{R^*(s)}{1 + HG^*(s)} \Rightarrow C^*(s) = \frac{R^*(s)}{1 + HG^*(s)} G^*(s) \tag{3.8.8}$$

Using z-transform in (3.8.8), we get

$$C(z) = \frac{R(z)}{1 + HG(z)} G(z) \tag{3.8.9}$$

EXERCISE 3.9

Find the expression that provides the output $Y(z)$ as a function of the input and the included system parameters, as shown in the following scheme:

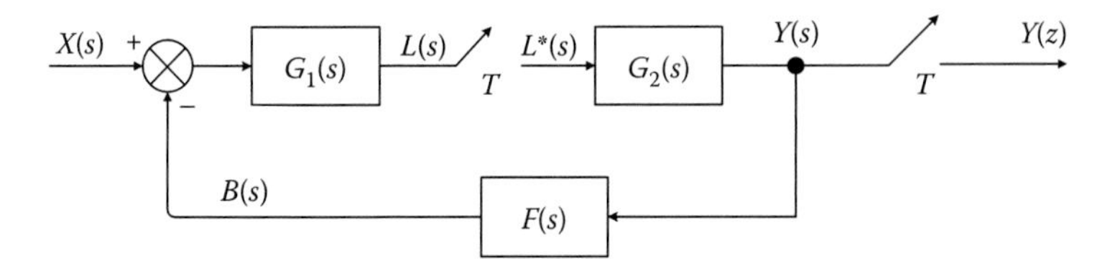

Solution

We compute $Y(z)$ from the system's model equations.

From the above block diagram, we have that

$$E(s) = X(s) - B(s) \tag{3.9.1}$$

$$B(s) = F(s)G_2(s)L^*(s) \tag{3.9.2}$$

$$L(s) = G_1(s)E(s) \tag{3.9.3}$$

$$\begin{aligned}(3.9.1),(3.9.3) \Rightarrow L(s) &= G_1(s)E(s) = G_1(s)(X(s) - B(s))\\ &\overset{(2)}{=} G_1(s)(X(s) - F(s)G_2(s)L^*(s)) \Rightarrow\end{aligned}$$

$$L(s) = G_1(s)X(s) - G_1(s)F(s)G_2(s)L^*(s) \tag{3.9.4}$$

By observing the expression (3.9.4), it is clear that both $L(s)$ and $L^*(s)$ are present; thus, to derive $L^*(s)$ as a common factor, we apply a starred Laplace transform as follows:

$$\begin{aligned}(3.9.4) \overset{*}{\Rightarrow} L^*(s) &= [G_1(s)X(s) - G_1(s)F(s)G_2(s)L^*(s)]^* =\\ &= [G_1(s)X(s)]^* - [G_1(s)F(s)G_2(s)L^*(s)]^* \Rightarrow\\ &= G_1X^*(s) - G_1FG_2{}^*(s)L^*(s) \Rightarrow\end{aligned}$$

$$L^*(s)(1 + G_1FG_2{}^*(s)) = G_1X^*(s) \tag{3.9.5}$$

However,

$$Y(s) = G_2L^*(s) \Rightarrow Y^*(s) = [G_2L^*(s)]^* = G_2^*(s)L^*(s) \tag{3.9.6}$$

$$(3.9.5),(3.9.6) \Rightarrow Y^*(s) = G_2^*(s)\frac{G_1X^*(s)}{1+G_1FG_2^*(s)} \tag{3.9.7}$$

Using z-transform in (3.9.7), we get

$$\Rightarrow Y(z) = G_2(z)\frac{G_1X(z)}{1+G_1FG_2(z)} \tag{3.9.8}$$

EXERCISE 3.10

Find the expression that provides the output $Y(z)$ as a function of the input and the included system parameters, as shown in the following scheme:

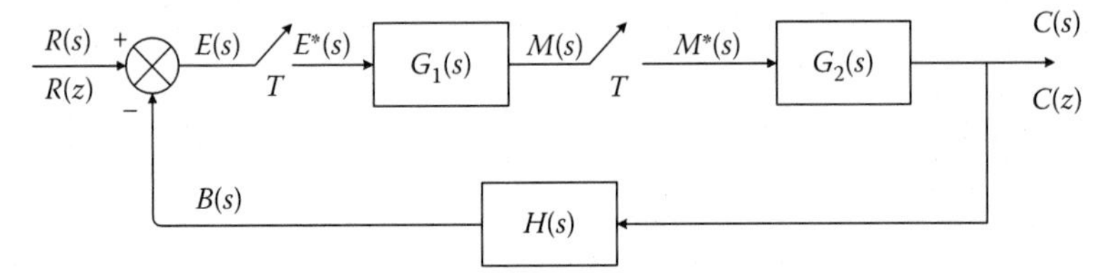

Solution

We compute $Y(z)$ from the system's model equations.

From the above block diagram, we have that

$$C(s) = G_2(s)M^*(s) \tag{3.10.1}$$

$$M(s) = G_1(s)E^*(s) \tag{3.10.2}$$

$$E(s) = R(s) - B(s) \tag{3.10.3}$$

$$B(s) = H(s)C(s) \tag{3.10.4}$$

$$E(s) = R(s) - H(s)C(s) \tag{3.10.5}$$

$$E(s) = R(s) - H(s)G_2(s)M^*(s) \tag{3.10.6}$$

Discretize Equation (3.10.6) as

$$C^*(s) = G_2^*(s)M^*(s) \tag{3.10.7}$$

$$(3.10.2)\overset{*}{\Rightarrow} M^*(s) = G_1{}^*(s)E^*(s) \Rightarrow M^*(s) = G_1{}^*(s)\left[R^*(s) - HG_2{}^*(s)M^*(s)\right]$$
$$\Rightarrow M^*(s) = G_1{}^*(s)R^*(s) - G_1{}^*(s)HG_2{}^*(s)M^*(s) \tag{3.10.8}$$

$$(3.10.6)\overset{*}{\Rightarrow} E^*(s) = R^*(s) - HG_2{}^*(s)M^*(s) \tag{3.10.9}$$

$$(3.10.7),(3.10.8) \Rightarrow \frac{C^*(s)}{R^*(s)} = \frac{G_1{}^*(s)G_2{}^*(s)}{1 + G_1{}^*(s)G_2H^*(s)} \tag{3.10.10}$$

Using z-transform in (3.10.10), we get

$$C(z) = \frac{G_2(z)G_1(z)R(z)}{1 + G_1(z)HG_2(z)} \tag{3.10.11}$$

EXERCISE 3.11

For the system of the following scheme, find the closed-loop transfer function $C(z)/R(z)$

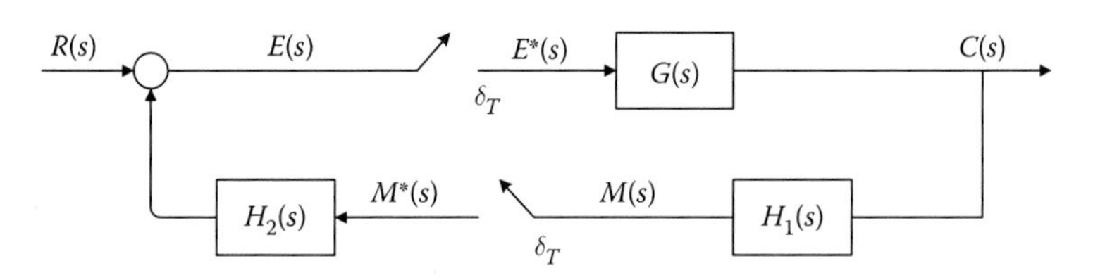

Solution

From the above block diagram, we have that

$$C(s) = G(s)E^*(s) \tag{3.11.1}$$

$$E(s) = R(s) - H_2(s)\,M^*(s) \tag{3.11.2}$$

$$M(s) = H_1(s)G(s)E^*(s) \tag{3.11.3}$$

Applying the starred Laplace transform at the above equations, it holds that

$$C^*(s) = G^*(s)E^*(s) \tag{3.11.4}$$

$$E^*(s) = R^*(s) - H_2^*(s)M^*(s) \tag{3.11.5}$$

$$M^*(s) = \left[G(s)H_1(s)\right]^*E^*(s) \tag{3.11.6}$$

$$(3.11.5),(3.11.6) \Rightarrow E^*(s) = R^*(s) - H_2^*(s)\left[G(s)H_1(s)\right]^* E^*(s) \Rightarrow$$

$$E^*(s) = \frac{R^*(s)}{1 + H_2^*(s)\left[G(s)H_1(s)\right]^*} \quad (3.11.7)$$

$$(3.11.4),(3.11.7) \Rightarrow C^*(s) = \frac{G^*(s)\,R^*(s)}{1 + H_2^*(s)\left[G(s)H_1(s)\right]^*} = \frac{G^*(s)\,R^*(s)}{1 + H^*(s)\left[GH_1(s)\right]^*}$$

Hence, the resultant transfer function is given by

$$\frac{C(z)}{R(z)} = \frac{G(z)}{1 + H(z)\,GH_1(z)} \quad (3.11.8)$$

EXERCISE 3.12

For the system of the following scheme, find the closed-loop transfer function $C(z)/R(z)$, using Mason's formula.

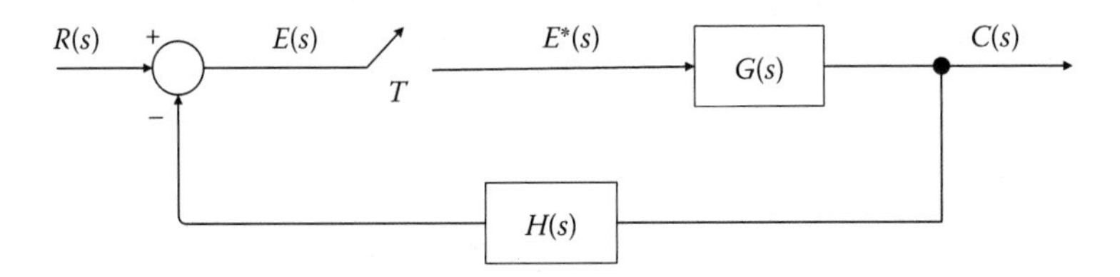

Solution

Design the S.F.G., which corresponds to the block diagram of the above scheme.

Observing that S.F.G. presents a discontinuity between $E(s)$ and $E^*(s)$, we redesign S.F.G. without discontinuities using the system model.

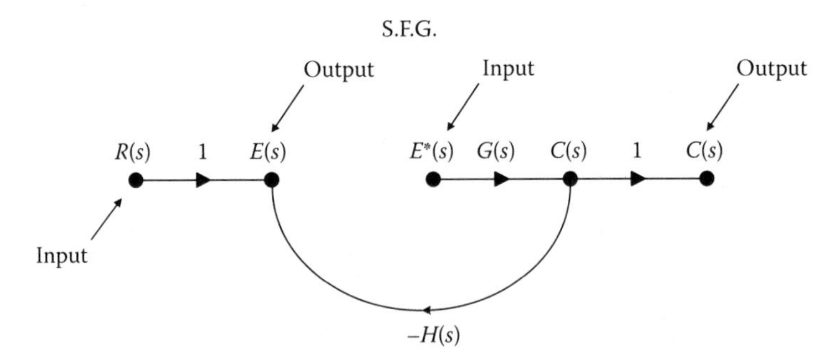

It holds that

$$E(s) = R(s) - H(s)G(s)E^*(s) \overset{*}{\Rightarrow} E^*(s) = R^*(s) - GH^*(s)E^*(s) \quad (3.12.1)$$

$$C(s) = G(s)E^*(s) \overset{*}{\Rightarrow} C^*(s) = G^*(s)E^*(s) \tag{3.12.2}$$

From Equations 3.12.1 and 3.12.2, design the equivalent S.F.G. as

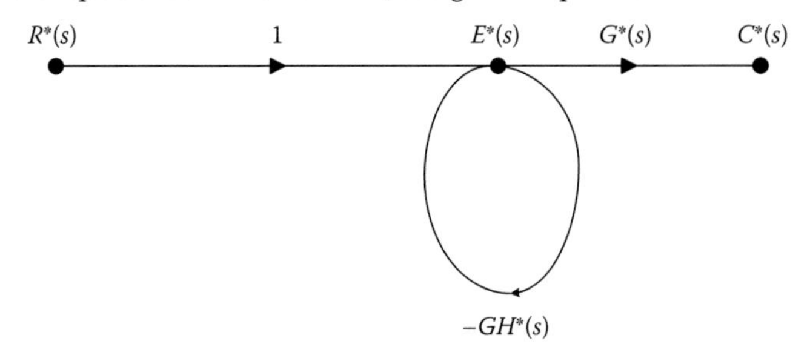

Applying Mason's formula, we have that

$$\frac{C^*(s)}{R^*(s)} = \frac{T_1\Delta_1}{\Delta} \tag{3.12.3}$$

where

$$T_1 = G^*(s)$$

$$\Delta = 1 - (-GH^*(s)) \quad \text{and} \quad \Delta_1 = 1 - (0) = 1$$

$$(3.12.3) \Rightarrow \frac{C^*(s)}{R^*(s)} = \frac{G^*(s)}{1 + GH^*(s)} \tag{3.12.4}$$

Thus, the desired transfer function is given by

$$\frac{C(z)}{R(z)} = \frac{G(z)}{1 + GH(z)} \tag{3.12.5}$$

The complex S.F.G. is formed by combining the initial and the equivalent diagram, which is presented at the following scheme:

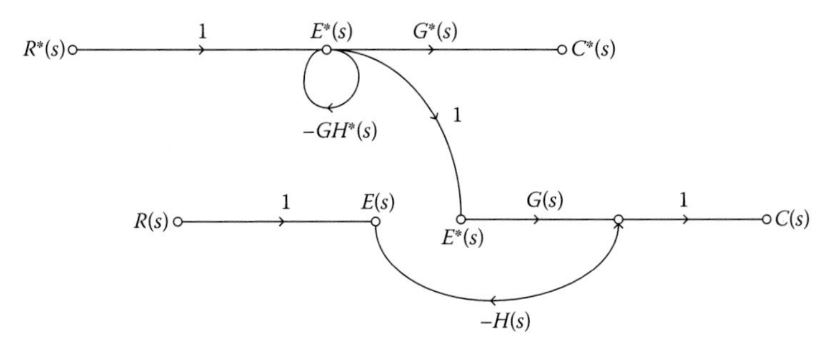

EXERCISE 3.13

For the system of the following scheme, find the closed-loop transfer function $C(z)/R(z)$, using Mason's formula.

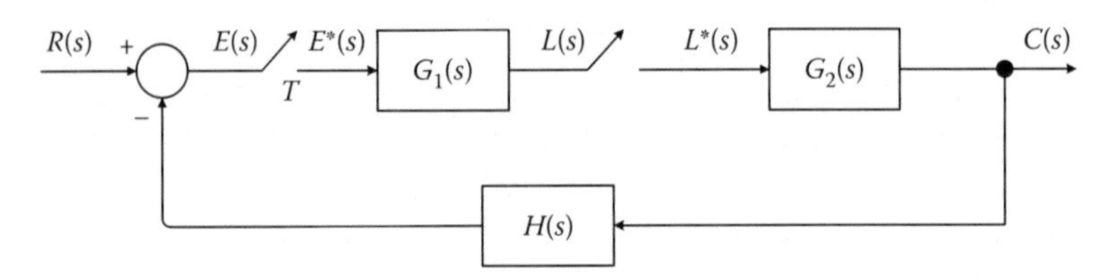

Solution

Design S.F.G., which corresponds to the block diagram of the above scheme.

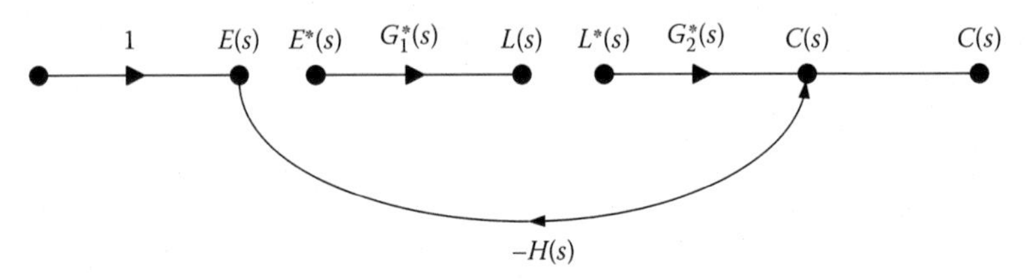

Observing that the S.F.G. presents a discontinuity between $E(s)$ and $E^*(s)$, and between $L(s)$ and $L^*(s)$, we redesign the S.F.G. without discontinuities using the system model.

$$E(s) = R(s) - H(s)G_2(s)L^*(s) \overset{*}{\Rightarrow} \\ E^*(s) = R^*(s) - HG_2{}^*(s)L^*(s) \tag{3.13.1}$$

$$C(s) = G_2(s)L^*(s) \overset{*}{\Rightarrow} C^*(s) = G_2{}^*(s)L^*(s) \tag{3.13.2}$$

$$L(s) = G_1(s)E^*(s) \overset{*}{\Rightarrow} L^*(s) = G_1{}^*(s)E^*(s) \tag{3.13.3}$$

From Equations 3.13.1, 3.13.2, and 3.13.3, design the equivalent S.F.G.

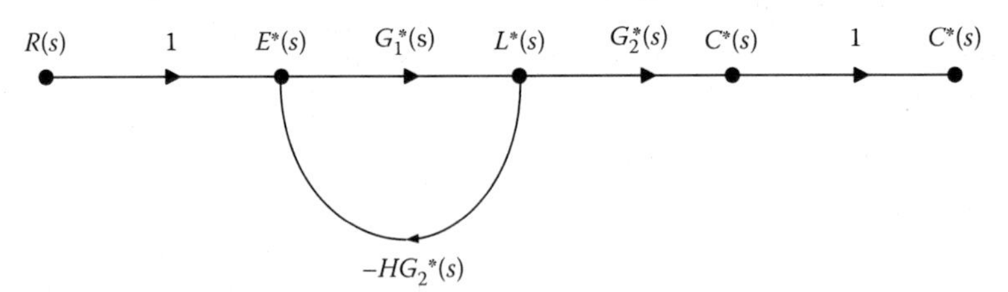

Applying Mason's formula, we get

$$\frac{C^*(s)}{R^*(s)} = \frac{T_1\Delta_1}{\Delta} \tag{3.13.4}$$

where

$$T_1 = G_1{}^*(s)G_2{}^*(s)$$

$$\Delta = 1-(-G^*{}_1(s)HG_2{}^*(s)) \quad \text{and} \quad \Delta_1 = 1-(0) = 1$$

Hence, substituting into the expression (3.13.4), it stems that

$$\frac{C^*(s)}{R^*(s)} = \frac{G_1{}^*(s)G_2{}^*(s)}{1+G_1{}^*(s)G_2H^*(s)} \tag{3.13.5}$$

The desired transfer function is presented as

$$\frac{C(z)}{R(z)} = \frac{G_1(z)G_2(z)}{1+G_1(z)G_2H(z)} \tag{3.13.6}$$

EXERCISE 3.14

For the system of the following scheme, find the closed-loop transfer function $C(z)/R(z)$, using Mason's formula.

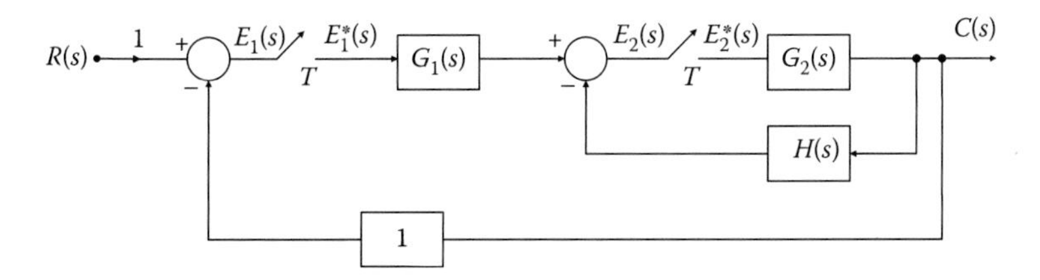

Solution

Design the S.F.G. which corresponds to the block diagram of the above scheme.

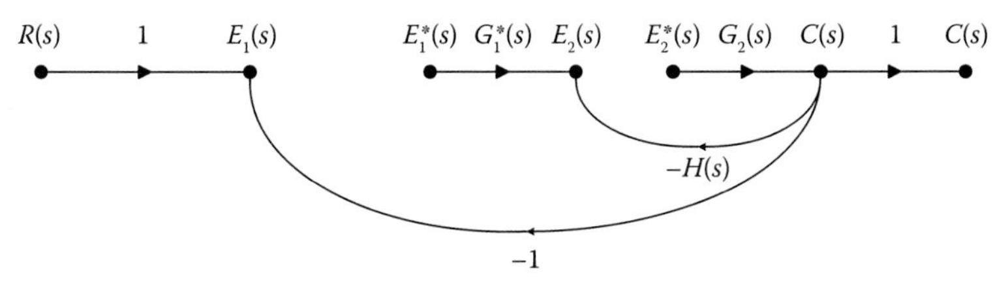

Observing that the S.F.G. presents a discontinuity between $E_1(s)$ and $E_1{}^*(s)$, and between $E_2(s)$ and $E_2{}^*(s)$, we redesign the S.F.G. without discontinuities using the system model.

It holds that

$$E_1(s) = R(s) - C(s) \Rightarrow E^*{}_1(s) = R^*(s) - C^*(s) \tag{3.14.1}$$

$$C(s) = G_2(s)E_2{}^*(s) \Rightarrow C^*(s) = G_2{}^*(s)E_2(s) \tag{3.14.2}$$

$$\begin{aligned}
&E_2(s) = G_1(s)E_1^*(s) - H(s)C(s) \Rightarrow \\
&E_2(s) = G_1(s)E_1^*(s) - H(s)G_2(s)E_2(s) \Rightarrow \\
&E_2^*(s) = G_1^*(s)E_1^*(s) - G_2H^*(s)E_2^*(s)
\end{aligned} \tag{3.14.3}$$

From Equations 3.14.1, 3.14.2, and 3.14.3, design the equivalent S.F.G.

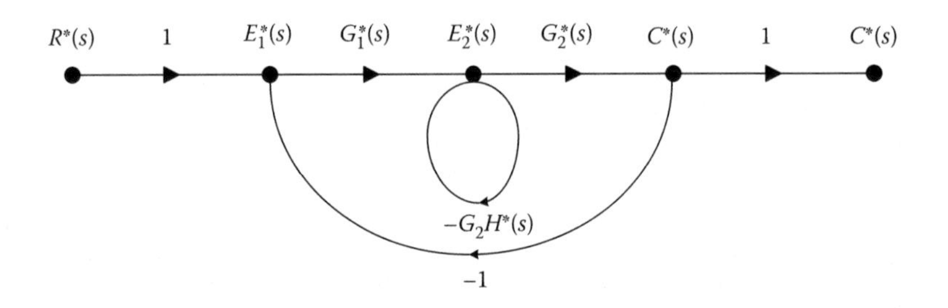

Applying Mason's formula, we have

$$\frac{C^*(s)}{R^*(s)} = \frac{T_1\Delta_1}{\Delta} \tag{3.14.4}$$

where

$$T_1 = G^*{}_1(s)G^*{}_2(s)$$

$$\Delta = 1 - (-G^*{}_1(s)G_2{}^*(s) - G_2H^*(s)) \quad \text{and} \quad \Delta_1 = 1 - (0) = 1$$

Hence, substituting into the expression (3.14.4), it yields

$$\frac{C^*(s)}{R^*(s)} = \frac{G_1{}^*(s)G_2{}^*(s)}{1 + G_2H^*(s) + G_1{}^*(s)G_2{}^*(s)} \tag{3.14.5}$$

The desired transfer function is presented as

$$\frac{C(z)}{R(z)} = \frac{G_1(z)G_2(z)}{1 + G_1(z)G_2(z) + Z[G_2H(s)]} \tag{3.14.6}$$

EXERCISE 3.15

For the system of the following scheme, find the closed-loop transfer function $C(z)/R(z)$, using Mason's formula.

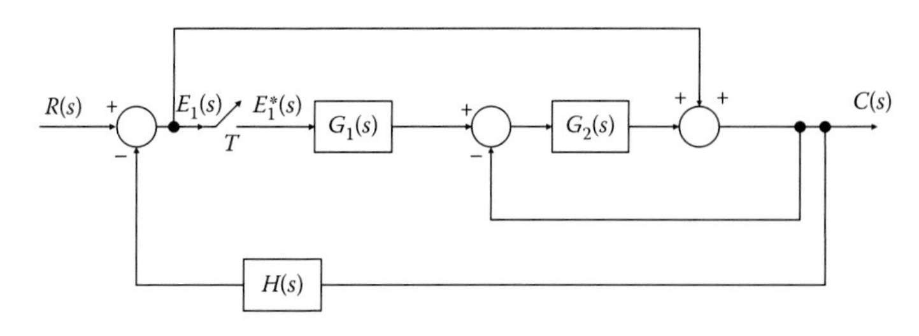

Solution

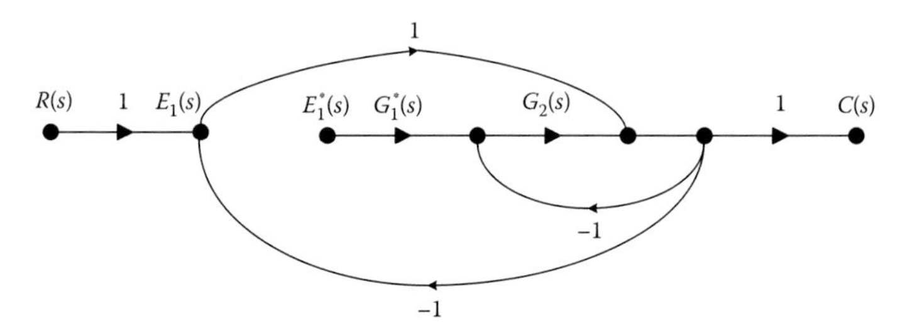

Design S.F.G. which corresponds to the block diagram of the above scheme.

Observing that the S.F.G. presents a discontinuity between $E_1(s)$ and $E_1{}^*(s)$, we redesign the S.F.G. without discontinuities using the system model.

It holds that

$$E_1(s) = R(s) - C(s) \tag{3.15.1}$$

$$\begin{aligned} C(s) &= E_1(s) + G_2(s)\big(G_1(s)E_1{}^*(s) - C(s)\big) \\ &= E_1(s) + G_1(s)G_2(s)E_1{}^*(s) - G_2(s)C(s) \end{aligned} \tag{3.15.2}$$

$$\begin{aligned} (3.15.1),(3.15.2) \Rightarrow C(s) &= R(s) - C(s) + G_1(s)G_2(s)E_1{}^*(s) - G_2(s)C(s) \\ &\Rightarrow 2C(s) + G_2(s)C(s) = R(s) + G_1G_2(s)E_1{}^*(s) \\ &\Rightarrow C(s) = \frac{R(s)}{2+G_2(s)} + \frac{G_1G_2(s)E_1{}^*(s)}{2+G_2(s)} \end{aligned} \tag{3.15.3}$$

$$\begin{aligned} E_1(s) = R(s) - C(s) &= R(s) - \frac{R(s)}{2+G_2(s)} - \frac{G_1G_2(s)E_1{}^*(s)}{2+G_2(s)} \\ \Rightarrow E_1(s) &= \frac{R(s)\big(1+G_2(s)\big)}{2+G_2(s)} - \frac{G_1G_2(s)E_1{}^*(s)}{2+G_2(s)} \end{aligned} \tag{3.15.4}$$

$$(3.15.3) \overset{*}{\Rightarrow} C^*(s) = \left(\frac{R(s)}{2+G_2(s)}\right)^* + \left(\frac{G_1G_2(s)}{2+G_2(s)}\right)^* E_1{}^*(s) \quad (3.15.5)$$

$$(3.15.4) \overset{*}{\Rightarrow} E_1{}^*(s) = \left(\frac{R(s)(1+G_2(s))}{2+G_2(s)}\right)^* + \left(\frac{G_1G_2(s)}{2+G_2(s)}\right)^* E_1{}^*(s) \quad (3.15.6)$$

From Equations 3.15.5 and 3.15.6, design the equivalent S.F.G. and observe that two inputs and one output appear.

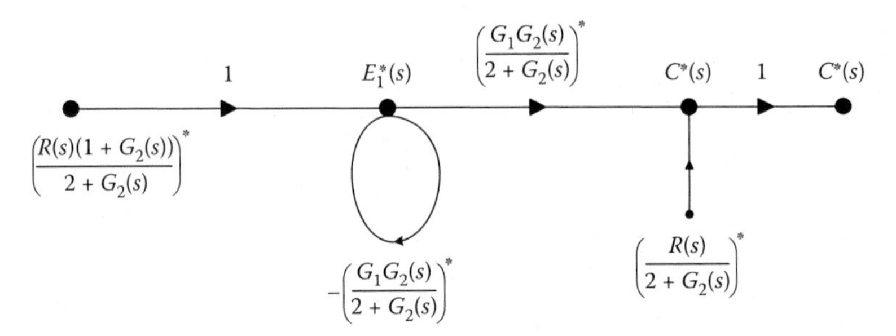

To compute the total output, we apply the superposition theorem, which states that

$$C^*(s) = C_1{}^*(s) + C_2{}^*(s) \Rightarrow C^*(s) = \frac{\left(\frac{G_1G_2(s)}{2+G_2(s)}\right)^*}{1+\left(\frac{G_1G_2(s)}{2+G_2(s)}\right)^*}\left(\frac{R(s)(1+G_2(s))}{2+G_2(s)}\right)^* + \left(\frac{R(s)}{2+G_2(s)}\right)^* \quad (3.15.7)$$

From the expression (3.15.7), the following expression is derived, which represents the z-transform of the system's output.

$$C(z) = \frac{Z\left[\frac{G_1G_2(s)}{2+G_2(s)}\right]}{1+Z\left[\frac{G_1G_2(s)}{2+G_2(s)}\right]} Z\left[\frac{R(s)(1+G_2(s))}{2+G_2(s)}\right] + Z\left[\frac{R(s)}{2+G_2(s)}\right] \quad (3.15.8)$$

The desired transfer function is presented as

$$\frac{C(z)}{R(z)} = \frac{Z\left[\frac{G_1G_2(s)}{2+G_2(s)}\right]}{1+Z\left[\frac{G_1G_2(s)}{2+G_2(s)}\right]} Z\left[\frac{R(s)(1+G_2(s))}{2+G_2(s)}\right]\frac{1}{R(z)} + Z\left[\frac{R(s)}{2+G_2(s)}\right]\frac{1}{R(z)} \quad (3.15.9)$$

EXERCISE 3.16

For the system of the following scheme, find the closed-loop transfer function $C(z)/R(z)$, using Mason's formula.

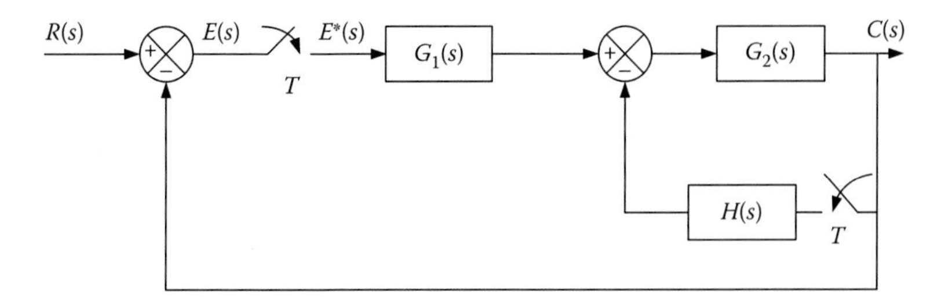

Solution

Design S.F.G. which corresponds to the block diagram of the above scheme.

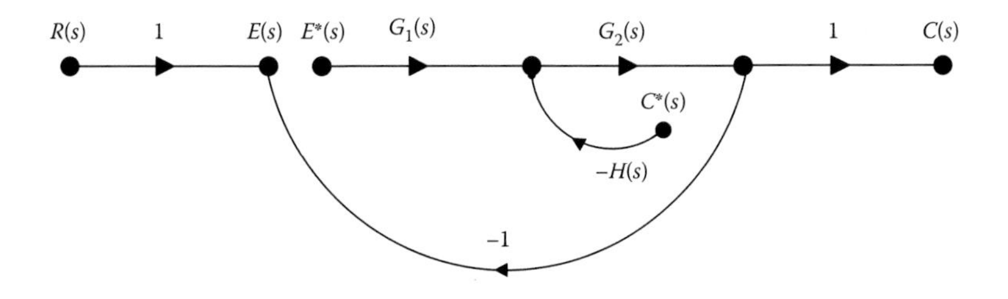

Observing that the S.F.G. presents a discontinuity between $E_1(s)$ and $E^*(s)$, and between $C(s)$ and $C^*(s)$, we redesign the complex S.F.G. without discontinuities using the system model.

It holds that

$$E(s) = R(s) - C(s) \tag{3.16.1}$$

$$\begin{aligned} C(s) &= (G_1(s)E^*(s) - H(s)C^*(s))G_2(s) \\ &= G_1(s)G_2(s)E^*(s) - G_2(s)H(s)C^*(s) \end{aligned} \tag{3.16.2}$$

To design the complex S.F.G., we use the pulse transform of Equations 3.16.1 and 3.16.2:

$$(3.16.2) \Rightarrow C^*(s) = G_1G_2^*(s)E^*(s) - G_2H^*(s)C^*(s) \tag{3.16.3}$$

$$\begin{aligned} (3.16.1) \Rightarrow E^*(s) &= R^*(s) - C^*(s) \\ &= R^*(s) - G_1G_2^*(s)E^*(s) + G_2H^*(s)C^*(s) \end{aligned} \tag{3.16.4}$$

The complex S.F.G. is given by

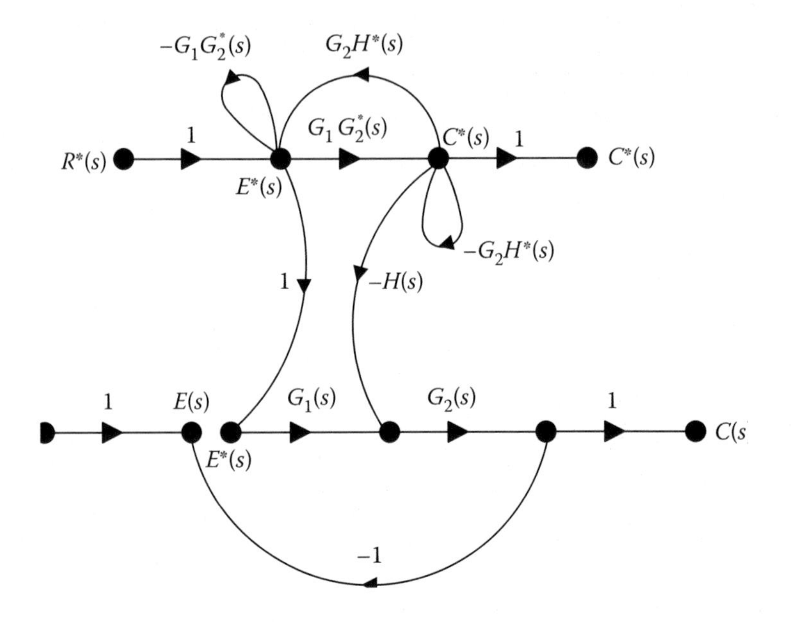

The transfer functions $C^*(s)/R^*(s)$, $E^*(s)/R^*(s)$, $C(s)/R^*(s)$ can be calculated with the aid of Mason's formula.

$$\frac{C^*(s)}{R^*(s)} = \frac{G_1G_2{}^*(s)}{1+G_1G_2{}^*(s)+G_2H^*(s)} \tag{3.16.5}$$

$$\frac{E^*(s)}{R^*(s)} = \frac{1\times(1-(-G_2H^*(s))}{1+G_1G_2{}^*(s)+G_2H^*(s)} = \frac{1+G_2H^*(s)}{1+G_1G_2{}^*(s)+G_2H^*(s)} \tag{3.16.6}$$

$$\frac{C(s)}{R^*(s)} = \frac{G_1(s)G_2(s)[1+G_2H^*(s)]-G_2(s)H(s)G_1G_2{}^*(s)}{1+G_1G_2{}^*(s)+G_2H^*(s)} \tag{3.16.7}$$

In z-domain, we get

$$\frac{E(z)}{R(z)} = \frac{1+G_2H(z)}{1+G_1G_2(z)+G_2H(z)} \tag{3.16.8}$$

$$\frac{C(z)}{R(z)} = \frac{G_1G_2(z)}{1+G_1G_2(z)+G_2H(z)} \tag{3.16.9}$$

EXERCISE 3.17

Provide the expression for the output $Y(z)$ for the following systems.

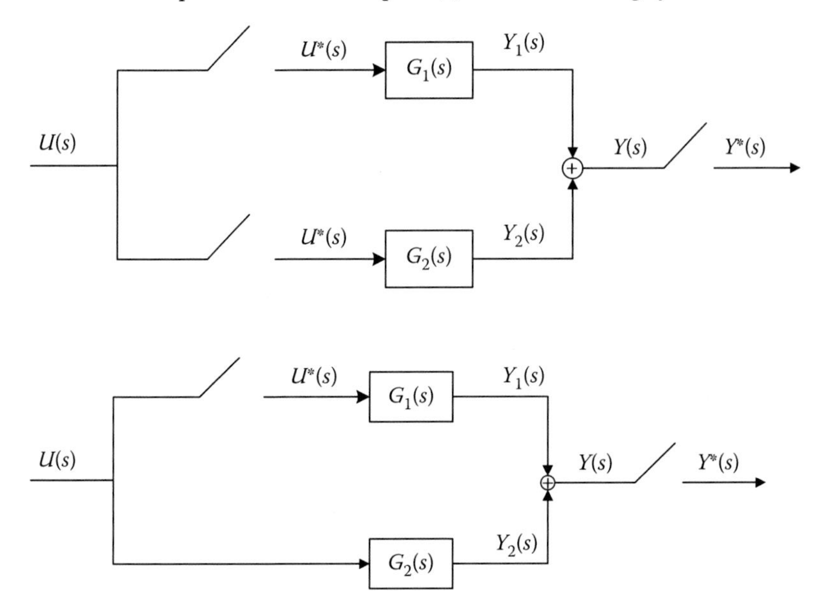

Solution

From the block diagram of the first scheme, the desired result is obtained as

$$Y_1(s)=G_1(s)U^*(s) \quad Y_2(s)=G_2(s)U^*(s)$$

$$Y(s)=Y_1(s)+Y_2(s)=G_1(s)U^*(s)+G_2(s)U^*(s)$$

$$Y^*(s)=G_1^*(s)U^*(s)+G_2^*(s)U^*(s)\overset{ZT}{\Rightarrow}$$

$$Y(z)=G_1(z)U(z)+G_2(z)U(z) \tag{3.17.1}$$

$$Y(z)=G_1(z)+G_2(z)$$

Similarly, we have for the second scheme

$$Y(s)=G_1(s)U^*(s)+G_2(s)U(s)$$

$$Y^*(s)=G_1^*(s)U^*(s)+(G_2(s)U(s))^*\overset{ZT}{\Rightarrow} \tag{3.17.2}$$

$$Y(z)=G_1(z)U(z)+G_2U(z)$$

EXERCISE 3.18

For the system of the following scheme, find the transfer function of $Y(z)/R(z)$.

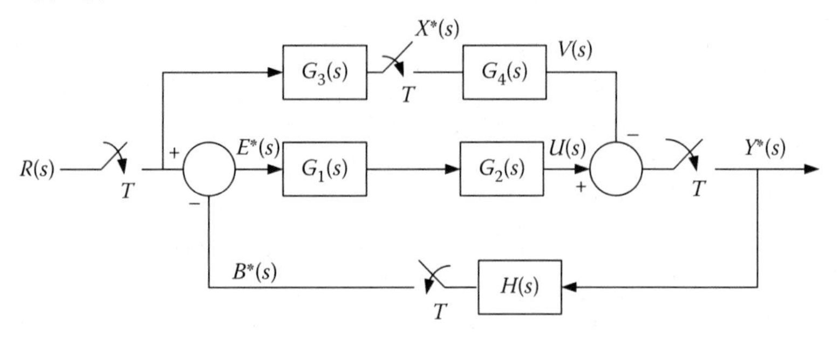

Solution

From the above block diagram, it holds that

$$E^*(s) = R^*(s) - B^*(s) \tag{3.18.1}$$

$$X^*(s) = (G_3(s)R^*(s))^* = G_3{}^*(s)R^*(s) \tag{3.18.2}$$

$$B^*(s) = (H(s)Y^*(s))^* = H^*(s)Y^*(s) \tag{3.18.3}$$

$$\begin{aligned} Y^*(s) &= (G_1(s)G_2(s)E^*(s))^* - (G_4(s)X^*(s))^* \Rightarrow \\ Y^*(s) &= (G_1(s)G_2(s))^*E^*(s) - G_4{}^*(s)X^*(s) \end{aligned} \tag{3.18.4}$$

$$(3.18.1) \Rightarrow E(z) = R(z) - B(z) \tag{3.18.5}$$

$$(3.18.2) \Rightarrow X(s) = G_3(z)R(z) \tag{3.18.6}$$

$$(3.18.3) \Rightarrow B(z) = H(z)Y(z) \tag{3.18.7}$$

$$(3.18.4) \Rightarrow Y(z) = z(G_1(s)G_2(s))E(z) - G_4(z)X(z) \tag{3.18.8}$$

Combining Equations 3.18.5, 3.18.6, 3.18.7, and 3.18.8, we get

$$Y(z)(1 + G_1G_2(z)H(z)) = (G_1G_2(z) - G_3(z)G_4(z))R(z)^* \Rightarrow$$

$$\frac{Y(z)}{R(z)} = \frac{G_1G_2(z) - G_3(z)G_4(z)}{(1 + G_1G_2(z)H(z)} \tag{3.18.9}$$

EXERCISE 3.19

Find the open-loop step response of the system of the following scheme.

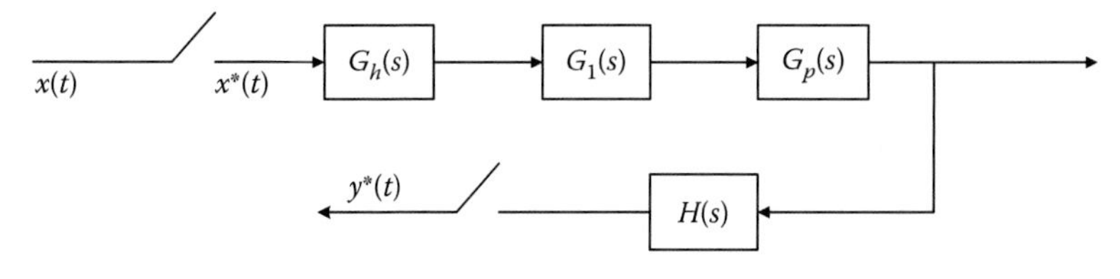

Let

$$T = 1s, G(s) = \left[G_h(s)G_1(s)G_p(s)H(s)\right] = \frac{9}{s(s^2+9)}$$

Solution

From the given block diagram, we have that

$$Y(z) = G(z)X(z) \text{ and } y(kT) = Z^{-1}[Y(z)] \tag{3.19.1}$$

where

$$G(z) = Z\left[G(s)\right] = Z\left[G_h(s)G_1(s)G_p(s)H(s)\right] \tag{3.19.2}$$

$$X(z) = Z[1(t)] = \frac{z}{z-1} \tag{3.19.3}$$

$$G(z) = Z[G(s)] = Z\left[\frac{9}{s(s^2+9)}\right] = Z\left[\frac{1}{s} - \frac{s}{s^2+9}\right] \tag{3.19.4}$$

$$(3.19.4) \Rightarrow G(z) = \frac{z}{z-1} - \frac{1-z^{-1}\cos 3T}{1-2z^{-1}\cos 3T + z^{-2}} = \frac{z}{z-1} - \frac{z^2 - z\cos 3T}{z^2 - 2z\cos 3T + 1}$$

$$= \frac{z^2 + z - z^2\cos 3T - z\cos 3T}{(z-1)(z^2 - 2z\cos 3T + 1)} = \frac{z(z+1)(1-\cos 3T)}{(z-1)(z^2 - 2z\cos 3T + 1)} \tag{3.19.5}$$

$$(3.19.1),(3.19.5) \Rightarrow Y(z) = X[z]G[z] = \frac{z}{z-1}\frac{z(z+1)(1-\cos 3T)}{(z-1)(z^2 - 2z\cos 3T + 1)}$$

$$= \frac{z^2(z+1)(1-\cos 3T)}{(z-1)^2(z^2 - 2z\cos 3T + 1)} \tag{3.19.6}$$

Applying the partial fraction expansion of

$$Y(z)=z\left[\frac{1}{(z-1)^2}+\frac{1}{2}\frac{1}{z-1}-\frac{1}{2}\frac{z+1}{z^2-2z\cos 3T+1}\right]$$
$$=\frac{z}{(z-1)^2}+\frac{1}{2}\frac{z}{z-1}-\frac{1}{2}\frac{z^2-z\cos 3T}{z^2-2z\cos 3T+1}-\frac{1}{2}\frac{1+\cos 3T}{\sin 3T}\frac{z\sin 3T}{z^2-2z\cos 3T+1}$$

$$\text{Hence, } y^*(t)=\left(\frac{t}{T}+\frac{1}{2}-\frac{1}{2}\cos 3t-\frac{1}{2}\frac{1+\cos 3T}{\sin 3T}\sin 3t\right) \tag{3.19.7}$$

For $T = 1\text{s}$, the desired expression is presented as

$$y^*(t)=(t+0.5-0.5\cos 3t-0.0354\sin 3t) \tag{3.19.8}$$

EXERCISE 3.20

Find the closed-loop pulse transfer function for the system of the following scheme.

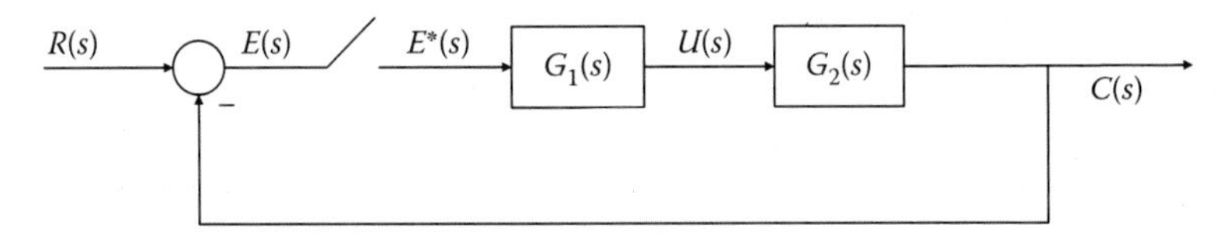

Let

$$G_1(s)=\frac{1-e^{-Ts}}{s},\quad G_2(s)=\frac{K}{s+1}$$

Solution

The pulse transfer function $G(z)$ arises as

$$G(z)=Z[G(s)]=Z\left[\frac{1-e^{-Ts}}{s}\frac{K}{s+1}\right]=(1-z^{-1})Z\left[\frac{K}{s(s+1)}\right]$$
$$=(1-z^{-1})Z\left[\frac{K}{s}-\frac{K}{s+1}\right]=(1-z^{-1})\left(\frac{K}{1-z^{-1}}-\frac{K}{1-e^{-T}z^{-1}}\right)\Rightarrow$$

$$G(z)=\frac{K(1-e^{-T})z^{-1}}{1-e^{-T}z^{-1}} \tag{3.20.1}$$

The closed-loop pulse transfer function $G_{cl}(z)$ is presented as

$$G_{cl}(z)=\frac{C(z)}{R(z)}=\frac{G(z)}{1+G(z)}=\frac{K(1-e^{-T})z^{-1}}{1+[K-(K+1)e^{-T}]z^{-1}} \tag{3.20.2}$$

EXERCISE 3.21

For the system of the scheme depicted in Exercise 3.20, let $G_1(s)=(1-e^{-Ts})/s$, $G_2(s)=10/(s+10)$, $H(s)=0.1$, $e^{-10T}=0.5$ and $T=0.07$ s. Find the step response of the system.

Solution

The closed-loop system output is

$$C(z)=\frac{G_1G_2(z)R(z)}{1+G_1G_2(z)H(z)} \tag{3.21.1}$$

where

$$R(z)=\frac{z}{z-1} \tag{3.21.2}$$

$$G_1G_2(z)=Z\left[\frac{1-e^{-Ts}}{s}\frac{10}{s+10}\right]=(1-z^{-1})Z\left[\frac{10}{s(s+10)}\right]=\frac{1-e^{-10T}}{z-e^{-10T}} \tag{3.21.3}$$

$$G_1G_2(z)H(z)=Z\left[\frac{1-e^{-Ts}}{s}\frac{10}{s+10}0.1\right]=(1-z^{-1})Z\left[\frac{1}{s(s+10)}\right]\Rightarrow$$
$$G_1G_2(z)H(z)=0.1\frac{1-e^{-10T}}{z-e^{-10T}} \tag{3.21.4}$$

Since $e^{-10T}=0.5$, it yields that

$$C(z)=\frac{0.5z}{(z-0.45)(z-1)} \tag{3.21.5}$$

The closed-loop step response is derived as

$$c(kT)=\sum_{i=1}^{n}\mathrm{Res}[C(z)z^{k-1}]_{z=z_i}=\left[(z-0.45)\frac{0.5z^k}{(z-0.45)(z-1)}\right]_{z=0.45}$$
$$+\left[(z-1)\frac{0.5z^k}{(z-0.45)(z-1)}\right]^{z=1}\Rightarrow$$

$$c(kT)=\frac{10}{11}(1-0.45^k) \tag{3.21.6}$$

$$\begin{aligned}c^*(t)&=0\delta(t)+0.5\delta(t-T)+0.725\delta(t-2T)+0.826\delta(t-3T)\\&+0.827\delta(t-4T)+0.894\delta(t-5T)+0.902\delta(t-6T)+\cdots\end{aligned} \tag{3.21.7}$$

EXERCISE 3.22

For the system of the scheme depicted in Exercise 3.20, let $G_1(s) = (1 - e^{-Ts})/s$, $G_2(s) = 1/(s(s+1))$, $T = 1$ s. Find the step response of the system.

Solution

From the given block diagram, it holds that

$$\begin{aligned} &C(s)=G(s)E^*(s), E(s)=R(s)-C(s) \\ &E^*(s)=R^*(s)-C^*(s)=R^*(s)-G^*(s)E^*(s) \\ &E^*(s)=\frac{R^*(s)}{1+G^*(s)} \end{aligned} \tag{3.22.1}$$

$$C(s)=G(s)\frac{R^*(s)}{1+G^*(s)} \tag{3.22.2}$$

where

$$G(s)=G_1(s)G_2(s) \tag{3.22.3}$$

From Equation 3.22.2, taking the inverse Laplace transform, we have that

$$\begin{aligned} c(t)=L^{-1}[C(s)]&=L^{-1}\left[G(s)\frac{R^*(s)}{1+G^*(s)}\right] \\ &=L^{-1}\left[\frac{1-e^{-Ts}}{s}\frac{1}{s(s+1)}\frac{R^*(s)}{1+G^*(s)}\right]\overset{T=1}{=}L^{-1}\left[\frac{1-e^{-s}}{s}\frac{1}{s(s+1)}\frac{R^*(s)}{1+G^*(s)}\right] \end{aligned} \tag{3.22.4}$$

$$\begin{aligned} &\text{Let} \quad X^*(s)=(1-e^{-s})\frac{R^*(s)}{1+G^*(s)}\overset{\text{ZT}}{\Rightarrow} \\ &X(z)=(1-z^{-1})\frac{R(z)}{1+G(z)} \end{aligned} \tag{3.22.5}$$

$$\text{where} \quad R(z)=Z[1(t)]=\frac{1}{1-z^{-1}} \tag{3.22.6}$$

$$\begin{aligned} \Rightarrow G(z)=Z[G(s)]=Z\left[\frac{1-e^{-Ts}}{s}\frac{1}{s(s+1)}\right]&=(1-z^{-1})Z\left[\frac{1}{s}\frac{1}{s(s+1)}\right] \\ &=\frac{0.368z^{-1}+0.264z^{-2}}{(1-0.368z^{-1})(1-z^{-1})} \end{aligned} \tag{3.22.7}$$

$(3.22.6),(3.22.7)\Rightarrow X(z)$

$$\begin{aligned} &=(1-z^{-1})\frac{1/(1-z^{-1})}{1+(0.368z^{-1}+0.264z^{-2})/(1-0.368z^{-1})(1-z^{-1})} \\ &=\frac{1-1.368z^{-1}+0.368z^{-2}}{1-z^{-1}+0.632z^{-2}} \end{aligned}$$

$$T=1, z=e^{Ts}=e^{s},$$

hence

$$X^*(s)=\frac{1-1.368e^{-s}+0.368e^{-2s}}{1-e^{-s}+0.632e^{-2s}} \tag{3.22.8}$$

$$\begin{aligned}
c(t)&=L^{-1}\left[\frac{1}{s^2(s+1)}X^*(s)\right]=\\
&=L^{-1}\left[\frac{1}{s^2(s+1)}\frac{1-1.368e^{-s}+0.368e^{-2s}}{1-e^{-s}+0.632e^{-2s}}\right]\\
&=L^{-1}\left[\frac{1}{s^2(s+1)}(1-0.368e^{-s}-0.632e^{-2s}-0.400e^{-3s}+0e^{-4s}+0.253e^{-5s}\right.\\
&\qquad\left.+0.253e^{-6s}\cdots)\right]
\end{aligned}$$

However,

$$L^{-1}\left[\frac{1}{s^2(s+1)}\right]=L^{-1}\left[\frac{1}{s^2}-\frac{1}{s}+\frac{1}{s+1}\right]=t-1+e^{-t}$$

$$\begin{aligned}
\text{thus } c(t)=t-1+e^{-t}&-0.368[(t-1)-1+e^{-(t-1)}]1(t-1)\\
&-0.0632[(t-2)-1+e^{-(t-2)}]1(t-2)\\
&-0.400[(t-3)-1+e^{-(t-3)}]1(t-3)\\
&-0.000[(t-4)-1+e^{-(t-4)}]1(t-4)\\
&+0.253[(t-5)-1+e^{-(t-5)}]1(t-5)\\
&+0.253[(t-6)-1+e^{-(t-6)}]1(t-5)+\cdots
\end{aligned}$$

Thereby, the step response of the system is

$$c(t)=\begin{cases}
t-1+e^{-t} & 0\le t<1\\
t-1+e^{-t}-0.368[(t-1)-1+e^{-(t-1)}]1(t-1) & 1\le t<2\\
t-1+e^{-t}-0.368[(t-1)-1+e^{-(t-1)}]1(t-1)- & \\
0.0632[(t-2)-1+e^{-(t-2)}]1(t-2) & 2\le t<3\\
\vdots &
\end{cases}$$

$$\begin{aligned}
&c(0)=0-1+1=0\\
&c(0.5)=0.5-1+0.607=0.107\\
&c(1.0)=0.368-0.368*0=0.368\\
&c(1.5)=0.723-0.368*0.107=0.684\\
&\vdots
\end{aligned}$$

EXERCISE 3.23

For the system of the scheme, find the transfer function $B(z)/M(z)$ using MATLAB® for $T = 0.01$ s.

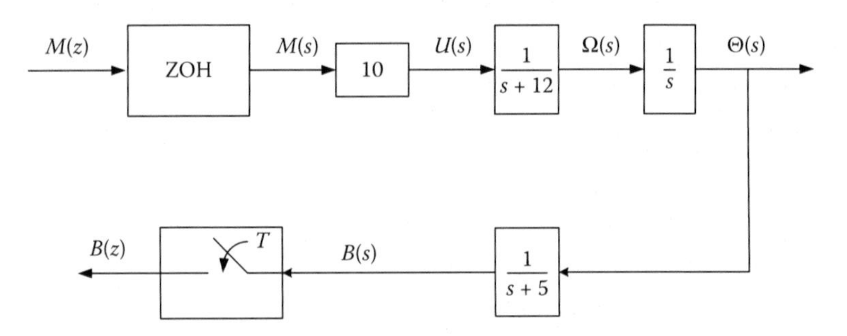

Solution

The desired transfer function is expressed as

$$\frac{B(z)}{M(z)} = Z\left\{\left(10 \cdot \frac{1}{s+12} \cdot \frac{1}{s} \cdot \frac{1}{s+5}\right) G_{zoh}(s)\right\} \tag{3.23.1}$$

The corresponding MATLAB code is given by

```
G1 = tf(10,[1 12]);
G2 = tf(1,[1 0]);
H = tf(1,[1 5]);
G = G1*G2*H;
Gd = c2d(G,0.01,'zoh')

Transfer function

1598e-006z^2 + 6126e-006z + 1468e-006
z^3- 2.838 z^2+ 2.682 z - 0.8437
Hence
```

$$\frac{B(z)}{M(z)} = \frac{1.598e-006z^2 + 6.126e-006z + 1.468e-006}{z^3 - 2.838z^2 + 2.682z - 0.8437} \tag{3.23.2}$$

4

Transfer Function Discretization

4.1 Introduction

There are two basic techniques for designing a digital filter for the automatic control of a system, which we will deal extensively in Chapter 8.

The first technique is called *discrete design* or *direct digital design* and is accomplished in two stages:

1. Discretization of the system that we want to control.
2. Construction of suitable digital controller using analytical methods of discrete-time systems.

The second technique is called simulation and is accomplished in the following steps:

1. Construction of the analog controller using continuous-time analytical methods.
2. Discretization of the analog controller.
3. Verification of the correct operation of the digital controller using discrete-time analytical methods.

In this chapter, we will describe how we can transform the transfer function $G(s)$ of an analog system in the s-domain to an equivalent transfer function $G(z)$ of a discrete system in the z-domain.

The $G(s)$ to $G(z)$ transform can be performed by *three different methods.*

The *first method* relies on the use of numerical methods for solving differential equations describing the given system and for converting them to difference equations. Thus, we convert the transfer function into an equivalent differential equation and we use numerical differential methods to solve the differential equation.

If we use *numerical integration methods* then the discretization process is as follows: Given the transfer function, we find the equivalent transfer function. Then, we integrate both parts of the differential equation and approach the integration regions numerically. We divide the integration region into several subspaces of length T, where T is the sampling period.

If we use *numerical differentiation methods*, then the discretization process is as follows: We calculate the equivalent differential equation of the transfer function and then approach the derivative of the output function numerically.

The *second method* is based on the response of an analog filter with a pole at the point $s = s_0$, when sampled over a time period T, is represented by the response of a discrete filter with a pole at the point $z = e^{s_0 T}$. Such a matching is used to equate the poles and zeros of $G(s)$ in s-domain with poles and zeros of $G(z)$ in z-domain.

The *third method* is based on the identification of response of continuous-time systems to specific inputs (step, impulse, and ramp functions), to those of discrete-time systems for the same inputs.

The equivalent discrete-time system must have *approximately the same dynamic characteristics* as the original continuous-time system.

This means that the discrete system is desired to have transmission and frequency response characteristics as close as possible to those of the original analog system.

In reality this cannot be achieved.

Using a specific discretization method is likely to have identical or nearly identical characteristics of the impulse response, while simultaneously having significant differences in the frequency response characteristics or vice versa.

The goal is to keep these important characteristics by selecting the appropriate discretization method.

Regarding the characteristics of the frequency response, it is important to note that an undesirable phenomenon appears during the discretization of the analog system displays, namely, the *frequency aliasing*, which occurs when the sampling frequency is too low to satisfy the sampling theorem. The selection of low sampling frequency results in a poor approximation of the analog system. Consequently, the satisfactory performance of the discrete system depends on the selection of the appropriate discretization method and sampling frequency.

4.2 Discretization Methods

The most important comparison points of the discretization methods to be mentioned are

- Ease of use
- Stability maintenance

- Impulse response maintenance
- Harmonic response maintenance

4.2.1 Impulse-Invariance Method or z-Transform Method

The equivalent discrete-time filter of the impulse response is the filter whose impulse response is identified with the impulse response of the continuous time filter $G(s)$, at time instances kT, k=0,1,2,3..., T where T denotes the sampling period.

The impulse response in the z-domain is the inverse z-transform of the transfer function $G(z)$. While in the s-domain the impulse response is the inverse Laplace transform of the transfer function $G(s)$.

Consider the systems of Figure 4.1.

The digital transfer function $G(z)$ arises from the corresponding analog $G(s)$, by implementing the following steps:

- From $G(s)$, we extract $g(t)$ in the time domain using inverse z-transform.
- From $g(t)$, via discretization, we extract the function $g(kT)$, where T is the sampling period.
- From $g(kT)$, using z-transform, we derive $G(z)$ in z-domain.

Thus,

$$G(s) \xrightarrow{L^{-1}} g(t) \xrightarrow{t=kT} g(kT) \xrightarrow{ZT} G(z) \tag{4.1}$$

or

$$G(z) = Z[g(kT)], \quad \text{where } g(kT) = [L^{-1}G(s)]_{t=kT} \tag{4.2}$$

Since the z-transform always projects a stable pole in the s-domain to a stable pole in the z-domain, we conclude that the discrete system will be stable if the original analog system is stable. Using this method, both the frequency and step responses are not preserved (frequency warping is observed due to overlap).

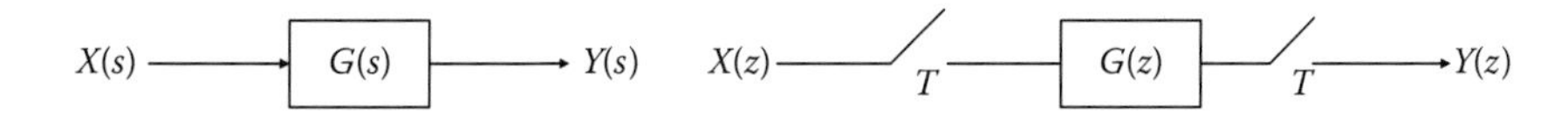

FIGURE 4.1
Analog and discrete system.

4.2.2 Step-Invariance Method or *z*-Transform Method with Sample and Hold

The aim of this method is to construct a discrete system $G(z)$ whose step response will consist of step response samples of the continuous system $G(s)$ at time instances kT, $k=0,1,2,3\ldots$, where T represents the sampling period.

Consider the system of Figure 4.2.

The digital transfer function $G(z)$ arises from the analog $G(s)$, using the expression (4.3) or (4.4).

$$G(z) = Z[G_h(s)\cdot G(s)] = Z\left[\frac{1-e^{Ts}}{s}\cdot G(s)\right]t \tag{4.3}$$

or

$$G(z) = (1-z^{-1})Z\left(\frac{G(s)}{s}\right) \tag{4.4}$$

If the original analog system $G(s)$ is stable then the equivalent discrete $G(z)$ obtained by the method of the invariance of the step response is stable. Using this method, neither the frequency (harmonic) nor the impulse responses are preserved.

At this point, the *zero-order hold (ZOH) circuit (filter)* is used with the transfer function $G_h(s)$. A D/A converter is actually a restraint network whose output is a partially continuous function. Figure 4.3 summarizes the ZOH operation, which is the preservation of the last sampled value of signal $f(t)$.

Consider the system $h_0(t)$ with an impulse response as given in Figure 4.4. The reconstructed signal is captured by the convolution of $h_0(t)$ with the postsampled signal, say $g^*(t)$, hence, $g(t) = g^*(t)^*h_0(t)$ and is sketched in Figure 4.5. Also, Figure 4.6 summarizes the first-order hold (FOH) operation.

Because the result of the convolution is stable for the duration between the value of the previous and next sample, this gives a reason for the name of this type of recovery filter, such as zero-order filter.

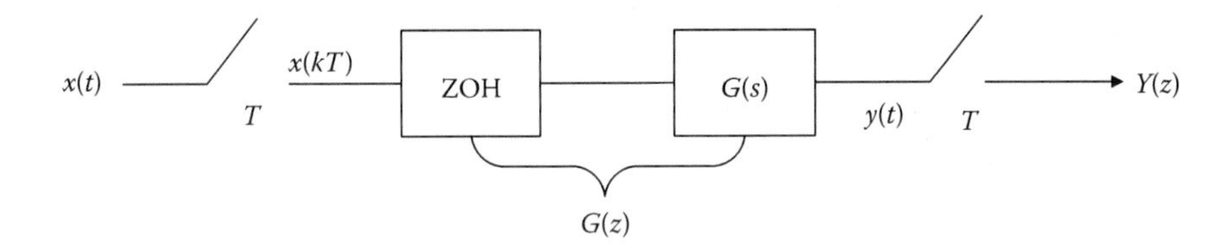

FIGURE 4.2
Sampled data system using ZOH.

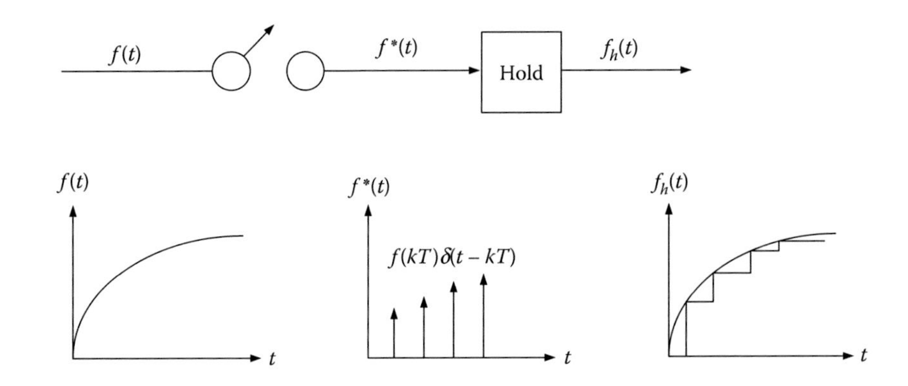

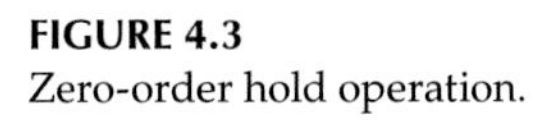

FIGURE 4.3
Zero-order hold operation.

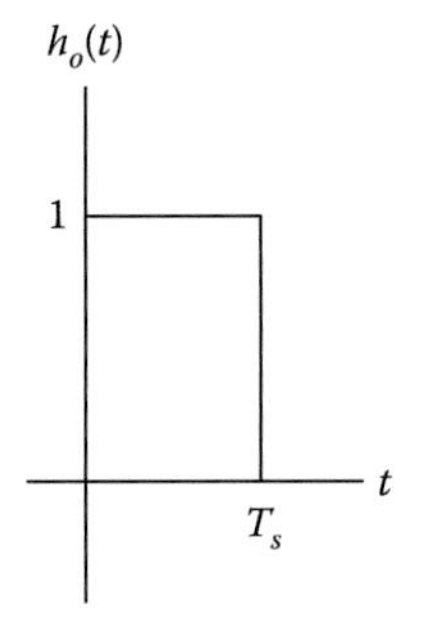

FIGURE 4.4
Impulse response $h_0(t)$.

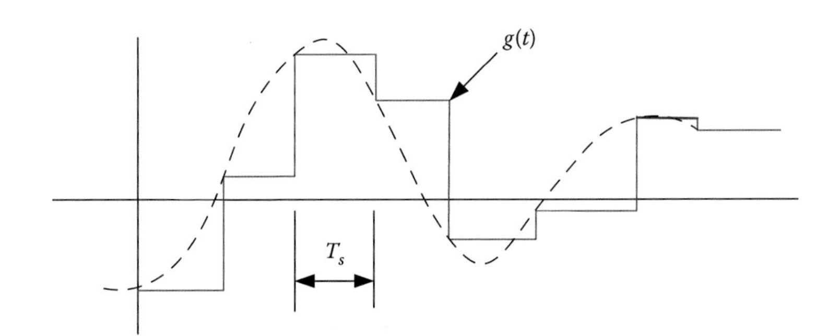

FIGURE 4.5
The reconstructed signal with ZOH.

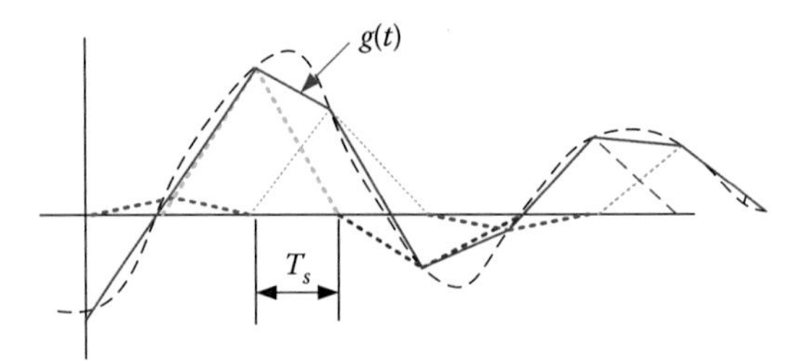

FIGURE 4.6
The reconstructed signal with FOH.

The impulse response, say $f_h(t)$, of the restraint network is a gate function, such that $f_h(t) = u(t) - u(t - T)$ or

$$F_h(s) = L\{u(t) - u(t-T)\} = L\{u(t)\} - L\{u(t-T)\}$$
$$\Rightarrow F_h(s) = \frac{1}{s} - \frac{1}{s}e^{-Ts} = \frac{1-e^{-Ts}}{s} \tag{4.5}$$

Apparently, since the input is the impulse function, the restraint system transfer function is given by

$$G_h(s) = \frac{1-e^{-Ts}}{s} \tag{4.6}$$

Furthermore, there are *higher order restraint filters* (first, second, etc.), which are used in practice.

For example, the output of a *first-order hold (FOH)* filter is not partially stable, as in ZOH, but partially linear with a slope $\{u(kT) - u[(k - 1)T]\}/T$ and it has a transfer function given by

$$G_{foh}(s) = \frac{T_s+1}{s}\left[\frac{1-e^{-Ts}}{s}\right]^2 \tag{4.7}$$

with output function $0 \le m \le \mathrm{T}$

$$y(kT+m) = u(kT) + \left[\frac{u(kT) - u[(k-1)T]}{T}\right] \tag{4.8}$$

4.2.3 Backward Difference Method

One way of calculating the derivative $\mathring{g}(t)$ at time instance $t = kT$ is by using the difference between the current and the previous sample divided by the sampling period, such that

$$\overset{\circ}{g}(t) = \frac{dg(t)}{dt} = \frac{g(k) - g(k-1)}{T} \tag{4.9}$$

The derivative of $\overset{\circ}{x}(t)$ in s-domain is $sX(s)$, while in z-domain is $z\left[\overset{\circ}{g}(t)\right] = z\left[\frac{g(k) - g(k-1)}{T}\right] = \frac{1 - z^{-1}}{T} \cdot G(z)$.

Comparing the above derivatives, the conversion of $G(s)$ in z-domain is presented as

$$G(z) = G(s)\Big|_{s=\frac{1-z^{-1}}{T}} \tag{4.10}$$

The expressions $s = ((1 - z^{-1})/T) \quad z = (1/(1 - sT))$ represent the projection from s-domain to z-domain and it is depicted in Figure 4.7.

This occurs by using the substitution $s = j\omega_s$ in the expression $z = (1/1 - sT)$.

$$z = \frac{1}{1 - sT}\Big|_{s=j\omega_s} = \frac{1}{1 - j\omega_s T} = \frac{1 + j\omega_s T}{1 + (\omega_s T)^2} = x + jy$$

But

$$\begin{aligned} x^2 + y^2 &= \left(\frac{1}{1 + (\omega_s T)^2}\right)^2 + \left(\frac{\omega_s T}{1 + (\omega_s T)^2}\right)^2 = \frac{1 + (\omega_s T)^2}{(1 + (\omega_s T)^2)^2} \\ &\Rightarrow x^2 + y^2 = \frac{1}{1 + (\omega_s T)^2} \Rightarrow x^2 + y^2 = x \Rightarrow x^2 - x + y^2 = 0 \\ &\Rightarrow x^2 - x + \frac{1}{4} - \frac{1}{4} + y^2 = 0 \Rightarrow \left(x - \frac{1}{2}\right)^2 + y^2 = \frac{1}{4} \end{aligned} \tag{4.11}$$

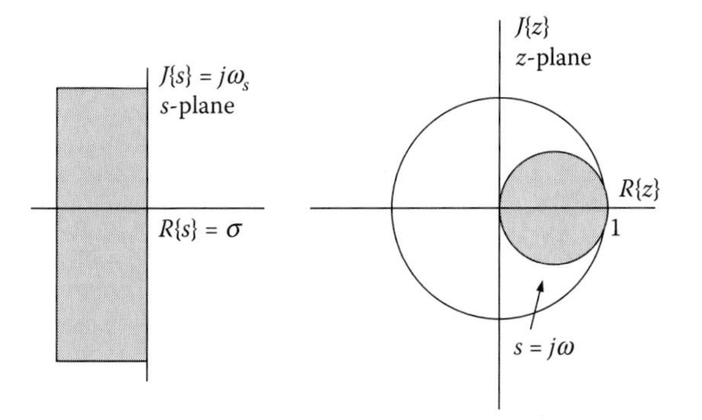

FIGURE 4.7
Projection from s-domain to z-domain using the backward difference method.

The expression (4.11) represents a circle equation with center (1/2) and radius (1/2). If we set $z = (1/1 - sT)$ where $s = \sigma + j\omega_s$, we get the same equation and the σ values define points within the circle (1/2, 1/2).

Using the backward difference method the stability is preserved but the harmonic response does not.

4.2.4 Forward Difference Method

Another way of calculating the derivative $\mathring{g}(t)$ at time instance $t = kT$ is by using the difference of the next sample and the current one, divided by the sampling period

$$\mathring{g}(t) = \frac{dg(t)}{dt} = \frac{g(k+1) - g(k)}{T} \tag{4.12}$$

The derivative of $\mathring{x}(t)$ in s-domain is $sX(s)$ while in z-domain is $z[\mathring{g}(t)] = z[(g(k+1) - g(k))/T] = ((z-1)/T) \cdot G(z)$. Comparing the above derivatives, the conversion of $G(s)$ in z-domain is presented as

$$G(z) = G(s)\Big|_{s=\frac{z-1}{T}} \tag{4.13}$$

Using the forward difference method the stability is not always preserved neither its harmonic response; thereby it is usually not preferred.

4.2.5 Bilinear or Tustin Method

Using this method, the conversion of $G(s)$ in z-domain is presented as

$$G(z) = G(s)\Big|_{s=\frac{2}{T}\frac{z-1}{z+1}} \tag{4.14}$$

The Tustin transform is defined by

$$s = \frac{2}{T}\frac{1-z^{-1}}{1+z^{-1}} \quad \text{and} \quad z = \frac{1+s(T/2)}{1-s(T/2)} \tag{4.15}$$

In fact, it represents one of the most popular methods because the stability of the analog system is preserved, while the impulse response can be also preserved in the case when the nonlinear relation between the analog and digital frequency is taken into account.

By substituting $s = j\omega_s$ in the expression (4.15) (to visualize the $j\omega_s$ axis in z-plane), we have

$$z = \left.\frac{1+s(T/2)}{1-s(T/2)}\right|_{s=j\omega_s} = \frac{1+j(\omega_s T/2)}{1-j(\omega_s T/2)} = \frac{1-(\omega_s T/2)^2 + j\omega_s T}{1+(\omega_s T/2)^2}$$

$$\Rightarrow z = x + jy = \frac{1-(\omega_s T/2)^2}{1+(\omega_s T/2)^2} + j\frac{\omega_s T}{1+(\omega_s T/2)^2}$$

But

$$x^2 + y^2 = \left(\frac{1-(\omega_s T/2)^2}{1+(\omega_s T/2)^2}\right)^2 + \left(\frac{\omega_s T}{1+(\omega_s T/2)^2}\right)^2 = \frac{1+(\omega_s T/2)^2}{1+(\omega_s T/2)^2} = 1 \quad (4.16)$$

The expression (4.16) $x^2 + y^2 = 1$ defines a circle with center 0 and radius 1, as Figure 4.8 shows.

Based on Figure 4.8, the left s half plane is presented within the unit circle in z-domain, using the Tustin method, therefore the discrete system *will always be stable*. The drawback of the Tustin method is the warping level in the ω frequency axis, which is caused by the nonlinear relation between the frequencies ω_s (in s-domain) and ω (in z-domain). Indeed, if we set $s = j\omega_s$ and $z = e^{j\omega}$ in the expression (4.15), we have that

$$j\omega_s = \frac{2}{T}\frac{1-e^{-j\omega}}{1+e^{-j\omega}} = \frac{2}{T}\frac{e^{-j(\omega/2)}\left(e^{j(\omega/2)} - e^{-j(\omega/2)}\right)}{e^{-j(\omega/2)}\left(e^{j(\omega/2)} + e^{-j(\omega/2)}\right)}$$

$$\Rightarrow j\omega_s = \frac{2}{T}\frac{2j\sin(\omega/2)}{2\cos(\omega/2)} = \frac{2}{T}j\tan\frac{\omega}{2} \quad \text{and hence}$$

$$\omega = 2\tan^{-1}\left(\frac{\omega_s T}{2}\right) \quad \text{and} \quad \omega_s = \frac{2}{T}\tan\frac{\omega}{2} \quad (4.17)$$

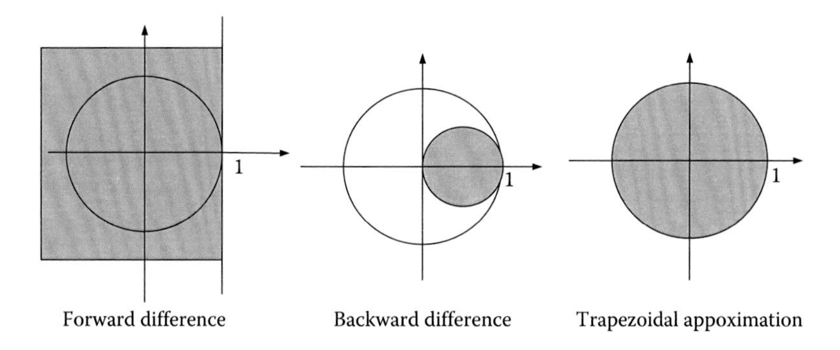

FIGURE 4.8
Stability regions using different methods.

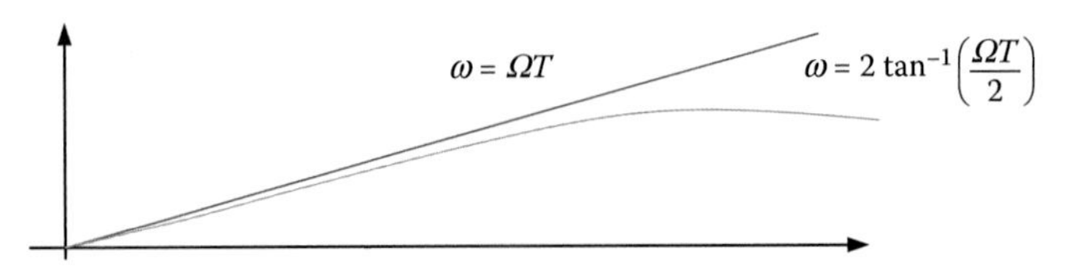

FIGURE 4.9
Nonlinear relation between ω_s and ω.

The nonlinear relation between ω_s and ω is referred as *frequency warping* and is presented in Figure 4.9.

When the specification of the digital filter is known, the most important analog frequencies are intentionally distorted (i.e., frequency prewarping) so as to provide the desired specifications for the digital filter. The usage limitation of the bilinear transform is the fact that the amplitude $|H(\omega_s)|$ of the analog filter *should be a partially linear function* of ω_s.

4.2.6 Frequency Prewarping Method

Even though the bilinear transformation method moves the left s half-plane within the unit circle in z-domain, preserving the stability, it causes distortion to the impulse response of the original system at the same time.

In various digital control and digital signal processing applications, the development of a digital filter $G(z)$ is desirable, which should tightly approach the frequency response of the continuous filter $G(s)$ with the constraint $0 \leq \omega < (\pi/T)((\pi/T) = \text{Nyquist frequency})$.

The frequency response of the continuous filter $G(s)$ is calculated with the substitution $s = j\omega$, while the frequency response is calculated by setting $z = e^{j\omega}$. To compare $G(j\omega)$ and $G(e^{j\omega})$ via the bilinear transform, we set $s = j\omega$ and $z = e^{j\omega}$ in the equation $s = (2/1)((1 + z^{-1})/(1 + z^{-1}))$, so we have

$$s = \frac{2}{T}\frac{1+z^{-1}}{1+z^{-1}} \Rightarrow j\omega_c = \frac{2}{T}\frac{1-e^{-j\omega_D T}}{1+e^{-j\omega_D T}} = \frac{2}{T}\frac{1-(1/e^{j\omega_D T})}{1+(1/e^{j\omega_D T})}$$
$$\Rightarrow j\omega_c = \frac{2}{T}\frac{e^{j(1/2)\omega_D T} - e^{-j(1/2)\omega_D T}}{e^{j(1/2)\omega_D T} + e^{-j(1/2)\omega_D T}} = \frac{2}{T}\frac{2j\sin(\omega_D T/2)}{2\cos(\omega_D T/2)}$$
$$\Rightarrow j\omega_c = j\frac{2}{T}\tan\frac{\omega_D T}{2} \Rightarrow \omega_c = \frac{2}{T}\tan\frac{\omega_D T}{2} \tag{4.18}$$

Hence, if $\omega_c = (2/T)\tan(\omega_D T/2)$ holds then $G(j\omega) = G(e^{j\omega})$.

For small ω_D values with respect to π/T, we have $j\omega_c \cong j(2/T)(\omega_D T/2) = j\omega_D$, thus the behavior of the digital filter satisfactorily approaches the frequency response of its analog counterpart.

If ω_D approaches the value π/T, then $j\omega_c = (2/T)\tan(\omega_D T/2) = j(2/T)\tan(\pi/2) \rightarrow j\infty$, therefore the ω_c frequency tends to infinity and the corresponding distortions become stronger. Hence, the distortions of frequency response can be greatly reduced in the case when the bilinear transform method is implemented using the constraint $\omega_c = (2/T)\tan(\omega_D T/2)$.

The implementation procedure of the Tustin transform method with frequency prewarping is as follows:

1. Given the transfer function $G(s)$, extract a new transfer function $G'(s)$ changing the poles and zeros of the original function as $(s+\alpha) \rightarrow (s+\alpha')|\alpha' = (2/T)\tan(aT/2)$ if α is a real root and $s^2 + 2j\omega_n s + \omega_n^2 \rightarrow s^2 + 2j\omega_n s + \omega_n'^2 \,|\, \omega_n' = (2/T)\tan(\omega_n T/2)$ for the complex roots.
2. Apply the Tustin method in $G'(s)$ substituting each s with $s = (2/T)((1-z^{-1})/(1+z^{-1}))$ into $G'(s)$.
3. Adapt the fixed term k of $G(z)$ using the expression $G(z)_{z=1} = G(s)$.

4.2.7 Matched Pole–Zero Method

A simple yet efficient way of deriving the equivalent discrete form of a continuous-time transfer function is the matched pole–zero method. If we calculate the z-transform of e(kT) samples for a continuous signal $e(t)$, then the poles of the discrete signal E(z) are connected to the poles of $E(s)$ according to the expression $z = e^{sT}$. The matched pole–zero method is based on the fact that we can use the z-transform and the expression $z = e^{sT}$ to locate the zeros of E(z).

According to this method, we separately consider the numerator and denominator of transfer function $Gc(s)$ of the continuous filter and we depict the poles of $Gc(s)$ at the poles of discrete-time transfer function $GD(z)$ and the zeros of $Gc(s)$ to the zeros of $GD(z)$.

The way how the zeros and poles that reach to infinity in s-domain ($s = \infty$) are depicted in z-domain using this method is noteworthy. The $j\omega$ axis from $\omega = 0$ to $\omega = (\pi/T)(= (\omega_s/2)$ Nyquist frequency) in s-domain, is depicted via $z = e^{sT}$ in the unit circle from $z = e^{j0} = 1$ to $z = e^{j(\pi T/T)} = -1$ in z-domain. So, if we choose ω_s as the cyclic frequency that satisfies the sampling theorem, then we can assume that the maximum achievable frequency is $\omega = (\omega_s/2)$ and not $\omega = \infty$.

Consequently, we can assume that the frequency response $G(j\omega)$ tends to zero as ω tends to π/T. Equivalently, it holds that $GD(z)$ (the discrete form of $Gc(s)$) tends to zero as z approaches to –1. Thereby, the point $z = -1$ in z-domain presents the maximum achievable frequency at $\omega = (\pi/T)$.

The pole–zero matching is accomplished as follows:

1. All finite poles of $Gc(s)$ are given in z-domain according to $z = e^{sT}$.
2. All finite zeros of $Gc(s)$ are given in z-domain according to $z = e^{sT}$.

3. a. The infinite zeros of $Gc(s)$, that is, the zeros at $s = \infty$, are presented in $GD(z)$ at $z = -1$. Therefore, we have a $z + 1$ factor in the numerator of $GD(z)$ for each infinite zero in $Gc(s)$. Similarly, the infinite poles of $Gc(s)$, if exist, are presented at $z = -1$. Hence, for each pole at $s = \infty$, we have a $z + 1$ factor in the denominator of $GD(z)$.

   b. If the numerator of $GD(z)$ has a smaller order than its denominator, we add powers of $z + 1$ at the numerator until the order between the numerator and denominator is equal.

4. Adapt the gain of $GD(z)$

   For low-pass filters, $GD(z)|\ z = 1 \equiv Gc(s)|\ s = 0$ should hold.

   For high-pass filters $GD(z)|\ z = -1 \equiv Gc(s)|\ s = \infty$ should hold.

If we use the pole–zero matching method to calculate $GD(z)$, certain difference equations arise in which the calculation of $y(kT + T)$ depends on $u(kT + T)$; so the output at the time instance $kT + T$ depends on its input value at the same time instance. Nonetheless, if a time delay is acceptable, that is, if the calculation of the output $y(kT + T)$ requires the input value $u(kT)$, then the numerator of $Gc(z) = (Y(z)/U(z))$ should have a smaller order than its denominator. In this case, the *modified pole–zero matching method* is used. the modified pole–zero matching method arises from the pole–zero matching method by ignoring Step 3(b).

Both methods preserve the stability of the original system because all the $s < 0$ points, via$z = e^{sT}$, are presented to the z points with $z < 1$, since $0 < e^{sT} < e^{0T} < 1$.

Let the analog transfer function having the following form:

$$G(s) = K_s \frac{(s + \mu_1)(s + \mu_2)\cdots(s + \mu_m)}{(s + \pi_1)(s + \pi_2)\cdots(s + \pi_n)}, \quad m \leq n \tag{4.19}$$

Then, based on the pole–zero matching method, $G(z)$ becomes

$$G(z) = K_z \frac{(z + 1)^{n-m}(z + z_1)(z + z_2)\cdots(z + z_m)}{(z + p_1)(z + p_2)\cdots(z + p_n)} \tag{4.20}$$

The z_i and p_i "are matched" with the corresponding μ_i and π_i according to

$$z_i = -e^{-\mu_i T} \quad \text{and} \quad p_i = -e^{-\pi_i T} \tag{4.21}$$

The n-m multiple zeros $(z+1)^{n-m}$ that arise in $G(z)$ represent the difference on the order between numerator and denominator polynomial.

The constant K_z (dc gain) is selected so as the $G(s)$ and $G(z)$ to be equal for $s=0$ and $z=1$ (where the behavior of systems in low frequencies is of particular interest), hence

$$G(z)(\text{for } z=1)=G(s) \quad (\text{for } s=0) \tag{4.22}$$

Therefore, K_z is easily computed as

$$K_z \cdot z^{n-m} \frac{(1+z_1)(1+z_2)\cdots(1+z_m)}{(1+p_1)(1+p_2)\cdots(1+p_m)} = K_s \frac{\mu_1\mu_2\cdots\mu_m}{\pi_1\pi_2\cdots\pi_n} \tag{4.23}$$

Alternatively, the constant K_z can be computed by

$$K_z = \frac{\lim\limits_{t\to\infty} y(t)}{\lim\limits_{n\to\infty} y(k)} = \frac{\lim\limits_{s\to 0} s(1/s)\cdot G(s)}{\lim\limits_{z\to 1}(z-1)(z/(z-1))G(z)} = \frac{G(0)}{G(1)} \tag{4.24}$$

4.3 Comparison of Discretization Methods

The relation between s and z for the Euler's forward, Euler's backward, and Tustin methods are

$$\begin{aligned} &\text{Euler's forward} \quad s \to \frac{z-1}{T}, \quad z \to 1+T_s \\ &\text{Euler's backward} \quad s \to \frac{z-1}{Tz}, \quad z \to \frac{1}{1-T_s} \\ &\text{Tustin} \quad s \to \frac{2}{T}\frac{z-1}{z+1}, \quad z \to \frac{1+(T/2)s}{1-(T/2)s} \end{aligned} \tag{4.25}$$

The latter methods are shown in Figure 4.10.

For each method, the stability region for continuous-time systems (left complex s half-plane) in z-domain is given in Figure 4.11.

In general, the Tustin method is the most accurate one, yet with marginal differences. The Tustin method with prewarping is the best choice if we desire accuracy at the response for a given frequency range.

Euler's backward method is used to discretize PID controllers in commercial digital control systems. Euler's forward method is less accurate but suitable for discretization of nonlinear differential equation modeling.

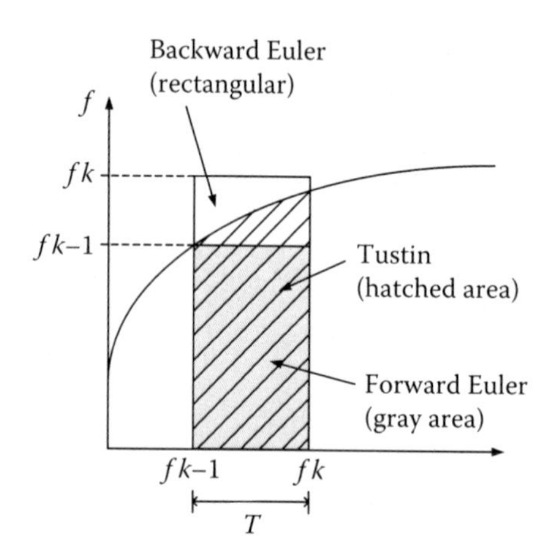

FIGURE 4.10

Euler's forward, Euler's backward, and Tustin methods.

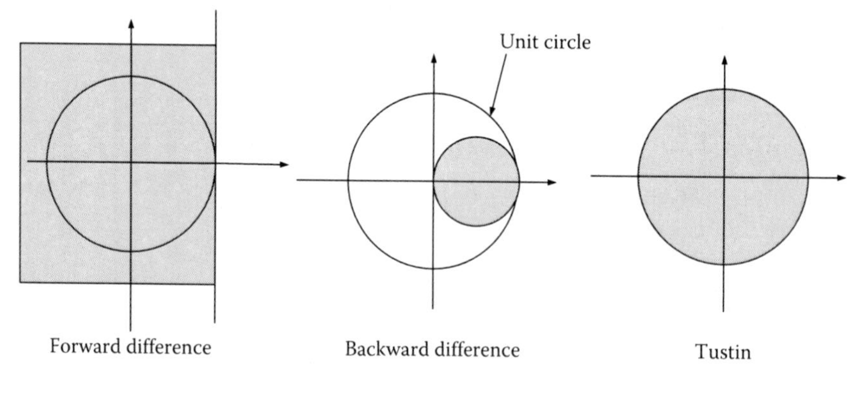

FIGURE 4.11
Stability regions of Euler's forward, Euler's backward, and Tustin methods.

FOH is typically a good choice if there is no special requirement. ZOH, although simpler, introduces a time delay/offset equal to half of the sampling time.

4.4 Formulas

The formula Tables 4.1 and 4.2 are discussed here.

TABLE 4.1

Discretization Methods

Discretization Method	Discrete Transfer Function
Invariant Impulse Response	$G(z) = Z[g(kT)]$ where $g(kT) = [L^{-1}G(s)]_{t=kT}$
Invariant Step Response	$G(z) = (1 - z^{-1})Z(G(s)/s)$
Backward Difference	$G(z) = G(s)\big\|_{s=\frac{1-z^{-1}}{T}}$ $s = \dfrac{1-z^{-1}}{T}$ and $z = \dfrac{1}{1-sT}$
Forward Difference	$G(z) = G(s)\big\|_{s=(z-1/T)}$
Bilinear Transform	$G(z) = G(s)\big\|_{s=(2/T)((z-1)/(z+1))}$ $s = \dfrac{2}{T}\dfrac{1-z^{-1}}{1+z^{-1}}$ and $z = \dfrac{1+s(T/2)}{1-s(T/2)}$
Pole–Zero Matching	$G(s) = K_s \dfrac{(s+\mu_1)(s+\mu_2)\cdots(s+\mu_m)}{(s+\pi_1)(s+\pi_2)\cdots(s+\pi_n)}, \quad m \leq n$ $G(z) = K_z \dfrac{(z+1)^{n-m}(z+z_1)(z+z_2)\cdots(z+z_m)}{(z+p_1)(z+p_2)\cdots(z+p_n)}$ $z_i = -e^{-\mu_i T}$ and $p_i = -e^{-\pi_i T}$ $K_z = \dfrac{\lim_{t\to\infty} y(t)}{\lim_{n\to\infty} y(k)} = \dfrac{\lim_{s\to 0} s(1/s)\cdot G(s)}{\lim_{z\to 1}(z-1)(z/(z-1))G(z)} = \dfrac{G(0)}{G(1)}$

TABLE 4.2

Discretization of Analog Transfer Functions of the Form $c/(s-b)^n$ Using the Invariant Impulse Response Method

$G(s)$	$G(z)\|H(z)$ where $a = e^{bT}$
$\dfrac{c}{s-b}$	$\dfrac{Tc}{1-az^{-1}}$
$\dfrac{c}{(s-b)^2}$	$\dfrac{T^2caz^{-1}}{(1-az^{-1})^2}$
$\dfrac{c}{(s-b)^3}$	$\dfrac{T^3caz^{-1}(1+az^{-1})}{2(1-az^{-1})^3}$
$\dfrac{c}{(s-b)^4}$	$\dfrac{T^4caz^{-1}(1+4az^{-1}a^2z^{-2})}{6(1-az^{-1})^4}$

4.5 Solved Exercises

EXERCISE 4.1

Consider a system with transfer function $G(s) = (1/s + \alpha)$. Find the discrete transfer function $G(z)$ using the invariant impulse response method.

Solution

If the input is described by the impulse function, the response is given as $g(t) = e^{-aT}$

From the definition of z-transform, we have

$$G(z) = \sum_{k=0}^{\infty} g(kT)z^{-k} = \sum_{k=0}^{\infty} e^{-akT} z^{-k} = \sum_{k=0}^{\infty} (e^{-aT} z^{-1})^k \tag{4.1.1}$$

or

$$G(z) = \frac{1}{1-e^{-aT}z^{-1}} = \frac{z}{z-e^{-aT}} = 1 + e^{-aT}z^{-1} + e^{-2aT}z^{-2} + \cdots \tag{4.1.2}$$

Observe from expressions (4.1.1) and (4.1.2) that the above method preserves the impulse response at the discrete time points $t = kT$.

Let's check whether the digital system, resulting from this method, preserves the step response of the analog system.

The step response of the analog filter is

$$Y(s) = \frac{1}{s} \cdot \frac{1}{s+a} = \frac{1/a}{s} - \frac{1/a}{s+\alpha} \overset{L^{-1}}{\Rightarrow} y(t) = \frac{1-e^{-aT}}{a} \tag{4.1.3}$$

The step response of the digital filter is

$$\begin{aligned} Y(z) &= \frac{z}{z-1} \cdot \frac{z}{z-e^{-\alpha T}} = \frac{1}{1-e^{-\alpha T}} \left(\frac{z}{z-1} - \frac{z-e^{-\alpha T}}{z-e^{-\alpha T}} \right) \\ &\overset{IZT}{\Rightarrow} y(k) = \frac{1}{1-e^{-\alpha T}} \left(1 - e^{-aT} e^{-akT} \right) u(k) \end{aligned} \tag{4.1.4}$$

It is obvious from the expressions (4.1.3) and (4.1.4) that the step response of the analog system differs from the corresponding response of the discrete system.

EXERCISE 4.2

Find the digital filter resulting from the conversion of the first-order analog low-pass filter $G(s) = (1/s + 1)$, using (a) the exponential method and (b) the FOH method.

Solution

a. Exponential method.
The desired transfer function is derived by

$$G(z)=(1-z^{-1})Z\left(\frac{G(s)}{s}\right) \quad (4.2.1)$$

$$(4.2.1)\Rightarrow G(z)=(1-z^{-1})Z\left[\frac{1}{s(s+1)}\right]=(1-z^{-1})Z\left[\frac{1}{s}-\frac{1}{s+1}\right]\Rightarrow$$

$$G(z)=(1-z^{-1})\left[\frac{z}{z-1}-\frac{z}{z-e^{-T}}\right]=(1-z^{-1})\left[\frac{(1-e^{-T})z}{(z-1)(z-e^{-T})}\right]$$

$$\Rightarrow G(z)=\frac{(1-e^{-T})z^{-1}}{1-e^{-T}z^{-1}} \quad (4.2.2)$$

For $T = 1$ s, we have

$$G(z)=\frac{0.632(z-1)}{(z-1)(z-0.368)}=\frac{0.632}{z-0.368} \quad (4.2.3)$$

For unit step input, where $X(z)=(z/z-1)$, we have

$$Y(z)=\frac{z}{z-1}\cdot\frac{0.632}{z-0.368} \quad (4.2.4)$$

with final value,

$$\lim_{k\to\infty} y(k)=\lim_{z\to 1} Y(z)(z-1)=1 \quad (4.2.5)$$

The final value of the step response of the continuous system is

$$\lim_{t\to\infty} y(t) \overset{F.V.T.}{=} \lim_{s\to 0} sY(s)=\lim_{s\to 0} s\cdot\frac{1}{s+1}\cdot\frac{1}{s}=1 \quad (4.2.6)$$

The inverse z-transform of $Y(z)$ is

$$\frac{Y(z)}{z}=\frac{1}{z-1}-\frac{1}{z-0.368}\overset{z^{-1}}{\Rightarrow} y(k)=1-(0.368)^{k},\quad k\geq 0 \quad (4.2.7)$$

where $y(0) = 0$
$y(1) = 1 - 0.368 = 0.632$
$y(2) = 1 - (0.368)^2 = 0.864$, etc.

The inverse z-transform of $Y(s)$ is

$$Y(s)=\frac{1}{s}-\frac{1}{s+1}\overset{L^{-1}}{\Rightarrow}y(t)=1-e^{-t} \tag{4.2.8}$$

It is clear that the values of the analog step response coincide with the $y(kT)$ values, at the discrete points $t = kT$. Consequently, this method preserves the step response.

Setting $A = (1 - e^{-T})$ and $B = e^{-T}$, and using the expression (4.2.2), the transfer function can be written as

$$\begin{aligned}G(z)&=\frac{(1-e^{-T})z^{-1}}{1-e^{-T}z^{-1}}=\frac{Az^{-1}}{1-Bz^{-1}}=\frac{Y(z)}{X(z)}\\&\Rightarrow Y(z)(1-Bz^{-1})=AX(z)\Rightarrow Y(z)-BY(z)z^{-1}=AX(z)z^{-1}\\&\Rightarrow y(k)-By(k-1)=Ax(k-1)\end{aligned} \tag{4.2.9}$$

Hence, the desired difference equation of the system is

$$y(k)=Ax(k-1)+By(k-1) \tag{4.2.10}$$

For $T = 1$ s, we have $A = 0.6321$ and $B = 0.3679$.
For unit input, the response is

$$y(k)=0.6321x(k-1)+0.3679y(k-1) \tag{4.2.11}$$

Construct the table of values and observe the response variation with respect of the discrete time.

k	0	1	2	3	4	5
$x(k)$	1	1	1	1	1	1
$y(k)$	0.6321	1	1	1	1	1

b. FOH.
Since

$$G(z)=Z[G_{h1}(s)G(s)]=(1-z^{-1})^2Z\left[\frac{(T_s+1)G(s)}{T_s^2}\right] \tag{4.2.12}$$

we proceed by computing the term

$$Z\left[\frac{(T_s+1)G(s)}{T_s^2}\right]=Z\left[\frac{1}{T}\left[\frac{(T_s+1)G(s)}{s^2}\right]\right]=\frac{1}{T}Z\left[\frac{(T_s+1)G(s)}{s^2}\right] \tag{4.2.13}$$

However

$$\frac{(T_s+1)G(s)}{s^2}=\left[\frac{1}{s^2}+\frac{T-1}{s}+\frac{1-T}{s+1}\right] \tag{4.2.14}$$

So

$$\begin{aligned}G(z)&=\frac{1}{T}(1-z^{-1})^2 Z\left[\frac{1}{s^2}+\frac{T-1}{s}+\frac{1-T}{s+1}\right]\\&=\frac{1}{T}(1-z^{-1})^2\left[\frac{Tz^{-1}}{(1-z^{-1})^2}+\frac{T-1}{(1-z^{-1})}+\frac{1-T}{1-e^{-T}z^{-T}}\right]\end{aligned} \tag{4.2.15}$$

Calculate the corresponding step responses using MATLAB®, as follows:

```
sysc=tf([1],[1 1])
Ts=0.1;
sysd=c2d(sysc,Ts,'foh')
step(sysc,'b',sysd,'r-')
```

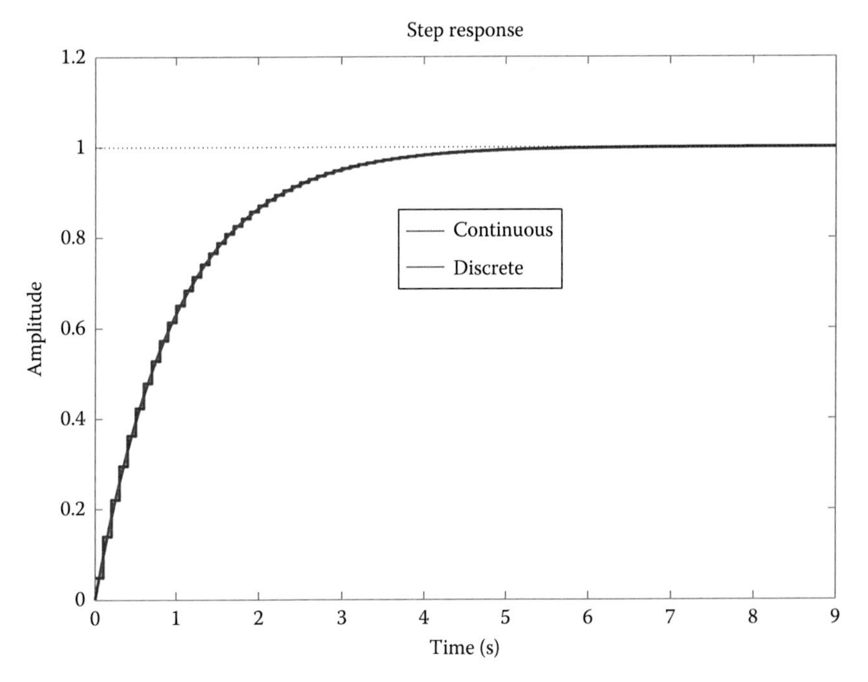

EXERCISE 4.3

Consider the transfer function $G_c(s)=(a/s(s+a))$. Plot the step responses of the analog and discretized system using ZOH. Let $a=0.1$ and $T=0.2s$.

Solution

Applying the invariance method of the step response we get

$$G_D(z)=z\left(\frac{1-e^{-Ts}}{s}\cdot\frac{a}{s(s+a)}\right)=(1-z^{-1})z\left(\frac{a}{s^2(s+a)}\right)$$

$$\Rightarrow G_D(z)=(1-z^{-1})z\left(\frac{1}{s^2}-\frac{1}{as}+\frac{1}{a}\frac{1}{s+a}\right)$$

$$=\frac{z-1}{z}\left(\frac{Tz}{(z-1)^2}-\frac{z}{a(z-1)}+\frac{1}{a}\frac{z}{z-e^{-aT}}\right)$$

$$\Rightarrow G_D(z)=\frac{(e^{-aT}+aT-1)z+(1-e^{-aT}-aTe^{-aT})}{a(z-1)(z-e^{-aT})} \tag{4.3.1}$$

For $a = 0.1$ and $T = 0.2s$ we have

$$G_c(s)=\frac{0.1}{s(s+0.1)} \tag{4.3.2}$$

and

$$G_D(z)=0.002\frac{z+0.98}{(z-1)(z-0.9802)} \tag{4.3.3}$$

Using MATLAB, estimate the corresponding step responses.

```
sysc=zpk([ ],[0  -0.1],0.1)
sysd=zpk([-0.98],[1 0.9802],0.002,0.2)
step(sysc,'r',sysd,'b'), axis([0 10 0 6])
```

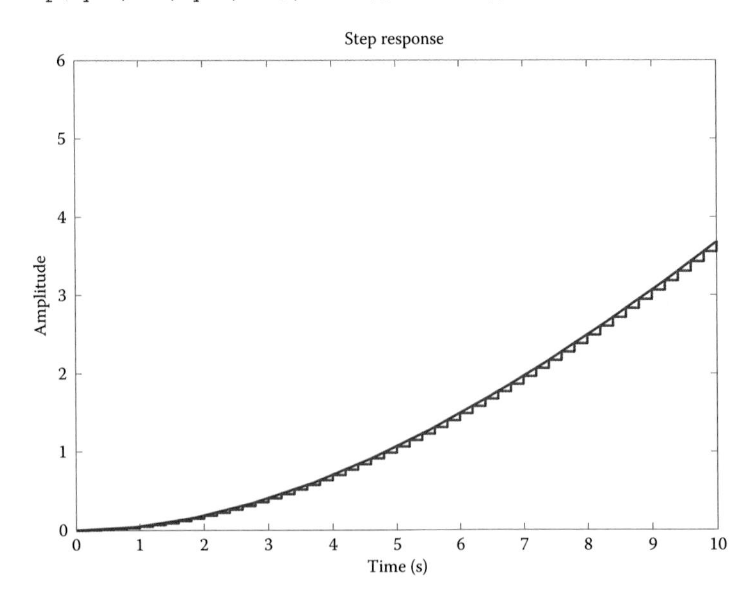

EXERCISE 4.4

Find the digital filter which is derived from the first-order analog low pass filter $G_P(s)=(1/s(s+1))$ using the exponential method.

Solution

The desired transfer function is derived by

$$G(z)=(1-z^{-1})Z\left(\frac{G_p(s)}{s}\right) \tag{4.4.1}$$

Hence

$$G(z)=(1-z^{-1})Z\left\{\frac{G_P(s)}{s}\right\}=(1-z^{-1})Z\left\{\frac{1}{s^2(s+1)}\right\}$$

$$\Rightarrow G(z)=(1-z^{-1})Z\left\{\frac{1}{s^2}-\frac{1}{s}+\frac{1}{s+1}\right\}=(1-z^{-1})\left(Z\left\{\frac{1}{s^2}\right\}-Z\left\{\frac{1}{s}\right\}+Z\left\{\frac{1}{s+1}\right\}\right)$$

$$=(1-z^{-1})\left(\frac{Tz^{-1}}{(1-z^{-1})^2}-\frac{1}{1-z^{-1}}+\frac{1}{1-e^{-T}z^{-1}}\right)$$

$$\Rightarrow G(z)=\frac{(T-1+e^{-T})z^{-1}+(1-e^{-T}-Te^{-T})z^{-2}}{(1-z^{-1})(1-e^{-T}z^{-1})} \tag{4.4.2}$$

EXERCISE 4.5

Find the digital filter which is derived from the first-order analog low-pass filter $G_P(s)=(ab/(s+a)(s+b))$, using the exponential method.

Solution

The desired transfer function is derived by

$$G_{HP}(z)=(1-z^{-1})Z\left\{\frac{G_P(s)}{s}\right\}=(1-z^{-1})Z\left\{\frac{ab}{s(s+a)(s+b)}\right\}$$

$$=(1-z^{-1})Z\left\{\frac{k_1}{s}-\frac{k_2}{s+a}+\frac{k_3}{s+b}\right\}$$

$$=(1-z^{-1})Z\left\{\frac{1}{s}+\frac{b/(a-b)}{s+a}+\frac{a/(b-a)}{s+b}\right\} \tag{4.5.1}$$

Hence

$$G_{HP}(z)=(1-z^{-1})\left(Z\left\{\frac{1}{s}\right\}+\frac{b}{a-b}Z\left\{\frac{1}{s+a}\right\}+\frac{a}{b-a}Z\left\{\frac{1}{s+b}\right\}\right)$$
$$=(1-z^{-1})\left(\frac{1}{1-z^{-1}}+\frac{b}{a-b}\frac{1}{1-e^{-aT}z^{-1}}+\frac{a}{b-a}\frac{1}{1-e^{-bT}z^{-1}}\right)$$
$$=\frac{(1-z^{-1})}{a-b}\left(\frac{a-b}{1-z^{-1}}+\frac{b}{1-e^{-aT}z^{-1}}-\frac{a}{1-e^{-aT}z^{-1}}\right) \quad (4.5.2)$$

EXERCISE 4.6

An integrating element with transfer function $G(s) = (1/s)$ is being sampled with period T_s. A ZOH circuit is applied in its input.

a. Define the pulse transfer function of the integrator.
b. Derive the difference equation of the integrator.
c. Derive the difference equation of the integrator for the bilinear transform.
d. For both cases, define the output signal values at time points $t = 0, T_s, 2T_s, 3T_s, 4T_s$, for step input and for $\delta(t)$ input.

Solution

a. The following scheme presents the continuous and sampled output of the integrator.

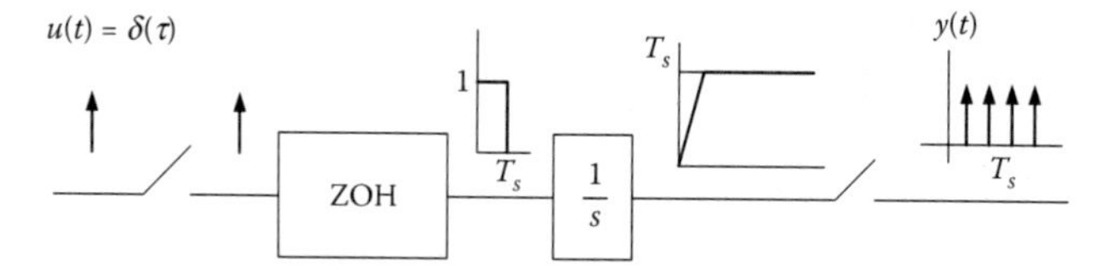

The sampled signal at the output of integrator can be written as

$$y(t)=0+T_s[\delta(t-T_s)+\delta(t-2T_s)+\delta(t-3T_s)+\cdots] \quad (4.6.1)$$

The z-transform of the signal is

$$Y(z)=0+T_s(z^{-1}+z^{-2}+z^{-3}+\cdots)=T_s z^{-1}(1+z^{-1}+z^{-2}+\cdots) \quad (4.6.2)$$

Substituting the geometric series at the right-most part of the expression (4.6.2) with a sum

$$Y(z)=\frac{T_s z^{-1}}{1-z^{-1}}=\frac{T_s}{z-1} \quad (4.6.3)$$

Since the z-transform of input signal is 1, the pulse transfer function becomes

$$G(z) = \frac{Y(z)}{U(z)} = \frac{T_s}{z-1} \tag{4.6.4}$$

b. From the expression (4.6.4), the difference equation is expressed as

$$y[nT_s] = T_s u[(n-1)T_s] + y[(n-1)T_s] \tag{4.6.5}$$

c. The pulse transfer function of an integrator, resulting from the bilinear method, is

$$G(z) = \frac{Y(z)}{U(z)} = \frac{T_s}{2}\frac{z+1}{z-1} \tag{4.6.6}$$

From the expression (4.6.6), the difference equation is expressed as

$$y[nT_s] = \frac{T_s}{2}u[nT_s] + \frac{T_s}{2}u[(n-1)T_s] + y[(n-1)T_s] \tag{4.6.7}$$

d. Let's calculate the output signal values at the points $n = 0, 1, 2, 3, 4$ for step input, using both methods.

n	0	1	2	3	4
y_{zoh}	0	T_s	$2\,T_s$	$3\,T_s$	$4\,T_s$
y_{Tustin}	$T_s/2$	$3\,T_s/2$	$5\,T_s/2$	$7\,T_s/2$	$9\,T_s/2$

For input $\delta(t)$

n	0	1	2	3	4
y_{zoh}	0	T_s	T_s	T_s	T_s
y_{Tustin}	$T_s/2$	T_s	T_s	T_s	T_s

EXERCISE 4.7

A proportional lag element with transfer function $G(s) = (1/1 + sT_1)$ is being sampled with period T_s. A ZOH circuit is applied in its input. Define the pulse transfer function of the integrator for step and impulse responses.

Solution

Design the system with transfer function $G(s)$ and ZOH in its input.

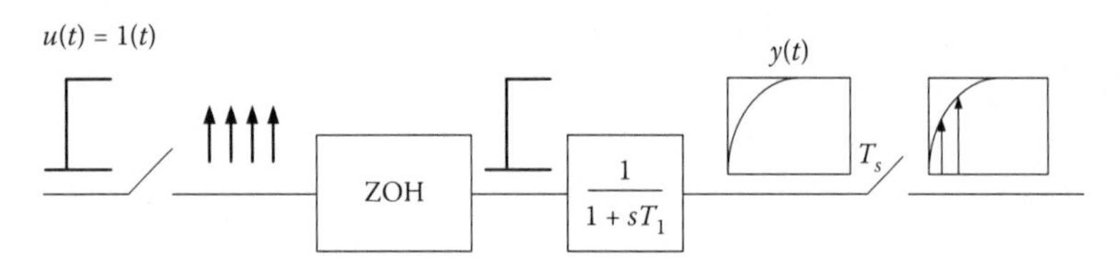

The signal $y(t)$ is analytically defined as

$$y(t) = 1 - e^{-t/T_1} \tag{4.7.1}$$

The mathematical expression of $y(t)$, after the sampling process, becomes

$$y^*(t) = 0 + (1 - e^{-(T_s/T_1)})\delta(t - T_s) + (1 - e^{-(2T_s/T_1)})\delta(t - 2T_s) + \cdots \tag{4.7.2}$$

Using the z-transform of $y^*(t)$ we have

$$\begin{aligned} Y(z) &= (1 - e^{-(T_s/T_1)})z^{-1} + (1 - e^{-(2T_s/T_1)})z^{-2} + (1 - e^{-(3T_s/T_1)})z^{-3} + \cdots \\ &= z^{-1}(1 + z^{-1} + z^{-2} + \cdots) - e^{-(T_s/T_1)}z^{-1}(1 + e^{-(T_s/T_1)}z^{-1} + e^{-(2T_s/T_1)}z^{-2} + \cdots) \end{aligned} \tag{4.7.3}$$

Substituting the geometric series at the right-most part of the expression (4.7.2) with a sum

$$Y(z) = \frac{z^{-1}}{1 - z^{-1}} - \frac{e^{-(T_s/T_1)}z^{-1}}{1 - e^{-(T_s/T_1)}z^{-1}} \tag{4.7.4}$$

The pulse transfer function is obtained by dividing the z-transform of output signal with the z-transform of input signal ($U(z) = (z/z - 1) = (1/1 - z^{-1})$).

$$G(z) = \frac{Y(z)}{U(z)} = z^{-1} - \frac{e^{-(T_s/T_1)}z^{-1}(1 - z^{-1})}{1 - e^{-(T_s/T_1)}z^{-1}} = \frac{(1 - e^{-(T_s/T_1)})z^{-1}}{1 - e^{-(T_s/T_1)}z^{-1}} = \frac{1 - e^{-(T_s/T_1)}}{z - e^{-(T_s/T_1)}} \tag{4.7.5}$$

for $\delta(t)$ as an input, the response of ZOH is a rectangular pulse of width 1 and length T_s. The output exponentially increases up to T_s and then it exponentially reduces up to T_1.

The z-transform of output signal, after the sampling process, is

$$Y(z)=0+(1-e^{-(T_s/T_1)})z^{-1}+(1-e^{-(T_s/T_1)})e^{-(T_s/T_1)}z^{-2}$$
$$+(1-e^{-(T_s/T_1)})e^{-(2T_s/T_1)}z^{-3}+\cdots$$
$$\Rightarrow Y(z)=(1-e^{-(T_s/T_1)})z^{-1}(1+e^{-(T_s/T_1)}z^{-1}+e^{-(2T_s/T_1)}z^{-2}+\cdots)$$
$$=\frac{(1-e^{-(T_s/T_1)})z^{-1}}{1-e^{-(T_s/T_1)}z^{-1}} \tag{4.7.6}$$

The z-transform of input signal is 1, so the pulse transfer function becomes

$$G(z)=\frac{Y(z)}{U(z)}=\frac{(1-e^{-(T_s/T_1)})z^{-1}}{1-e^{-(T_s/T_1)}z^{-1}}=\frac{1-e^{-(T_s/T_1)}}{z-e^{-(T_s/T_1)}} \tag{4.7.7}$$

Observing the expressions (4.7.5) and (4.7.7), it is clear that the pulse transfer function is identical with the two given input signals.

EXERCISE 4.8

a. Find the discretized with ZOH transfer function $G(s)=(V(s)/U(s))$ for a cruise control system, where u is the input, v is the speed, and b is the damping coefficient.
b. Repeat the procedure for the system of the second scheme, where the transfer function is $G(s)=(Y(s)/U(s))$.

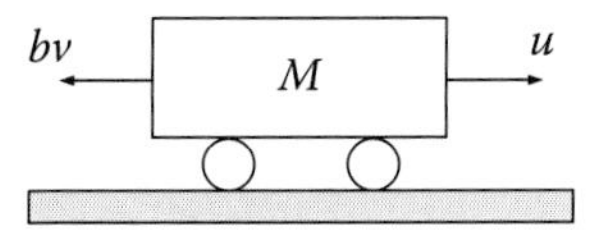

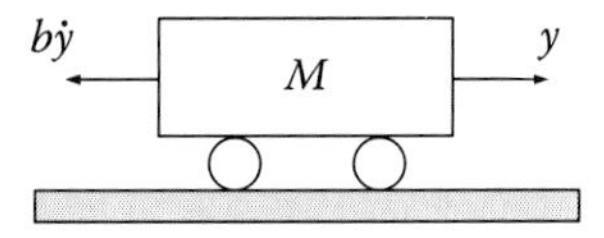

Solution

a. The transfer function in the first case is

$$G(s)=\frac{V(s)}{U(s)}=\frac{1}{Ms+b}=\frac{K/\tau}{s+1/\tau} \tag{4.8.1}$$

where $K=1/b$ and $\tau=M/b$
We know that

$$G(z)=(1-z^{-1})Z\left[\frac{G(s)}{s}\right] \tag{4.8.2}$$

$$\frac{G(s)}{s}=\frac{K/\tau}{s(s+1/\tau)}=(K/\tau)\left[\frac{\tau}{s}-\frac{\tau}{s+1/\tau}\right] \tag{4.8.3}$$

Thereby

$$G(z)=(1-z^{-1})Z\left[\left(\frac{K}{\tau}\right)\left[\frac{\tau}{s}-\frac{\tau}{s+1/\tau}\right]\right]$$

$$=(1-z^{-1})Z\left[K\left[\frac{1}{s}-\frac{1}{s+1/\tau}\right]\right]$$

$$\Rightarrow G(z)=\frac{z-1}{z}K\left[\frac{z}{z-1}-\frac{z}{z-e^{-T/\tau}}\right]=K\left[1-\frac{z-1}{z-e^{-T/\tau}}\right] \tag{4.8.4}$$

or

$$G(z)=K\left[\frac{z-e^{-T/\tau}-z+1}{z-e^{-T/\tau}}\right]=K\left[\frac{1-e^{-T/\tau}}{z-e^{-T/\tau}}\right] \tag{4.8.5}$$

b. The transfer function in the second case is

$$G(s)=\frac{Y(s)}{U(s)}=\frac{1}{s(Ms+b)}=\frac{K/\tau}{s(s+1/\tau)} \tag{4.8.6}$$

$$\frac{G(s)}{s}=\frac{K/\tau}{s^2(s+1/\tau)}=K\left[\frac{1}{s^2}-\frac{\tau}{s}+\frac{\tau}{s+1/\tau}\right] \tag{4.8.7}$$

Hence

$$G(z)=\frac{z-1}{z}K\left[\frac{z}{(z-1)^2}-\frac{\tau z}{z-1}+\frac{\tau z}{z-e^{-T/\tau}}\right] \tag{4.8.8}$$

Making the appropriate simplifications, it yields

$$G(z)=K\left[\frac{1}{z-1}-\tau+\frac{\tau(z-1)}{z-e^{-T/\tau}}\right]$$

$$=K\left[\frac{(1-\tau+\tau e^{-T/\tau})z+(\tau-e^{-T/\tau}(\tau+1)}{(z-1)(z-e^{-T/\tau})}\right] \tag{4.8.9}$$

EXERCISE 4.9

The second-order transfer function is $G(s)=(1/(1+sT_1)(1+sT_2))$. The input and output are being sampled with period T_s. A ZOH circuit is applied in its input. Define the pulse transfer function.

Solution

Design the system with transfer function $G(s)$ and ZOH in its input.

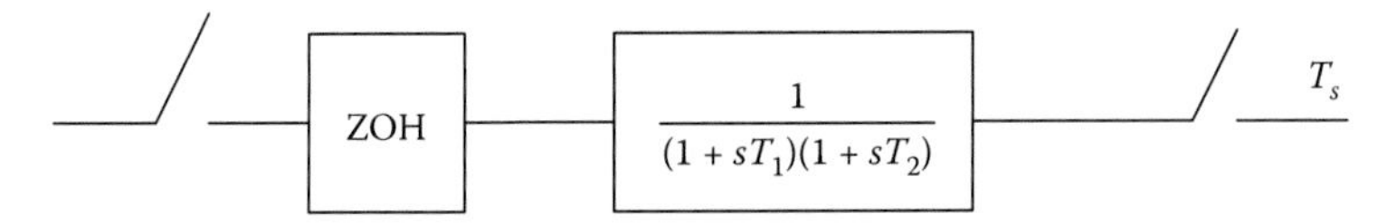

Applying a partial fraction expansion to $G(s)$, the block diagram of the sampled system is presented as

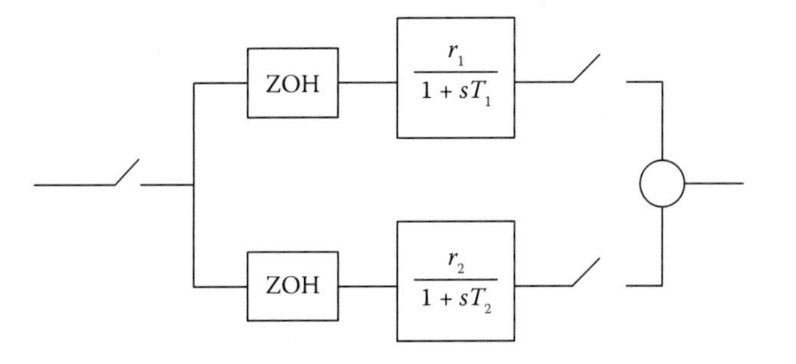

$$\text{where } r_1 = \frac{T_1}{T_1 - T_2}; \quad r_2 = \frac{T_2}{T_2 - T_1} \tag{4.9.1}$$

The pulse transfer function of the first-order transfer function with transfer function $(1/(1+sT_1))$, assuming ZOH, is $((1-e^{-T_s/T})/(z-e^{-T_s/T}))$. The total pulse transfer function is

$$G(z) = \frac{T_1}{T_1 - T_2}\frac{1-e^{-(T_s/T_1)}}{z-e^{-(T_s/T_1)}} + \frac{T_2}{T_2 - T_1}\frac{1-e^{-(T_s/T_2)}}{z-e^{-(T_s/T_2)}} = K\frac{z+a}{(z-e^{-(T_s/T_1)})(z-e^{-(T_s/T_2)})} \tag{4.9.2}$$

where K and a are provided from the expressions (4.9.3) and (4.9.4).

$$K = \frac{T_1(1-e^{-(T_s/T_1)}) - T_2(1-e^{-(T_s/T_2)})}{T_1 - T_2} \tag{4.9.3}$$

$$a = \frac{T_2 e^{-(T_s/T_1)}(1-e^{-(T_s/T_2)}) - T_1 e^{-(T_s/T_2)}(1-e^{-(T_s/T_1)})}{T_1(1-e^{-(T_s/T_1)}) - T_2(1-e^{-(T_s/T_2)})} \tag{4.9.4}$$

EXERCISE 4.10

Using the backward difference method, convert the transfer function $G(s) = (\alpha/s + b)$ into a digital form and define the difference equation for computer emulation.

Solution

It holds that

$$G(z) = G(s)\big|_{s=((1-z^{-1})/T)}$$

$$\Rightarrow G(z) = \frac{a}{((1-z^{-1})/T)+b} = \frac{aT}{(1+bT)-z^{-1}} = \frac{Y(z)}{X(z)} \quad (4.10.1)$$

Hence

$$Y(z)(1+bT) - z^{-1}Y(z) = X(z)(aT) \quad (4.10.2)$$

The difference equation is

$$y(k)(1+bT) - y(k-1) = aTx(k) \quad \text{or}$$

$$y(k) = \left(\frac{1}{1+bT}\right) y(k-1) + aTx(k) \quad (4.10.3)$$

EXERCISE 4.11

Convert the analog controller $G(s) = (4/(s+4))$ into a digital form using the Tustin method.

Solution

It holds that

$$G(z) = G(s)\big|_{s=(2/T)(z-1/z+1)} \quad (4.11.1)$$

consequently

$$G(z) = \frac{4}{(2/T)(1-z^{-1}/1+z^{-1})+4} = \frac{4T}{2+4T} \cdot \frac{1+z^{-1}}{1+z^{-1}((4T-2)/(4T+2))} = \frac{Y(z)}{X(z)} \quad (4.11.2)$$

The resultant difference equation is

$$y(k) = -\frac{4T-2}{4T+2} y(k-1) + \frac{4T}{4T+2}[x(k) + x(k-1)] \quad (4.11.3)$$

For $T = 0.1$ s, we have

$$y(k) = \frac{2}{3} y(k-1) + \frac{1}{6}(x(k) + x(k-1)), \quad k \geq 0 \quad (4.11.4)$$

EXERCISE 4.12

Consider the transfer function $G_c(s) = (1/(s+1)(s+4))$ which has poles at points $s = -1$ and $s = -4$, thus it is stable. Calculate the equivalent discrete-time transfer function $G_D(z)$, using the backward difference method.

Solution

We set $s = ((1 - z^{-1})/T)$ in $G_c(s)$.

Thus

$$G_D(z) = G_c(s)\Big|_{s=((1-z^{-1})/T)} = \frac{1}{\left(\left(\dfrac{1-z^{-1}}{T}\right)+1\right)\left(\left(\dfrac{1-z^{-1}}{T}\right)+4\right)}$$

$$= \frac{T^2}{(1+T-z^{-1})(1+4T-z^{-1})} = \frac{T^2z^2}{[(1+T)z-1][(1+4T)z-1]}$$

$$\Rightarrow G_D(z) = \frac{T^2(1+T)^{-1}(1+4T)^{-1}z^2}{\left(z-\left(\dfrac{1}{1+T}\right)\right)\left(z-\left(\dfrac{1}{1+4T}\right)\right)} \quad (4.12.1)$$

The poles of $G_D(z)$ are $z_1 = (1/(1+T))$ and $z_2 = (1/(1+4T))$.
Since $T > 0$ we have $|z_1| = |1/(1+T)| < 1$ and $|z_2| = |1/(1+4T)| < 1$
Consequently, the arising system is also stable.
For $T = 1$, we have

$$G_D(z) = \frac{1}{10}\frac{z^2}{(z-(1/2))(z-(1/5))} \quad (4.12.2)$$

Therefore, the poles are located at $z_1 = 1/2$ and $z_2 = 1/5$ with $|z_1| 0.5 < 1$ and $|z_2| = 0.2 < 1$; hence the arising system (setting $s = ((1-z^{-1})/T)$) is stable.

Design the step response of the analog and discrete system (for $T = 1$ s) in the following scheme.

We import the above functions in MATLAB using the *zpk* command and then we estimate the step response using the *step* command.

```
zero1=[];
pole1=[-1 -4];
k1=[1];
 sysc=zpk(zero1,pole1,k1)
 zero2=[0 0];
 pole2=[0.5 0.2];
 k2=[1/10];
 T=1;
 sysd=zpk(zero2,pole2,k2,T)
step(sysc,'b',sysd,'r')
```

Observe that the step responses of the continuous-time and discrete-time system are almost identical, using the backward difference method.

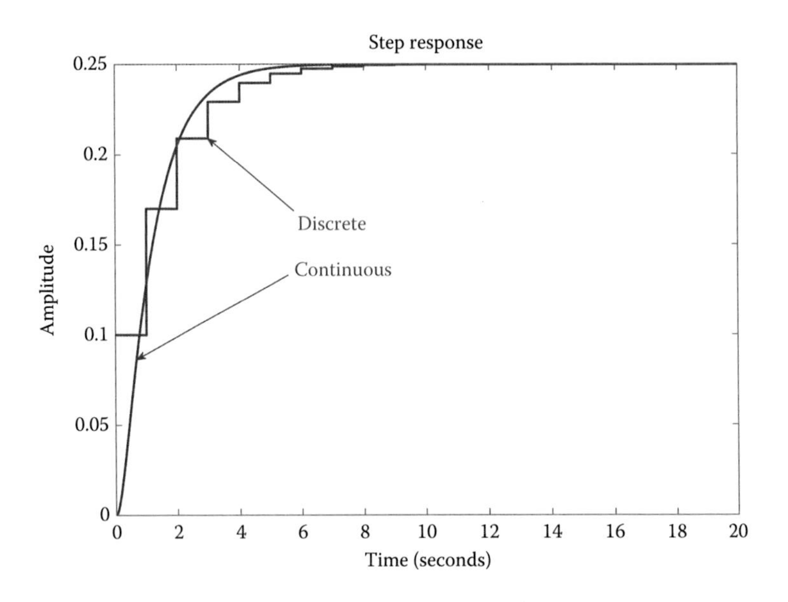

EXERCISE 4.13

Consider the transfer function $G_c(s) = 10((0.5s+1)/(0.1s+1))$ which has a pole at point $s = -10$, so it is stable. Calculate the equivalent discrete-time transfer function $G_D(z)$, using the bilinear method.

Solution

We set $s = (2/T)((1 - z^{-1})/(1 + z^{-1}))$ in $G_C(s)$.

Hence

$$
\begin{aligned}
G_D(z) = G_c(s)\big|_{s=(2/T)((1-z^{-1})/(1+z^{-1}))} &= 10\frac{0.5(2/T)((1-z^{-1})/(1+z^{-1}))+1}{0.1(2/T)((1-z^{-1})/(1+z^{-1}))+1} \\
&= 10\frac{(1-z^{-1}+T(1+z^{-1})/T(1+z^{-1}))}{(0.2(1-z^{-1})+T(1+z^{-1})/T(1+z^{-1}))} \\
&= \frac{10(1-z^{-1}+T+Tz^{-1})}{0.2-0.2z^{-1}+T+Tz^{-1}} \\
&= 10\frac{1+T+(-1+T)z^{-1}}{0.2+T+(-0.2+T)z^{-1}} \\
\Rightarrow G_D(z) &= \frac{(1+T)z+(T-1)}{(0.2+T)z+(T-0.2)}
\end{aligned}
\tag{4.13.1}
$$

If $T = 0.025$ s, then

$$
G_D(z) = \frac{10.25z-9.75}{0.225z-0.175} = \frac{45.55z-43.33}{z-0.7778} \tag{4.13.2}
$$

The function $G_D(z)$ has a pole at $z = 0.7778$ with $|z| = |0.7778| < 1$, so the stability is preserved.

The digital controller $GD(z)$ has the optimum performance when the circular sampling frequency ωs is at least 20 times higher than the bandwidth circular frequency ωb. Let $\omega b = 10$ rad/s. and, hence, we choose $\omega s > 20\omega b$, $\omega s = 25\omega b = 250$ rad/s.

So, $T = (2\pi/\omega_s) = 0{,}025$ s.

Implementing the latter example in MATLAB, we have

```
sysc=tf(10*[0.5 1],[0.1 1]);
 sysd=c2d(sysc,0.025,'tustin');
step(sysc,'r+',sysd,'b-')
```

Design the step responses of the analog and discrete system (for $T = 1$ s) in the following scheme. Observe that the step responses are almost identical.

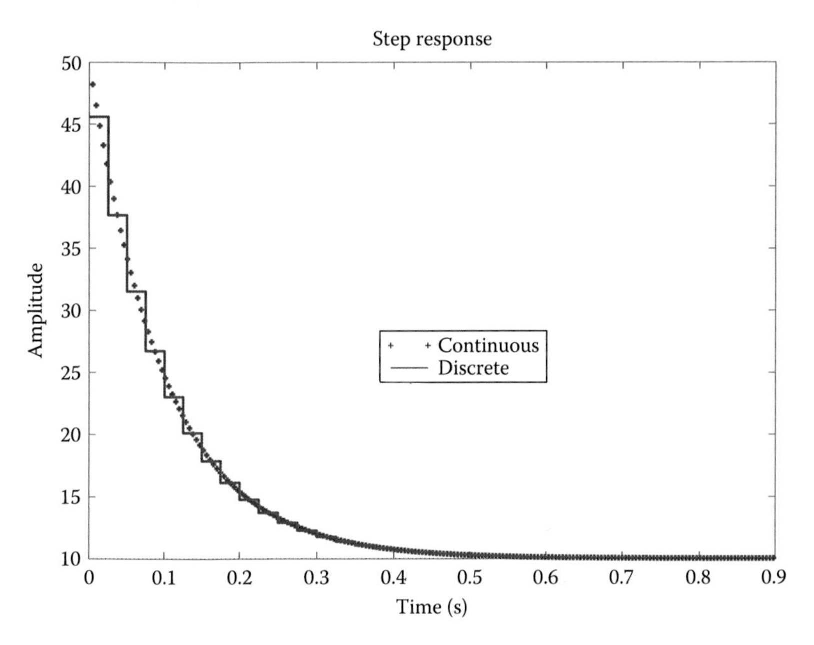

EXERCISE 4.14

Find the equation for the digital filter, derived from the conversion of the analog filter with transfer function $H(s) = (s+1)/((s+1)^2 + 4)$ using the pole–zero matching method, for $T = 1$ s.

Solution

$H(s)$ is written as

$$H(s) = \frac{s+1}{(s+1+2j)(s+1-2j)} \tag{4.14.1}$$

Since the order of denominator and numerator is $n = 2$ and $m = 1$, respectively, it is assumed that there is $n - m = 2 - 1 = 1$ zero to infinity. Note that every zero of $G(s)$ tend to infinity, they are substituted with zeros $z = -1$ in z-domain. It stems that

$$H(z)=\frac{(1-z^{-1}e^{-j\pi})*(1-z^{-1}e^{-T})}{1-2z^{-1}e^{-T}\cos(2T)+e^{-2T}z^{-2}} \tag{4.14.2}$$

Explanation of the term $1 - z^{-1}e^{-j\pi}$: The s-domain frequency, $\omega_s \to \infty$ (when zero tends to infinity) is equivalent to z-domain frequency, $\omega \to \pi$, and such a relation is given by

$$1-z^{-1}e^{j\omega_s T}=1-z^{-1}e^{j(\omega/T)T}=1-z^{-1}e^{j\pi}=1-z^{-1}(-1)=1+z^{-1} \tag{4.14.3}$$

$$(4.14.2),(4.14.3)\Rightarrow H(z)=\frac{(1+z^{-1})(1-z^{-1}e^{-T})}{1-2z^{-1}e^{-T}\cos(2T)+e^{-2T}z^{-2}}$$

$$\Rightarrow H(z)=\frac{(z+1)(z-e^{-T})}{z^2-2ze^{-T}\cos(2T)+e^{-2T}} \tag{4.14.4}$$

For $T = 1$ s, we get

$$(4.14.4)\Rightarrow H(z)=\frac{(z+1)(z-0.3678)}{z^2+0.306z+0.135} \tag{4.14.5}$$

Yet

$$H(z)=\frac{Y(z)}{X(z)}\overset{(4.14.5)}{=}\frac{z^2+0.6322z-0.3678}{z^2+0.306z+0.135} \tag{4.14.6}$$

The expression (4.14.6) should be multiplied with the dc gain factor as

$$H(z)=\frac{Y(z)}{X(z)}=K_{dc}\cdot\frac{z^2+0.6322z-0.3678}{z^2+0.306z+0.135} \tag{4.14.7}$$

where K_{dc} is calculated as

$$K_{dc}=\frac{\lim_{t\to\infty}y(t)}{\lim_{k\to\infty}y(k)} \tag{4.14.8}$$

$$\lim_{t\to\infty} y(t) \overset{F.V.T.}{\underset{L.T}{=}} \lim_{s\to 0} s\left(\frac{1}{s}\right)H(s) = \lim_{s\to 0} H(s) = 0.2$$

$$\lim_{k\to\infty} y(k) \overset{F.V.T.}{\underset{Z.T}{=}} \lim_{z\to 1}(z-1)\frac{z}{z-1}H(z) = \lim_{z\to 1} zH(z) = \frac{1.2644}{1.441} = 0.877$$

$$(4.14.8) \Rightarrow K_{dc} = \frac{0.2}{0.877} = 0.23 \tag{4.14.9}$$

$$(4.14.7), (4.14.9) \Rightarrow H(z) = \frac{Y(z)}{X(z)} = 0.23\frac{1+0.6322z^{-1}-0.3678z^{-2}}{1+0.306z^{-1}+0.135z^{-2}}$$

$$\Rightarrow (1+0.306z^{-1}+0.135z^{-2})Y(z) = 0.23(1+0.6322z^{-1}-0.3678z^{-2})X(z)$$

$$\Rightarrow Y(z) = -0.135z^{-2}Y(z) - 0.306z^{-1}Y(z) + 0.23X(z) + 0.1454z^{-1}X(z) - 0.084z^{-2}X(z) \tag{4.14.10}$$

Thereby, the difference equation of the digital filter becomes

$$y(k) = -0.135y(k-2) - 0.306y(k-1) + +0.23x(k) + 0.1454x(k-1) - 0.0846x(k-2) \tag{4.14.11}$$

EXERCISE 4.15

Convert the analog filter with transfer function $H_\alpha(s) = (1/((s+0.1)^2+9))$ into a digital filter by using: (a) the backward difference method and (b) the invariant impulse method.

Solution

a. Using the backward difference method and the transform $s = ((1-z^{-1})/T)$, we have that

$$H(z) = H_\alpha(s)\big|_{s=((1-z^{-1})/T)} = \frac{1}{\left(\left(\frac{1-z^{-1}}{T}\right)+0.1\right)^2+9}$$

$$\Rightarrow H(z) = \frac{T^2/(1+0.2T+9.01T^2)}{1-\left(\frac{2(1+0.1T)}{1+0.2T+9.01T^2}\right)z^{-1}+\left(\frac{1}{1+0.2T+9.01T^2}\right)z^{-2}} \tag{4.15.1}$$

For $T = 0.1$ s, the poles are

$$P_{1,2} = 0.91 \pm j0.27 = 0.949e^{\pm j16.5^\circ} \tag{4.15.2}$$

Thus, the poles are located near to the unit circle (exactly the same occurs if we choose $T \le 0.1$).

b. $H_\alpha(s)$ can be described as fraction expansion

$$H_\alpha(s) = \frac{1/2}{s+0.1-j3} + \frac{1/2}{s+0.1+j3} \tag{4.15.3}$$

$H(z) = H_\alpha(s)\big|_{s=((1-z^{-1})/T)}$, therefore

$$H(z) = Z[H_\alpha(s)] \overset{(4.15.2)}{=} Z\left[\frac{1/2}{s+0.1-j3}\right] + Z\left[\frac{1/2}{s+0.1+j3}\right]$$

$$\Rightarrow H(z) = \frac{1/2}{1-e^{-0.1T}e^{j3T}z^{-1}} + \frac{1/2}{1-e^{-0.1T}e^{-j3T}z^{-1}}$$

$$\Rightarrow H(z) = \frac{1-(e^{-0.1T}\cos 3T)z^{-1}}{1-(2e^{-0.1T}\cos 3T)z^{-1}+e^{-0.2T}z^{-1}} \tag{4.15.4}$$

EXERCISE 4.16

Convert the analog filter with transfer function $G(s)=((s+0.1)/((s+0.1)^2+9))$ into a digital filter using the invariant impulse method.

Solution

$$G(s) = \frac{s+0.1}{(s+0.1)^2+9} \tag{4.16.1}$$

Using the invariant impulse method, we get

$$G(z) = Z(G(s)) = Z\left[\frac{s+0.1}{(s+0.1)^2+9}\right] = \frac{z^2 - e^{-0.1T}\cos 3Tz}{z^2 - 2e^{-0.1T}\cos 3Tz + e^{-0.2T}} \tag{4.16.2}$$

$$(4.16.1) \Rightarrow |G(j\omega_s)| = \frac{|j\omega_{s+0.1}|}{|(j\omega_{s+0.1})^2+9|} = \frac{\sqrt{0.01+\omega_s^2}}{\sqrt{(9.01-\omega_s^2)^2+0.04\omega_s^2}} \tag{4.16.3}$$

$$(4.16.2) \Rightarrow |G(e^{j\omega})| = \frac{|e^{j2\omega} - e^{-0.01}\cos 0.3e^{j\omega}|}{|e^{j2\omega} - 2e^{-0.01}\cos 0.3e^{j\omega} + e^{-0.02}|}$$

$$\Rightarrow |G(e^{j\omega})| = \frac{|\cos 2\omega + j\sin 2\omega - 0.946(\cos\omega + j\sin\omega)|}{|\cos 2\omega + j\sin 2\omega - 1.8916(\cos\omega + j\sin\omega) + 9802|}$$

$$\Rightarrow |G(e^{j\omega})| = \frac{\sqrt{(\cos 2\omega - 0.946\cos\omega)^2 + (\sin 2\omega - 0.946\sin\omega)^2}}{\sqrt{(0.9802+\cos 2\omega - 1.8916\cos\omega)^2 + (\sin 2\omega - 1.8916\sin\omega)^2}} \tag{4.16.4}$$

EXERCISE 4.17

Convert the analog filter with transfer function $G(s) = 2/((s+1)(s+2))$ into a digital filter, using (a) the backward difference method and (b) the invariant impulse response method.

Solution

a. Backward difference method.

$$G(s) = \frac{2}{(s+1)(s+2)} \tag{4.17.1}$$

Using this method, we have

$$G(z) = G(s)\Big|_{s=((1-z^{-1})/T)} = \frac{2}{(2-z^{-1})(3-z^{-1})} = \frac{2z^2}{(2z-1)(3z-1)} \tag{4.17.2}$$

Setting $z = e^{j\omega}$ in the expression (4.17.2), calculate the discrete frequency response as

$$G(e^{j\omega}) = G(z)\Big|_{z=e^{j\omega}} = \frac{2e^{j2\omega}}{(2e^{j\omega}-1)(3e^{j\omega}-1)} \tag{4.17.3}$$

The amplitude and phase of the digital frequency response of the discrete system are

$$\left.\begin{aligned} |G(e^{j\omega})| &= \frac{2}{\sqrt{(2\cos\omega-1)^2+4\sin^2\omega}\sqrt{(3\cos\omega-1)^2+9\sin^2\omega}} \\ \varphi(\omega) &= 2\omega - \tan^{-1}\frac{2\sin\omega}{2\cos\omega-1} - \tan^{-1}\frac{3\sin\omega}{3\cos\omega-1} \end{aligned}\right\} \tag{4.17.4}$$

Setting $s = j\omega$ in the expression (4.17.1), calculate the analog frequency response as

$$G(j\omega) = G(s)\Big|_{s=j\omega} = \frac{2}{(j\omega+1)(j\omega+2)} \tag{4.17.5}$$

The amplitude and phase of the digital frequency response of the analog system are

$$\left.\begin{aligned} |G(j\omega)| &= \frac{2}{\sqrt{1+\omega^2}\sqrt{4+\omega^2}} \\ \theta(\omega) &= -\tan^{-1}\omega - \tan^{-1}\frac{\omega}{2} \end{aligned}\right\} \tag{4.17.6}$$

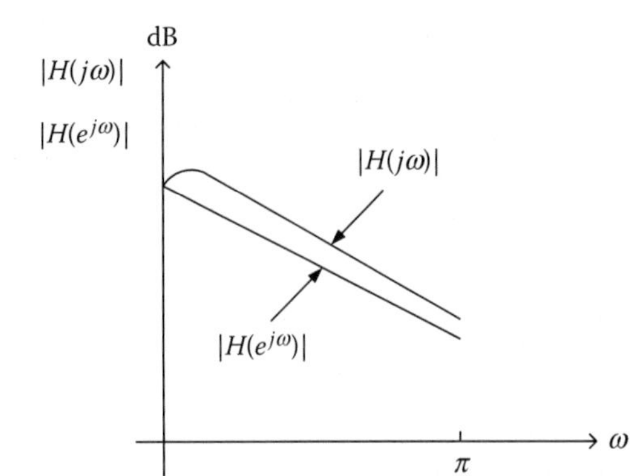

Using the backward difference method, the analog system is transferred to z-domain with negligible differences.

b. Invariant impulse response method.

$$G(s)=\frac{2}{(s+1)(s+2)}=\frac{2}{s+1}-\frac{2}{s+2} \tag{4.17.7}$$

Taking the z-transform of $G_c(s)$, it yields

$$G(z)=\{[L^{-1}\{G(s)\}]_{t=kT}\}=Z\left\{L^{-1}\left\{\frac{2}{s+1}-\frac{2}{s+2}\right\}\Bigg|_{t=kT}\right\}$$

$$=2\frac{z}{z-e^{-1T}}-2\frac{z}{z-e^{-2T}}=\frac{2z}{z-e^{-2T}}-\frac{2z}{z-e^{-2T}}$$

$$=\frac{2z(z-e^{-2T})-2z(z-e^{-T})}{(z-e^{-T})(z-e^{-2T})}=\frac{2z^2-2e^{-2T}z-2z^2+2ze^{-T}}{(z-e^{-T})(z-e^{-2T})}$$

$$\Rightarrow G(z)=\frac{2ze^{-T}-2e^{-2T}z}{(z-e^{-T})(z-e^{-2T})} \tag{4.17.8}$$

So, $G(z)$ is formed as

$$G(z)=\frac{(2e^{-T}-2e^{-2T})z}{(z-e^{-T})(z-e^{-2T})}=\frac{2(e^{-T}-e^{-2T})z^{-1}}{(1-e^{-T}z^{-1})(1-e^{-2T}z^{-1})} \tag{4.17.9}$$

If $T=0.1$ then

$$G(z)=0.1722\frac{z}{(z-0.9048)(z-0.8187)} \tag{4.17.10}$$

Using MATLAB, estimate the corresponding impulse responses as follows:

```
sysc=zpk([ ],[-1 -2],2)
sysd=zpk([0],[0.9048 0.8187],0.1722,0.1)
impulse(sysc,'r',sysd,'b')
```

Observe that the impulse response is preserved, at the sampling time instances, as expected.

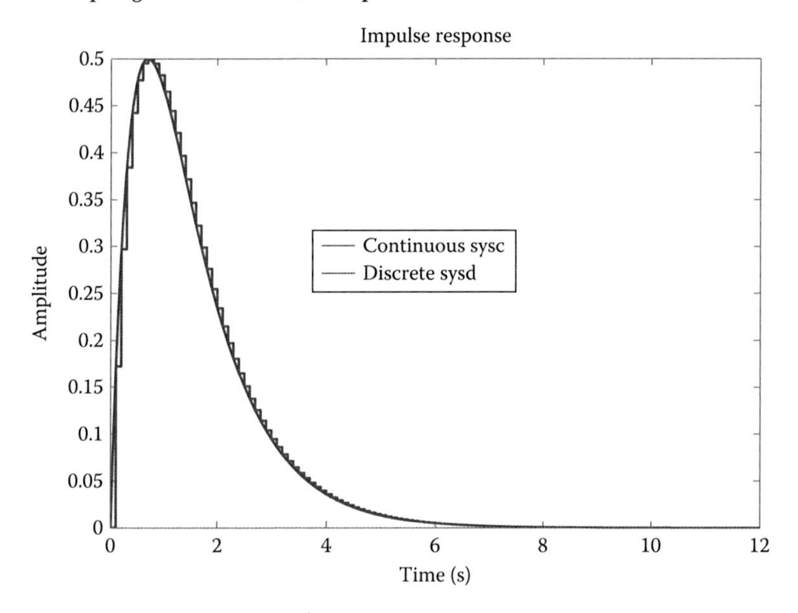

EXERCISE 4.18

Consider the transfer function $G_c(s) = 3/(s+3)$ which has a pole at point $s = -3$, so it is stable. Calculate the equivalent discrete-time transfer function $G_D(z)$, using the forward difference method.

Solution

We set $s = ((1-z^{-1})/Tz^{-1})$.

Hence

$$G(z) = G(s)\big|_{s=((1-z^{-1})/Tz^{-1})} = \frac{3}{((1-z^{-1})/Tz^{-1})+3} = \frac{3Tz^{-1}}{1-z^{-1}+3Tz^{-1}}$$
$$= \frac{3T}{z-1+3T} = \frac{3T}{z-(1-3T)} \tag{4.18.1}$$

The pole of $G_D(z)$ is at the point $z_1 = 1 - 3T$. The system is stable only for values of T, where it holds that $|z_1| = |1 - 3T| < 1$.

For $T = (1/6)$ we have $|z_1| = |1-(3/6)| = 0{,}5 < 1$, so the arising system is stable, setting $s = ((1-z^{-1})/Tz^{-1})$.

For $T = 5/6$ we have $|z_1| = |1-3(5/6)| = 1.5 > 1$, so the arising system is stable, setting $s = ((1-z^{-1})/Tz^{-1})$.

We calculate the corresponding functions and step responses using MATLAB as

```
zero1=[];
pole1=[-3];
k1=[3];
sysc=zpk(zero1,pole1,k1)
 T=1/6;
 zero2=[];
pole2=[0.5];
 k2=[1/2];
sysd1=zpk(zero2,pole2,k2,T)
 T1=5/6;
zero3=[];
pole3=[-3/2];
k3=[2.5];
sysd2=zpk(zero3,pole3,k3,T1)
step(sysc,'b',sysd1,'r',sysd2,'g')
```

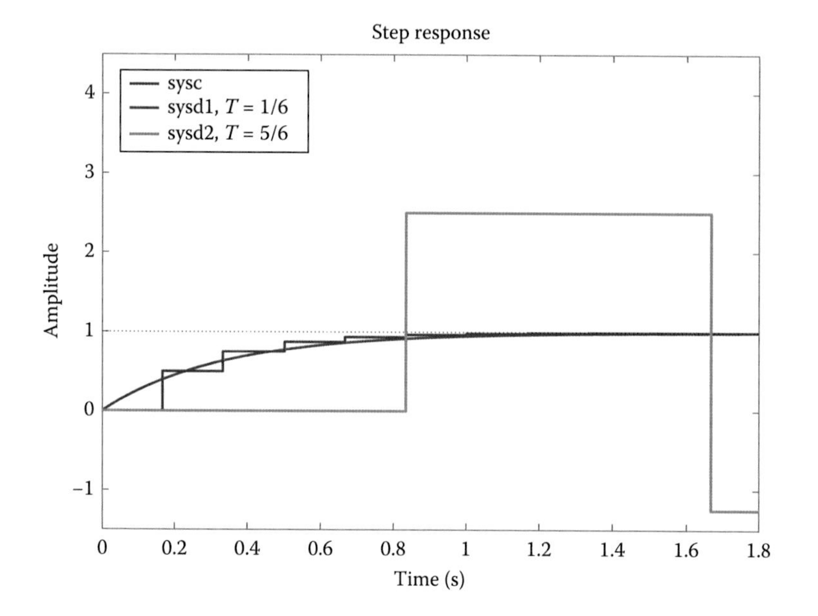

EXERCISE 4.19

Consider the transfer function $G_c(s)=((200\pi)^2/(s^2+200\pi\cdot s+(200\pi)^2))$. Calculate the equivalent discrete-time transfer function $G_D(z)$, using the bilinear transform method with frequency change at $T=0.002$ s.

Solution

The poles of $G_c(s)$ are in the points $s=-100\pi\pm j100\pi\sqrt{3}$.

We have

$$2J\omega_n=200\pi\Rightarrow\omega_n=100\pi/J \tag{4.19.1}$$

$$\omega_n^2 = (200\pi)^2 \Rightarrow \omega_n = 200\pi \quad \text{and} \quad \zeta = \frac{100\pi}{\omega_n} = 0.5 \tag{4.19.2}$$

Hence

$$\omega_n' = \frac{2}{T}\tan\frac{\omega_n T}{2} = \frac{2}{0.002}\cdot\tan\frac{200\pi\cdot 0.002}{2}$$
$$= 1000\tan\frac{200\pi}{1000} = 726.54\ \text{rad/s} \tag{4.19.3}$$

Also, for complex roots we have

$$s^2 + 2j\omega_n s + \omega_n{}^2 \rightleftarrows s^2 + 2j\omega_n^* s + \omega_n^{*2} \tag{4.19.4}$$

and $G_c'(s)$ is given by

$$G_c'(s) = \frac{k}{s^2 + 2\cdot 0.5\cdot 726.54s + (726.54)^2} = \frac{k}{s^2 + 726.54s + (726.54)^2} \tag{4.19.5}$$

Applying the Tustin transform method in $G_c'(s)$, that is, we set

$$s = \frac{2}{T}\frac{z-1}{z+1} = \frac{2}{0{,}002}\frac{z-1}{z+1} = 10^3\frac{z-1}{z+1} \tag{4.19.6}$$

$G_D(z)$ is presented as

$$G_D(z) = G_c'(s)\Big|_{s=10^3((z-1)/(z+1))}$$
$$= \frac{k}{(10^3(z-1)/(z+1))^2 + 726{,}54\cdot 10^3((z-1)/(z+1)) + (726{,}54)^2} \tag{4.19.7}$$

$$G_D(z) = \frac{k(z+1)^2}{10^6(z-1)^2 + 726{,}54\cdot 10^3(z-1)(z+1) + (726{,}54)^2(z+1)^2}$$
$$\Rightarrow G_D(z) = \frac{k(z+1)^2}{2254405z^2 - 944273z + 801321} \tag{4.19.8}$$

To compute the constant k, we set $G_D(1) = G_C(0)$.
Hence

$$G_D(1) = 1 \Rightarrow \frac{4k}{2111453.1} = 1 \Rightarrow k = 527863.28 \tag{4.19.9}$$

Thus

$$G_D(z) = \frac{0,2341(z+1)^2}{z^2 - 0.4189z + 0.3554} \tag{4.19.10}$$

For sampling period $T = 0.001$ s, we get

$$\omega_n^* = 649.839 \quad \text{and} \quad G_D'(z) = \frac{0.0738(z+1)^2}{z^2 - 1.25052 + 0.5457} \tag{4.19.11}$$

Implementing this example in MATLAB, design the Bode diagrams for both the continuous-time and discrete-time systems for sampling periods $T = 0.002$ s and $T = 0.001$ s, whereas we can draw some useful outcomes.

```
num=[(200*pi)^2];
den=[1 200*pi  (200*pi)^2];
 sys=tf(num,den);
 [numd,dend]=c2dm(num,den,0.002,'prewarp',726.54)
 sysd=tf(numd,dend,0.002);
[numd1,dend1]=c2dm(num,den,0.001,'prewarp',649.8394)
 sys=tf(num,den);
sysd=tf(numd,dend,0.002);
 sysd1=tf(numd1,dend1,0.001);
 bode(sys,'r',sysd,'b',sysd1,'g')
```

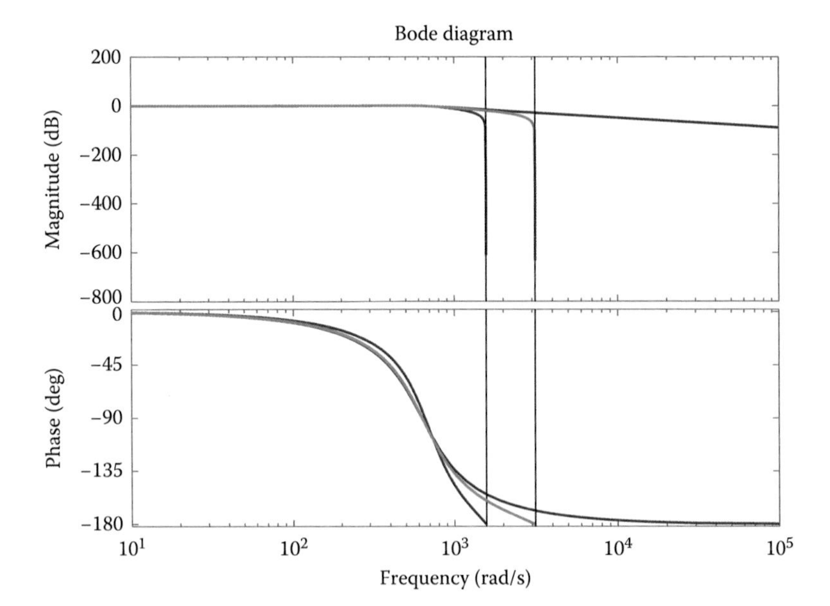

EXERCISE 4.20

Consider the system of the following scheme. Find the open-loop and closed-loop step response.

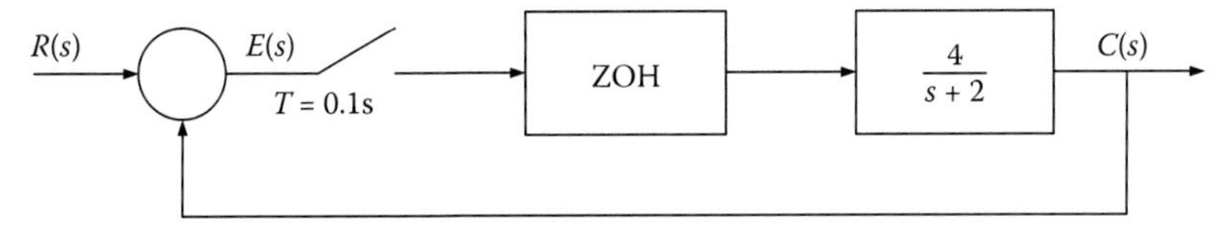

Solution

The output of the discrete system in z-domain is given by

$$C(z) = \frac{G(z)}{1+G(z)}R(z) \tag{4.20.1}$$

where $G(z)$ denotes the transfer function controlled by ZOH.

$$G(z) = Z\left[\frac{1-e^{-T_s}}{s}\frac{4}{s+2}\right] = \frac{z-1}{z}Z\left[\frac{4}{s(s+2)}\right]$$
$$= \frac{z-1}{z}\frac{2(1-e^{-2T})}{(z-1)(z-e^{-2T})} = \frac{0.3625}{z-0.8187}, T = 0.1s \tag{4.20.2}$$

The closed-loop transfer function is

$$G_{cl}(z) = \frac{G(z)}{1+G(z)} = \frac{0.3625}{z-0.4562} \tag{4.20.3}$$

For step input $R(z) = Z(1/s) = (z/(z-1))$ so

$$(4.20.1) \Rightarrow C(z) = \frac{0.3625z}{(z-1)(z-0.4562)} = \frac{0.667z}{z-1} + \frac{-0.667z}{z-0.4562}$$
$$\Rightarrow c(kT) = 0.667[1-(0.4562)^k] \tag{4.20.4}$$

The expression (4.20.4) provides the open-loop step response. We also calculate the step response of the analog system for comparison reasons.

$$G_a(s) = \frac{G_p(s)}{1+G_p(s)} = \frac{4}{s+6}$$
$$\Rightarrow C_a(s) = \frac{4}{s(s+6)} = \frac{0.667}{s} + \frac{-0.667}{s+6}$$
$$\Rightarrow c_a(t) = 0.667(1-e^{-6t}) \tag{4.20.5}$$

We formulate the subsequent table of values, where it is clear that the step response of the analog and discrete system is preserved.

kT	$C(kT)$	$c_a(t)$
0	0	0
0.1	0.363	0.300
0.2	0.528	0.466
0.3	0.603	0.557
0.4	0.639	0.606
0.5	0.654	0.634
0.6	0.661	0.648
.	.	.
.	.	.
.	.	.
1.0	0.666	0.665

The continuous-time system output is expressed as

$$C(s) = G(s)\left[\frac{R(z)}{1+G(z)}\right]_{z=e^{Ts}} = \frac{4(1-e^{-Ts})}{s(s+2)}\left[\frac{(z/z-1)}{1+G(z)}\right]_{z=\varepsilon^{Ts}}$$
$$= \frac{4}{s(s+2)}\left[\frac{1}{1+G(z)}\right]_{z=e^{Ts}} = C_1(s)\left[\frac{1}{1+G(z)}\right]_{z=e^{Ts}} \tag{4.20.6}$$

where

$$\frac{1}{1+G(z)} = \frac{1}{1+(0.3625/(z-0.8187))} = \frac{z-0.8187}{z-0.4562}$$
$$= 1 - 0.363z^{-1} - 0.165z^{-2} - \cdots \tag{4.20.7}$$

$$C_1(s) = \frac{4}{s(s+2)} = \frac{2}{s} - \frac{2}{s+2}$$
$$\Rightarrow c_1(t) = 2(1-e^{-2t}) \tag{4.20.8}$$

$$(4.20.6) \overset{(4.20.7),(4.20.8)}{\Rightarrow} C(s) = C_1(s)[1-0.363e^{-T_s} - 0.165e^{-2T_s} - \cdots]$$
$$c(t) = 2(1-e^{-2t}) - 0.363(2)(1-e^{-2(t-T)})u(t-T)$$
$$-0.165(2)(1-\varepsilon^{-2(t-2T)})u(t-2T) - \cdots \tag{4.20.9}$$

For $T = 0.1\ s$

$$c(3T) = c(0.3) = 2(1-\varepsilon^{-0.6}) - 0.363(2)(1-\varepsilon^{-0.4})$$
$$-0.165(2)(1-\varepsilon^{-0.2}) = 0.603 \tag{4.20.10}$$

EXERCISE 4.21

Consider the transfer function $G_c(s) = k_c((s + a)/(s + b))$. Discretize it by using the pole–zero matching method. Provide some conclusions regarding the impact of sampling time onto the deviation between the continuous-time and discrete-time step responses. Repeat the process for $G_c(s) = (a/(s+a))$.

Solution

$G_c(s)$ has a zero at $s = -a$ and a pole at $s = -b$.

The zero point at $s = -a$ will appear at $z = e^{-aT}$ in z-domain, while the pole at $s = -b$ will appear at $z = e^{-bT}$ in z-domain.

So

$$G_D(z) = k_D \frac{z - e^{-aT}}{z - e^{-bT}} \tag{4.21.1}$$

To calculate k_D, we solve the equation $G_c(0) = G_D$

$$G_c(0) = G_D(1) \Rightarrow k_c \frac{a}{b} = k_D \frac{1 - e^{-aT}}{1 - e^{-bT}} \Leftrightarrow k_D = k_c \frac{a}{b} \frac{1 - e^{-bT}}{1 - e^{-aT}} \tag{4.21.2}$$

Therefore

$$G_D(z) = k_c \frac{a}{b} \frac{1 - e^{-bT}}{1 - e^{-aT}} \cdot \frac{z - e^{-aT}}{z - e^{-bT}} \tag{4.21.3}$$

Let $a = 1$, $b = 2$ and $k_c = 1$. Then, the transfer function becomes

$$G_c(s) = \frac{s+1}{s+2} \tag{4.21.4}$$

If $G_c(s)$ is being discretized with sampling period $T = 0.1$, then

$$k_D = 1 \frac{1}{2} \frac{1 - e^{-0.2}}{1 - e^{-0.1}} = 0.9524 \tag{4.21.5}$$

and

$$G_D(z) = 0.9524 \frac{z - e^{-0.1}}{z - e^{-0.2}} \tag{4.21.6}$$

The deviation between the two step responses is quite important. Using the corresponding MATLAB commands, we estimate functions $G_c(s)$ and $G_D(z)$

```
sysc=tf([1 1],[1 2])
 sysd=c2d(sysc,0.1,'matched')
 step(sysc,'b',sysd,'r')
```

However, if $G_c(s)$ is being discretized with smaller sampling period $T = 0.001$, then the deviation is emphatically reduced. We add the command

```
sysd=c2d(sysc,0.001,'matched')
```

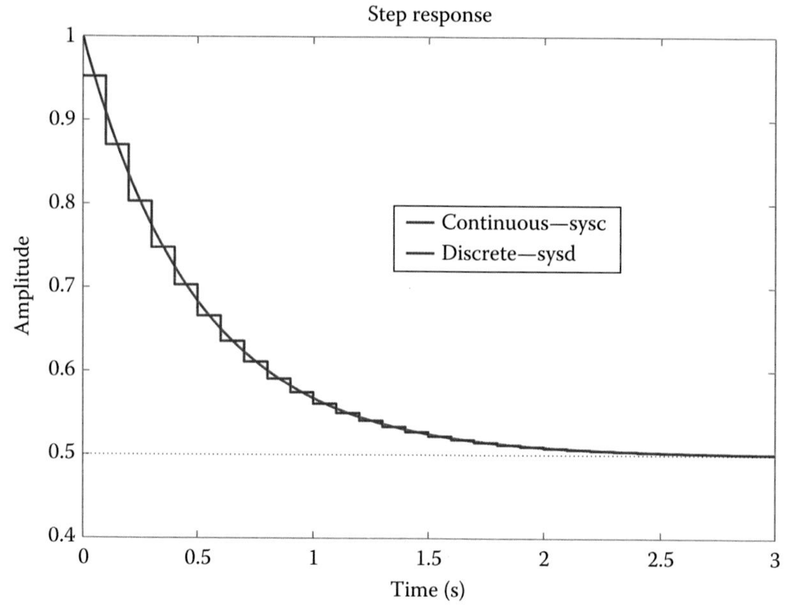

Since the denominator of $G_c(s)$ has a higher order than the numerator, then we have to set the $z+1$ parameter in the numerator.

If $G_c(s) = (a/(s+a))$, then the equivalent function $G_D(z)$ is given by

$$G_D(z) = k_D \frac{a(z+1)}{z - e^{-aT}} \tag{4.21.7}$$

And since $G_c(0) = G_D(1)$, we have

$$G_c(0) = G_D(1) \Leftrightarrow \frac{a}{0+a} = k_D \frac{a(1+1)}{1-e^{-aT}} \Leftrightarrow k_D = \frac{1-e^{-aT}}{2a} \tag{4.21.8}$$

Setting $A = e^{-aT}$ and $k_D = ((1-e^{-aT})/2a)$, we have

$$G_D(z) = k_D \frac{a(z+1)}{(z-A)} = \frac{(1-A)}{2a} \frac{a(z+1)}{(z-A)} = \frac{(1-A)(z+1)}{2(z-A)} \tag{4.21.9}$$

Setting $G_D(z) = (Y(z)/U(z))$, it yields

$$\frac{Y(z)}{U(z)} = \frac{(1-A)(z+1)}{2(z-A)} = \frac{(1-A)(1+z^{-1})}{2(1-Az^{-1})}$$

$$\Leftrightarrow \frac{Y(z)}{U(z)} = \frac{(1-A)+(1-A)z^{-1}}{2-2Az^{-1}}$$

$$\Leftrightarrow 2Y(z) - 2Az^{-1}Y(z) = (1-A)U(z) + (1-A)z^{-1}U(z)$$

$$\Leftrightarrow Y(z) = Az^{-1}Y(z) + \frac{(1-A)}{2}U(z) + \frac{(1-A)}{2}z^{-1}U(z) \tag{4.21.10}$$

Using the inverse z-transform, it holds

$$y(k) = Ay(k-1) + \frac{(1-A)}{2}u(k) + \frac{(1-A)}{2}u(k-1)$$

$$\Leftrightarrow y(k) = Ay(k-1) + \frac{(1-A)}{2}[u(k) + u(k-1)] \tag{4.21.11}$$

Observe that the output $y(k)$ depends on the input value $u(k)$ at the time instance k.

If $a = 5$, the continuous-time transfer function will be $G_c(s) = (5/(s+5))$ and the transfer function $G_D(z)$ of the equivalent discrete system will be expressed as

$$G_D(z) = k_D \frac{5(z+1)}{z - e^{-5T}} \tag{4.21.12}$$

If the sampling period is $T = 1/15$, we have

$$k_D = \frac{1 - e^{-5(1/15)}}{10} = 0.02835 \tag{4.21.13}$$

and

$$G_D(z) = 0.02835 \frac{5(z+1)}{z - 0.7165} = 0.1417 \frac{(z+1)}{z - 0.7165} \tag{4.21.14}$$

We estimate the corresponding step responses using MATLAB

```
sysc=zpk([],[-5],5);
sysd=zpk([-1],[0.7165],0.1417,1/15);
```

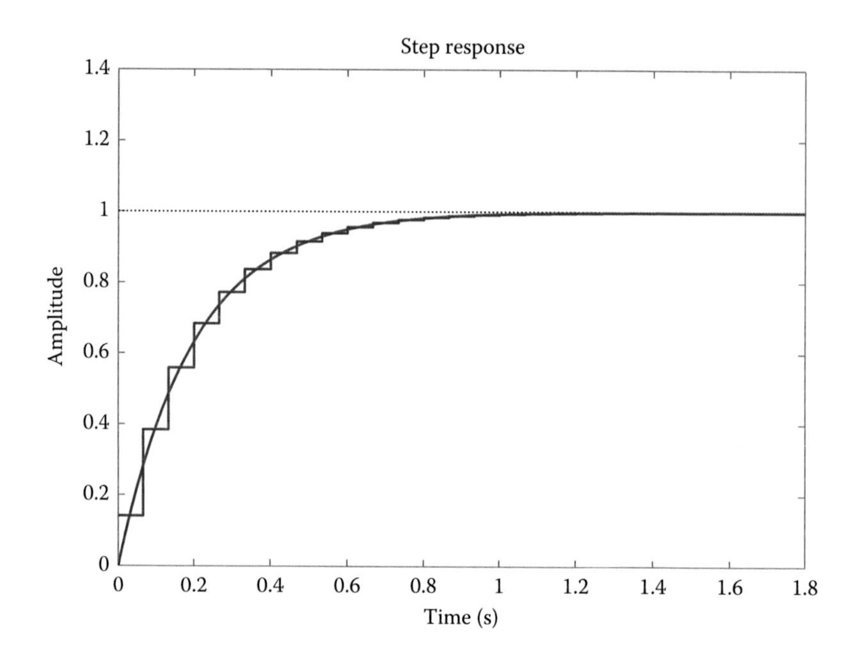

However, if we use the modified method for the same transfer function $Gc(s) = (a/(s + a))$, we will have

$$G_D(z) = k_D \frac{a}{z - e^{-aT}}$$

where

$$k_D = \frac{1 - e^{-aT}}{a}$$

If $a = 5$, then the continuous-time transfer function will be $Gc(s) = (5/(s + 5))$ and the equivalent discrete-time transfer function $G_D(z)$ will be presented as

$$G_D(z) = k_D \frac{5}{z - e^{-5T}} = \frac{1 - e^{-5T}}{5} \frac{5}{z - e^{-5T}} = \frac{1 - e^{-5T}}{z - e^{-5T}} \tag{4.21.15}$$

For sampling period $T = 1/15$ then

$$G_D(z) = 0.2835 \frac{1}{z - 0.7165} \tag{4.21.16}$$

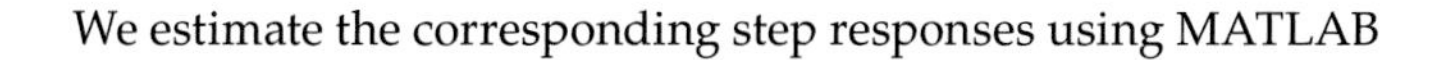

We estimate the corresponding step responses using MATLAB

```
sysc=zpk([],[-5],5);
sysd=zpk([],[0.7165],0.2835,1/15)
step(sysc,'r',sysd,'b')
```

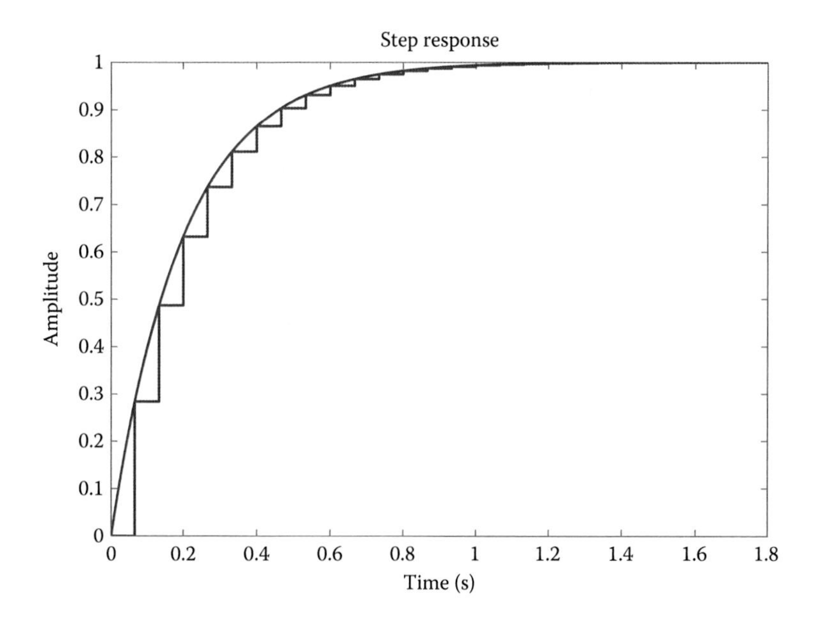

Using the same method, we compute the corresponding difference equation

$$G_D(z) = \frac{Y(z)}{U(z)} \Leftrightarrow \frac{Y(z)}{U(z)} = k_D \frac{a}{z - e^{-aT}} = \frac{1 - e^{-aT}}{a} \frac{a}{z - e^{-aT}} = \frac{1 - e^{-aT}}{z - e^{-aT}} \quad (4.21.17)$$

Setting $A = e^{-aT}$, the above equation becomes

$$\begin{aligned} \frac{Y(z)}{U(z)} = \frac{1-A}{z-A} \Leftrightarrow \frac{Y(z)}{U(z)} &= \frac{(1-A)z^{-1}}{1-Az^{-1}} \\ &\Leftrightarrow (1 - Az^{-1})Y(z) = (1-A)z^{-1}U(z) \\ &\Leftrightarrow Y(z) - Az^{-1}Y(z) = (1-A)z^{-1}U(z) \\ &\Leftrightarrow Y(z) = Az^{-1}Y(z) + (1-A)z^{-1}U(z) \end{aligned} \quad (4.21.18)$$

Deriving the inverse transform of the latter equation, we have that

$$\underline{y(k)} = Ay(k-1) + (1-A)\underline{u(k-1)} \quad (4.21.19)$$

Observe that the output value $y(k)$ depends on the input value $u(k-1)$ at the time instance $k-1$.

EXERCISE 4.22

Consider the system with transfer function $H(s)=((s-1)/(s^2+3s+2))$.

Discretize the system using the forward and backward Euler difference methods, and using the ZOH method, for sampling periods $T=0.1$ s. and $T=1$ s. Compare the discretized frequency responses and derive useful conclusions.

Solution

The forward Euler method uses the expression $s=(z-1)/T$, while the backward Euler method uses $s=(z-1)/Tz$.

The MATLAB code that extracts the resultant pulse transfer functions is given as follows. For the discretization via the forward and backward Euler methods, the MATLAB command **bilin.m** ([Gz] = BILIN(Gs,VERS,METHOD,AUG)) has been used, which performs a bilinear transform in the frequency domain, such that $s=((az+d)/(cz+b))$ (if vers = 1) or $z=((d-bs)/(cs-a))$ (if vers = −1), so as $Gs=C(Is-A)^{-1}B+D \Leftrightarrow Gz=Cb(Iz-Ab)^{-1}\,Bb+Db$

```
%T=0.1s
H=tf([1 -1],conv([1 1],[1 2]))
Hf=tf(bilin(ss(H),1,'fwdrec',.1)) % Forward Euler
Hb=tf(bilin(ss(H),1,'bwdrec',.1)) % Backward Euler
Hz=c2d(H,.1)% ZOH
%T=1s
Hf1=tf(bilin(ss(H),1,'fwdrec',1)) % Forward Euler
Hb1=tf(bilin(ss(H),1,'bwdrec',1)) % Backward Euler
Hz1=c2d(H,1)% ZOH
```

We have the following transfer functions:

$$H_{FE,0.1}(z)=\frac{0.1z-0.11}{z^2-1.7z+0.72} \qquad H_{FE,1}(z)=\frac{z-2}{z^2+z}$$

$$H_{BE,0.1}(z)=\frac{0.0682z^2-0.0758z}{z^2-1.742z+0.758} \qquad H_{BE,1}(z)=\frac{-0.167z}{z^2-0.833z+0.167}$$

$$H_{ZOH,0.1}(z)=\frac{0.0816z-0.09}{z^2-1.724z+0.741} \qquad H_{ZOH,1}(z)=\frac{0.0328z-0.306}{z^2-0.503z+0.0498}$$

Design the frequency responses of the discretized systems along with the corresponding ones of the analog system

```
subplot(121)
bode(H,Hf,Hb,Hz) %T=0.1s
subplot(122)
bode(H,Hf1,Hb1,Hz1) %T=1s
```

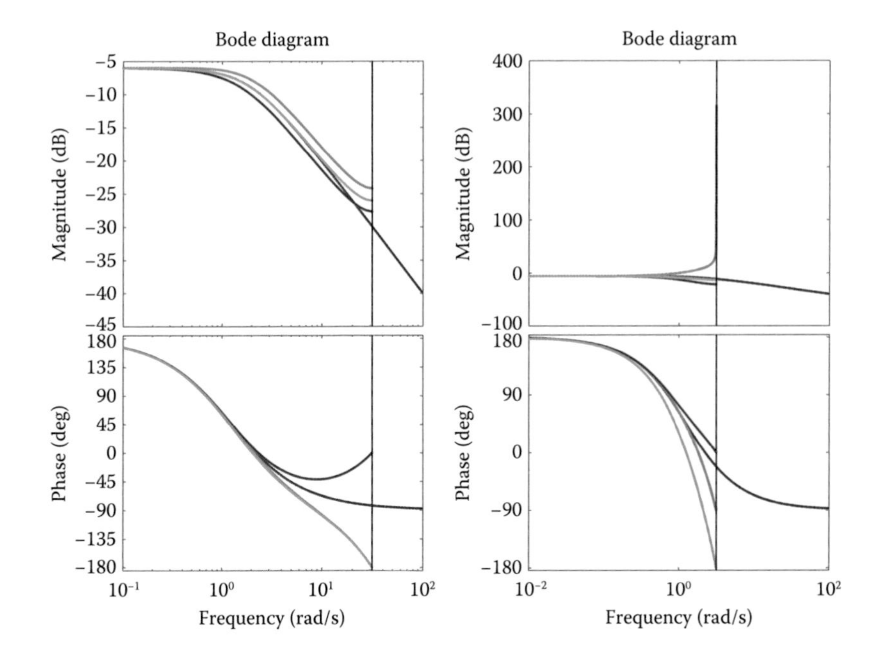

Observe that the quality of the approach worsens with increasing sampling period.

EXERCISE 4.23

Consider a continuous-time process with transfer function $P(s)=(10/(s+10))$. Discretize it with the aid of ZOH and sampling period $T_s=0.01$ s. Using w-transform with the Tustin method, calculate the analog transfer function and derive relevant conclusions.

Solution

The continuous-time transfer function with ZOH is given by

$$P(z)=\mathbf{Z}\left\{\frac{1-e^{-sT_s}}{s}P(s)\right\}=\mathbf{Z}\left\{\frac{1-e^{-0.1s}}{s}\frac{10}{s+10}\right\} \tag{4.23.1}$$

The corresponding MATLAB code is

```
numPs=10;
 denPs=[1 10];
 Ts=0.1;
 [numPz,denPz]=c2dm(numPs,denPs,Ts,'zoh');
```

Hence

$$P(z)=\frac{0.63212}{z-0.36788} \tag{4.23.2}$$

Using the **d2cm.m** command, we transfer from z-domain to w-domain

```
[numPw,denPw]=d2cm(numPz,denPz,Ts,'tustin');
```

The result is

$$P(w) = 9.241\frac{1-0.05w}{w+9.241} \tag{4.23.3}$$

Obviously, the expression (4.23.3) seems different than $P(s) = (10/(s+10))$ The latter difference can be better revealed, by taking a frequency-domain analysis.

```
omega= logspace(-1,2,200);
[Mags,Phases]=bode(numPs,denPs,omega);
[Magw,Phasew]=bode(numPw,denPw,omega);
 loglog(omega,[Mags,Magw]),grid;
semilogx(omega,Phases,omega,Phasew),grid;
```

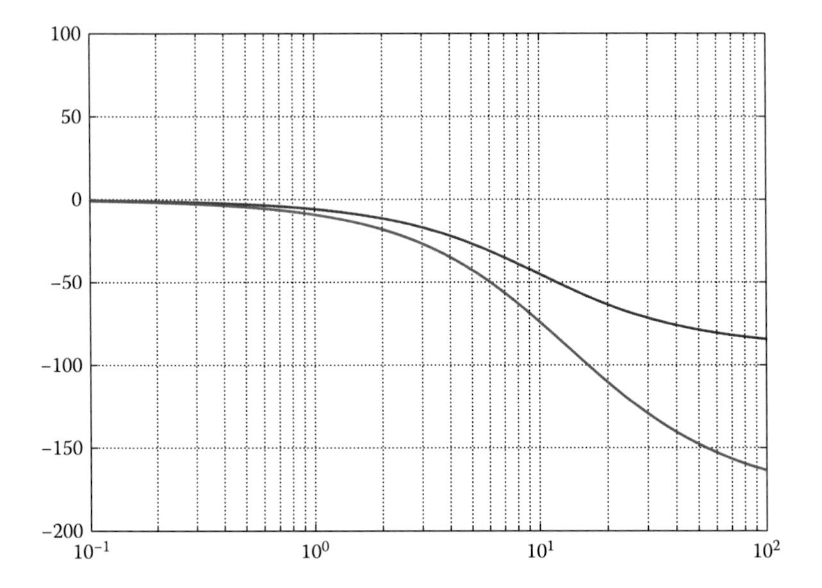

It can be seen that the frequency response of $P(w)$ tightly approaches the corresponding response of $P(s)$ in low frequency regions, while the deviation starts to grow in higher frequency regions.

EXERCISE 4.24

Consider the system of the following scheme with transfer function $G(s) = (3/(s+1)(s+3))$, which is being discretized with ZOH and a digital controller having the transfer function $D(z) = 10/3$, $H(s) = 1$.

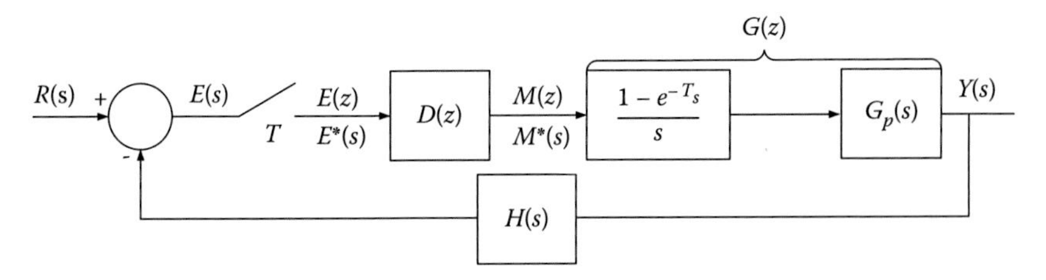

Derive a function that creates and illustrates the closed-loop step response of the above scheme, and utilize it to illustrate the step responses of various sampling periods. Also, compare them with the corresponding ones of the analog system.

Solution

The following function, namely, cl_stepzoh.m, creates and illustrates the closed-loop step response of the given discrete system:

```
function [y,u,t] = cl_stepzoh (cn,cd,pn,pd,T,N)
[A,B,C,D] = tf2ss(pn,pd) ;
[Ad,Bd] = c2d(A,B,T/10) ;
[pnd,pdd] = ss2tf(Ad,Bd,C,D) ;
order_c = max(length(cn),length(cd)) - 1 ;
order_pd = max(length(pnd),length(pdd)) - 1 ;
xc = zeros(1,order_c) ;
xp = zeros(1,order_pd) ;
y = 0 ;
r = 1 ;
u0 = [ ] ;
y0 = [ ] ;
for i=1:N
  if length(xc)==0,
     [ucsim,xc] = filter(cn,cd,r-y) ;
  else
     [ucsim,xc] = filter(cn,cd,r-y,xc) ;
  end
      % ucsim = u(i-i)
      % xc now = xc(i)
  up = ucsim*ones(1,10) ;
  [ypsim,xp] = filter(pnd,pdd,up,xp) ;
  y = xp(1) ;  % assumes that the discretized plant is
               % strictly proper !!!
  u0 = [u0 up] ;
  y0 = [y0 ypsim] ;
end
t0 = [0:10*N-1]*T/10 ;
if nargout == 0
   subplot(211), plot(t0,y0), xlabel('time'), ylabel('output')
   subplot(212), plot(t0,u0), xlabel('time'), ylabel('input')
   subplot(111)
   return
else
   y = y0 ;
   u = u0 ;
```

```
    t = t0 ;
end
```

where u denotes the system input, y is the system step response using ZOH, *pn* and *pd* are the system numerator and denominator, respectively, *cn* and *cd* are the numerator and denominator of digital controller, respectively, T is the sampling period, and N stands for the number of samples for y and u.

```
% Estimate the transfer function for T=0.05
Gnum = [0 0 3] ;Gden = [1 4 3] ;
Dnum = 10/3 ;Dden = 1 ;
[nol,dol] = series(Dnum,Dden,Gnum,Gden);
[ncl,dcl] = cloop(nol,dol,-1);
T = 0.05 ;
% Create the step response using the function cl_stepzoh.m
[y,u,t] = cl_stepzoh (Dnum,Dden,Gnum,Gden,T,ceil(5/T)) ;
% Illustrate of the system step response and its input for
T=0.05 sec.
cl_stepzoh (Dnum,Dden,Gnum,Gden,T,ceil(5/T))
% Create the closed-loop step response of the analog system
tc = t;
yc = step(ncl,dcl,tc);
FontWeight = 'normal';  % or FontWeight = 'bold';
% Illustrate the system step response and input
plot(t,y,'-',t,u,'-','Linewidth',3)
grid
title('Step response: y and u, T = 0.05','FontSize',20,...
    'FontWeight',FontWeight)
xlabel('Time (sec.)','FontSize',20,'FontWeight',FontWeight)
set(gca,'FontWeight','bold','FontSize',15)
```

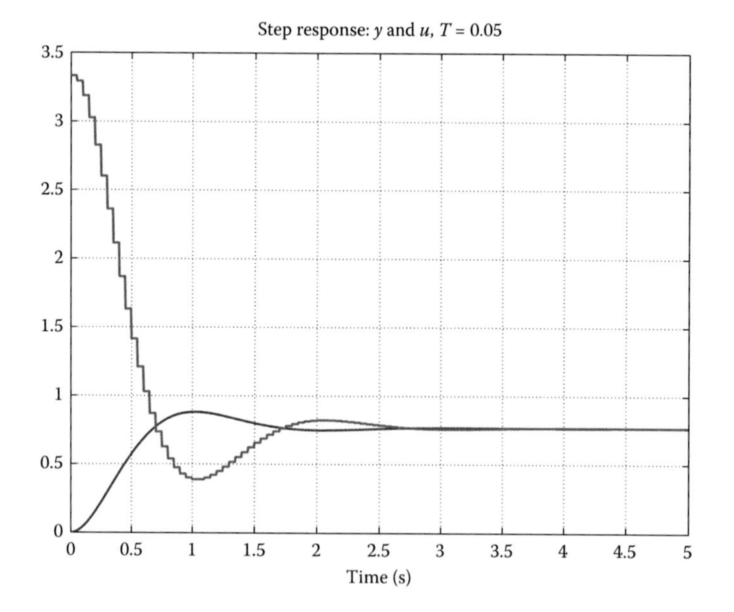

```
% Compare the closed-loop step responses of the analog and
discretized system for T = 0.05
plot(t,y,'-',tc,yc,'--',t,0.9231*ones(1,length(t)),'--',...
    t,0.7692*ones(1,length(t)),'--','Linewidth',3)
axis([0 5 0 1])
grid
title('Stepresponse: T = 0.05','FontSize',20,'FontWeight',
FontWeight)
xlabel('Time (sec.)','FontSize',20,'FontWeight',FontWeight)
set(gca,'FontWeight','bold','FontSize',15)
```

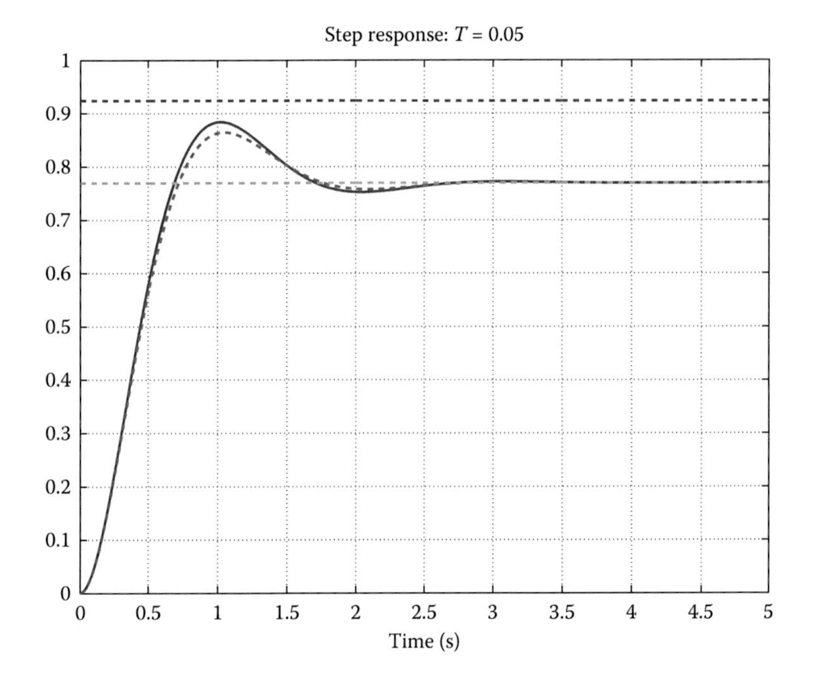


```
% Compare the closed-loop step responses of the analog and
discretized system for T = 0.15
T = 0.15 ;
[y,u,t] = cl_stepzoh (Dnum,Dden,Gnum,Gden,T,ceil(5/T)) ;
plot(t,y,'-',tc,yc,'--',t,0.9231*ones(1,length(t)),'--',...
    t,0.7692*ones(1,length(t)),'--','Linewidth',3)
axis([0 5 0 1])
grid
title('Step response: T = 0.15','FontSize',20,'FontWeight',
FontWeight)
xlabel('Time (sec.)','FontSize',20,'FontWeight',FontWeight)
set(gca,'FontWeight','bold','FontSize',15)
```

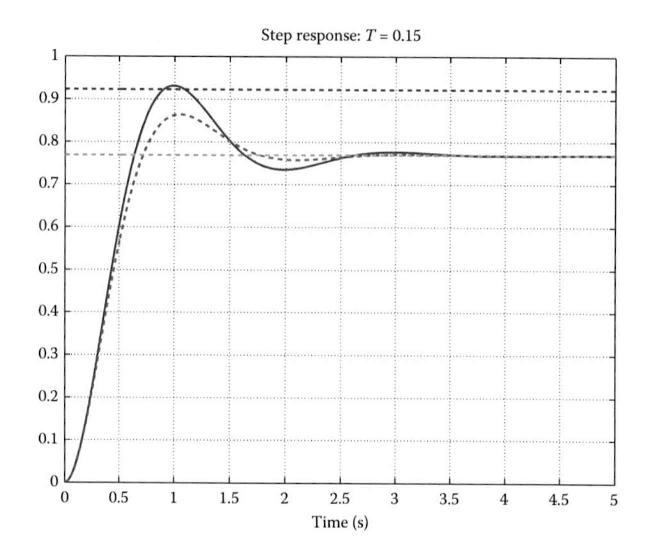


```
% Compare the closed-loop step responses of the analog and
discretized system for T= 0.25
T = 0.25;
[y,u,t] = cl_stepzoh (Dnum,Dden,Gnum,Gden,T,ceil(5/T)) ;
plot(t,y,'-',tc,yc,'--',t,0.9231*ones(1,length(t)),'--',...
    t,0.7692*ones(1,length(t)),'--','Linewidth',3)
axis([0 5 0 1])
grid
title('Step response: T = 0.25','FontSize',20,'FontWeight',
FontWeight)
xlabel('Time (sec.)','FontSize',20,'FontWeight',FontWeight)
set(gca,'FontWeight','bold','FontSize',15)
```

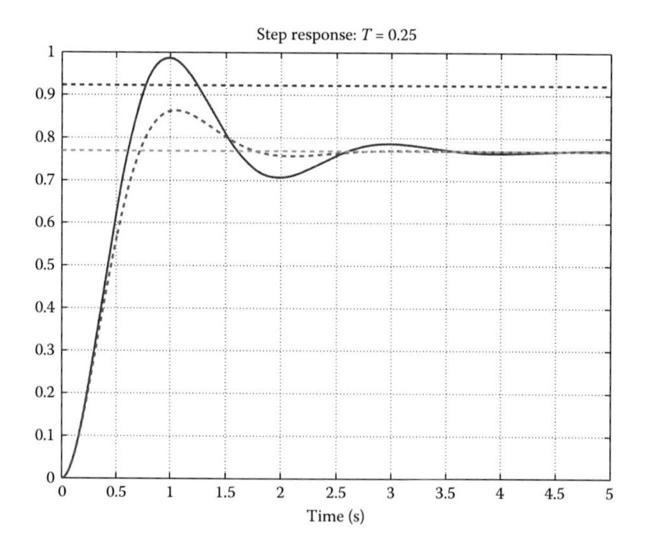

EXERCISE 4.25

Find the transform in z-domain for the system of the following scheme, where the two subsystems are serially connected.

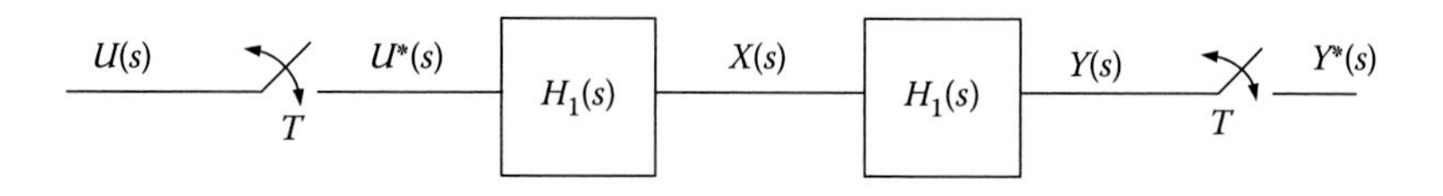

Repeat the above process in the case when the two subsystems are being separated by a sampler.

$$H_1(s)=\frac{1}{s+2} \quad \text{and} \quad H_2(s)=\frac{2}{s+4}, T=0.06s$$

Solution

Apparently, we have two different pulse transfer functions in these two cases. In particular, in the first case, the total transfer function is given in a form of $H(z) = z(H_1(s)\cdot H_2(s))$, while in the second case is in a form of $H(z) = z(H_1(s))\cdot z(H_2(s))$.

The corresponding MATLAB code is

```
% Estimation of the transfer function when the two
subsystems are serially connected
num_s=[0 2];
den_s=conv([1 2],[1 4]);
SYS_s=tf(num_s,den_s)
T=0.06;
num_z1=[exp(-2*T)-exp(-4*T) 0];
den_z1=[conv([1 -exp(-2*T)],[1 -exp(-4*T)])];
SYS_z1=tf(num_z1,den_z1,T)
```

The following transfer function arises

$$H(z)=z(H_1(z)H_2(z))=\frac{0.1003z}{z^2-1.674z+0.6977} \tag{4.25.1}$$

```
% Estimation of the transfer function when the two
subsystems are separated by a sampler
num_z2= [2 0 0];
SYS_z2=tf(num_z2,den_z1,T)
```

The following transfer function arises

$$H(z)=z(H_1(z))\cdot z(H_2(z))=\frac{2z^2}{z^2-1.674z+0.6977} \tag{4.25.2}$$

It is obvious that: $z(H_1(z)H_2(z)) \neq z(H_1(z))\cdot z(H_2(z))$

EXERCISE 4.26

Using LabVIEW, discretize an analog integrator with the aid of the forward, trapezoidal and backward methods, and indicate the impact of each differential integrator in a sine input signal.

Solution

In this simulation example, three algorithms are used to compute the discrete integrator: forward, trapezoidal, and backward. In the following scheme, the block diagram for the simulation is provided. LabVIEW Simulation Module is used.

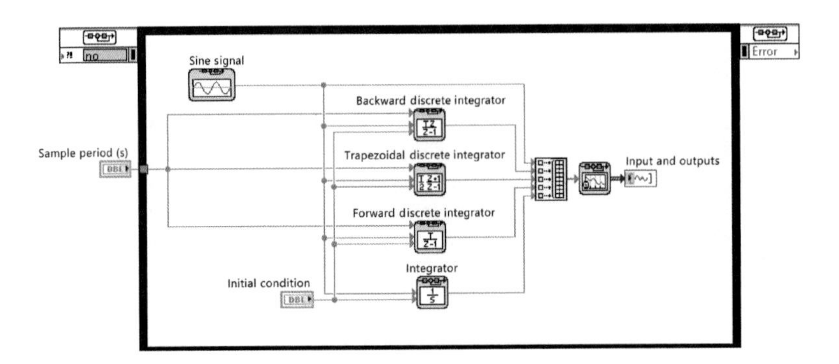

The front panel of the simulation scenario for sampling period $T = 0.6$ s is presented as follows:

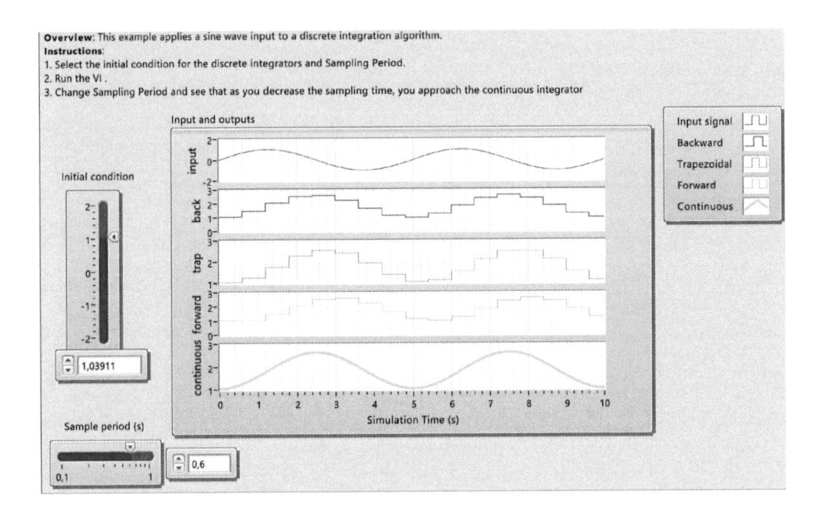

Reducing the sampling period to $T = 0.1$ s, observe the reduction impact (from $T = 0.6$ s to $T = 0.1$ s) at the integrators' performance when the input is a sine signal of frequency 0.2 Hz.

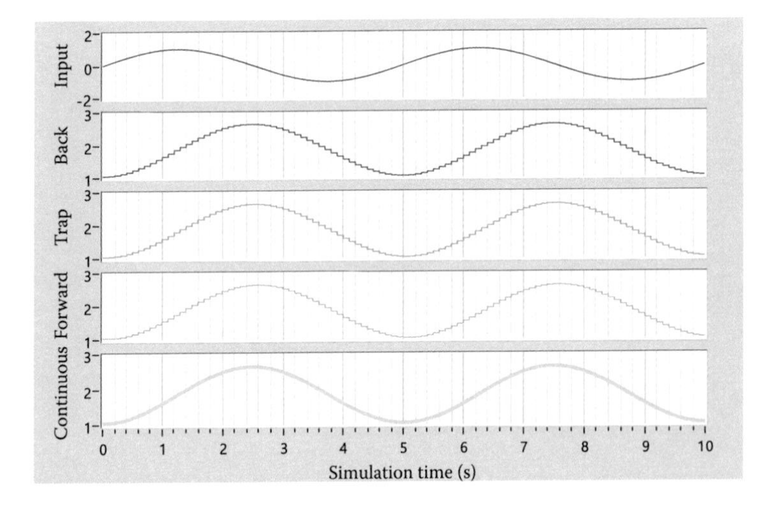

EXERCISE 4.27

Using LabVIEW, provide the responses of FOH and ZOH circuits for a sine signal.

Solution

Using LabVIEW control design and simulation module, we can create and simulate linearly, nonlinearly, and discrete control systems. From the block diagram of a vi, in functions palette, we find the control design and simulation module.

a. The front panel and block diagram of vi, which will be used to show the ZOH operation via the simulation module, are presented as follows: Selecting a sampling period $T = 0.11$ s, observe that the sine signal with frequency 0.5 Hz and the signal after the impact of ZOH closely match. When the sampling period increases, the quantization error also increases.

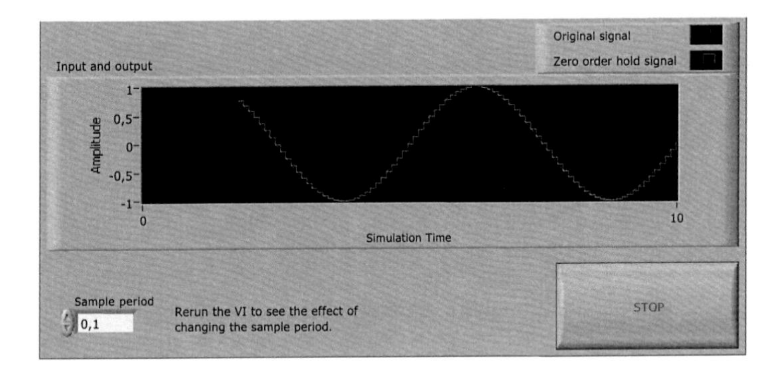

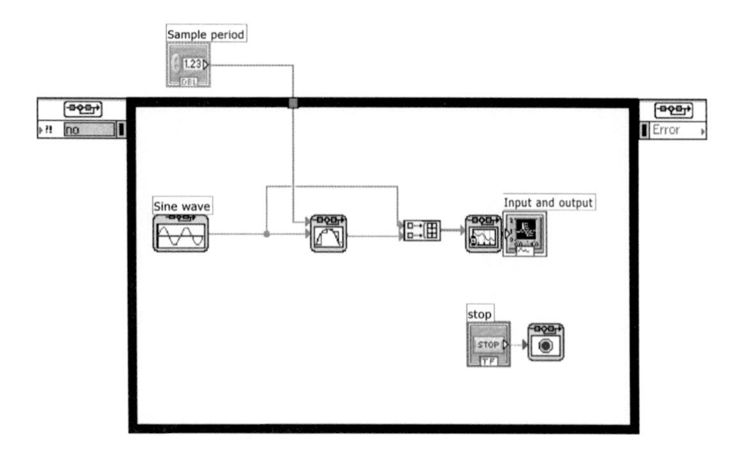

b. The front panel and block diagram of vi, which will be used to show the FOH operation via the simulation module, are presented as follows:

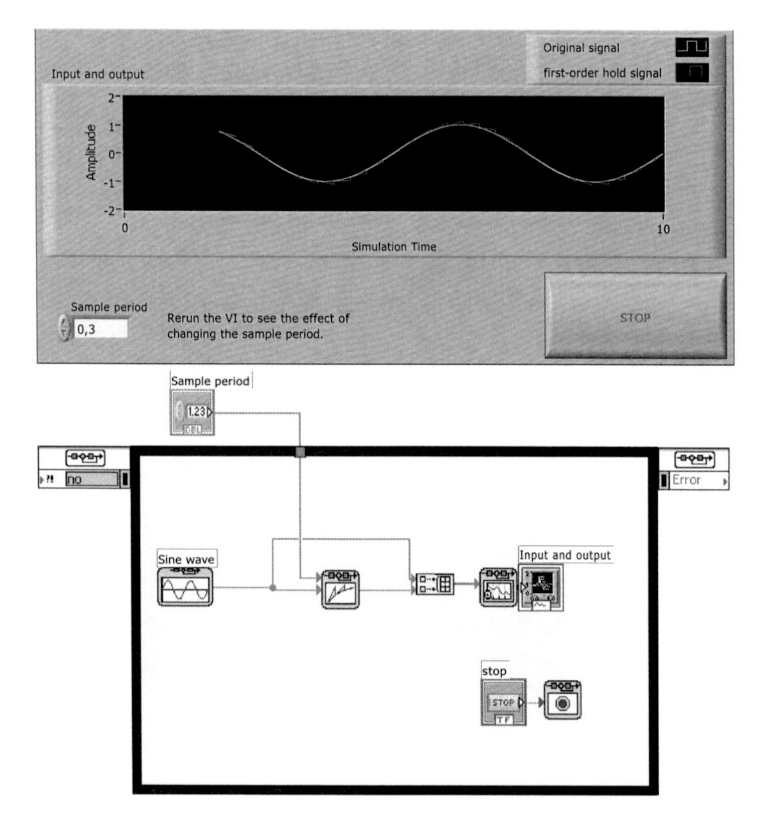

Changing the sampling period to $T = 0.3$ s, observe the negative impact of its increase onto the signal after the FOH operation.

EXERCISE 4.28

Consider the transfer function $G(s)=(4/(s+2))$. Discretize it utilizing all the discretization methods for $T=1$ and 0.1 s, by using the LabVIEW software platform.

Solution

We will use *LabVIEW Control Design and Simulation Module* so as to discretize the transfer function of the analog system.

Using the *Continuous to Discrete* **Conversion.VI,** we can discretize an analog transfer function using various methods (e.g., ZOH—Tustin—Prewarp—Forward—Backward—Z-Transform—FOH—Matched) and illustrate the step response, zero-pole graph, and impulse response of the resultant digital system, by comparing it with the original analog system.

The resultant transfer function of the system, discretized by *ZOH*, is presented as

$$G(z)=\frac{1.72933}{z-0.135335} \tag{4.28.1}$$

In the following, the front panel with the elements of analog system and the step response of analog and discrete system are presented using *ZOH* for $T=1$ s. Relevant deviations can be observed.

If we select $T=0.1$ s, the step response of discretized system will be much closer to the corresponding response of analog system.

By experimenting with all the given discretization methods, the following transfer functions of discrete system arise:

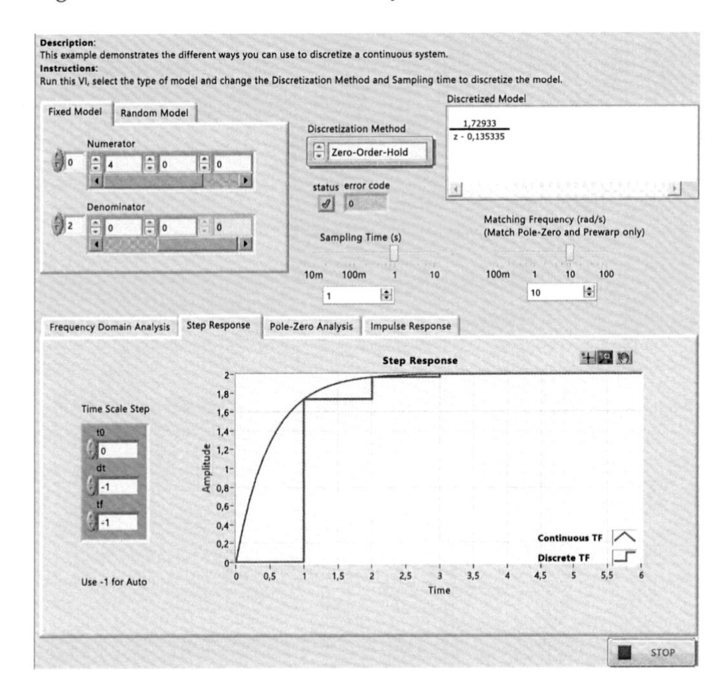

Tustin (Bilinear)

$$G(z) = \frac{1.181818z + 0.181818}{z - 0.818182} \quad (4.28.2)$$

Prewarp

$$G(z) = \frac{1.196997z + 0.196997}{z - 0.803003} \quad (4.28.3)$$

Forward

$$G(z) = \frac{0.4}{z - 0.8} \quad (4.28.4)$$

Backward

$$G(z) = \frac{0.33333z + 0.11022E - 16}{z - 0.833333} \quad (4.28.5)$$

z-Transform

$$G(z) = \frac{4z}{z - 0.818731} \quad (4.28.6)$$

First-Order Hold

$$G(z) = \frac{0.187308z + 0.175231}{z - 0.818731} \quad (4.28.7)$$

Matched Pole–Zero

$$G(z) = \frac{0.347651z}{z - 0.818731} \quad (4.28.8)$$

0	**Zero-Order-Hold (default)**	
1	Tustin (bilinear)	$s \to \frac{2(z-1)}{T(z+1)}$
2	Prewarp	$s \to \frac{(z-1)}{T^{*}(z+1)}$, where $T^{*} = \frac{2\tan(wT/2)}{w}$
3	Forward	$s \to \frac{(z-1)}{T}$
4	Backward	$s \to \frac{(z-1)}{zT}$
5	***z*-Transform – Impulse Invariant Transform**	
6	First-Order-Hold	
7	Matched Pole–Zero	

NOTE: The front panel and block diagram, which are respectively presented below, have been designed to discretize an analog system with ZOH. By appropriately setting $G(s) = (4/(s+2))$, we have a pulse transfer function identical to the expression (4.28.1).

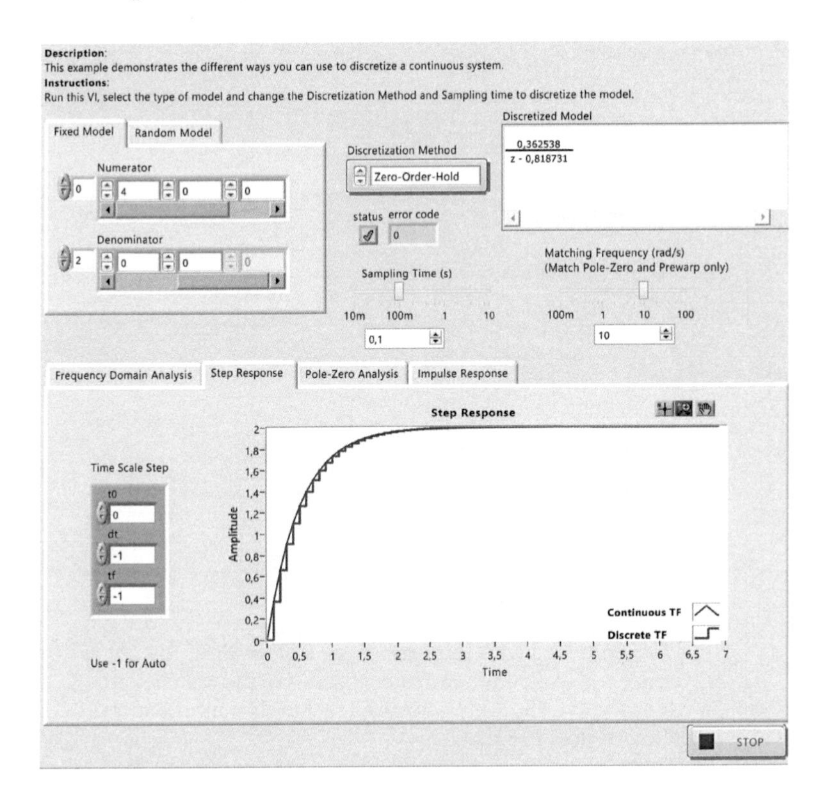

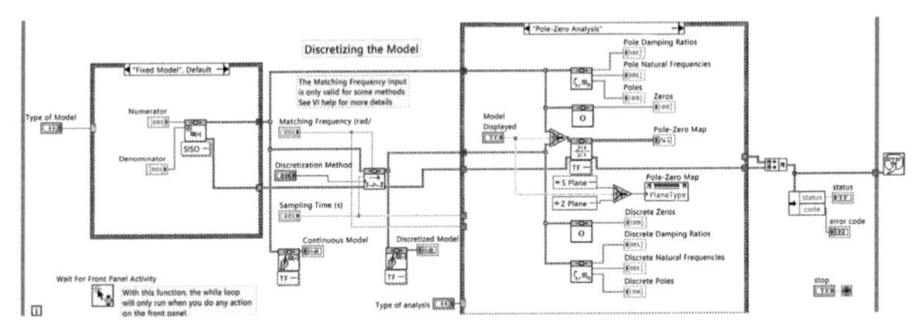

EXERCISE 4.29

Consider the transfer function $G(s) = (2s/(s+1)^2(s+2))$. (a) Discretize it using all the discretization methods given in the comprehensive control (CC) program and theoretically prove the simulation results.

(b) To discretize with the sample equivalence method, design the system open-loop step response and verify the results.

Solution

a. With the Enter command of the program, the transfer function $G(s)$ is imported.

function enter

Apply Ok Cancel Help

Filters:

enter

Enters a tf using polynomial coefficients. The numerator and denominator each have polynomial factors: an*s^n+...+a1*s+a0. The coefficients are entered high to low: an,...,a1,a0.

Output parameters:

Tf: f

Input parameters:

numerator factors = 1

1st factor, order: 1 a1,a0: 2,0

2nd factor, order: an,...,a0

3rd factor, order: an,...,a0

4th factor, order: an,...,a0

denominator factors = 2

1st factor, order: 1 a1,a0: 1,2

2nd factor, order: 2 a2,a1,a0: 1,2,1

3rd factor, order: an,...,a0

4th factor, order: an,...,a0

Alternatively, we can write

```
CC>G=enter(1,1,2,0,2,1,1,2,2,1,2,1)
```

1. Discretize $G(s)$ with the *forward rectangle method*.
   Using the Convert command, discretize the transfer function $G(s)$. We choose the forward rectangle method for sampling period $T = 1$ s.

function convert

Apply Ok Cancel Help

Conversions:

convert

Convert analog system to digital.

Output parameters:

Digital system: gd

Input parameters:

Analog system: g

Convert option:

1: forward rectangle

2: backward rectangle

3: bilinear

4: Tustin prewarped

5: not used

6: pole-zero-map

7: sample equivalence

8: zoh equivalence

9: first-order-hold equivalence

10: slewer-hold equivalence

Sample period: 1

Prewarped freq: 1

Delay: Auto 0

Alternatively, we can write

CC>gd=convert(G,1,1)

The transfer function in z-domain arises as

$$gd(z) = \frac{2(z-1)}{z^2(z+1)} \tag{4.29.1}$$

To prove the expression (4.29.1), it suffices to set $s = ((z-1)/T) = z-1$ in transfer function $G(s)$, hence

$$gd(z) = \frac{2(z-1)}{(z-1+1)^2(z-1+2)} = \frac{2(z-1)}{z^2(z+1)} \tag{4.29.2}$$

2. Discretize $G(s)$ with the *backward rectangle method.*
Using the Convert command, discretize the transfer function $G(s)$. We choose the forward rectangle method for sampling period $T = 1$ s. We write the command

```
CC>gd=convert(G,2,1)
```

The transfer function in z-domain arises as

$$gd(z) = \frac{0.1667z^2(z-1)}{(z-0.3333)(z^2-z+0.25)} \tag{4.29.3}$$

To prove the expression (4.29.3), it suffices to set $s = (z-1/zT) = 1-z^{-1}, T = 1$ in $G(s)$ where $s = ((z-1)/zT) = 1-z^{-1}, T = 1$ hence

$$gd(z) = \frac{2(1-z^{-1})}{(1-z^{-1}+1)^2(1-z^{-1}+2)} \Rightarrow gd(z) = \frac{0.166z^2(z-1)}{(z-0.5)^2(z-0.33)} \tag{4.29.4}$$

3. Discretize $G(s)$ with the *bilinear method.*
Using the Convert command, discretize the transfer function $G(s)$. We choose the Bilinear method for sampling period $T = 1$ s. We write the command

```
CC>gd=convert(G,3,1)
```

The transfer function in z-domain arises as

$$gd(z) = \frac{0.111(z-1)(z+1)^2}{z(z-0.333)^2} \tag{4.29.5}$$

To prove the expression (4.29.5), it suffices to set $s=(2(1-z^{-1})/T(1+z^{-1}))=(2(z-1)/T(z+1)), T=1$ in $G(s)$ where $s=(2(1-z^{-1})/T(1+z^{-1}))=(2(z-1)/T(z+1)), T=1$ hence

$$gd(z)=\frac{4\frac{z-1}{z+1}}{[2(z-1/z+1)+1]^2[2(z-1/z+1)+2]}$$

$$\Rightarrow gd(z)=\frac{0.111(z-1)(z+1)^2}{z(z-0.333)^2} \quad (4.29.6)$$

4. Discretize $G(s)$ with the *pole–zero map method.*
   Using the Convert command, discretize the transfer function $G(s)$. We choose the pole–zero map method for sampling period $T = 1$ s. We write the command

```
CC>gd=convert(G,6,1)
```

The transfer function in z-domain arises as

$$gd(z)=\frac{0.086(z-1)(z+1)^2}{(z-0.368)^2(z-0.135)} \quad (4.29.7)$$

To prove the expression (4.29.7), it suffices to match the poles and zeros of pulse transfer function with the corresponding ones of $G(s)$, according to the expressions $z_i = e^{-\mu_i T} = e^{-\mu_i}$ and $p_i = e^{-\pi_i T} = e^{-\pi_i}$, and to compute the dc gain factor so as $G(s)$ and $gd(z)$ to be equal for $s = 0$ and $z = 1$.

$$gd(z)=\frac{K_{dc}(1+z)^2(z-1)}{(z-e^{-T})(z-e^{-T})(z-e^{-2T})} \quad (4.29.8)$$

We set $T = 1$ and compute K_{dc} according to $gd(z)|_{z=1} = G(s)|_{s=0}$. It yields that $K_{dc} = 0.086$. The final form of the function is

$$gd(z)=\frac{0.086(z-1)(z+1)^2}{(z-0.368)^2(z-0.135)} \quad (4.29.9)$$

5. Discretize $G(s)$ with the *sample equivalence* (or *sampled inverse Laplace transform*) *method.*
   Using the Convert command, discretize the transfer function $G(s)$. We choose the sample equivalence method for sampling period $T = 1$ s. We write the command

```
CC>gd=convert(G,7,1)
```

The transfer function in z-domain arises as

$$gd(z) = \frac{0.1944z(z-1.248)}{(z-0.3679)^2(z-0.1353)} \tag{4.29.10}$$

To prove the expression (4.29.10), it suffices to compute the inverse Laplace transform of $G(s)$, set $t = kT$ and then calculate the discretized transfer function, using tabulated values (from the given tables).

$$L^{-1}[G(s)] = L^{-1}\left[\frac{2s}{(s+1)^2(s+2)}\right] = e^{-t}(4-2t) - 4e^{-2t} \tag{4.29.11}$$

$$\begin{aligned} g(kT) &= e^{-kT}(4-2kT) - 4e^{-2kT} \\ &\Rightarrow g(kT) = 4e^{-kT} - 2kTe^{-kT} - 4e^{-2kT} \end{aligned} \tag{4.29.12}$$

$$\begin{aligned} gd(z) &= 4\frac{z}{(z-e^{-T})} - 2\frac{Tze^{-T}}{(z-e^{-T})^2} - 4\frac{z}{(z-e^{-2T})} \\ &\xrightarrow{T=1} gd(z) = \frac{0.194z(z-1.24)}{(z-0.368)^2(z-0.135)} \end{aligned} \tag{4.29.13}$$

6. Discretize $G(s)$ with the *ZOH equivalence method.*
   Using the Convert command, discretize the transfer function $G(s)$. We choose the ZOH equivalence method for sampling period $T = 1$ s. We write the command

```
CC>gd=convert(G,8,1)
```

The transfer function in z-domain arises as

$$gd(z) = \frac{0.2707(z+0.2642)(z-1)}{(z-0.3679)^2(z-0.1353)} \tag{4.29.14}$$

The proof of the expression (4.29.14) is given as follows

$$\begin{aligned} gd(z) &= (1-z^{-1})Z\left[\frac{G(s)}{s}\right] \\ &= (1-z^{-1})Z\left[\frac{2}{(s+1)^2} - \frac{2}{(s+1)} + \frac{2}{(s+2)}\right] \\ &\xrightarrow{T=1} gd(z) = \frac{z-1}{z}\left(\frac{2ze^{-1}}{(z-e^{-1})^2} - \frac{2z}{z-e^{-1}} + \frac{2z}{z-e^{-2}}\right) \\ &\Rightarrow gd(z) = \frac{0.270(z-1)(0.270z+0.071)}{(z-0.368)^2(z-0.135)} \end{aligned} \tag{4.29.15}$$

7. Discretize $G(s)$ with the *first-order hold equivalence method.*
   Using the Convert command, discretize the transfer function $G(s)$. We choose the FOH equivalence method for sampling period $T = 1$ s. We write the command

```
CC>gd=convert(G,9,1)
```

The transfer function in z-domain arises as

$$gd(z) = \frac{0.3996(z+0.3679)(z-1)}{(z-0.3679)(z-0.1353)} \tag{4.29.16}$$

8. Discretize $G(s)$ with the *Tustin prewarped method.*
   Using the Convert command, discretize the transfer function $G(s)$. We choose the Tustin prewarped method for sampling period $T = 1$ s. We write the command

```
CC>gd=convert(G,4,1)
```

The transfer function in z-domain arises as

$$gd(z) = \frac{0.1193(z-1)(z+1)^2}{(z-0.2934)^2(z+0.4425)} \tag{4.29.17}$$

b. To design the system open-loop step response with transfer function $gd(z) = (0.1944z(z-1.248)/((z-0.3679)^2(z-0.1353)))$, we type the command

```
CC>time
```

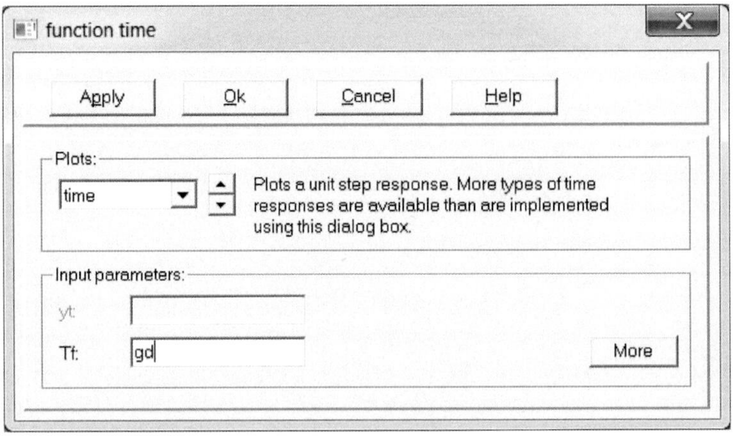

function time

Apply Ok Cancel Help

Plots:

time

Plots a unit step response. More types of time responses are available than are implemented using this dialog box.

Input parameters:

yt:

Tf: gd

More

The following graph represents the resultant step response

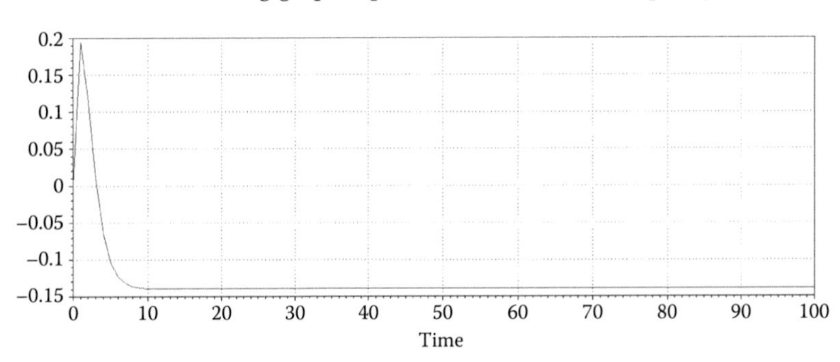

The theoretical proof follows:

$$Y(z) = gd(z) * U(z) = \frac{0.194z(z-1.24)}{(z-0.368)^2(z-0.135)} \cdot \frac{z}{z-1}$$

$$\Rightarrow Y(z) = \frac{1.159z^2}{(z-0.368)^2} - \frac{1.64z}{(z-0.368)} + \frac{0.62z}{(z-0.135)} - \frac{0.14z}{(z-1)}$$

$$\xrightarrow{Z^{-1}[\,],\,T=1} y(n) = e^{-n}(1.15n - 0.49) + 0.62e^{-2n} - 0.14 \quad (4.29.18)$$

function izt

Apply | Ok | Cancel | Help

Displays:
izt — Display the Inverse z-Transform.

Output parameters:
None

Input parameters:
Input tf: y

Region of convergence:
- 'causal' (default)
- 'anti-causal'
- 'stable'
- 'roc' (as entered below:)

roc:

From $y(n)$, we provide a table of values, which are identical to the values of the above graph.

n	$Y(n)$
0	−0.010
1	0.180
2	0.110
3	0.009
4	−0.060
5	−0.100
6	−0.120
7	−0.133
8	−0.137
9	−0.138
10	−0.139

The IZT command for $Y(z)$ would provide the same result.

$$CC > y = gd * z/(z-1)y(n) = -0.1395 + (1.164n + 0.6774)(0.3679)^n + 0,6261(0.1353)^n \tag{4.29.19}$$

In the same graph, we can add the step response of the closed-loop step response with transfer function $G_{cl}(z) = (gd(z)/(1+gd(z)))$ and derive some useful results. From the menu, plot options, select Add new line and add the closed-loop transfer function. The following graph arises, which jointly shows the two step responses.

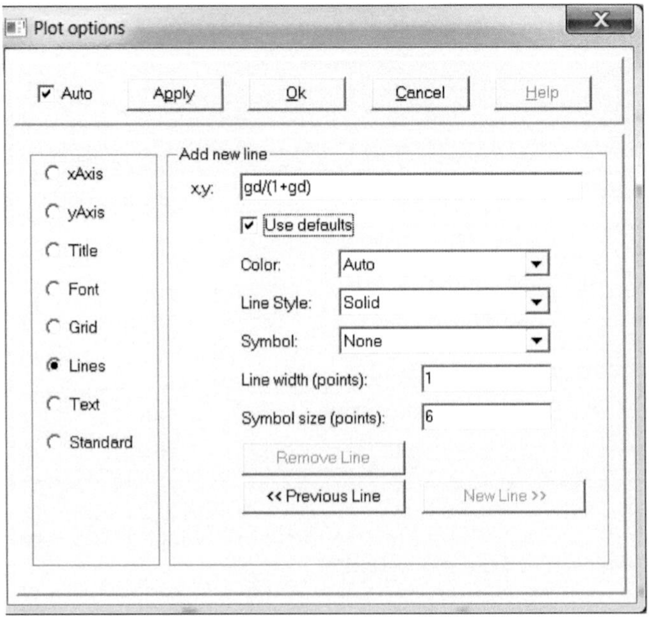

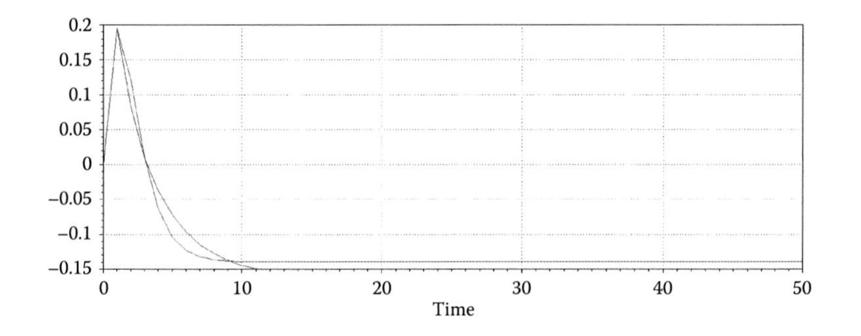
0.2
0.15
0.1
0.05
0
−0.05
−0.1
−0.15
0
10
20
30
40
50
Time

5

State-Space Representation

5.1 Introduction

A large class of systems (linear, nonlinear, time-varying, nontime-varying, etc.) is analyzed using the methodology of the *state space,* where the system is described by a set of *first-order difference equations,* describing the state variables.

The *state variables* are the smallest number of variables describing the future response of a system when the current state of the system, the inputs, and the equations that describe its function are known. The state variables may not always be observed or measured, however, they affect the behavior of the system. They determine how the system evolves and somehow "save" its previous behavior.

By *state* of a system we refer to past, present, and future of the system. From a mathematical viewpoint, the state of the system is expressed by the state variables which represent a new "dimension" in the study of systems. The state variables are quantities related to the internal structure of the system and provide important information about it; which other traditional methods fail to give us. Thus, this method of analysis provides additional information about the system, not limited to the study of the transfer function, or system response time.

Definition 5.1

The state variables $x_1(n), x_2(n), \ldots x_k(n)$ are defined as a (minimum) number of variables, such that we know

1. Their values at time instance n_0
2. The system input for $n \geq n_0$
3. The mathematical model

to determine the system status at any time $n \geq n_0$.

The *state differential equation* gives the relationship between the system inputs, system state, and rate of change.

The *advantages* of the system description using the state equations method in comparison to the conventional methods are

- The simulation and scheduling in computer systems is quite easy since they represent a linear difference equations system.
- They facilitate the solution of control problems, such as stability and optimized control.
- Besides the linear systems, they are also able to describe nonlinear systems, which cannot be performed using the transfer function.
- There is the possibility of describing the state of the entire system each time; unlike the transfer function, which connects the input with the output.

5.2 Discrete-Time State-Space Equations

The state equations of a discrete-time system is a first-order difference equation system of the form

$$x(k+1) = Ax(k) + Bu(k) \tag{5.1}$$

The system output vector $y(k)$ is

$$y(k) = Cx(k) + Du(k) \tag{5.2}$$

The matrices A, B, C, D are called *state-space matrices*. Specifically, the matrix A is a square matrix of dimension nxn, it is called the *state matrix* and represents the physical (actual) system; the matrix B of dimension nxr is called the *input matrix*; the matrix C of dimension mxn is called the *output matrix*; and the matrix D of dimension mxr is called the *feedforward matrix*.

In the case of a single input–single output (SISO) system, where $r = m = 1$, the system is described by the difference equations

$$\left.\begin{aligned} &x(k+1) = Ax(k) + bu(k) \\ &y(k) = c^T x(k) + du(k) \\ &x(0) = x_o \end{aligned}\right\} \tag{5.3}$$

where

c is a column vector with n elements
b is a column vector with n elements

d is a scalar and
$x(0) = x_o$ is a column vector of the initial conditions of state variables.

Is it possible to learn everything about the dynamic behavior of a system, described by state equations, using only measurements of the input and output? In other words, when and how can we estimate the system state vector?

In practice, it is impossible to measure all the system states. But if the system mathematical model is available, then we can calculate (estimate) the state vector, by using the already measured inputs and outputs.

5.2.1 Eigenvalues and Eigenvectors

The elements of the system matrix depend on the components comprising the system.

Consider an n order system with column vectors $x = X_i$ $(i = 1,2,\ldots,n)$ and real or complex values for the parameter λ, which satisfy the equation

$$Ax = \lambda x \Rightarrow (\lambda I - A)x = 0 \tag{5.4}$$

The matrix $(\lambda I_n - A)$ is called *characteristic matrix* of the system. The values of parameter λ satisfying $(\lambda I_n - A)x = 0$ represent a column vector and are called *eigenvalues* or *characteristic values* of the system, which arise from the solution of the linear system.

Eliminating the determinant of the characteristic matrix, the *characteristic polynomial* of the system is revealed, such as

$$P(\lambda) = \det(\lambda I - A) = \lambda^n + a_{n-1}\lambda^{n-1} + \cdots + a_1\lambda + a_0 = 0 \tag{5.5}$$

The roots of $P(\lambda)$, namely, its eigenvalues, denote the *poles* of the closed system. The characteristic equation of the system is given by

$$[zI - A] = 0 \tag{5.6}$$

5.3 Solution of State Equations

1. Solution in time domain

   The general solution of state vector is

$$x(k) = A^k x(0) + \sum_{j=0}^{k-1} A^{k-j-1}Bu(j), \quad k \geq 1 \tag{5.7}$$

   where $A^k = \Phi(k)$ = the transition matrix.

The exponential matrix A^k is denoted as $\Phi(k)$ and called *state transition matrix,* which represents the response of the system only by the influence of the initial conditions (i.e., the free response of the system).

The *output vector* is defined as

$$y(k)=C\Phi(k)x(0)+\sum_{j=0}^{k-1}C\Phi(k-j-1)Bu(j)+Du(k), \quad k\geq 1 \tag{5.8}$$

2. Solution in z-domain

The *general solution of state vector* is

$$x(k)=IZT(X(z))=IZT[z(zI-A)^{-1}x(0)+(zI-A)^{-1}BU(z)] \tag{5.9}$$

The *output vector* is defined as

$$y(k)=IZT(Y(z))=IZT[(C(zI-A)^{-1}B+D)U(z)] \tag{5.10}$$

The *state transition matrix* is

$$\Phi(k)=A^k=IZT[z(zI-A)^{-1}], \quad k\geq 1 \tag{5.11}$$

The pulse transfer function is presented as

$$H(z)=\frac{Y(z)}{U(z)}=C(zI-A)^{-1}B+D \tag{5.12}$$

In Figure 5.1, the transforms between the model at state space and the discrete transfer function are presented.

5.4 State-Space Representation

5.4.1 Direct Form

Consider a system described by the transfer function

$$H(z)=\frac{Y(z)}{U(z)}=\frac{b_1z^{-1}+b_2z^{-2}+\cdots+b_mz^{-m}}{1+\alpha_1z^{-1}+\alpha_2z^{-2}+\cdots+\alpha_nz^{-n}} \tag{5.13}$$

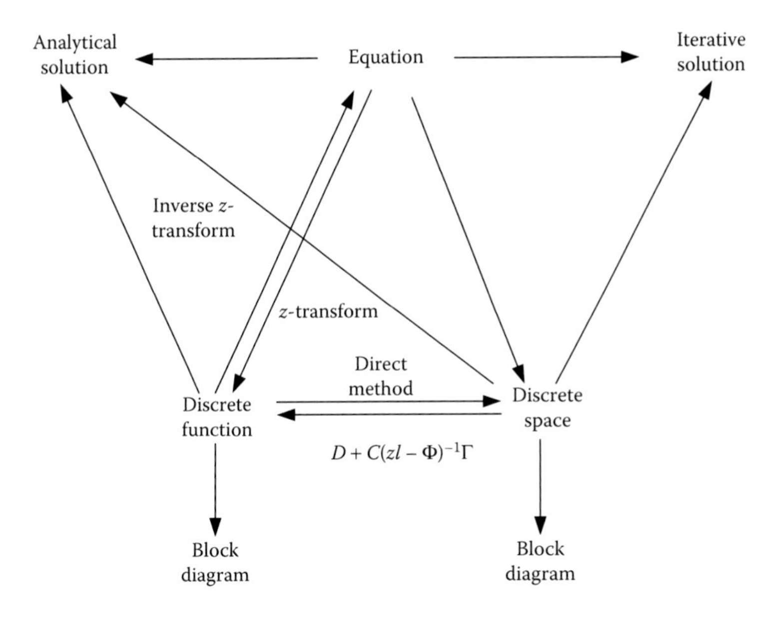

FIGURE 5.1
Transforms between the model at state space and the discrete transfer function.

Solving with respect to $Y(z)$, we have

$$Y(z) = -\left[a_1 z^{-1} + \cdots + a_n z^{-n}\right]Y(z) + \left[b_1 z^{-1} + \cdots + b_n z^{-n}\right]U(z)$$

or

$$Y(z) = (b_1 U(z) - a_1 Y(z))z^{-1} + \left(b_2 U(z) - a_2 Y(z)\right)z^{-2} + \cdots \tag{5.14}$$

The block diagram of the discrete system in *direct form* is given in Figure 5.2.

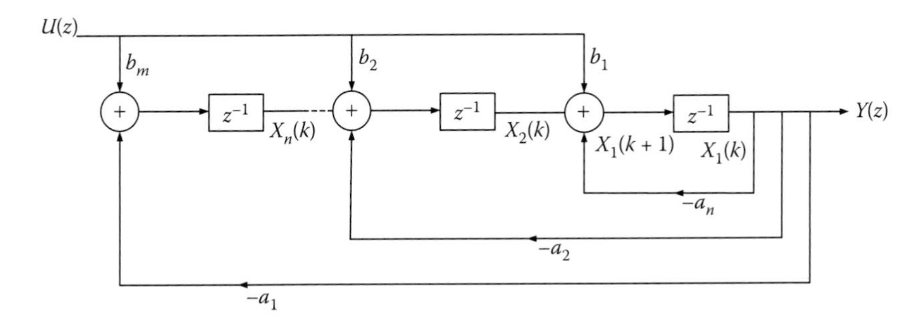

FIGURE 5.2
Block diagram of the discrete system in direct form.

The state equations are

$$\left.\begin{aligned} x_1(k+1) &= -a_1x_1(k)+x_2(k)+b_1U(k) \\ x_2(k+1) &= -a_2x_1(k)+x_3(k)+b_2U(k) \\ &\cdots \qquad\qquad \cdots \\ x_n(k+1) &= -a_nx_1(k)+b_nU(k) \end{aligned}\right\} \tag{5.15}$$

The state equations and the system output are given in *direct form* as

$$\left.\begin{bmatrix} x_1(k+1) \\ x_2(k+1) \\ \vdots \\ x_n(k+1) \end{bmatrix} = \begin{bmatrix} -a_1\ 1\ldots 0 \\ -a_2\ 0\ 1\ldots 0 \\ \\ -a_n\ 0\ldots 1 \end{bmatrix} = \begin{bmatrix} x_1(k) \\ x_2(k) \\ \vdots \\ x_n(k) \end{bmatrix} + \begin{bmatrix} b_1 \\ b_2 \\ \\ b_n \end{bmatrix} u(k)\right\} \tag{5.16}$$

$$y(n) = [1 \quad 0\cdots 0]x(k)$$

5.4.2 Canonical Form

Consider a system described by the transfer function of the expression (5.13). The digital diagram is generated after a proper disintegration of the quotient of the transfer function as shown below. The technique is the separation of the diagram construction process in two steps which are connected through a new variable $W(z)$.

$$\frac{Y(z)}{U(z)} = \frac{\text{numerator}}{\text{denominator}} \Rightarrow \frac{Y(z)}{\text{numerator}} = \frac{U(z)}{\text{denominator}} = W(z) \tag{5.17}$$

Step 1: Structure of $U(z)/\text{denominator} = W(z)$

$$\frac{U(z)}{\text{denominator}} = W(z) = \frac{U(z)}{1+a_1z^{-1}+\cdots+a_nz^{-n}} \tag{5.18}$$

The solution with respect to $W(z)$ is

$$W(z) = -\alpha_1z^{-1}W(z) - a_2z^{-2}W(z)+\cdots+U(z) \tag{5.19}$$

Step 2: Structure of $Y(z)/\text{numerator} = W(z)$

$$Y(z) = b_1z^{-1}W(z)+b_2z^{-2}W(z)+\cdots+b_mz^{-m}W(z) \tag{5.20}$$

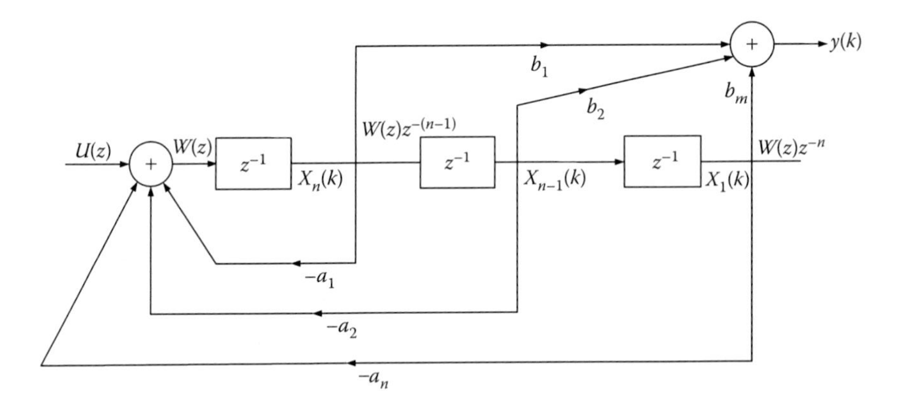

FIGURE 5.3
Block diagram in canonical form.

The block diagram that reflects the discrete system in canonical form, which corresponds to the expressions (5.19) and (5.20), is given in Figure 5.3.

The state equations are given by

$$\left.\begin{aligned} &x_1(k+1) = x_2(k) \\ &x_2(k+1) = x_3(k) \\ &\vdots \quad\quad \vdots \quad\quad\quad\quad \vdots \quad\quad\quad\quad \vdots \\ &x_n(k+1) = u(k) - a_1x_n(k) - a_2x_{n-1}(k) - \cdots - a_nx_1(k) \\ &\text{and} \\ &y(k) = b_mx_1(k) + b_{m-1}x_2(k) + \cdots + b_1x_n(k) \end{aligned}\right\} \tag{5.21}$$

The state equations and the system output in *canonical form* can be expressed in vectors/matrices as

$$\left.\begin{aligned} &\begin{bmatrix} x_1(k+1) \\ x_2(k+1) \\ \vdots \\ x_n(k+1) \end{bmatrix} = \begin{bmatrix} 0 & 1 & 0\ldots0 \\ 0 & 0 & 1\ldots0 \\ & \vdots & \\ -a_n & \ldots - a_1 & \end{bmatrix} \begin{bmatrix} x_1(k) \\ x_2(k) \\ \vdots \\ x_n(k) \end{bmatrix} + \begin{bmatrix} 0 \\ 0 \\ \vdots \\ 1 \end{bmatrix} u(k) \\ &y(k) = [b_mb_{m-1}\ldots b_1] \;\; x(k) \end{aligned}\right\} \tag{5.22}$$

5.4.3 Controllable Canonical Form

Consider a system described by the transfer function of the expression

$$H(z) = \frac{Y(z)}{U(z)} = \frac{b_0z^m + b_1z^{m-1} + \cdots + b_m}{z^n + \alpha_1z^{n-1} + \alpha_2z^{n-2} + \cdots + \alpha_n} \tag{5.23}$$

We study the case when $m = n$.

$$\text{Let} \quad \frac{\overline{X(z)}}{U(z)} = \frac{1}{z^n + \alpha_1 z^{n-1} + \alpha_2 z^{n-2} + \cdots + \alpha_n} \tag{5.24}$$

The output $Y(z)$ can be written in terms of $X(z)$

$$Y(z) = (b_0 z^m + b_1 z^{m-1} + \cdots + b_n)\overline{X(z)} \tag{5.25}$$

or in the time-domain, such as

$$y(k) = b_0\,\overline{x}(k+n) + b_1\,\overline{x}(k+n-1) + \cdots + b_n\,\overline{x}(k) \tag{5.26}$$

The block diagram that reflects the discrete system in controllable canonical form is presented in Figure 5.4.

The state equations are given by

$$\begin{aligned} x_1(k+1) &= x_2(k) \\ x_2(k+1) &= x_3(k) \\ \vdots \qquad & \vdots \qquad\qquad \vdots \qquad\qquad \vdots \\ x_n(k+1) &= u(k) - a_1 x_n(k) - a_2 x_{n-1}(k) - \cdots - a_n x_1(k) \end{aligned} \tag{5.27}$$

$$\begin{aligned} y(k) = {} & (b_n - a_n b_0) x_1(k) + (b_{n-1} - a_{n-1} b_0) x_2(k) \\ & + \cdots + (b_1 - a_1 b_0) x_n(k) + b_0 u(k) \end{aligned} \tag{5.28}$$

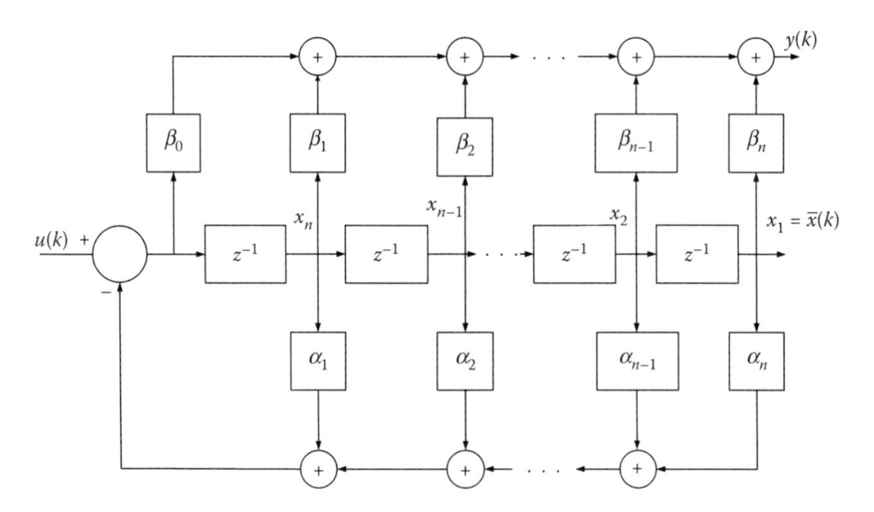

FIGURE 5.4
Block diagram in controllable canonical form.

The matrices at state space are provided by

$$\left.\begin{aligned} A &= \begin{bmatrix} 0 & 1 & 0\ldots0 \\ 0 & 0 & 1\ldots0 \\ & \vdots & \\ -a_n\ldots & -a_1 & \end{bmatrix} \quad B = \begin{bmatrix} 0 \\ 0 \\ \vdots \\ 1 \end{bmatrix} \\ C &= [b_n - a_n b_0 \quad b_{n-1} - a_{n-1} b_0 \quad b_1 - a_1 b_0] \quad D = b_0 \end{aligned}\right\} \tag{5.29}$$

5.4.4 Observable Canonical Form

The expression (5.23) can be written as

$$(z^n + \alpha_1 z^{n-1} + \alpha_2 z^{n-2} + \cdots + \alpha_n)Y(z) = (b_0 z^m + b_1 z^{m-1} + \cdots + b_m)U(z)$$

or

$$Y(z) = b_0 U(z) - z^{-1}(a_1 Y(z) - b_1 U(z)) - \cdots z^{-n}(a_n Y(z) - b_n U(z)) \tag{5.30}$$

The block diagram that reflects the discrete system in observable canonical form is presented in Figure 5.5.

The state equations are

$$\left.\begin{aligned} x_n(k+1) &= x_{n-1}(k) - a_1(x_n(k) + b_0 u(k)) + b_1 u(k) \\ x_{n-1}(k+1) &= x_{n-2}(k) - a_2(x_n(k) + b_0 u(k)) + b_2 u(k) \\ \vdots \quad & \quad \vdots \qquad\qquad \vdots \qquad\qquad \vdots \\ x_1(k+1) &= -a_n(x_n(k) + b_0 u(k)) + b_n u(k) \\ y(k) &= x_n(k) + b_0 u(k) \end{aligned}\right\} \tag{5.31}$$

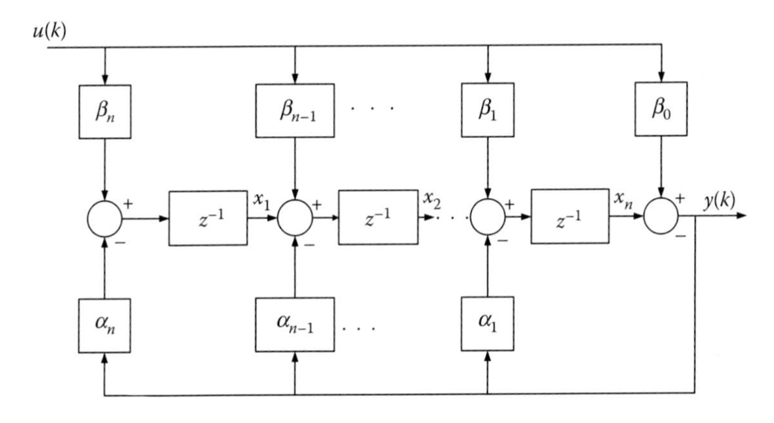

FIGURE 5.5
Block diagram in observable canonical form.

The matrices at state space are provided by

$$\left.\begin{aligned} A &= \begin{bmatrix} 0 & 0 & 0\ldots & 0 & -a_n \\ 1 & 0 & 0\ldots & 0 & -a_{n-1} \\ 0 & 1 & 0\ldots & 0 & -a_{n-2} \\ \ldots & & & & \\ 0 & 0 & 0\ldots & 1 & -a_1 \end{bmatrix} \quad B = \begin{bmatrix} b_n - a_n b_0 \\ b_{n-1} - a_{n-1} b_0 \\ \vdots \\ b_1 - a_1 b_0 \end{bmatrix} \\ C &= \begin{bmatrix} 0 & 0 & 0\ldots 1 \end{bmatrix} \quad D = b_0 \end{aligned}\right\} \tag{5.32}$$

5.4.5 Jordan Canonical Form

The transfer function of the expression (5.23) can be rewritten in a partial fraction expansion as

$$\frac{Y(z)}{U(z)} = b_0 + \frac{r_1}{z-\lambda_1} + \frac{r_2}{z-\lambda_2} + \cdots + \frac{r_n}{z-\lambda_n} \tag{5.33}$$

The parallel implementation of the transfer function of the latter expression is presented in Figure 5.6.

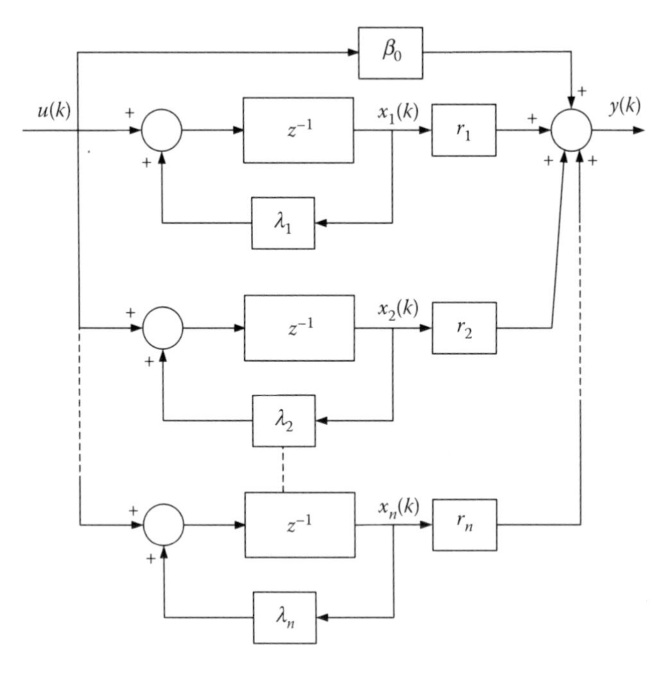

FIGURE 5.6
Block diagram in Jordan canonical form.

Assuming the outputs of the delay elements as state variables, it can easily be shown that the model at state space is represented by the following matrices:

$$\left.\begin{aligned} A &= \begin{bmatrix} \lambda_1 & 0 & 0\ldots & 0 & 0 \\ 0 & \lambda_2 & 0\ldots & 0 & 0 \\ 0 & 0 & \lambda_3 & 0 & 0 \\ \cdots & & & & \\ 0 & 0 & \ldots & 0 & \lambda_n \end{bmatrix} \quad B = \begin{bmatrix} 1 \\ 1 \\ \vdots \\ 1 \end{bmatrix} \\ C &= [r_1 \quad r_2 \quad r_3 \cdots r_n] \quad D = b_0 \end{aligned}\right\} \tag{5.34}$$

5.5 Controllability and Observability

The *controllability* of a system refers to whether it is possible to move a system from a given initial state to any final state in finite time (Kalman, 1960).

Consider the system described in the state space by Equations 5.1 and 5.2. This system will be controllable when $S = n$. Matrix S is called *controllability matrix* and is derived by

$$S = [B \vdots AB \vdots A^2B \vdots \ldots A^{n-1}B] \tag{5.35}$$

An example of a noncontrollable system is presented in Figure 5.7, where we can see that regardless of the level of influence of the input u_1, the variable x_2 remains unaffected.

The *observability* of a system refers to whether each position $x(k)$ can be determined by observing the output $y(k)$ at a finite time.

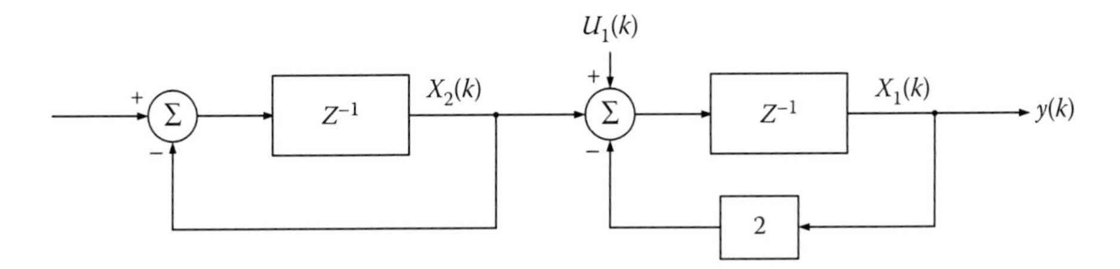

FIGURE 5.7
Noncontrollable system.

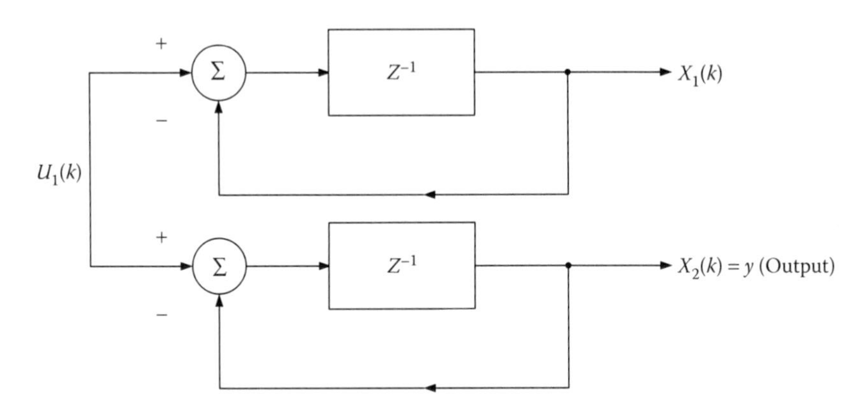

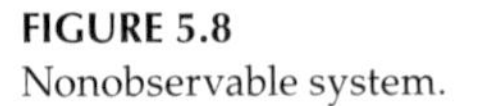
FIGURE 5.8
Nonobservable system.

Such a system is observable when $R = n$. Matrix R is called *observability matrix* and is given by

$$R = \begin{pmatrix} C^T \\ C^T A \\ \vdots \\ C^T A^{n-1} \end{pmatrix} \tag{5.36}$$

A typical example of a nonobservable system is presented in Figure 5.8, where we cannot acquire information of the variable $x_1(k)$, regardless of how much time we observe the output $y(k)$.

Overall, a system is controllable when we can control the system operation process, given an initial state, while a system is observable when all the provided information about the system state must be recovered from knowledge of the obtained measurements.

In a sense, the observability is a tradeoff to the concept of controllability, which is related to the acquisition of certain controlling tools that allow us to achieve a desired state.

How the properties of controllability and observability are involved in the design and analysis of digital control systems?

In the automatic control theory, concepts of controllability and observability are related to the design of the controller, which solves the problem of regulation through the positioning of the hedged system poles at desired positions.

The digital control engineer selects the appropriate measuring tools (transducers, sensors, etc.) and decides the variables to be measured and at what point, so that the system is observable. Also, the engineer decides how many directional elements are required in order for the system to be controllable.

5.6 State-Space Discretization

The transformation of a continuous-time system to an equivalent discrete-time system, in the state space, is achieved by two methods.

- The first is based on numerical integration and differentiation techniques for solving the differential equation $\dot{x}(t) = Ax(t) + Bu(t)$.
- The second method is based on the sampling of input signal $u(t)$ and holding the sample values $u(kT)$ for a time period equal to the sampling period T (hold equivalence).

5.6.1 Discretization of Continuous-Time Systems in the State Space with Numerical Integration Methods

Consider the following continuous-time system

$$\dot{x}(t) = Ax(t) + Bu(t) \tag{5.37}$$

$$y(t) = Cx(t) + Du(t) \tag{5.38}$$

A particular method to calculate the equivalent discrete system, corresponding to a continuous one, is to integrate the differential equation of the state space and then to approach the factors that contain integrals using numerical integration methods.

Hence, we have that

$$\dot{x}(t) = Ax(t) + Bu(t) \Rightarrow \int_{t_0}^{t} \frac{dx(t)}{dt} dt = \int_{t_0}^{t} [Ax(t) + Bu(t)]dt$$

$$\Rightarrow x(t) = x(t_0) + \int_{t_0}^{t} [Ax(t) + Bu(t)]dt \tag{5.39}$$

$$x(kT + T) = x(kT) + \int_{kT}^{kT+T} [Ax(t) + Bu(t)]dt \tag{5.40}$$

- Euler's forward method

  The resultant system, using this method, is expressed as

$$x((k+1)T) = \tilde{A}\,x(kT) + \tilde{B}\,u(kT) \tag{5.41}$$

$$y(kT) = \tilde{C}x(kT) + \tilde{D}u(kT) \tag{5.42}$$

where

$$\tilde{A} = I + AT,\ \tilde{B} = BT,\ \tilde{C} = C,\ \tilde{D} = D$$

- Euler's backward method
  The resultant system, using this method, is expressed as

$$\overline{x}(\kappa T + T) = \tilde{A}\overline{x}(kT) + \tilde{B}u(kT) \tag{5.43}$$

$$y(kT) = \tilde{C}\overline{x}(kT) + \tilde{D}u(kT) \tag{5.44}$$

where

$$\tilde{A} = [I - AT]^{-1},\ \tilde{B} = [I - AT]^{-1}BT$$

$$\tilde{C} = C[I - AT]^{-1},\ \tilde{D} = C[I - AT]^{-1}BT + D$$

- Trapezoidal method
  The resultant system, using this method, is expressed as

$$\overline{x}\ (kT + T) = \tilde{A}\overline{x}(kT) + \tilde{B}u(kT) + \tilde{\tilde{B}}u(kT - T) \tag{5.45}$$

$$y(kT) = \tilde{C}\overline{x}(kT) + \tilde{D}u(kT) + \tilde{\tilde{D}}u(kT - T) \tag{5.46}$$

where

$$\tilde{A} = \left[I - \frac{AT}{2}\right]^{-1}\left[I + \frac{AT}{2}\right], \tilde{B} = \left[I - \frac{AT}{2}\right]^{-1}\frac{BT}{2}$$

$$\tilde{\tilde{B}} = \left[I - \frac{AT}{2}\right]^{-1}\frac{BT}{2} \quad \tilde{C} = C\left[I - \frac{AT}{2}\right]^{-1}\left[I + \frac{AT}{2}\right],$$

$$\tilde{D} = C\left[I - \frac{AT}{2}\right]^{-1}\frac{BT}{2} + D,$$

$$\tilde{\tilde{D}} = C\left[I - \frac{AT}{2}\right]^{-1}\frac{BT}{2}$$

5.6.2 Discretization of Continuous Time Systems in State-Space with Numerical Differentiation Methods

Consider the continuous-time system of Equations 5.37 and 5.38. A particular method used to approach the derivative of state vector $\dot{x}(t)$.

- Euler's forward method

  The resultant system, using this method, is expressed as

$$x(kT+T)=\tilde{A}x(kT)+\tilde{B}u(kT) \tag{5.47}$$

$$y(kT)=\tilde{C}x(kT)+\tilde{D}u(kT) \tag{5.48}$$

  where

$$\tilde{A}=(I+AT),\tilde{B}=BT,\tilde{C}=C,\tilde{D}=D$$

- Euler's backward method

  The resultant system, using this method, is expressed as

$$\overline{x}(kT+T)=\tilde{A}\overline{x}(kT)+\tilde{B}u(kT) \tag{5.49}$$

$$y(kT)=C[I-AT]^{-1}\overline{x}(kT)+(C[I-AT]^{-1}BT+D)u(kT) \tag{5.50}$$

  where

$$\begin{aligned}\tilde{A}&=[I-AT]^{-1}, \quad \tilde{B}=[I-AT]^{-1}BT,\\ \tilde{C}&=C[I-AT]^{-1}, \quad \tilde{D}=C[I-AT]^{-1}BT+D\end{aligned}$$

- Trapezoidal method

  The resultant system, using this method, is expressed as

$$\overline{x}(kT+T)=\tilde{A}\overline{x}(kT)+\tilde{B}u(kT)+\tilde{\tilde{B}}u(kT-T) \tag{5.51}$$

$$y(kT)=\tilde{C}\overline{x}(kT)+\tilde{\tilde{C}}u(kT-T)+\tilde{D}u(kT) \tag{5.52}$$

where

$$\tilde{A} = \left[I - \frac{AT}{2}\right]^{-1}\left[I + \frac{AT}{2}\right],$$

$$\tilde{B} = \left[I - \frac{AT}{2}\right]^{-1}\frac{BT}{2},$$

$$\tilde{\tilde{B}} = \left[I - \frac{AT}{2}\right]^{-1}\frac{BT}{2},$$

$$\tilde{C} = C\left[I - \frac{AT}{2}\right]^{-1}\left[I + \frac{AT}{2}\right],$$

$$\tilde{\tilde{C}} = C\left[I - \frac{AT}{2}\right]^{-1}\frac{BT}{2},$$

$$\tilde{D} = C\left[I - \frac{AT}{2}\right]^{-1}\frac{BT}{2} + D$$

5.6.3 Discretization with the Zero-Order Hold Method

The resultant system, using this method, is expressed as

$$x[(k+1)T] = \tilde{A}x(kT) + \tilde{B}u(kT) \tag{5.53}$$

$$y(kT) = \tilde{C}x(kT) + \tilde{D}u(kT) \tag{5.54}$$

where

$$\tilde{A} = e^{AT}, \quad \tilde{B} = \left[\int_0^T e^{Aw}dw\right]B, \quad \tilde{C} = C, \tilde{D} = D$$

or $\tilde{B} = (e^{AT} - I)A^{-1}B$ if A is invertible.

5.6.4 Discretization with the First-Order Hold Method

The resultant system, using this method, is expressed as

$$x((k+1)T) = \tilde{A}x(kT) + \tilde{B}u(kT) + \tilde{\tilde{B}}u((k-1)T) \tag{5.55}$$

$$y(kT) = \tilde{C}x(kT) \tag{5.56}$$

where

$$\tilde{A} = e^{AT}, \quad \tilde{B} = \left[\int_0^T e^{Aw} \left[2 - \frac{w}{T} \right] dw \right] B,$$

$$\tilde{\tilde{B}} = \left[\int_0^T e^{Aw} \left[\frac{w}{T} - 1 \right] dw \right] B, \quad \text{and} \quad \tilde{C} = C$$

5.7 Formula Tables

The formula Tables 5.1 through 5.3 are discussed here.

5.8 Solved Exercises

EXERCISE 5.1

Consider the block diagram of the following scheme.

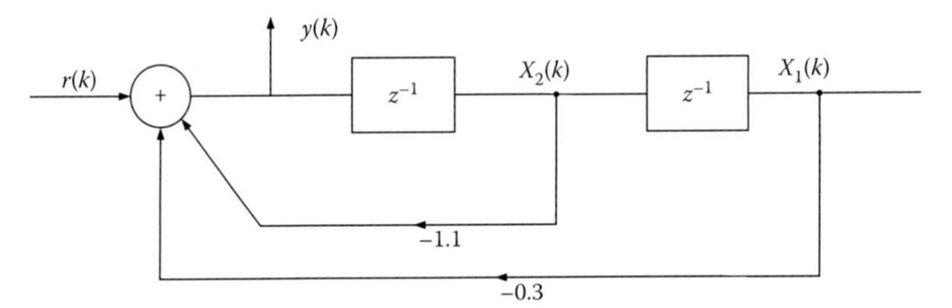

a. Derive the state equations of the given system.
b. Derive the transfer function $H(z) = Y(z)/R(z)$.

Solution

a. Assume the state variables $x_1(k)$ and $x_2(k)$ at the output of the two delay elements, thus the desired state equations are presented as

$$\begin{aligned} &x_1(k+1) = x_2(k) \\ &x_2(k+1) = -0.3x_1(k) - 1.1x_2(k) + r(k) \\ &\text{and} \\ &y(k) = -0.3x_1(k) - 1.1x_2(k) + r(k) \end{aligned} \tag{5.1.1}$$

TABLE 5.1

State Equations of Linear Discrete Systems

#	System Category	State-Space Description	Matrix Dimensions
1	Multiple input–multiple output systems 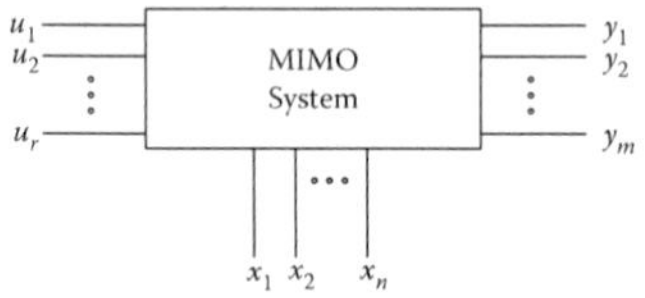 (MIMO System; inputs $u_1, u_2, \dots, u_r$; outputs $y_1, y_2, \dots, y_m$; states $x_1, x_2, \dots, x_n$)	$x(k+1) = Ax(k) + Bu(k)$ $y(k) = Cx(k) + Du(k)$ $x(0) = x_0$	$A:(n \quad x \quad n)$ $B:(n \quad x \quad r)$ $C:(m \quad x \quad n)$ $D:(m \quad x \quad r)$ $r, m > 1$
2	Multiple input–single output systems 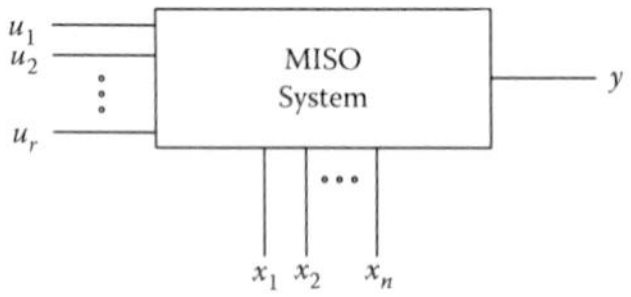 (MISO System; inputs $u_1, u_2, \dots, u_r$; output y; states $x_1, x_2, \dots, x_n$)	$x(k+1) = Ax(k) + Bu(k)$ $y(k) = c^T x(k) + du(k)$ $x(0) = x_0$	$A:(n \quad x \quad n)$ $B:(n \quad x \quad r)$ $c:(n \quad x \quad 1)$ $d:(m \quad x \quad 1)$ $r > 1, \ m = 1$
3	Single input–multiple output systems 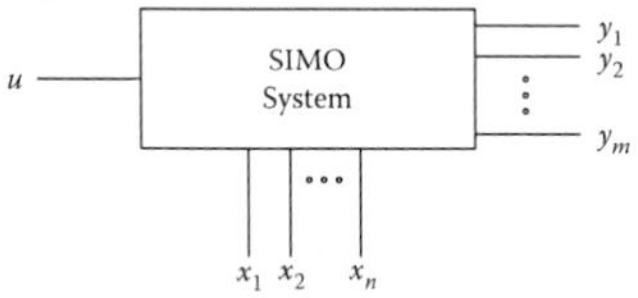 (SIMO System; input u; outputs $y_1, y_2, \dots, y_m$; states $x_1, x_2, \dots, x_n$)	$x(k+1) = Ax(k) + bu(k)$ $y(k) = Cx(k) + du(k)$ $x(0) = x_0$	$A:(n \quad x \quad n)$ $b:(n \quad x \quad 1)$ $C:(m \quad x \quad n)$ $d:(m \quad x \quad 1)$ $r = 1$ $m > 1$
4	Single input–single output systems 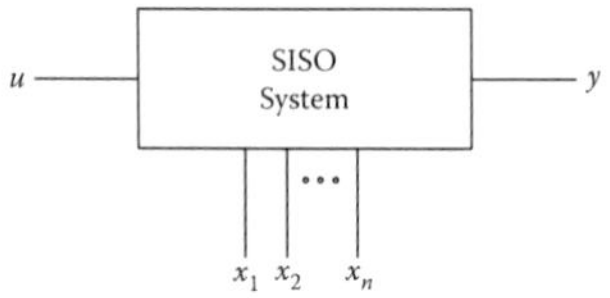 (SISO System; input u; output y; states $x_1, x_2, \dots, x_n$)	$x(k+1) = Ax(k) + bu(k)$ $y(k) = c^T x(k) + du(k)$ $x(0) = x_0$	$A:(n \quad x \quad n)$ $b:(n \quad x \quad 1)$ $c:(n \quad x \quad 1)$ $d:(scalar)$

Hence, the state equations can be written in vector form as

$$
\begin{aligned}
\begin{bmatrix} x_1(k+1) \\ x_2(k+1) \end{bmatrix} &= \begin{bmatrix} 0 & 1 \\ -0.3 & -1.1 \end{bmatrix} \begin{bmatrix} x_1(k) \\ x_2(k) \end{bmatrix} + \begin{bmatrix} 0 \\ 1 \end{bmatrix} r(k) \\
y(k) &= [-0.3 \quad -1.1] \begin{bmatrix} x_1(k) \\ x_2(k) \end{bmatrix} + r(k)
\end{aligned}
\tag{5.1.2}
$$

TABLE 5.2

State-Space Description of Dynamic Systems in Various Forms

1 Transfer function of a form

$$H(z)=\frac{Y(z)}{U(z)}=\frac{b_1z^{-1}+b_2z^{-2}+\cdots+b_mz^{-m}}{1+\alpha_1z^{-1}+\alpha_2z^{-2}+\cdots+\alpha_nz^{-n}}$$

Direct form

⇓

$$\begin{bmatrix}x_1(k+1)\\x_2(k+1)\\\vdots\\x_n(k+1)\end{bmatrix}=\begin{bmatrix}-a_1\;1\ldots0\\-a_2\;0\quad1\ldots0\\\\-a_n\;0\ldots1\end{bmatrix}=\begin{bmatrix}x_1(k)\\x_2(k)\\\vdots\\x_n(k)\end{bmatrix}+\begin{bmatrix}b_1\\b_2\\\\b_n\end{bmatrix}u(k)$$

$$y(n)=\begin{bmatrix}1 & 0\cdots0\end{bmatrix}x(k)$$

2 Transfer function of a form

$$H(z)=\frac{Y(z)}{U(z)}=\frac{b_1z^{-1}+b_2z^{-2}+\cdots+b_mz^{-m}}{1+\alpha_1z^{-1}+\alpha_2z^{-2}+\cdots+\alpha_nz^{-n}}$$

Canonical form

⇓

$$\begin{bmatrix}x_1(k+1)\\x_2(k+1)\\\vdots\\x_n(k+1)\end{bmatrix}=\begin{bmatrix}0 & 1 & 0\cdots0\\0 & 0 & 1\cdots0\\\vdots\\-a_n\cdots-a_1\end{bmatrix}\begin{bmatrix}x_1(k)\\x_2(k)\\\vdots\\x_n(k)\end{bmatrix}+\begin{bmatrix}0\\0\\\vdots\\1\end{bmatrix}u(k)$$

$$y(k)=[b_mb_{m-1}\cdots b_1]x(k)$$

3 Transfer function of a form

$$H(z)=\frac{Y(z)}{U(z)}=\frac{b_0z^m+b_1z^{m-1}+\cdots+b_m}{z^n+\alpha_1z^{n-1}+\alpha_2z^{n-2}+\cdots+\alpha_n}$$

Controllable canonical form

$$A=\begin{bmatrix}0 & 1 & 0\ldots0\\0 & 0 & 1\ldots0\\\vdots\\-a_n\ldots-a_1\end{bmatrix}\quad B=\begin{bmatrix}0\\0\\\vdots\\1\end{bmatrix}$$

$$C=[b_n-a_nb_0\quad b_{n-1}-a_{n-1}b_0\quad b_1-a_1b_0]\quad D=b_0$$

4 Transfer function of a form

$$H(z)=\frac{Y(z)}{U(z)}=\frac{b_0z^m+b_1z^{m-1}+\cdots+b_m}{z^n+\alpha_1z^{n-1}+\alpha_2z^{n-2}+\cdots+\alpha_n}$$

Observable canonical form

(Continued)

TABLE 5.2 (*Continued*)

State-Space Description of Dynamic Systems in Various Forms

	$$A=\begin{bmatrix}0 & 0 & 0\ldots & 0 & -a_n\\ 1 & 0 & 0\ldots & 0 & -a_{n-1}\\ 0 & 1 & 0\ldots & 0 & -a_{n-2}\\ \ldots & & & & \\ 0 & 0 & 0\ldots & 1 & -a_1\end{bmatrix}\quad B=\begin{bmatrix}b_n-a_nb_0\\ b_{n-1}-a_{n-1}b_0\\ \vdots\\ b_1-a_1b_0\end{bmatrix}$$ $$C=\begin{bmatrix}0 & 0 & 0\ldots & 1\end{bmatrix}\quad D=b_0$$
5	Transfer function of a form $$H(z)=\frac{Y(z)}{U(z)}=\frac{b_0z^m+b_1z^{m-1}+\cdots+b_m}{z^n+\alpha_1z^{n-1}+\alpha_2z^{n-2}+\cdots+\alpha_n}$$ Jordan canonical form $$A=\begin{bmatrix}\lambda_1 & 0 & 0\ldots & 0 & 0\\ 0 & \lambda_2 & 0\ldots & 0 & 0\\ 0 & 0 & \lambda_3 & 0 & 0\\ \ldots & & & & \\ 0 & 0 & \ldots & 0 & \lambda_n\end{bmatrix}\quad B=\begin{bmatrix}1\\ 1\\ \vdots\\ 1\end{bmatrix}$$ $$C=[r_1\quad r_2\quad r_3\ldots r_n]\quad D=b_0$$

b. It holds that

$$H(z)=C^T(zI-A)^{-1}b+D \tag{5.1.3}$$

Calculation of $(zI-A)$

$$(zI-A)=z\begin{bmatrix}1 & 0\\ 0 & 1\end{bmatrix}-\begin{bmatrix}0 & 1\\ -0.3 & -1.1\end{bmatrix}=\begin{bmatrix}z & -1\\ 0.3 & z+1\end{bmatrix} \tag{5.1.4}$$

Calculation of $(zI-A)^{-1}$

$$(zI-A)^{-1}=\begin{pmatrix}z & 1\\ 0.3 & z+1.1\end{pmatrix}^{-1}=\frac{1}{z^2+1.1z+0.3}\begin{bmatrix}z+1.1 & 1\\ -0.3 & z\end{bmatrix} \tag{5.1.5}$$

$$(5.1.3),(5.1.5)\Rightarrow H(z)=C^T(zI-A)^{-1}b+D$$

$$=[-0.3\quad -1.1]\frac{1}{z^2+1.1z+0.3}\begin{bmatrix}z+1.1 & 1\\ -0.3 & z\end{bmatrix}\begin{bmatrix}0\\ 1\end{bmatrix}+1\Rightarrow$$

After some algebraic manipulations, the transfer function is

TABLE 5.3

Transfer Matrix—State Equation Solution—Controllability—Observability

#	Illustration	Formula
1	Transfer Function $H(z)$	$H(z)=C(zI-A)^{-1}B+D$
2	State Equation Solution in Time Domain	$x(k)=A^k x(0)+\sum_{j=0}^{k-1} A^{k-j-1}Bu(j),\quad k\geq 1$ $y(k)=C\Phi(k)x(0)+\sum_{j=0}^{k-1} C\Phi(k-j-1)Bu(j)+Du(k),\quad k\geq 1$
3	State Equation Solution in Z Domain	$x(k)=IZT[z(zI-A)^{-1}x(0)+(zI-A)^{-1}BU(z)]$ $y(k)=IZT(Y(z))=IZT[(C(zI-A)^{-1}B+D)U(z)]$
4	Transition Matrix	$\Phi(k)=A^k=IZT[z(zI-A)^{-1}],\quad k\geq 1$
5	State Vector Controllability	$S=\left\|B \;\vdots\; AB \;\vdots\; \cdots \;\vdots\; A^{n-1}B\right\|$ TAΞH $S=n$
6	State Vector Observability	$R^T=[C^T \;\vdots\; A^TC^T \;\vdots\; \cdots \;\vdots\; (A^T)^{n-1}C^T]$ TAΞH $R^T=n$

$$H(z)=\frac{z^2}{z^2+z+0.3} \tag{5.1.6}$$

EXERCISE 5.2

a. Derive the transition matrix $\Phi(k)$ of the discrete-time system $x(k+1)=Ax(k)+bu(k)$ with

$$A=\begin{bmatrix}0 & 1\\ -0.16 & -1\end{bmatrix},\quad b=\begin{bmatrix}1\\1\end{bmatrix},\quad x(0)=\begin{bmatrix}1\\1\end{bmatrix}$$

b. Calculate the state vector $x(k)$ for step input.

Solution

a. It holds that

$$\Phi(k) = A^k = z^{-1}\left\{z(ZI - A)^{-1}\right\}$$

where $(ZI - A)^{-1} = \begin{bmatrix} z & -1 \\ 0.16 & z+1 \end{bmatrix}^{-1}$

$$\Rightarrow (ZI - A)^{-1} = \begin{bmatrix} \dfrac{z+1}{(z+0.2)(z+0.8)} & \dfrac{1}{(z+0.2)(z+0.8)} \\ \dfrac{-0{,}16}{(z+0.2)(z+0.8)} & \dfrac{z}{(z+0.2)(z+0.8)} \end{bmatrix}$$

$$\Rightarrow (ZI - A)^{-1} = \begin{bmatrix} \dfrac{4z/3}{z+0.2} - \dfrac{z/3}{z+0.8} & \dfrac{5z/3}{z+0.2} - \dfrac{5z/3}{z+0.8} \\ \dfrac{-0{,}8z/3}{z+0.2} + \dfrac{0{,}8z/3}{z+0.8} & \dfrac{-z/3}{z+0.2} + \dfrac{4z/3}{z+0.8} \end{bmatrix} \quad (5.2.1)$$

$$\Rightarrow \Phi(k) = \begin{bmatrix} \dfrac{4(-0.2)^k}{3} - \dfrac{(-0.8)^k}{3} & \dfrac{5(-0.2)^k}{3} - \dfrac{5(-0.8)^k}{3} \\ \dfrac{-0.8(-0.2)^k}{3} + \dfrac{0{,}8(-0.8)^k}{3} & -\dfrac{(-0.2)^k}{3} + \dfrac{4(-0.8)^k}{3} \end{bmatrix}$$

b. To calculate the state vector $x(k)$, we have

$$X(z) = (ZI - A)^{-1}(zX(0) + BU(z)), \quad U(z) = \frac{z}{z-1}$$

where

$$zX(0) + BU(z) = \begin{bmatrix} z \\ -z \end{bmatrix} + \begin{bmatrix} \dfrac{z}{z-1} \\ \dfrac{z}{z-1} \end{bmatrix} = \begin{bmatrix} \dfrac{z^2}{z-1} \\ \dfrac{-z^2+2z}{z-1} \end{bmatrix} \Rightarrow$$

$$X(z) = \begin{bmatrix} \dfrac{(z^2+2)z}{(z+0.2)(z+0.8)(z-1)} \\ \dfrac{(-z^2+1.84z)z}{(z+0.2)(z+0.8)(z-1)} \end{bmatrix} = \begin{bmatrix} \dfrac{-17z/6}{z+0.2} + \dfrac{22z/9}{z+0.8} + \dfrac{25z/18}{z-1} \\ \dfrac{3{,}4z/6}{z+0.2} - \dfrac{17{,}6z/9}{z+0.8} + \dfrac{7z/18}{z-1} \end{bmatrix}$$

$$\Rightarrow x(k) = Z^{-1}\{X(z)\} = \begin{bmatrix} -17(-0.2)^k/6 + 22(-0.8)^k/9 + 25/18 \\ 3.4(-0.2)^k/6 - 17.6(-0.8)^k/9 + 7/18 \end{bmatrix} \quad (5.2.2)$$

EXERCISE 5.3

Consider a discrete-time system with the difference equation $y(k+2) + 5y(k+1) + 3y(k) = u(k+1) + 2u(k)$

Calculate (a) its transfer function and (b) the state-space model.

Solution

a. Applying the z-transform of the given difference equation, assuming zero initial conditions, we get

$$z[y(k+2)+5y(k+1)+3y(k)] = z[u(k+1)+2u(k)]$$
$$\Rightarrow H(z) = \frac{Y(z)}{U(z)} = \frac{z+2}{z^2+5z+3} \tag{5.3.1}$$

b. Define the state variables as

$$\begin{aligned} x_1(k) &= y(k) \\ x_2(k) &= x_1(k+1) - u(k) \end{aligned} \tag{5.3.2}$$

Inserting the state variables into the difference equation, we have that

$$\begin{aligned} x_1(k+1) &= x_2(k) + u(k) \\ x_2(k+1) &= -3x_1(k) - 5x_2(k) - 3u(k) \end{aligned} \tag{5.3.3}$$

Therefore, $x(k+1) = Ax(k) + bu(k)$ with

$$x(k) = \begin{bmatrix} x_1(k) \\ x_2(k) \end{bmatrix}, \quad A = \begin{bmatrix} 0 & 1 \\ -3 & -5 \end{bmatrix}, \quad b = \begin{bmatrix} 1 \\ -3 \end{bmatrix} \tag{5.3.4}$$

EXERCISE 5.4

Illustrate the digital controller (using the canonical form method) with the transfer function

$$H(z) = \frac{1+\alpha z^{-1}}{1+\beta z^{-1}} \cdot \frac{1+\gamma z^{-1}+\delta z^{-2}}{1+\varepsilon z^{-1}+\zeta z^{-2}}$$

Solution

Observe that

$$H(z) = H_1(z) \cdot H_2(z) \tag{5.4.1}$$

where

$$H_1(z) = \frac{1+\alpha z^{-1}}{1+\beta z^{-1}} \tag{5.4.2}$$

and

$$H_2(z) = \frac{1+\gamma z^{-1}+\delta z^{-2}}{1+\varepsilon z^{-1}+\zeta z^{-2}} \tag{5.4.3}$$

We select the canonical form method and design the digital diagram of the first-order system $H_1(z)$ as follows. Insert the variable $W(z)$ and reach to the expressions (5.4.5) and (5.4.6).

$$H_1(z) = \frac{1+\alpha z^{-1}}{1+\beta z^{-1}} = \frac{Y_1(z)}{U(z)} \Rightarrow \frac{Y_1(z)}{1+\alpha z^{-1}} = \frac{U(z)}{1+\beta z^{-1}} = W(z) \tag{5.4.4}$$

It holds:

$$\frac{U(z)}{1+\beta z^{-1}} = W(z) \Rightarrow u(k) = w(k) + \beta w(k-1) \Rightarrow$$

$$w(k) = u(k) - \beta w(k-1) \tag{5.4.5}$$

and

$$\frac{Y_1(z)}{1+\alpha z^{-1}} = W(z) \Rightarrow y_1(k) = w(k) + \alpha w(k-1) \tag{5.4.6}$$

The following block diagram illustrates the first-order digital system.

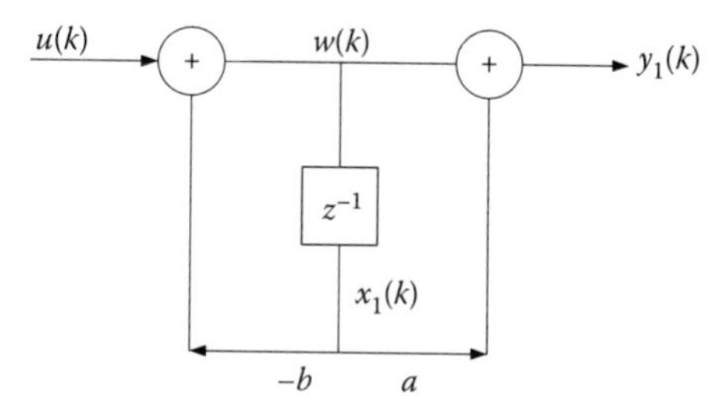

To illustrate the second-order system, utilize the intermediate variable $P(z)$ resulting to the expressions (5.4.8) and (5.4.9).

$$H_2(z) = \frac{1+\gamma z^{-1}+\delta z^{-2}}{1+\varepsilon z^{-1}+\zeta z^{-2}} = \frac{Y(z)}{Y_1(z)} \Rightarrow \frac{Y(z)}{1+\gamma z^{-1}+\delta z^{-2}}$$
$$= \frac{Y_1(z)}{1+\varepsilon z^{-1}+\zeta z^{-2}} = P(z) \tag{5.4.7}$$

$$\frac{Y_1(z)}{1+\varepsilon z^{-1}+\zeta z^{-2}} = P(z) \Rightarrow p(k) = -\varepsilon p(k-1) - \zeta p(k-2) + y_1(k) \tag{5.4.8}$$

and

$$\frac{Y(z)}{1+\gamma z^{-1}\delta z^{-2}} = P(z) \Rightarrow y(k) = p(k) + \gamma p(k-1) + \delta p(k-2) \qquad (5.4.9)$$

The following block diagram illustrates the second-order digital system.

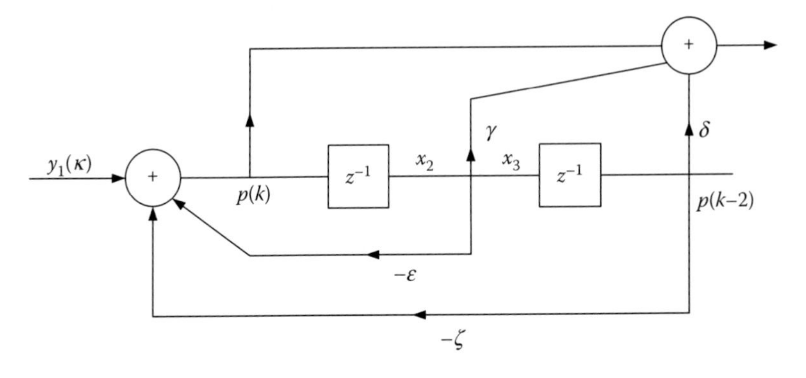

The illustration of the total system is achieved by the connection (in series) of the above two diagrams.

Assuming the state variables $x_1(k)$, $x_2(k)$, and $x_3(k)$ at the output of the delay elements, the system model at state space arise

$$\begin{aligned}
&x_1(k+1) = -\beta x_1(k) + u(k)\\
&\text{with } \; y_1(k) = (\alpha - \beta)x_1(k) + u(k)\\
&x_2(k+1) = (\alpha - \beta)x_1(k) - \varepsilon x_2(k) - \zeta x_3(k) + u(k)\\
&x_3(k+1) = x_2(k)\\
&y(k) = (\alpha - \beta)x_1(k) + (\gamma - \varepsilon)x_2(k) + (\delta - \zeta)x_3(k) + u(k)
\end{aligned} \qquad (5.4.10)$$

Consequently,

$$\begin{bmatrix} x_1(k+1)\\ x_2(k+1)\\ x_3(k+1)\end{bmatrix} = \begin{bmatrix} -\beta & 0 & 0\\ \alpha-\beta & -\varepsilon & -\zeta\\ 0 & 1 & 0\end{bmatrix}\begin{bmatrix} x_1(k)\\ x_2(k)\\ x_3(k)\end{bmatrix} + \begin{bmatrix}1\\1\\0\end{bmatrix}u(k) \qquad (5.4.11)$$

$$y(k) = [\alpha-\beta \quad \gamma-\varepsilon \quad \delta-\zeta]\begin{bmatrix} x_1(k)\\ x_2(k)\\ x_3(k)\end{bmatrix} + [1]u(k) \qquad (5.4.12)$$

EXERCISE 5.5

Consider the discrete-time system with transfer function $H(z) = -1 + 2z^{1} - z^{-2} - 2z^{-3}$

Derive the state-space model.

Solution

The input–output relation at the space of z-complex frequency is

$$\begin{aligned} &Y(z) = H(z)U(z) = (-1 + 2z^1 - z^{-2} - 2z^{-3})U(z) \\ &\Rightarrow Y(z) = -U(z) + \frac{2}{z}U(z) - \frac{1}{z^2}U(z) - \frac{2}{z^3}U(z) \end{aligned} \tag{5.5.1}$$

Set

$$\left.\begin{aligned} &X_1(z) = \frac{1}{z}U(z) \\ &X_2(z) = \frac{1}{z^2}U(z) = \frac{1}{z}X_1(z) \\ &X_3(z) = \frac{1}{z^3}U(z) = \frac{1}{z}\frac{1}{z^2}U(z) = \frac{1}{z}X_2(z) \end{aligned}\right\} \tag{5.5.2}$$

From the expressions (5.5.1) and (5.5.2), it holds that

$$Y(z) = -U(z) + 2X_1(z) - X_2(z) - 2X_3(z) \tag{5.5.3}$$

From the following expressions in (5.5.4), using the inverse z-transform, we get the expressions in (5.5.5).

$$\left.\begin{aligned} zX_1(z) &= U(z) \\ zX_2(z) &= X_1(z) \\ zX_3(z) &= X_2(z) \end{aligned}\right\} \tag{5.5.4}$$

$$\left.\begin{aligned} x_1[k+1] &= u[k] \\ x_2[k+1] &= x_1[k] \\ x_3[k+1] &= x_2[k] \end{aligned}\right\} \tag{5.5.5}$$

$$\text{and } y[n] = -u[k] + 2x_1[k] - x_2[k] - 2x_3[k] \tag{5.5.6}$$

The vector-form state equations are given by

$$\left.\begin{aligned} &\begin{bmatrix} x_1[k+1] \\ x_2[k+1] \\ x_3[k+1] \end{bmatrix} = \begin{bmatrix} 0 & 0 & 0 \\ 1 & 0 & 0 \\ 0 & 2 & 0 \end{bmatrix} \begin{bmatrix} x_1[k] \\ x_2[k] \\ x_3[k] \end{bmatrix} + \begin{bmatrix} 1 \\ 0 \\ 0 \end{bmatrix} u[k] \\ &y[k] = \begin{bmatrix} 2 & -1 & -2 \end{bmatrix} \begin{bmatrix} x_1[k] \\ x_2[k] \\ x_3[k] \end{bmatrix} - u[k] \end{aligned}\right\} \tag{5.5.7}$$

EXERCISE 5.6

Consider the discrete-time system with transfer function $H(z)=Y(z)/U(z)=(z^{-1}+2z^{-2})/(1+4z^{-1}+3z^{-2})$.

a. Derive the state-space model in controllable canonical form, observable canonical form, and diagonal canonical form.
b. Calculate the transition state matrix for the observable canonical form.

Solution

a. Using the formula table, we have
Controllable canonical form

$$\begin{bmatrix} x_1(k+1) \\ x_2(k+1) \end{bmatrix} = \begin{bmatrix} 0 & 1 \\ -3 & -4 \end{bmatrix} \begin{bmatrix} x_1(k) \\ x_2(k) \end{bmatrix} + \begin{bmatrix} 0 \\ 1 \end{bmatrix} u(k)$$
$$y(k) = \begin{bmatrix} 2 & 1 \end{bmatrix} \begin{bmatrix} x_1(k) \\ x_2(k) \end{bmatrix} \tag{5.6.1}$$

Observable canonical form

$$\begin{bmatrix} x_1(k+1) \\ x_2(k+1) \end{bmatrix} = \begin{bmatrix} 0 & -3 \\ 1 & -4 \end{bmatrix} \begin{bmatrix} x_1(k) \\ x_2(k) \end{bmatrix} + \begin{bmatrix} 2 \\ 1 \end{bmatrix} u(k)$$
$$y(k) = \begin{bmatrix} 0 & 1 \end{bmatrix} \begin{bmatrix} x_1(k) \\ x_2(k) \end{bmatrix} \tag{5.6.2}$$

Diagonal canonical form

$$H(z) = \frac{Y(z)}{U(z)} = \frac{z^{-1}+2z^{-2}}{1+4z^{-1}+3z^{-2}} = \frac{z^{-1}+2z^{-2}}{(1+z^{-1})(1+3z^{-1})}$$
$$\Rightarrow H(z) = \frac{(1/2)z^{-1}}{1+z^{-1}} + \frac{(1/2)z^{-1}}{1+3z^{-1}} \tag{5.6.3}$$

$$\begin{bmatrix} x_1(k+1) \\ x_2(k+1) \end{bmatrix} = \begin{bmatrix} -1 & 0 \\ 0 & -3 \end{bmatrix} \begin{bmatrix} x_1(k) \\ x_2(k) \end{bmatrix} + \begin{bmatrix} 1 \\ 1 \end{bmatrix} u(k)$$
$$y(k) = \begin{bmatrix} 1/2 & 1/2 \end{bmatrix} \begin{bmatrix} x_1(k) \\ x_2(k) \end{bmatrix} \tag{5.6.4}$$

b. The system of the observable canonical form can be described by the matrices of the expression (5.6.2).
The transition state matrix is provided by

$$\Phi(k) = Z^{-1}((zI-A)^{-1}) \tag{5.6.5}$$

$$(zI-A)^{-1}=\frac{Adj(zI-A)}{|zI-A|}=\begin{bmatrix} z+1 & 0 \\ 0 & z+3 \end{bmatrix}^{-1}=\begin{bmatrix} \frac{1}{z+1} & 0 \\ 0 & \frac{1}{z+3} \end{bmatrix} \tag{5.6.6}$$

Hence

$$\Phi(k)=Z^{-1}\begin{bmatrix} \frac{1}{z+1} & 0 \\ 0 & \frac{1}{z+3} \end{bmatrix}=Z^{-1}\begin{bmatrix} \frac{1}{1+z^{-1}} & 0 \\ 0 & \frac{1}{1+3z^{-1}} \end{bmatrix} \tag{5.6.7}$$

$$\Rightarrow \Phi(k)=\begin{bmatrix} (-1)^k & 0 \\ 0 & (-3)^k \end{bmatrix}$$

EXERCISE 5.7

Considering the following state-space model, design the illustration diagram.

$$\begin{cases} x(k+1)=\begin{bmatrix} 0 & 1 \\ -6 & -5 \end{bmatrix}x(k)+\begin{bmatrix} -3 \\ 10 \end{bmatrix}u(k) \\ y(k)=\begin{bmatrix} 1 & 0 \end{bmatrix}x(k)+u(k) \end{cases} \quad x(k)=\begin{bmatrix} x_1(k) \\ x_2(k) \end{bmatrix}$$

Solution

From the given state-space model of the expression (5.7.1) below, design the block diagram as follows.

$$x(k+1)=\begin{bmatrix} 0 & 1 \\ -6 & -5 \end{bmatrix}x(k)+\begin{bmatrix} -3 \\ 10 \end{bmatrix}u(k) \tag{5.7.1}$$

$$y(k)=\begin{bmatrix} 1 & 0 \end{bmatrix}x(k)+u(k) \tag{5.7.2}$$

The resultant illustration diagram of the discrete-time system becomes

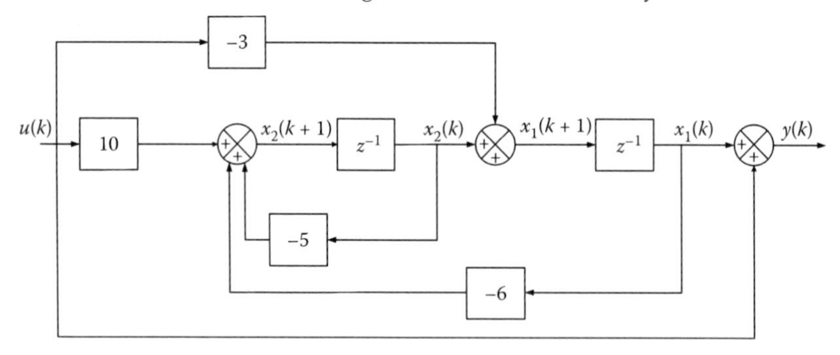

EXERCISE 5.8

Derive the transition matrix of the discrete-time system with a state-space model. Is the system controllable?

$$\begin{pmatrix} x_1(k+1) \\ x_2(k+1) \end{pmatrix} = \begin{pmatrix} 0 & 1 \\ -0.16 & -1 \end{pmatrix} \begin{pmatrix} x_1(k) \\ x_2(k) \end{pmatrix} + \begin{pmatrix} 0 \\ 1 \end{pmatrix} u(k)$$

$$y(k) = \begin{pmatrix} 2 & 1 \end{pmatrix} \begin{pmatrix} x_1(k) \\ x_2(k) \end{pmatrix}$$

Solution

The transition matrix of the discrete system is computed by

$$\Phi(k) = A^k = IZT[z(zI - A)^{-1}], \quad k \geq 1 \tag{5.8.1}$$

The state-space model is

$$\begin{pmatrix} x_1(k+1) \\ x_2(k+1) \end{pmatrix} = \begin{pmatrix} 0 & 1 \\ -0.16 & -1 \end{pmatrix} \begin{pmatrix} x_1(k) \\ x_2(k) \end{pmatrix} + \begin{pmatrix} 0 \\ 1 \end{pmatrix} u(k) \tag{5.8.2}$$

$$y(k) = \begin{pmatrix} 2 & 1 \end{pmatrix} \begin{pmatrix} x_1(k) \\ x_2(k) \end{pmatrix} \tag{5.8.3}$$

Hence

$$A = \begin{pmatrix} 0 & 1 \\ -0.16 & -1 \end{pmatrix}, \quad b = \begin{pmatrix} 0 \\ 1 \end{pmatrix} \quad \text{and} \quad c^T = \begin{pmatrix} 2 & 1 \end{pmatrix} \tag{5.8.4}$$

Calculation of $(zI - A)^{-1}$

$$\begin{aligned}
(zI - A)^{-1} &= \left(\begin{pmatrix} z & 0 \\ 0 & z \end{pmatrix} - \begin{pmatrix} 0 & 1 \\ -0.16 & -1 \end{pmatrix} \right)^{-1} \\
&= \begin{bmatrix} z & -1 \\ 0.16 & z+1 \end{bmatrix}^{-1} = \frac{1}{z(z+1)+0.16} \begin{bmatrix} z+1 & 1 \\ -0.16 & z \end{bmatrix} \\
&= \frac{1}{(z+0.2)(z+0.8)} \begin{bmatrix} z+1 & 1 \\ -0.16 & z \end{bmatrix} \\
&= \begin{bmatrix} \dfrac{z+1}{(z+0.2)(z+0.8)} & \dfrac{1}{(z+0.2)(z+0.8)} \\ \dfrac{-0.16}{(z+0.2)(z+0.8)} & \dfrac{z}{(z+0.2)(z+0.8)} \end{bmatrix}
\end{aligned} \tag{5.8.5}$$

To derive the transition matrix $\Phi(k)$, it suffices to compute the inverse z-transform of each term of the expression (5.8.1), such that

$$\Phi(k)=A^k=Z^{-1}[(zI-A)^{-1}z]$$
$$=Z^{-1}\begin{bmatrix}\dfrac{z(z+1)}{(z+0.2)(z+0.8)} & \dfrac{z}{(z+0.2)(z+0.8)}\\ \dfrac{-0.16z}{(z+0.2)(z+0.8)} & \dfrac{z^2}{(z+0.2)(z+0.8)}\end{bmatrix} \quad (5.8.6)$$
$$=Z^{-1}\begin{bmatrix}\dfrac{\frac{4}{3}}{z+0.2}+\dfrac{-\frac{1}{3}}{z+0.8} & \dfrac{\frac{5}{3}}{z+0.2}+\dfrac{-\frac{5}{3}}{z+0.8}\\ \dfrac{-\frac{0.8}{3}}{z+0.2}+\dfrac{\frac{0.8}{3}}{z+0.8} & \dfrac{-\frac{1}{3}}{z+0.2}+\dfrac{\frac{4}{3}}{z+0.8}\end{bmatrix}$$

Hence

$$\Phi(k)=A^k=Z^{-1}[(zI-A)^{-1}z]$$
$$=\begin{bmatrix}\frac{4}{3}(-0.2)^k-\frac{1}{3}(-0.8)^k & \frac{5}{3}(-0.2)^k-\frac{5}{3}(-0.8)^k\\ -\frac{0.8}{3}(-0.2)^k+\frac{0.8}{3}(-0.8)^k & -\frac{1}{3}(-0.2)^k+\frac{4}{3}(-0.8)^k\end{bmatrix} \quad (5.8.7)$$

The controllability matrix is given by

$$S=[B\vdots AB\vdots A^2B\vdots\ldots A^{n-1}B]=\begin{bmatrix}0 & 1\\ 1 & -1\end{bmatrix} \quad (5.8.8)$$

$$\det[B\ AB]=\det\begin{bmatrix}0 & 1\\ 1 & -1\end{bmatrix}=0-1=-1\neq 0$$

The order of matrix S is 2, so the system is controllable.

EXERCISE 5.9

Considering the system of Exercise 5.7, find the discrete output $y(k)$ for step input and $x(0)=\begin{pmatrix}x_1(0)\\ x_2(0)\end{pmatrix}=\begin{pmatrix}1\\ 1\end{pmatrix}$.

Solution

The general solution of state equations in time domain is

$$\left.\begin{aligned}x(k)&=A^kx(0)+\sum_{j=0}^{k-1}A^{k-j-1}Bu(j),\quad k\geq 1\\ y(k)&=C\Phi(k)x(0)+\sum_{j=0}^{k-1}C\Phi(k-j-1)Bu(j)+Du(k),\quad k\geq 1\end{aligned}\right\} \quad (5.9.1)$$

Thus

$$x(k) = \Phi(k)x(0) + \sum_{j=0}^{k-1} \Phi(k-j-1)Bu(j), \quad k = 1,2,3,\cdots \tag{5.9.2}$$

$$\begin{aligned} x(k) &= \Phi(k)x(0) + \sum_{j=0}^{k-1} \Phi(k-j-1)Bu(j) \\ &= \begin{bmatrix} \frac{4}{3}(-0.2)^k - \frac{1}{3}(-0.8)^k & \frac{5}{3}(-0.2)^k - \frac{5}{3}(-0.8)^k \\ -\frac{0.8}{3}(-0.2)^k + \frac{0.8}{3}(-0.8)^k & -\frac{1}{3}(-0.2)^k + \frac{4}{3}(-0.8)^k \end{bmatrix} \begin{bmatrix} 1 \\ 1 \end{bmatrix} \\ &\quad + \sum_{j=0}^{k-1} \begin{bmatrix} \frac{4}{3}(-0.2)^j - \frac{1}{3}(-0.8)^j & \frac{5}{3}(-0.2)^j - \frac{5}{3}(-0.8)^j \\ -\frac{0.8}{3}(-0.2)^j + \frac{0.8}{3}(-0.8)^j & -\frac{1}{3}(-0.2)^j + \frac{4}{3}(-0.8)^j \end{bmatrix} \begin{bmatrix} 0 \\ 1 \end{bmatrix} \end{aligned} \tag{5.9.3}$$

The desired discrete output will be

$$y(k) = Cx(k) = \begin{bmatrix} 2 & 1 \end{bmatrix} x(k) \quad k = 0,1,2,3\cdots \tag{5.9.4}$$

EXERCISE 5.10

Derive the discrete model at the state space of a continuous-time system, which is described by the differential equation:

$$\frac{d^2y}{dt^2} + 3\frac{dy}{dt} + 2y = \frac{du}{dt} + 3u$$

and is being discretized with ZOH.

Solution

Derive the transfer function of the continuous-time system from its corresponding differential equation

$$\begin{aligned} &L\left[\frac{d^2y}{dt^2} + 3\frac{dy}{dt} + 2y\right] = L\left[\frac{du}{dt} + 3u\right] \\ &s^2Y(s) + 3sY(s) + 2Y(s) = sU(s) + 3U(s) \\ &G(s) = \frac{Y(s)}{U(s)} = \frac{s+3}{s^2+3s+2} \end{aligned} \tag{5.10.1}$$

Write the system state equations in controllable canonical form

$$\begin{aligned} \frac{d}{dt}\begin{pmatrix} x_1(t) \\ x_2(t) \end{pmatrix} &= \begin{pmatrix} 0 & 1 \\ -2 & -3 \end{pmatrix}\begin{pmatrix} x_1(t) \\ x_2(t) \end{pmatrix} + \begin{pmatrix} 0 \\ 1 \end{pmatrix} u(t) \\ y(t) &= \begin{pmatrix} 3 & 1 \end{pmatrix}\begin{pmatrix} x_1(t) \\ x_2(t) \end{pmatrix} \end{aligned} \tag{5.10.2}$$

Calculate e^{At} and e^{AT}.

$$\varphi(t)=e^{At}=L^{-1}\left[(sI-A)^{-1}\right]=L^{-1}\left[\left[\begin{pmatrix} s & 0 \\ 0 & s \end{pmatrix}-\begin{pmatrix} 0 & 1 \\ -2 & -3 \end{pmatrix}\right]^{-1}\right]$$

$$=L^{-1}\left[\begin{pmatrix} s & -1 \\ 2 & s+3 \end{pmatrix}^{-1}\right]=L^{-1}\left[\frac{1}{s(s+3)+2}\begin{pmatrix} s+3 & 1 \\ -2 & s \end{pmatrix}\right]$$

$$=L^{-1}\begin{bmatrix} \dfrac{s+3}{s^2+3s+2} & \dfrac{1}{s^2+3s+2} \\ \dfrac{-2}{s^2+3s+2} & \dfrac{s}{s^2+3s+2} \end{bmatrix}=L^{-1}\begin{bmatrix} \dfrac{2}{s+1}-\dfrac{1}{s+2} & \dfrac{1}{s+1}-\dfrac{1}{s+2} \\ \dfrac{-2}{s+1}+\dfrac{2}{s+2} & \dfrac{-1}{s+1}+\dfrac{2}{s+2} \end{bmatrix}$$

$$\Rightarrow \varphi(t)=\begin{bmatrix} 2e^{-t}-e^{-2t} & e^{-t}-e^{-2t} \\ -2e^{-t}+2e^{-2t} & -e^{-t}+2e^{-2t} \end{bmatrix} \tag{5.10.3}$$

From the expression (5.10.3), calculate the matrix Φ.

$$\Phi=\begin{bmatrix} 2e^{-T}-e^{-2T} & e^{-T}-e^{-2T} \\ -2e^{-T}+2e^{-2T} & -e^{-T}+2e^{-2T} \end{bmatrix} \tag{5.10.4}$$

The matrix Γ will be

$$\Gamma=\int_0^T e^{Aq}Bdq=\int_0^T \begin{bmatrix} 2e^{-q}-e^{-2q} & e^{-q}-e^{-2q} \\ -2e^{-q}+2e^{-2q} & -e^{-q}+2e^{-2q} \end{bmatrix}\begin{bmatrix} 0 \\ 1 \end{bmatrix}dq$$

$$=\int_0^T \begin{bmatrix} e^{-q}-e^{-2q} \\ -e^{-q}+2e^{-2q} \end{bmatrix}dq=\begin{bmatrix} -e^{-q}+\frac{1}{2}e^{-2q} \\ e^{-q}-e^{-2q} \end{bmatrix}_0^T \tag{5.10.5}$$

$$\Rightarrow \Gamma=\begin{bmatrix} -e^{-T}+\frac{1}{2}e^{-2T} \\ e^{-T}-e^{-2T} \end{bmatrix}-\begin{bmatrix} -1+\frac{1}{2} \\ 1-1 \end{bmatrix}=\begin{bmatrix} \frac{1}{2}-e^{-T}+\frac{1}{2}e^{-2T} \\ e^{-T}-e^{-2T} \end{bmatrix}$$

Finally, we get

$$\begin{aligned} x(k+1)&=\Phi x(k)+\Gamma u(k) \\ &=\begin{bmatrix} 2e^{-T}-e^{-2T} & e^{-T}-e^{-2T} \\ -2e^{-T}+2e^{-2T} & -e^{-T}+2e^{-2T} \end{bmatrix}\begin{bmatrix} x_1(k) \\ x_2(k) \end{bmatrix} \\ &+\begin{bmatrix} \frac{1}{2}-e^{-T}+\frac{1}{2}e^{-2T} \\ e^{-T}-e^{-2T} \end{bmatrix}u(k) \end{aligned} \tag{5.10.6}$$

$$y(k)=\begin{pmatrix} 3 & 1 \end{pmatrix}\begin{bmatrix} x_1(k) \\ x_2(k) \end{bmatrix}$$

EXERCISE 5.11

Derive the discrete model of a continuous-time state space described by the transfer function $G(s)=Y(s)/U(s)=1/s(s+2)$, which is being discretized with ZOH (T = 1 s). Derive the transfer function of the discretized system.

Solution

Provide the state equations of the continuous-time system in a controllable canonical form

$$\begin{aligned}\frac{d}{dt}\begin{pmatrix}x_1(t)\\x_2(t)\end{pmatrix}&=\begin{pmatrix}0&1\\0&-2\end{pmatrix}\begin{pmatrix}x_1(t)\\x_2(t)\end{pmatrix}+\begin{pmatrix}0\\1\end{pmatrix}u(t)\\ y(t)&=\begin{pmatrix}1&0\end{pmatrix}\begin{pmatrix}x_1(t)\\x_2(t)\end{pmatrix}\end{aligned} \quad (5.11.1)$$

Calculate e^{At} and e^{AT}.

$$\varphi(t)=L^{-1}[(sI-A)^{-1}]=\begin{bmatrix}1&\frac{1}{2}(1-e^{-2t})\\0&e^{-2t}\end{bmatrix} \quad (5.11.2)$$

From the expression (5.11.2), compute the matrix Φ.

$$\Phi=\varphi(T)=\begin{bmatrix}1&\frac{1}{2}(1-e^{-2T})\\0&e^{-2T}\end{bmatrix} \quad (5.11.3)$$

The matrix Γ will be

$$\begin{aligned}\Gamma&=\int_0^T e^{Aq}Bdq=\int_0^T\begin{bmatrix}1&\frac{1}{2}(1-e^{-2q})\\0&e^{-2q}\end{bmatrix}\begin{bmatrix}0\\1\end{bmatrix}dq\\&=\begin{bmatrix}\frac{1}{2}(T+\frac{1}{2}(e^{-2T}-1))\\\frac{1}{2}(1-e^{-2T})\end{bmatrix}\end{aligned} \quad (5.11.4)$$

Finally, we have

$$\begin{aligned}x(k+1)&=\Phi x(k)+\Gamma u(k)\\&=\begin{bmatrix}1&\frac{1}{2}(1-e^{-2T})\\0&e^{-2T}\end{bmatrix}\begin{bmatrix}x_1(k)\\x_2(k)\end{bmatrix}+\begin{bmatrix}\frac{1}{2}(T+\frac{1}{2}(e^{-2T}-1))\\\frac{1}{2}(1-e^{-2T})\end{bmatrix}u(k)\\ y(k)&=(1\quad 0)\begin{bmatrix}x_1(k)\\x_2(k)\end{bmatrix}\end{aligned} \quad (5.11.5)$$

for T = 1 s., the latter equations become

$$x(k+1)=\begin{bmatrix}1 & 0.4323\\0 & 0.1353\end{bmatrix}\begin{bmatrix}x_1(k)\\x_2(k)\end{bmatrix}+\begin{bmatrix}0.2838\\0.4323\end{bmatrix}u(k)$$
$$y(k)=(1 \quad 0)\begin{bmatrix}x_1(k)\\x_2(k)\end{bmatrix} \tag{5.11.6}$$

The transfer function is given by $G(z) = C(zI - \Phi)^{-1}\Gamma$, thus

$$\begin{aligned}G(z)&=C(zI-\Phi)^{-1}\Gamma\\&=(1 \quad 0)\begin{pmatrix}z-1 & -0.4323\\0 & z-0.1353\end{pmatrix}^{-1}\begin{pmatrix}0.2838\\0.4323\end{pmatrix}\\ \Rightarrow G(z)&=\frac{0.2838z^{-1}+0.1485z^{-2}}{1-1.1353z^{-1}+0.1353z^{-2}}\end{aligned} \tag{5.11.7}$$

The same transfer function is obtained if we directly apply ZOH in the given transfer function

$$\begin{aligned}G(z)&=(1-z^{-1})Z\left[\frac{G(s)}{s}\right]=(1-z^{-1})Z\left[\frac{1}{s^2(s+2)}\right]\\&=\frac{0.2838z^{-1}+0.1485z^{-2}}{1-1.1353z^{-1}+0.1353z^{-2}}\end{aligned} \tag{5.11.8}$$

EXERCISE 5.12

Given the following state-space model of a discrete system, check its observability.

$$\begin{bmatrix}x_1(k+1)\\x_2(k+1)\end{bmatrix}=\begin{bmatrix}1.1 & -0.3\\1 & 0\end{bmatrix}\begin{bmatrix}x_1(k)\\x_2(k)\end{bmatrix}+\begin{bmatrix}1\\-0.5\end{bmatrix}u(k)$$
$$y(k)=(1 \quad -0.5)\begin{pmatrix}x_1(k)\\x_2(k)\end{pmatrix}$$

Solution

The observability matrix is presented as

$$R=\begin{pmatrix}C^T\\C^TA\\\vdots\\C^TA^{n-1}\end{pmatrix} \tag{5.12.1}$$

$$\begin{aligned}
&C^T = (1 \quad -0.5)\\
&C^T A = (1 \quad -0.5)\begin{bmatrix} 1.1 & -0.3 \\ 1 & 0 \end{bmatrix} = (0.6 \quad -0.3)\\
&\begin{bmatrix} C^T \\ C^T A \end{bmatrix} = \begin{bmatrix} 1 & -0.5 \\ 0.6 & -0.3 \end{bmatrix}
\end{aligned} \tag{5.12.2}$$

$$\det\begin{bmatrix} C^T \\ C^T A \end{bmatrix} = \det\begin{bmatrix} 1 & -0.5 \\ 0.6 & -0.3 \end{bmatrix} = 0$$

The order of matrix R is 1, hence the system is not observable.

EXERCISE 5.13

Given the following state-space model of a discrete system, check its controllability and observability. If the system is being discretized does it remain controllable and observable? Derive the transfer function of the analog and discrete system.

$$\frac{d}{dt}\begin{bmatrix} x_1(t) \\ x_2(t) \end{bmatrix} = \begin{bmatrix} 0 & 1 \\ -1 & 0 \end{bmatrix}\begin{bmatrix} x_1(t) \\ x_2(t) \end{bmatrix} + \begin{bmatrix} 0 \\ 1 \end{bmatrix} u(t)$$

$$y(t) = (1 \quad 0)\begin{pmatrix} x_1(t) \\ x_2(t) \end{pmatrix}$$

Solution

a. Controllability matrix

$$S = \begin{bmatrix} b & Ab \end{bmatrix} = \begin{bmatrix} 0 & 1 \\ 1 & 0 \end{bmatrix} \tag{5.13.1}$$

The order of matrix S is 2, hence the continuous-time system is controllable.
Observability matrix

$$R = \begin{bmatrix} c \\ cA \end{bmatrix} = \begin{bmatrix} 1 & 0 \\ 0 & 1 \end{bmatrix} \tag{5.13.2}$$

The order of matrix R is 2, hence the continuous-time system is observable.

b. The discrete-time system arises by computing the following matrices

$$\left.\begin{aligned} &\varphi(t)=e^{At}=L^{-1}\left[(sI-A)^{-1}\right] \\ &\Phi=\varphi(t)\,|_{t=T}=e^{At}\,|_{t=T}=e^{AT} \\ &\Gamma=\int_0^T e^{At}Bdq \end{aligned}\right\} \tag{5.13.3}$$

$$\varphi(t)=e^{AT}=L^{-1}\left[(sI-A)^{-1}\right]=L^{-1}\left\{\begin{bmatrix} s & -1 \\ 1 & s \end{bmatrix}^{-1}\right\}$$

$$\Rightarrow \varphi(t)=L^{-1}\left\{\begin{bmatrix} \dfrac{s}{s^2+1} & \dfrac{1}{s^2+1} \\ \dfrac{-1}{s^2+1} & \dfrac{s}{s^2+1} \end{bmatrix}\right\}=\begin{bmatrix} \cos T & \sin T \\ -\sin T & \cos T \end{bmatrix} \tag{5.13.4}$$

$$\Gamma=\int_0^T e^{At}Bdq=\int_0^T \begin{bmatrix} \cos q & \sin q \\ -\sin q & \cos q \end{bmatrix} dq \begin{bmatrix} 0 \\ 1 \end{bmatrix}=\begin{bmatrix} 1-\cos T \\ \sin T \end{bmatrix} \tag{5.13.5}$$

Calculate the controllability matrix

$$S=\begin{bmatrix} \Gamma & A\Gamma \end{bmatrix}=\begin{bmatrix} 1-\cos T & \cos T-\cos^2 T+\sin^2 T \\ \sin T & -\sin T+2\sin T\cos T \end{bmatrix} \tag{5.13.6}$$

$$\begin{aligned} \det(S) &= \begin{vmatrix} (1-\cos T) & (\cos T-\cos^2 T+\sin^2 T) \\ \sin T & (-\sin T+2\sin T\cos T) \end{vmatrix} \\ &= \begin{vmatrix} 1-\cos T & 1-\cos^2 T+\sin^2 T \\ \sin T & 2\sin T\cos T \end{vmatrix} \end{aligned}$$

$$\Rightarrow \det(S) = \begin{vmatrix} 1-\cos T & 1-\cos^2 T+\sin^2 T-2\cos T(1-\cos T) \\ \sin T & 0 \end{vmatrix} \tag{5.13.7}$$

$\det(S) = -2\sin T(1-\cos T) = 0$ the system is not controllable for $T = n\pi, n = 0,1,2,\ldots$

Calculate the observability matrix

$$R = \begin{bmatrix} c \\ cA \end{bmatrix} = \begin{bmatrix} 1 & 0 \\ \cos T & \sin T \end{bmatrix} \tag{5.13.8}$$

$\det(R) = \sin T = 0$: Nonobservable system for $T = n\pi$, $n = 0,1,2,\ldots$

c. The transfer function of the analog system is

$$G(s) = C(sI-A)^{-1}B + D = \begin{bmatrix} 1 & 0 \end{bmatrix} \begin{bmatrix} s & -1 \\ 1 & s \end{bmatrix}^{-1} \begin{bmatrix} 0 \\ 1 \end{bmatrix}$$

$$\Rightarrow G(s) = \begin{bmatrix} 1 & 0 \end{bmatrix} \begin{bmatrix} \frac{s}{s^2+1} & \frac{1}{s^2+1} \\ \frac{-1}{s^2+1} & \frac{s}{s^2+1} \end{bmatrix} \begin{bmatrix} 0 \\ 1 \end{bmatrix} = \frac{1}{s^2+1} \tag{5.13.9}$$

The transfer function of the discretized system is

$$\begin{aligned} G(z) &= C(zI-G)^{-1}H \\ &= \begin{bmatrix} 1 & 0 \end{bmatrix} \begin{bmatrix} z-\cos T & -\sin T \\ \sin T & z-\cos T \end{bmatrix}^{-1} \begin{bmatrix} 1-\cos T \\ \sin T \end{bmatrix} \\ &= \frac{1}{(z-\cos T)^2 + \sin^2 T} \begin{bmatrix} 1 & 0 \end{bmatrix} \begin{bmatrix} z-\cos T & \sin T \\ -\sin T & z-\cos T \end{bmatrix} \begin{bmatrix} 1-\cos T \\ \sin T \end{bmatrix} \end{aligned}$$

$$\Rightarrow G(z) = \frac{(z-\cos T)(1-\cos T) + \sin^2 T}{(z-\cos T)^2 + \sin^2 T} \tag{5.13.10}$$

EXERCISE 5.14

The model of a discrete system in the state space is given as follows

$$\begin{aligned} &\bar{x}_{k+1} = A\bar{x}_k + B\bar{u}_k \quad A = \begin{pmatrix} 1 & 1 \\ 0 & 3 \end{pmatrix}, \quad B = \begin{pmatrix} 1 \\ 2 \end{pmatrix}, \quad C = (0 \quad 1), \quad D = 0, \quad \bar{x}_0^{\,T} = (1 \quad 1) \\ &y_k = C\bar{x}_k \end{aligned}$$

a. Derive the system transfer function.
b. Calculate the step response.

Solution

a. The discrete-time transfer function is derived by

$$
\begin{aligned}
G(z) &= C\cdot(zI-A)^{-1}\cdot B+D \\
&= (0\ \ 1)\cdot\left[\begin{pmatrix} z & 0 \\ 0 & z\end{pmatrix}-\begin{pmatrix} 1 & 1 \\ 0 & 3\end{pmatrix}\right]^{-1}\cdot\begin{pmatrix}1\\2\end{pmatrix} \\
&= (0\ \ 1)\cdot\begin{pmatrix} z-1 & -1 \\ 0 & z-3\end{pmatrix}^{-1}\cdot\begin{pmatrix}1\\2\end{pmatrix} \\
&= \frac{1}{(z-1)(z-3)}\cdot(0\ \ 1)\cdot\begin{pmatrix} z-3 & 1 \\ 0 & z-1\end{pmatrix}\cdot\begin{pmatrix}1\\2\end{pmatrix} \\
&= \frac{1}{(z-1)(z-3)}\cdot(0\ \ z\text{-}1)\cdot\begin{pmatrix}1\\2\end{pmatrix} \\
&= \frac{2\cdot(z-1)}{(z-1)(z-3)} \Rightarrow
\end{aligned}
$$

$$G(z)=\frac{2}{z-3} \tag{5.14.1}$$

b. The step response is given as

$$
\left.
\begin{aligned}
Y(z)=\frac{2}{z-3}\cdot\frac{z}{z-1}\Rightarrow\frac{Y(z)}{z}=\frac{2}{(z-3)\cdot(z-1)}\Rightarrow\frac{Y(z)}{z}=\frac{A}{z-3}+\frac{B}{z-1} \\
A=\left.\frac{2}{z-1}\right|_{z=3}=\frac{2}{3-1}=1 \text{ and } B=\left.\frac{2}{z-3}\right|_{z=1}=\frac{2}{1-3}=-1
\end{aligned}
\right\}
$$

$$\Rightarrow\frac{Y(z)}{z}=\frac{1}{z-3}-\frac{1}{z-1}$$

$$\Rightarrow Y(z)=\frac{z}{z-3}-\frac{z}{z-1}$$

Consequently

$$
\begin{aligned}
&y(k)=Z^{-1}\{Y(z)\}=Z^{-1}\left\{\frac{z}{z-3}-\frac{z}{z-1}\right\}\Rightarrow \\
&y(k)=3^k-u(k)
\end{aligned}
\tag{5.14.2}
$$

The system step response is described as *total* and includes a term which corresponds to the transition state (3^k) and a term corresponding to the steady state $u(k)$.

EXERCISE 5.15

(a) Derive the discrete-time transfer function described by the following matrices and (b) Evaluate the system output when it is influenced, or not influenced, from the initial conditions, for a sampling period $T = 0.3$ s and a rectangular pulse as an input signal of frequency 1 Hz, using LabVIEW.

$$A = \begin{bmatrix} 1 & 0 & 0 \\ 1 & 1 & 0 \\ 0 & 1 & 0 \end{bmatrix} \quad B = \begin{bmatrix} 1 \\ 0 \\ 0 \end{bmatrix} \quad C = \begin{bmatrix} 0 & 0 & 1 \end{bmatrix}$$

Solution

a. The discrete-time transfer function is calculated as

$$G(z) = C \cdot (zI - A)^{-1} \cdot B + D \tag{5.15.1}$$

Calculation of matrix $zI - A$ and its determinant

$$\left.\begin{aligned} zI - A &= \begin{bmatrix} z-1 & 0 & 0 \\ -1 & z-1 & 0 \\ 0 & -1 & z \end{bmatrix} \\ D &= z^3 - 2z^2 + z = z(z-1)^2 \end{aligned}\right\} \tag{5.15.2}$$

Calculation of matrix $(zI - A)^{-1}$

$$(zI - A)^{-1} = \begin{bmatrix} \dfrac{1}{z-1} & 0 & 0 \\ \dfrac{1}{(z-1)^2} & \dfrac{1}{z-1} & 0 \\ \dfrac{1}{z(z-1)^2} & \dfrac{1}{z(z-1)} & \dfrac{1}{z} \end{bmatrix} \tag{5.15.3}$$

$$(5.15.1) \Rightarrow G(z) = \frac{1}{z^3 - 2z^2 + z} \tag{5.15.4}$$

b. In the following block diagram, which has been designed using the LabVIEW Simulation Module, the given discrete system is implemented in the state space. The simulation scenario assumes a rectangular pulse as an input signal with frequency 1 Hz and studies the system response with or without the initial conditions' effect (which correspond to the state variables) and is also determined by the sampling period. The results are depicted in the front panel and correspond to zero and nonzero initial conditions.

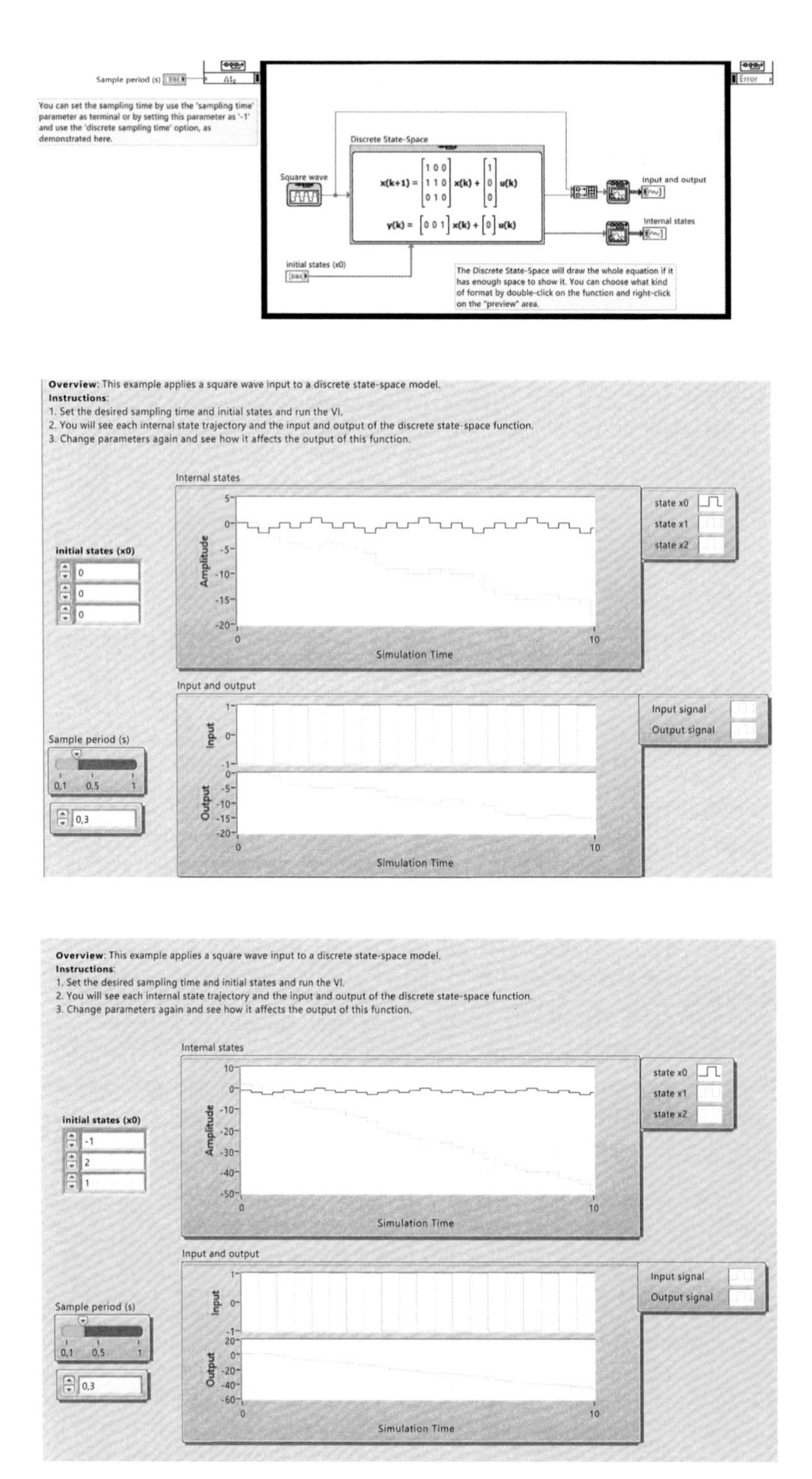
Sample period (s)
You can set the sampling time by use the 'sampling time' parameter as terminal or by setting this parameter as '-1' and use the 'discrete sampling time' option, as demonstrated here.
Discrete State-Space
Square wave
Input and output
Internal states
initial states (x0)
The Discrete State-Space will draw the whole equation if it has enough space to show it. You can choose what kind of format by double-click on the function and right-click on the "preview" area.
Error
Overview: This example applies a square wave input to a discrete state-space model.
Instructions:
1. Set the desired sampling time and initial states and run the VI.
2. You will see each internal state trajectory and the input and output of the discrete state-space function.
3. Change parameters again and see how it affects the output of this function.
Internal states
state x0
state x1
state x2
Amplitude
Simulation Time
Input and output
Input signal
Output signal
Input
Output
0,3

EXERCISE 5.16

Consider the discrete system described by the transfer function

$$G(z) = \frac{Y(z)}{U(z)} = \frac{3 + 2z^{-1}}{2 + 6z^{-1} + 4z^{-2}}$$

a. Derive the state equations in direct and canonical form.
b. Compute the system transfer function.

Solution

a. Direct form

From the given transfer function, we provide the expression (5.16.1) below, whereon the illustration diagram of the discrete system in direct form is presented

$$Y(z) = \left(-3z^{-1} - 2z^{-2}\right)Y(z) + \left(1.5 + z^{-1}\right)U(z) \tag{5.16.1}$$

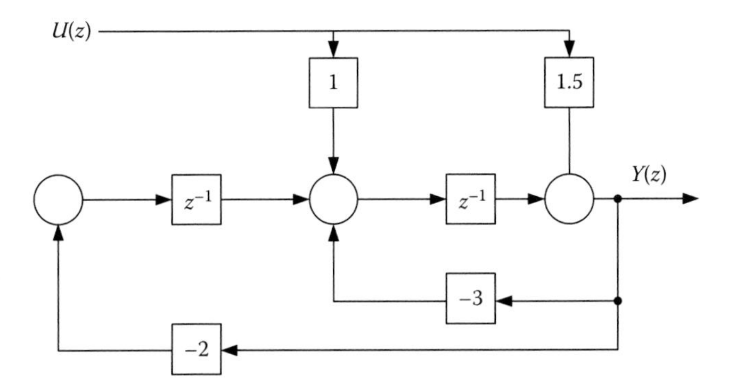

Assuming the outputs of the delay elements as state variables, the subsequent model of expressions (5.16.2) and (5.16.3) arises

$$\left.\begin{aligned} x_1(k+1) &= u(k) + x_2(k) - 3\left[1.5u(k) + x_1(k)\right] = -3.5u(k) - 3x_1(k) + x_2(k) \\ x_2(k+1) &= -2\left[1.5u(k) + x_1(k)\right] = -3u(k) - 2x_1(k) \end{aligned}\right\} \tag{5.16.2}$$

$$y(k) = 1.5u(k) + x_1(k) \tag{5.16.3}$$

The state-space matrices are

$$A = \begin{pmatrix} -3 & 1 \\ -2 & 0 \end{pmatrix} \quad B = \begin{pmatrix} -3.5 \\ -3 \end{pmatrix}, C = \begin{pmatrix} 1 & 0 \end{pmatrix}, D = 1.5 \tag{5.16.4}$$

Canonical form
Using the intermediate variable $W(z)$, we reach to the expressions (5.16.6) and (5.16.7), whereon the illustration diagram of the discrete system in canonical form is presented

$$\frac{Y(z)}{3+2z^{-1}}=\frac{U(z)}{2+6z^{-1}+4z^{-2}}=W(z) \tag{5.16.5}$$

$$\begin{aligned}&(5.16.5)\Rightarrow 2W(z)=-6z^{-1}W(z)-4z^{-2}W(z)+U(z)\\&W(z)=-3z^{-1}W(z)-2z^{-2}W(z)+.5U(z)\end{aligned} \tag{5.16.6}$$

$$Y(z)=3W(z)+2z^{-1}W(z) \tag{5.16.7}$$

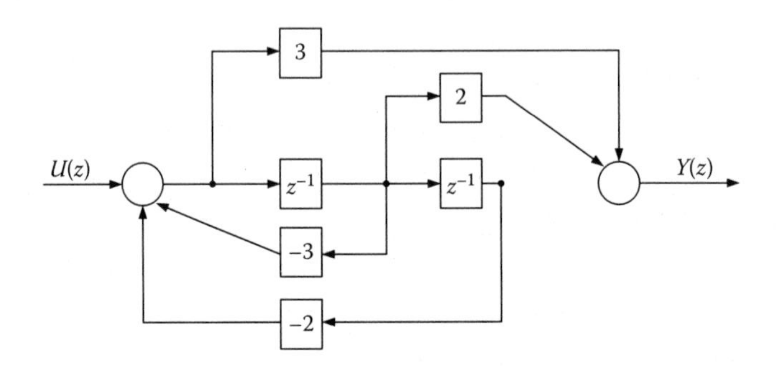

b. Calculation of transfer function

$$G(z)=C(zI-A)^{-1}B+D \tag{5.16.8}$$

Calculate the matrix $(zI - A)^{-1}$, from the corresponding equations in direct form

$$(zI-A)^{-1}=\begin{pmatrix}z+3 & 1\\ -2 & z\end{pmatrix}^{-1}=\frac{\begin{pmatrix}z & 1\\ -2 & z+3\end{pmatrix}}{z^2+3z+2} \tag{5.16.9}$$

$$(5.16.8)\Rightarrow G(z)=C(zI-A)^{-1}B+D=\frac{-3.5z-3}{z^2+3z+2}+1.5=\frac{1.5z^2+z}{z^2+3z+2} \tag{5.16.10}$$

The transfer function of the expression (5.16.10) can be written as

$$G(z)=\frac{1.5z^2+z}{z^2+3z+2}=\frac{1.5+z^{-1}}{1+3z^{-1}+2z^{-2}}=\frac{3+2z^{-1}}{2+6z^{-1}+4z^{-2}} \tag{5.16.11}$$

Observe that the transfer function of the latter expression is identical to the given one.

EXERCISE 5.17

Consider a discrete system with the transfer function

$$G(z) = \frac{Y(z)}{U(z)} = \frac{2 + 4z^{-1}}{1 + 5z^{-1} + 6z^{-2}}$$

a. Derive the state equations in direct form.
b. Compute the system transfer function.

Solution

a. Direct form

From the given transfer function, we derive the expression (5.17.1), whereon the illustration diagram of the discrete system in direct form is presented

$$Y(z) = \left(-5z^{-1} - 6z^{-2}\right)Y(z) + \left(2 + 4z^{-1}\right)U(z) \tag{5.17.1}$$

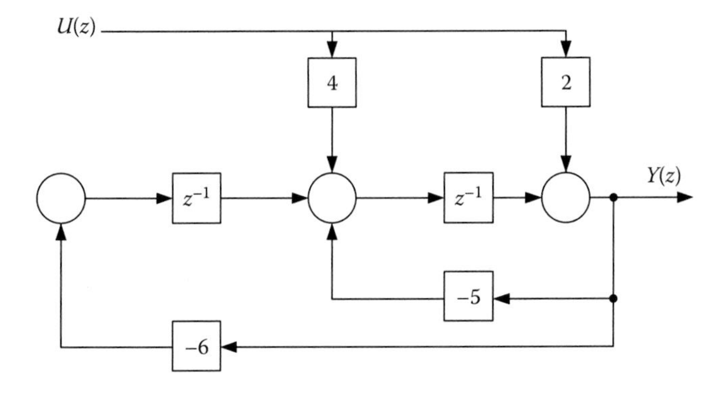

Assuming the outputs of the delay elements as state variables, the subsequent model of expressions (5.17.2) and (5.17.3) arises

$$\left.\begin{aligned} x_1(k+1) &= 4u(k) + x_2(k) - 5\left[2u(k) + x_1(k)\right] \\ &= -6u(k) - 5x_1(k) + x_2(k) \\ x_2(k+1) &= -6\left[2u(k) + x_1(k)\right] = -12u(k) - 6x_1(k) \end{aligned}\right\} \tag{5.17.2}$$

$$y(k) = 2u(k) + x_1(k) \tag{5.17.3}$$

The state-space matrices are

$$A=\begin{pmatrix}-5 & 1\\ -6 & 0\end{pmatrix}\quad B=\begin{pmatrix}-6\\ -12\end{pmatrix}, C=\begin{pmatrix}1 & 0\end{pmatrix}, D=2 \tag{5.17.4}$$

b. Verification of the results

$$G(z)=C(zI-A)^{-1}B+D$$

$$(zI-A)^{-1}=\begin{pmatrix}z+5 & -1\\ 6 & z\end{pmatrix}^{-1}=\frac{\begin{pmatrix}z & 1\\ -6 & z+5\end{pmatrix}}{z^2+5z+6}$$

$$\begin{aligned}G(z)&=C(zI-A)^{-1}B+D=\frac{-14z-58}{z^2+5z+6}+2\\ &=\frac{2z^2+4z}{z^2+5z+6}=\frac{2+4z^{-1}}{1+5z^{-1}+6z^{-2}}\end{aligned} \tag{5.17.5}$$

Observe that the transfer function of the latter expression is identical to the given one.

EXERCISE 5.18

Provide a state-space model for the system described by the following difference equation:

$$y(k+2)=u(k)+1.7y(k+1)-0.72y(k)$$

Solution

We select the state variables

$$x_2(k)=y(k+1)=x_1(k+1) \tag{5.18.1}$$

Hence

$$\left.\begin{aligned}x_1(k+1)&=x_2(k)\\ x_2(k+1)&=y(k+2)=u(k)+1.7x_2(k)-0.72x_1(k)\end{aligned}\right\} \tag{5.18.2}$$

The associated discrete model is given by

$$\begin{bmatrix}x_1(k+1)\\ x_2(k+1)\end{bmatrix}=\begin{bmatrix}0 & 1\\ -0.72 & 1.7\end{bmatrix}\begin{bmatrix}x_1(k)\\ x_2(k)\end{bmatrix}+\begin{bmatrix}0\\ 1\end{bmatrix}u(k) \tag{5.18.3}$$

$$y(k)=\begin{bmatrix}1 & 0\end{bmatrix}\begin{bmatrix}x_1(k)\\ x_2(k)\end{bmatrix} \tag{5.18.4}$$

EXERCISE 5.19

Derive the transfer function

$$D(z) = \frac{U(z)}{E(z)} = \frac{3 + 3.6z^{-1} + 0.6z^{-2}}{1 + 0.1z^{-1} - 0.2z^{-2}}$$

in direct form, canonical form, in series, and parallel form.

Solution

a. Calculation in direct form
   From the given transfer function, we have

$$u(k) = 3e(k) + 3.6e(k-1) + 0.6e(k-2) - 0.1u(k-1) + 0.2u(k-2) \tag{5.19.1}$$

The system state equations are

$$\begin{bmatrix} x_1(k+1) \\ x_2(k+1) \end{bmatrix} = \begin{bmatrix} -0.1 & 1 \\ 0.2 & 0 \end{bmatrix} \begin{bmatrix} x_1(k) \\ x_2(k) \end{bmatrix} + \begin{bmatrix} 3.6 - 0.3 \\ 0.6 + 0.6 \end{bmatrix} e(k) \tag{5.19.2}$$

$$u(k) = \begin{bmatrix} 1 & 0 \end{bmatrix} \begin{bmatrix} x_1(k) \\ x_2(k) \end{bmatrix} + 3e(k) \tag{5.19.3}$$

The illustration diagram in direct form follows

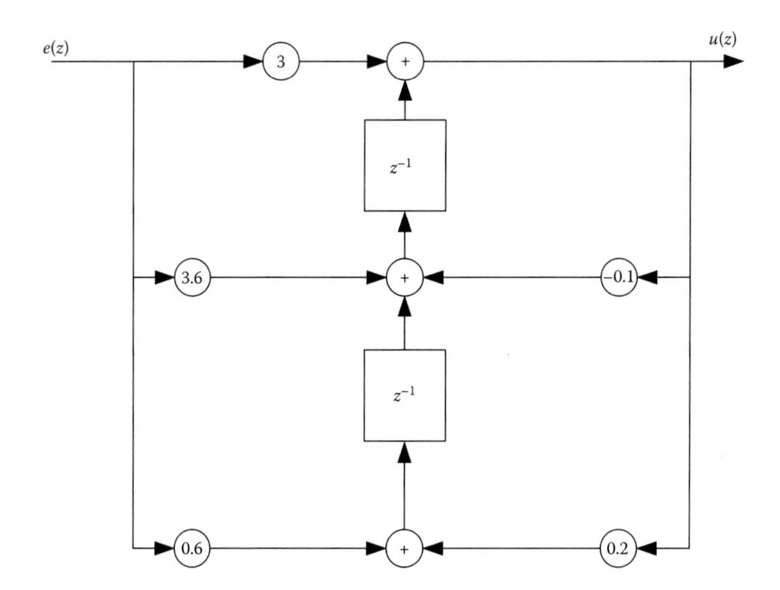

b. Calculation in canonical form
It holds that

$$w(k) = e(k) - 0.1w(k-1) + 0.2w(k-2) \tag{5.19.4}$$

$$u(k) = 3w(k) + 3.6w(k-1) + 0.6w(k-2) \tag{5.19.5}$$

The illustration diagram in canonical form follows

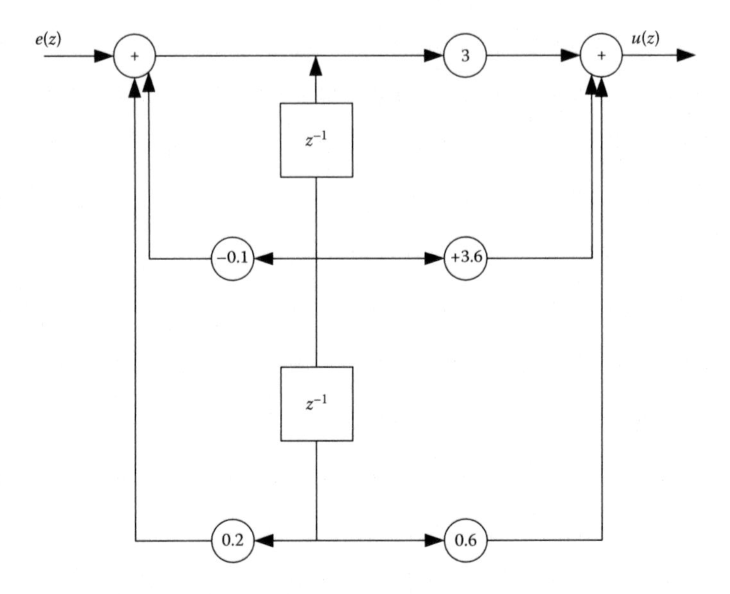

c. Calculation in series

$$D(z) = \frac{3 + 3.6z^{-1} + 0.6z^{-2}}{1 + 0.1z^{-1} - 0.2z^{-2}} = \frac{3(z+1)(z-0.2)}{(z+0.5)(z-0.4)}$$
$$\Rightarrow D(z) = \frac{3(1+z^{-1})(1-0.2z^{-1})}{(1+0.5z^{-1})(1-0.4z^{-1})} \tag{5.19.6}$$

The system state equations are

$$\begin{bmatrix} x_1(k+1) \\ x_2(k+1) \end{bmatrix} = \begin{bmatrix} -0.5 & 0 \\ 1-0.5 & 0.4 \end{bmatrix} \begin{bmatrix} x_1(k) \\ x_2(k) \end{bmatrix} + \begin{bmatrix} 3 \\ 3 \end{bmatrix} e(k) \tag{5.19.7}$$

$$u(k) = [1-0.5 \quad 0.2+0.4] \begin{bmatrix} x_1(k) \\ x_2(k) \end{bmatrix} + 3e(k) \tag{5.19.8}$$

The illustration diagram in series follows

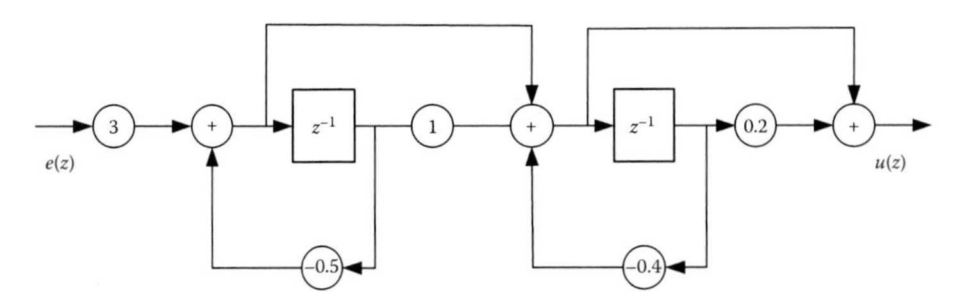

d. Calculation in parallel form

The transfer function is presented as

$$D(z)=\frac{3(z+1)(z-0.2)}{(z+0.5)(z-0.4)}=-3-\frac{1}{1+0.5z^{-1}}+\frac{7}{1-0.4z^{-1}} \tag{5.19.9}$$

The system state equations are

$$\begin{bmatrix} x_1(k+1) \\ x_2(k+1) \end{bmatrix}=\begin{bmatrix} -0.5 & 0 \\ 0 & 0.4 \end{bmatrix}\begin{bmatrix} x_1(k) \\ x_2(k) \end{bmatrix}+\begin{bmatrix} 1 \\ 1 \end{bmatrix}e(k) \tag{5.19.10}$$

$$u(k)=[0.5 \quad 7\cdot 0.4]\begin{bmatrix} x_1(k) \\ x_2(k) \end{bmatrix}+(-3-1+7)e(k) \tag{5.19.11}$$

The illustration diagram in parallel form follows

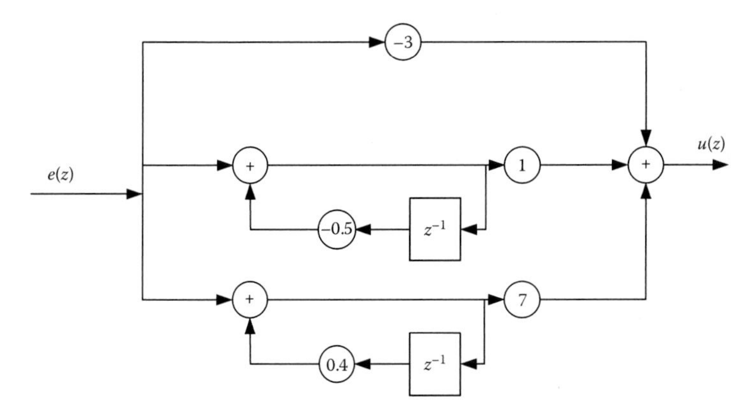

EXERCISE 5.20

Consider the system

$$\dot{x}(t)=\begin{bmatrix} 0 & 1 \\ 0 & 0 \end{bmatrix}x(t)+\begin{bmatrix} 0 \\ 1 \end{bmatrix}u(t)$$

$$y(t)=\begin{bmatrix} 1 & 0 \end{bmatrix}x(t)+\begin{bmatrix} 0 \end{bmatrix}u(t)$$

Discretize the system by using (a) the forward difference method, (b) the backward difference method, and (c) the trapezoidal method. Compare the step responses of the analog and discretized system for all the above cases.

Solution

a. The state-space matrices are

$$A=\begin{bmatrix}0 & 1\\ 0 & 0\end{bmatrix}, B=\begin{bmatrix}0\\ 1\end{bmatrix}, C=\begin{bmatrix}1 & 0\end{bmatrix}, D=\begin{bmatrix}0\end{bmatrix}$$

The resultant system using the forward difference method is given as

$$\left.\begin{aligned} x((k+1)T) &= \tilde{A}x(kT)+\tilde{B}u(kT) \\ y(kT) &= \tilde{C}x(kT)+\tilde{D}u(kT) \end{aligned}\right\}$$

where

$$\tilde{A}=I+AT, \tilde{B}=BT, \tilde{C}=C, \tilde{D}=D$$

Using MATLAB®, we develop the function **eulerforw.m**, which has as an input the matrices a, b, c, d of the continuous system and the desired sampling period T, which calculates the corresponding matrices ad, bd, cd, dd of the equivalent discrete system according to the forward difference method.

```
function [ad,bd,cd,dd]=eulerforw(a,b,c,d,T)
[s1 s2]=size(a);
I=eye(s1);
ad=I+a*T;
bd=b*T;
cd=c;
dd=d;
```

Thereby, the step responses of the analog and discretized system are designed and we can observe that they are almost identical.

```
a=[0 1;0 0];
b=[0;1];
c=[1 0];
d=[0];
T=0.1;
[ad,bd,cd,dd]=eulerforw(a,b,c,d,T)
sys=ss(a,b,c,d);
sysd=ss(ad,bd,cd,dd,T);
step(sys,'b',sysd,'r')
```

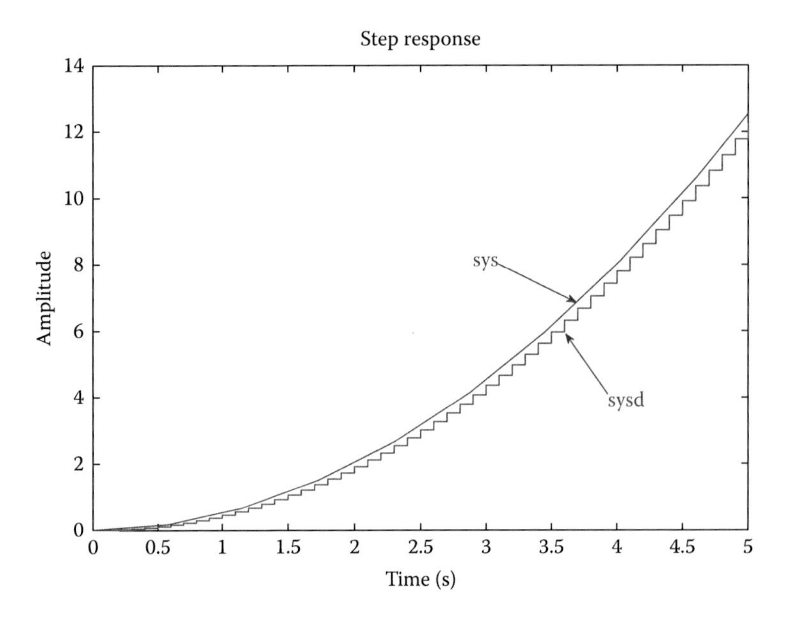

b. The resultant system using the backward difference method is given as

$$\left.\begin{array}{l}\bar{x}(\kappa T+T)=\tilde{A}\bar{x}(kT)+\tilde{B}u(kT)\\ y(kT)=\tilde{C}\bar{x}(kT)+\tilde{D}u(kT)\end{array}\right\}$$

where

$$\tilde{A}=[I-AT]^{-1},\quad \tilde{B}=[I-AT]^{-1}BT$$

$$\tilde{C}=C[I-AT]^{-1},\quad \tilde{D}=C[I-AT]^{-1}BT+D$$

Using MATLAB, we develop the function **eulerback.m**, which has as an input the matrices a, b, c, d of the continuous system and the desired sampling period T, which calculates the corresponding matrices ad, bd, cd, dd of the equivalent discrete system according to the backward difference method.

```
function [ad,bd,cd,dd]=eulerback(a,b,c,d,T);
[s1 s2]=size(a);
I=eye(s1);
ad=inv(I-a*T);
bd=inv(I-a*T)*b*T;
cd=c*inv(I-a*T);
dd=c*inv(I-a*T)*b*T+d;
```

Thereby, the step responses of the analog and discretized system are designed and we can observe that they are almost identical.

```
a=[0 1;0 0];
b=[0;1];
c=[1 0];
d=[0];
T=0.1;
[ad,bd,cd,dd]=eulerback(a,b,c,d,T)
sys=ss(a,b,c,d);
sysd=ss(ad,bd,cd,dd,T);
step(sys,'b',sysd,'r')
```

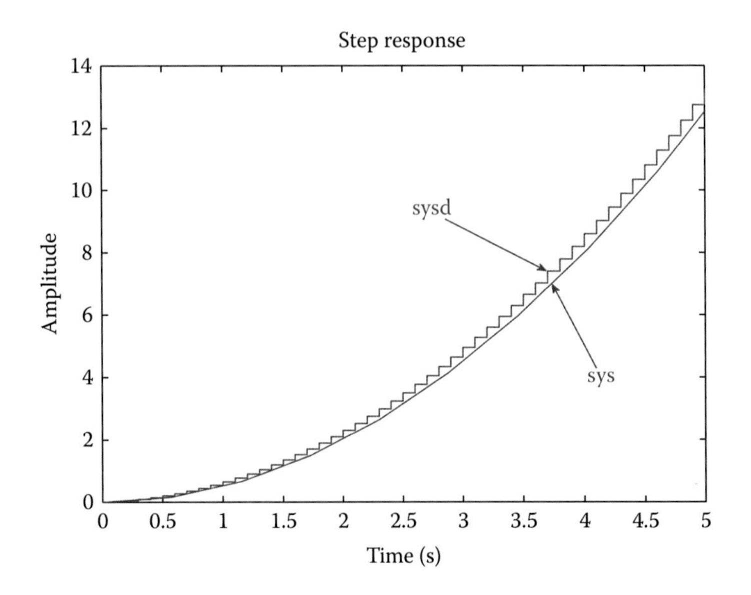

c. Utilizing the trapezoidal method, we assume that

$$\int_{kT}^{kT+T} x(t) = \frac{T}{2}[x(kT) + x(kT+T)] \quad \text{and} \quad \int_{kT}^{kT+T} u(t) = \frac{T}{2}[u(kT) + u(kT+T)]$$

The resultant discrete system, derived from the corresponding continuous one, is given by

$$\left.\begin{aligned} \bar{x}(kT+T) &= \tilde{A}\bar{x}(kT) + \tilde{B}u(kT) + \tilde{\tilde{B}}u(kT-T) \\ y(kT) &= \tilde{C}\bar{x}(kT) + \tilde{D}u(kT) + \tilde{\tilde{D}}u(kT-T) \end{aligned}\right\}$$

where

$$\tilde{A}=\left[I-\frac{AT}{2}\right]^{-1}\left[I+\frac{AT}{2}\right],$$

$$\tilde{B}=\left[I-\frac{AT}{2}\right]^{-1}\frac{BT}{2},$$

$$\tilde{\tilde{B}}=\left[I-\frac{AT}{2}\right]^{-1}\frac{BT}{2}$$

$$\tilde{C}=C\left[I-\frac{AT}{2}\right]^{-1}\left[I+\frac{AT}{2}\right],$$

$$\tilde{D}=C\left[I-\frac{AT}{2}\right]^{-1}\frac{BT}{2}+D,$$

$$\tilde{\tilde{D}}=C\left[I-\frac{AT}{2}\right]^{-1}\frac{BT}{2}$$

Using MATLAB, we develop the function **trap.m**, which has as an input the matrices a, b, c, d of the continuous system and the desired sampling period T, which calculates the corresponding matrices ad, bd, cd, dd of the equivalent discrete system according to the trapezoidal method.

```
function [ad,bd,cd,dd]=trap(a,b,c,d,T)
[s1 s2]=size(a);
I=eye(s1);
ad=(I+a*T/2)*inv(I-a*T/2);
bd=inv(I-a*T/2)*b*sqrt(T);
cd=sqrt(T)*c*inv(I-a*T/2);
dd=d+c*(inv(I-a*T/2)*b*T/2;
```

Thereby, the step responses of the analog and discretized system are designed and we can observe that they are almost identical.

```
a=[0 1;0 0];
b=[0;1];
c=[1 0];
d=[0];
T=0.1;
[ad,bd1,bd2,cd,dd1,dd2]=trap2(a,b,c,d,T)
sys=ss(a,b,c,d);
sysd=ss(ad,bd,cd,dd,T);
step(sys,'b',sysd,'r')
```

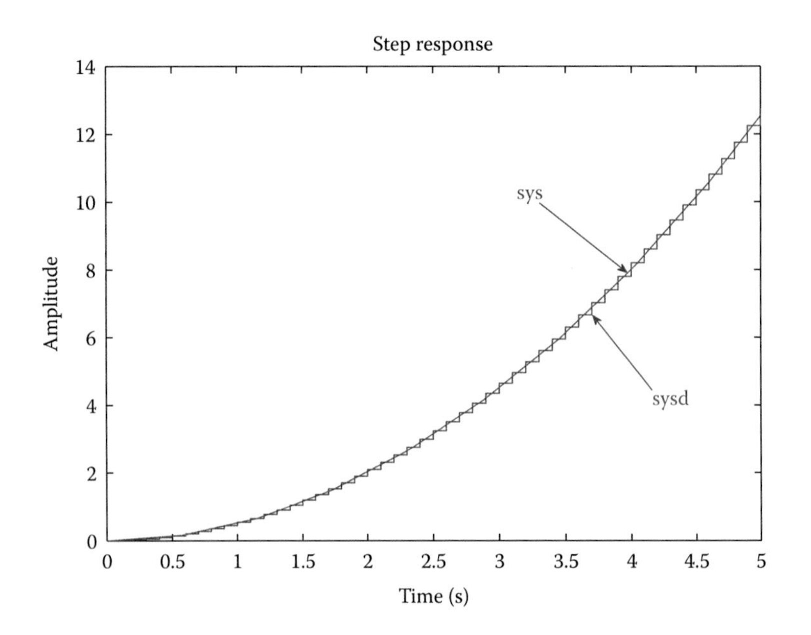

EXERCISE 5.21

Consider the system $\dot{x}(t)=\begin{bmatrix}0 & 1\\ -6 & 5\end{bmatrix}x(t)+\begin{bmatrix}0\\1\end{bmatrix}u(t)$

$$y(t)=\begin{bmatrix}1 & 0\\ 0 & 1\end{bmatrix}x(t)+\begin{bmatrix}0\\0\end{bmatrix}u(t)$$

Discretize the system using (a) ZOH and (b) FOH. Compare the step responses of the analog and discretized system in both cases.

Solution

a. Utilizing the ZOH method and using MATLAB, the step responses of the analog and discretized system are designed. Observe that both responses are almost identical.

```
A=[0 1;-6 5];
B=[0;1];
C=[1 0;0 1];
D=[0;0];
[AD,BD,CD,DD]=c2dm(A,B,C,D,T,'zoh')
 sys=ss(A,B,C,D);
sysd=ss(AD,BD,CD,DD,T);
step(sys,'b',sysd,'r')
```

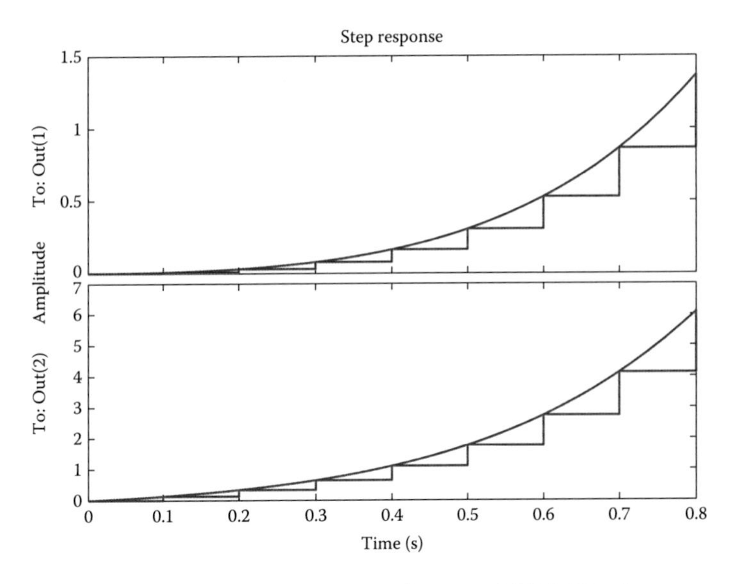

b. Utilizing the FOH method and using MATLAB, the step responses of the analog and discretized system are designed. Observe that both responses are almost identical.

```
A=[0 1;-6 5];
B=[0;1];
C=[1 0;0 1];
D=[0;0];
[AD,BD,CD,DD]=c2dm(A,B,C,D,T,'foh')
 sys=ss(A,B,C,D);
sysd=ss(AD,BD,CD,DD,T);
 step(sys,'b',sysd,'r')
```

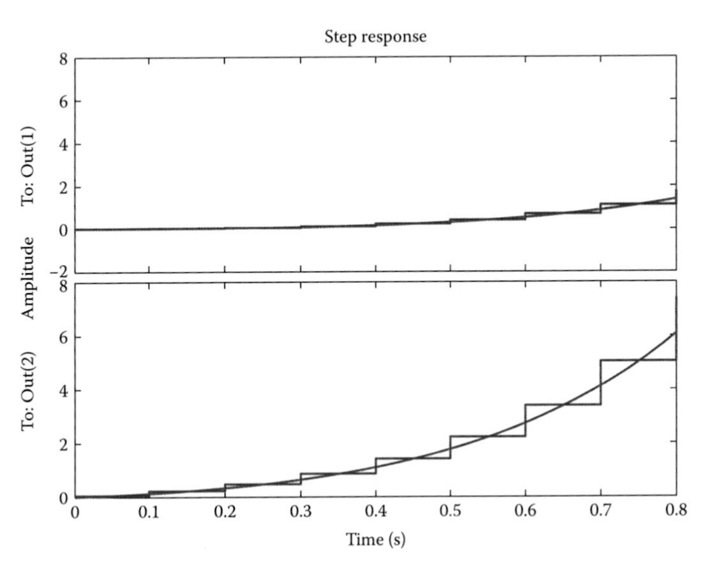

EXERCISE 5.22

Consider the loop transfer function $G(s) = 2s/((s+1)^2(s+2))$. (a) Discretize it using thesample equivalence method (or sampled inverse Laplace transform) and evaluate $G(z)$ in the state space in canonical form and (b) Verify the results using the simulation program Comprehensive Control—CC.

Solution

a. To calculate the discretized system with the aid of sample equivalence method, it suffices to compute the inverse Laplace transform of $G(s)$, substitute $t = kT$, and after using relevant tabulated values, to compute the discretized transfer function.

$$L^{-1}[G(s)] = L^{-1}\left[\frac{2s}{(s+1)^2(s+2)}\right] = e^{-t}(4-2t) - 4e^{-2t} \tag{5.22.1}$$

$$\begin{aligned} g(kT) &= e^{-kT}(4-2kT) - 4e^{-2kT} \Rightarrow \\ g(kT) &= 4e^{-kT} - 2kTe^{-kT} - 4e^{-2kT} \end{aligned} \tag{5.22.2}$$

$$\begin{aligned} gd(z) &= 4\frac{z}{\left(z-e^{-T}\right)} - 2\frac{Tze^{-T}}{\left(z-e^{-T}\right)^2} - 4\frac{z}{\left(z-e^{-2T}\right)} \xrightarrow{T=1} \\ gd(z) &= \frac{0.194z(z-1.248)}{(z-0.368)^2(z-0.135)} \end{aligned} \tag{5.22.3}$$

The transfer function of the digital system, without feedback, can be written in fractionized form as

$$\begin{aligned} F_5(z) &= \frac{0.194z(z-1.248)}{(z-0.368)^2(z-0.135)} \\ &= \frac{-0.272}{(z-0.368)^2} + \frac{0.733}{(z-0.368)} + \frac{-0.537}{(z-0.135)} \end{aligned} \tag{5.22.4}$$

From the above, the following block diagram in canonical form arises

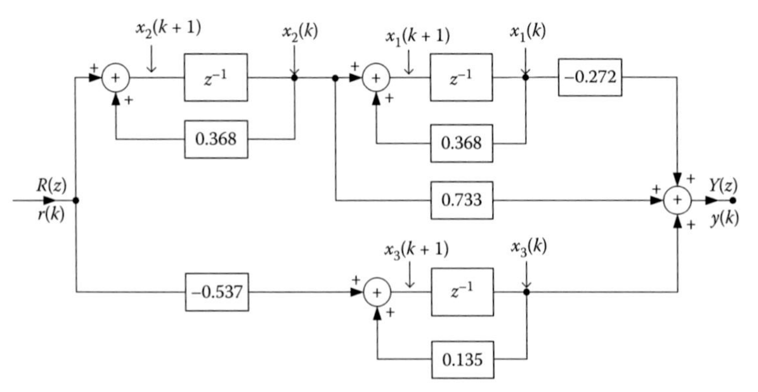

Block diagram of canonical form

From the above block diagram, we have the subsequent equations in the state space

$$\begin{aligned}
x_1(k+1) &= x_2(k) + 0.368 * x_1(k) \\
x_2(k+1) &= r(k) + 0.368 * x_2(k) \\
x_3(k+1) &= r(k) + 0.135 * x_3(k) \\
y(k) &= -0.272 * x_1(k) + 0.733 * x_2(k) - 0.537 * x_3(k)
\end{aligned} \quad (5.22.5)$$

In vector form, we get

$$\begin{bmatrix} x_1(k+1) \\ x_2(k+1) \\ x_3(k+1) \end{bmatrix} = \begin{bmatrix} 0.368 & 1 & 0 \\ 0 & 0.368 & 0 \\ 0 & 0 & 0.135 \end{bmatrix} \begin{bmatrix} x_1(k) \\ x_2(k) \\ x_3(k) \end{bmatrix} + \begin{bmatrix} 0 \\ 1 \\ 1 \end{bmatrix} r(k) \quad (5.22.6)$$

$$y(k) = \begin{bmatrix} -0.272 & 0.733 & -0.537 \end{bmatrix} \begin{bmatrix} x_1(k) \\ x_2(k) \\ x_3(k) \end{bmatrix} + 0 \cdot r(k) \quad (5.22.7)$$

The system state matrices with loop transfer function $gd(z)$ resulted from using fraction expansion in canonical form. We did not use the formulas of the canonical form, controllable canonical form, observable canonical form, or diagonal canonical form. Nevertheless, the resulting expressions will always be the same with respect to the system stability, controllability, and observability, regardless of the applied method.

The above four matrices are defined, respectively, as A, B, C, D, which will be used to describe the system hereinafter, that is, $P = [A, B, C, D]$.

Evaluation of the closed system stability using the matrix A (for unit-feedback): We take the system characteristic equation and if its roots are within the unit circle of complex z-plane, then the system is stable.

$$C.E.: \; p(z) = 0 \Rightarrow |\lambda * I - A| = 0 \Rightarrow$$

$$\begin{vmatrix} \lambda - 0.368 & -1 & 0 \\ 0 & \lambda - 0.368 & 0 \\ 0 & 0 & \lambda - 0.135 \end{vmatrix} = 0 \Rightarrow$$

$$(\lambda - 0.368)^2 * (\lambda - 0.135) =$$

$$\left. \begin{matrix} \lambda_{1,2} = 0.135 \Rightarrow |\lambda_{1,2}| < 1 \\ \lambda_3 = 0.368 \Rightarrow |\lambda_3| < 1 \end{matrix} \right\} \Rightarrow \text{STABLE SYSTEM}$$

Evaluation of system controllability
It holds that

$$AB=\begin{bmatrix}1\\0.368\\0.135\end{bmatrix},\quad A^2B=\begin{bmatrix}0.736\\0.135\\0.018\end{bmatrix} \tag{5.22.8}$$

Thus, the controllability matrix of the state vector is

$$S=\begin{bmatrix}B & AB & A^2B\end{bmatrix}\Rightarrow S=\begin{bmatrix}0 & 1 & 0.736\\1 & 0.368 & 0.135\\1 & 0.135 & 0.018\end{bmatrix} \tag{5.22.9}$$

The determinant of matrix S equals to −0.054, that is, nonzero, hence rank(S) = 3; so, the system is controllable.
Similarly, the observability matrix of the state vector is

$$R^T=\begin{bmatrix}-0.272 & -0.099 & -0.037\\0.733 & -1.138*10^{-4} & -0.099\\-0.537 & -0.073 & -0.01\end{bmatrix}\Rightarrow\left|R^T\right|=-0.002\neq 0 \tag{5.22.10}$$

The determinant of matrix R^T is nonzero, hence rank(R^T) = 3; so, the system is observable.
Transition from state matrices to the system transfer function: Based on $G(z)=C[z-IA]^{-1}B+D$, we obtain the transfer function of the discretized system as

$$G(z)=\frac{0.193718z^2-0.241970z-1.410^{-5}}{(z-0.368)^2(z-0.135)}\cong gd(z) \tag{5.22.11}$$

The deviations between $G(z)$ and $gd(z)$ are due to the rounding of the involved mathematical calculations.

b. *Verification of the results using the simulator COMPREHENSIVE CONTROL—CC*
Insert the transfer function into the program Comprehensive Control—CC, such that

```
CC>G=enter(1,1,2,0, 2,1,1,2,2,1,2,1)
```

Using the command *Convert*, the transfer function $G(s)$ is being discretized. We select the Sample Equivalence method for sampling period T = 1 s. We write

```
CC>gd=convert(G,7,1)
```

The loop transfer function in z-domain arises

$$gd(z) = \frac{0.1944z(z-1.248)}{(z-0.3679)^2(z-0.1353)} \tag{5.22.12}$$

We provide the matrices A, B, C, D to the program as

```
CC>A=(0.368,1,0;0,0.368,0;0,0,0.135)
CC>B=(0;1;1)
CC>C=(-0.272,0.733,-0.537)
CC>D=0
```

We develop the system in the state space using the command **pack** as

```
CC>p=pack(A,B,C,D)
CC>p
 p.a =
   0,3680000           1           0
           0   0,3680000           0
           0           0   0,1350000
 p.b =
           0
           1
           1
 p.c =
  -0,2720000   0,7330000  -0,5370000
 p.d = 0
```

Compute the system poles

```
CC>poles(p)
 ans =
   0,1350000
   0,3680000
   0,3680000
```

Compute the controllability matrix

```
CC>y=conmat(p)
CC>y
 y =
           0           1   0,7360000
           1   0,3680000   0,1354240
           1   0,1350000   0,0182250
```

Compute the order of controllability matrix

```
CC>rank(y)
```

ans = 3 ⇒ Controllable system.

Compute the observability matrix

```
CC>y1=obsmat(p)
CC>y1
 y1 =
```

```
  -0,2720000        0,7330000        -0,5370000
  -0,1000960       -2,256000e-003 -0,0724950
  -0,0368353       -0,1009262        -9,786825e-003
CC>rank(y1)
```

ans = 3 ⇒ Observable system.

Derive the transfer function, by using the above matrices via the command **faddeeva** as:

```
CC>g=faddeeva(p)
CC>g
               0,196z^2-0,2455z+0,0004128
 g(z)  =  ─────────────────────────────────
              z^3-0,871z^2+0,2348z-0,01828
```

We note that the simulation results verify the corresponding theoretical ones. Hence, it is confirmed that the Comprehensive Control program is an efficient simulator, which assists us in deriving our results and useful outcomes accurately and quite fast.

6

Stability of Digital Control Systems

6.1 Stability

The stability is a structural systemic property directly related to the type of system response. The response may be bounded or asymptotically tend to zero. Otherwise, the system response would take emphatically high values, which would remove the system by its modeling limits or cause damage to the system itself.

A system is stable if, for finite input, the output is also finite. This fundamental principle is known as *bounded input–bounded output (BIBO) stability criterion.*

The output of a stable system is within acceptable limits while the corresponding output of an unstable system theoretically tends to infinity.

The stability of a discrete control system is directly connected with *the positions of roots* of the characteristic equation (poles) of the transfer function.

- When the poles are inside the unit circle ($|z| = 1$), then the response of the various disturbance signals appear decreasing.
- When there are poles on the circumference of the unit circle or outside it, then the response with respect to a disturbance input appears stable or increasing.

A linear time invariant discrete system is *stable* if the poles of the closed-loop system are *inside the unit circle* (i.e., they have real parts between -1 and 1), while it is *unstable* if at least one pole is located outside the unit circle.

If the system characteristic equation has roots in the circumference of the unit cycle with all other roots being located inside, then the steady-state output will operate unabated oscillations of finite amplitude when its input is a finite function. Such behavior makes the system *marginally stable.* All the above are illustrated in Figures 6.1 and 6.2.

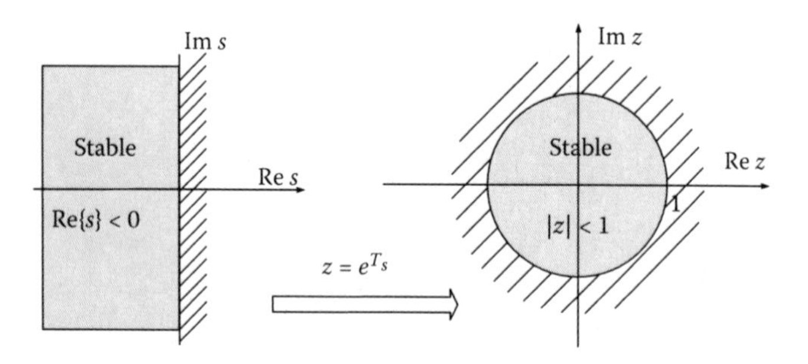

FIGURE 6.1
Stability of analog and discrete system.

The most *prevalent techniques* for determining the stability of a discrete-time system are

- Unit-circle criterion
- Routh criterion using the bilinear mobius transformation
- Jury criterion
- Root locus method
- Nyquist stability criterion
- Bode stability criterion

The stability analysis of digital control systems is similar to the stability analysis of analog systems and all the known methods can be applied to digital control systems with some modifications.

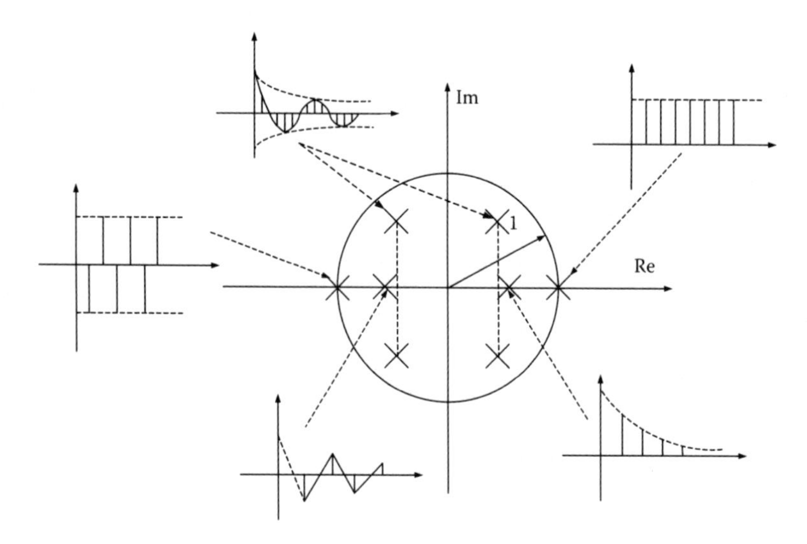

FIGURE 6.2
Time responses as a function of the poles location on the unit circle.

6.2 Unit-Circle Criterion

The key relation between analog and digital domain is $z = e^{sT}$
or $z = e^{(\sigma + j\omega)T} = e^{\sigma T} \cdot e^{j\omega T} = e^{\sigma T} \cdot e^{j\omega T} \cdot e^{\pm j2k\pi} \Rightarrow$

$$z = e^{\sigma T} \times e^{j(\omega T \pm j2k\pi)} \tag{6.1}$$

Hence, the variable z is a vector of length $e^{\sigma T}$ and phase $\omega T \pm j2k\pi$. As we know from the analog control system theory, a system is stable when its poles are located on the left half complex plane s, that is, when $\mathrm{Re}\{s\} = \sigma < 0$ with marginal stability $\sigma = 0$. By moving into the z-domain, observe that for $\sigma = 0 \Rightarrow |z| = |e^{0T}| = 1$ and $\sigma \to -\infty$, $|z| = |e^{-\infty T}| \to 0$.

Thereby, the key relation between stability and poles location is transferred from the left half-plane s into the unit circle with center being the intersection of complex z-plane axes (Figure 6.3).

6.3 Routh Criterion Using the Bilinear Mobius Transformation

Routh stability criterion (Routh criterion—1875) is a method of determining whether any polynomial has all its roots in the left complex half-plane.

Möbius transform (Möbius transformation—in honor of August Ferdinand Möbius):

$$w = \frac{z+1}{z-1} \Rightarrow z = \frac{w+1}{w-1} \tag{6.2}$$

illustrates the unit circle of the z-plane in the left-half w-plane.

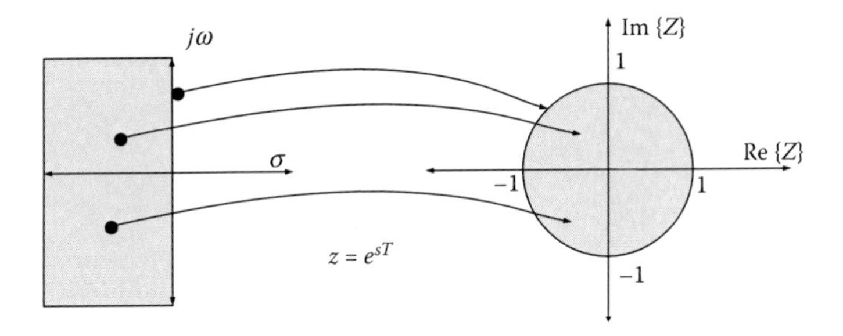

FIGURE 6.3
Relation between the left s half-plane with the interior of unit circle in z-plane.

Proof: Assume that

$$w = \alpha + j\beta, \quad z = x + jy \Rightarrow$$

$$w = \alpha + j\beta = \frac{z+1}{z-1} = \frac{x+jy+1}{x+jy-1}\cdot\left(\frac{x-1-jy}{x-1-jy}\right)$$

$$= \frac{x^2+y^2-1}{(x-1)^2+y^2} - j\frac{2y}{(x-1)^2+y^2}$$

$$\alpha \le 0 \Rightarrow x^2+y^2-1 \le 0 \Rightarrow x^2+y^2 \le 1 \quad (6.3)$$

From the expression (6.3), we conclude that the left w half-plane is illustrated within the unit circle of z-plane.

Consider the characteristic equation

$$P(z) = a_0 z^n + a_1 z^{n-1} + \cdots + a_{n-1} z + a_n = 0 \quad (6.4)$$

With the aid of the bilinear transform, the characteristic equation becomes

$$a_0\left(\frac{1+w}{1-w}\right)^n + a_1\left(\frac{1+w}{1-w}\right)^{n-1} + \cdots + a_{n-1}\left(\frac{1+w}{1-w}\right) + a_n = 0 \quad (6.5)$$

or

$$Q(w) = b_0 w^n + b_1 w^{n-1} + \cdots b_{n-1} z + b_n = 0 \quad (6.6)$$

Hence, we transform $P(z) = 0$ into $Q(w) = 0$ and study the stability of the discrete control system using the Routh criterion similar to the continuous-time control systems. The Möbious transform is illustrated in Figure 6.4.

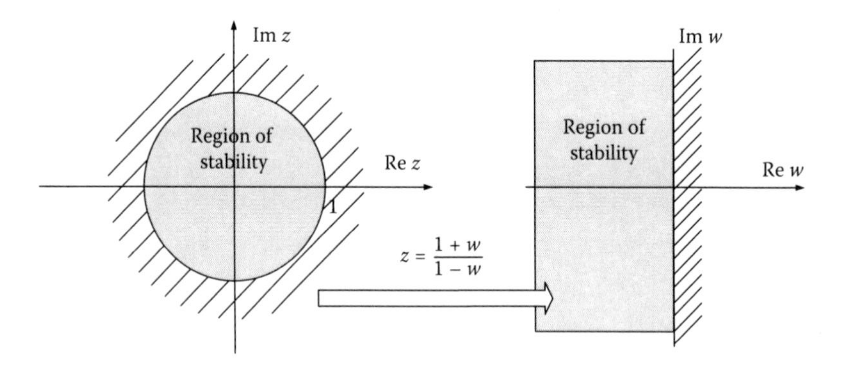

FIGURE 6.4
z- and w-domains via the Möbious bilinear transform.

6.4 Jury Criterion (Eliahu I. Jury 1955)

Let the characteristic equation of a sampled data system be

$$\alpha(z) = \alpha_0 z^n + \alpha_1 z^{n-1} + \cdots + \alpha_{n-1} z + \alpha_n = 0 \tag{6.7}$$

We form the *Jury table*

a_0	a_1	a_2	a_{n-1}	a_n	$B_n = \dfrac{a_n}{a_0}$
a_n	a_{n-1}		a_1	a_0	
a_0^{n-1}		a_1^{n-1}	a_{n-1}^{n-1}		$B_{n-1} = \dfrac{a_{n-1}^{n-1}}{a_0^{n-1}}$
a_{n-1}^{n-1}		a_{n-2}^{n-1}	a_0^{n-1}		
–					$B_{n-1} = \dfrac{a_{n-1}^{n-1}}{a_0^{n-1}}$
–					
a_0^0					

where

$$\alpha_i^{k-1} = \alpha_i^k - \beta_k \alpha_{k-i}^k \quad \text{and} \quad B_k = \frac{\alpha_k^k}{\alpha_0^k} \tag{6.8}$$

The third row of the above table is obtained by multiplying the second row with $B_n = (\alpha_n/\alpha_0)$ and subtracting the result from the first row. Thus, the last element of the third row becomes zero.

The latter process is being repeated until the $2n + 1$ row is reached, which includes a single term.

Jury stability criterion: If $\alpha_0 > 0$ then the polynomial $\alpha(z)$ has all its roots inside the unit circle when all $\alpha_0^k k = 0,1,...,n-1$ are positive.

NOTE: If all $\alpha_0^k, k = 1,2\ldots,$ are positive, it can be shown that the condition $\alpha_0 > 0$ is equivalent to the following two conditions:

$$\begin{aligned} &a(1) > 0 \\ &(-1)^n a(-1) > 0 \end{aligned} \tag{6.9}$$

The latter conditions represent necessary stability conditions; therefore they can be used prior to the formation of the Jury table.

6.5 Root Locus Method

The *root locus method* (root locus analysis—Evans 1948) is a graphical method that serves as a means of formulating the locus in the complex plane, whereupon the roots of the system characteristic equation are driven as changing the parameter of a certain value.

Since the transfer function of a digital control system is the ratio of two polynomials of z, the same rules can be applied as for the formulation of the root locus in s-domain.

When formulating the root locus, we can find the absolute and relative stability of the system. The absolute stability requires that all the roots of the denominator of the transfer function (poles) are within the unit circle.

The relative stability is determined by the location of roots within the circle relative to locus of fixed attenuation, fixed frequency, and fixed J.

Consider the loop transfer function

$$G(z)H(z) = K\frac{P(z)}{Q(z)} \tag{6.10}$$

where $P(z)$ and $Q(z)$ are polynomials of the complex variable z.

The roots of the characteristic equation are the poles of the closed system and are obtained by

$$Q(z) + KP(z) = 0 \tag{6.11}$$

where K denotes the system variable.

The position of poles of the transfer function in complex z-plane affects the transient response of the system and determines its stability. From the expression (6.11), observe that any change in the value of constant K results to a location shift for the poles in complex plane.

The *root locus diagram* is a graphical illustration of the location of poles of the closed system in z-plane, as one its parameters changes, say K. In the root locus diagram, we receive satisfactory information for the stability and overall behavior of a system.

6.5.1 Rules for Approximate Establishment of Root Locus

Some rules that apply on the approximate establishment of the root locus for the characteristic equation of a discrete control system.

RULE 1: The poles of the loop transfer function (let them be in a form of $G(z)H(z)$) are the *points of departure* of root locus.

RULE 2: The zeros of loop transfer function and the infinity when $m < n$ are the *points of arrival* of root locus.

RULE 3: The number of independent branches of the locus equals to max(n, m) where m and n are the number of zeros and poles of the loop transfer function, respectively.

RULE 4: The root locus presents *symmetry* with respect to the real axis (horizontal axis).

RULE 5: The intersection of asymptotic lines with the horizontal axis is given by

$$\sigma_\alpha = \frac{\sum_{i=1}^{n}(p_i) - \sum_{j=1}^{m}(z_j)}{n-m} \tag{6.12}$$

where

$\sum_{i=1}^{n}(p_i) =$ the algebraic sum of the values of poles of the transfer function.

$\sum_{j=1}^{m}(z_j) =$ the algebraic sum of the values of zeros of the loop transfer function.

RULE 6: For high z-values, the root locus asymptotically approaches the straight lines forming angles with the horizontal axis

$$\sphericalangle \varphi_\alpha = \frac{(2\rho+1)\pi}{n-m}, \quad \begin{matrix} \rho = 0,1,\ldots,|n-m|-1 \\ K \geq 0 \end{matrix} \tag{6.13}$$

RULE 7: A part of the real axis can be a part of root locus if, for $K \geq 0$, the number of poles and zeros that are located to the right of this part is *odd*.

RULE 8: The separation and arrival points of the branches from and to the horizontal axis are called *break away points* of root locus and are expressed as

$$(6.11) \Rightarrow K = -\frac{Q(z)}{P(z)} \tag{6.14}$$

Each root of the equation $(dK/dz) = 0$ is an *accepted break away point* if it satisfies the condition $1 + G(z)H(z) = 0$ or $\boxed{|G(z)H(z)| = 1}$ for some real value of K.

RULE 9: The intersection points of the root locus and the circumference of the unit circle (where $|z| = 1$) are the points where the system goes from stable to instable and are calculated from the algebraic Jury stability criterion or from the algebraic Ruth stability criterion using the Möbius transform.

RULE 10: The departure angles of root locus for a complex pole or the arrival angles for a complex zero are calculated as

$$\sphericalangle \varphi_d = (2\rho+1)\pi - \left(\sum_{i=1}^{n} \varphi_{p_i} - \sum_{j=1}^{m} \varphi_{z_j} \right) \tag{6.15}$$

where

$\sum_{i=1}^{n} \varphi_{p_i}$ = the algebraic sum of the angles of poles with respect to the reference complex pole (or zero).

$\sum_{j=1}^{m} \varphi_{z_j}$ = the algebraic sum of the angles of zeros with respect to the reference complex pole (or zero).

6.6 Nyquist Stability Criterion

The *Nyquist stability criterion* (Nyquist stability criterion—1932) is based on the graphical representation of the open-loop transfer function for a particular closed path in the complex frequency domain and provides information not only on the stability of the closed systems but for their *relative stability* as well. The special closed road is called *Nyquist path* or Nyquist plot and includes the right complex half plane. In Figure 5.7, Nyquist path Γ_C is presented.

The Nyquist stability criterion studies the stability of the closed-loop system, when the open-loop transfer function is considered as known. To apply the Nyquist stability criterion in discrete-time systems, it suffices to set $z = e^{j\omega T}$ in the open-loop transfer function and to design the polar diagram with the circular frequency ω as a parameter.

Consider the discrete system of Figure 6.5.

Its transfer function is presented as

$$G_{cl}(z) = \frac{Y(z)}{U_c(z)} = \frac{H(z)}{1+H(z)} \tag{6.16}$$

The characteristic polynomial is given by

$$1+H(z) \tag{6.17}$$

It holds that:

$$N = Z - P \tag{6.18}$$

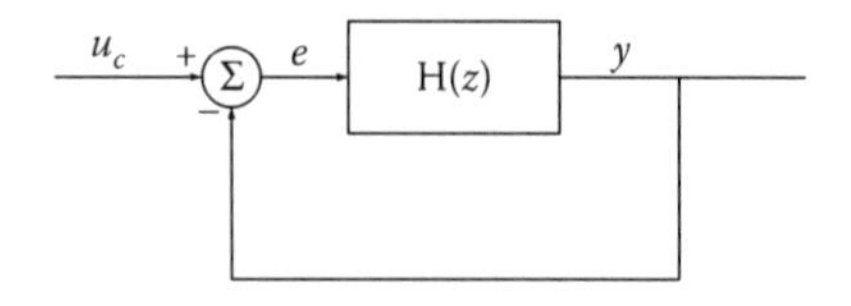

FIGURE 6.5
Closed discrete system.

where

Z = the roots of the characteristic equation $1 + H(z) = 0$, except the unit circle.

N = the number of encirclements of the point $-1 + j0$ clockwise to $H(z)$.

P = the number of poles of $H(z)$ outside the unit circle.

Based on Nyquist criterion, to preserve the stability of the closed system, then $P = 0$ should hold, thus

$$N = Z \tag{6.19}$$

Figure 6.6 illustrates a typical Nyquist path.

Nyquist stability criterion: If the open-loop system is stable, then the stability of the closed-loop system is determined by the case when the point $-1 + j0$ is surrounded by the Nyquist diagram of $H(e^{j\omega T})$ for ωT, from 0 to π.

The gain margin k_g is defined as the quantity arising from the expression (6.20) and it is the inverse value of gain $|H(e^{j\omega T})|$ into the frequency for which the phase angle tends to –180°:

$$k_g = \frac{1}{H(e^{j\omega_C T})} \tag{6.20}$$

where ω_C = the critical frequency where the Nyquist diagram of $H(e^{j\omega T})$ intersects the axis Re{GH}, that is

$$\arg(H(e^{j\omega_C T}) = -\pi \tag{6.21}$$

A closed system is stable if $k_g > 0$.

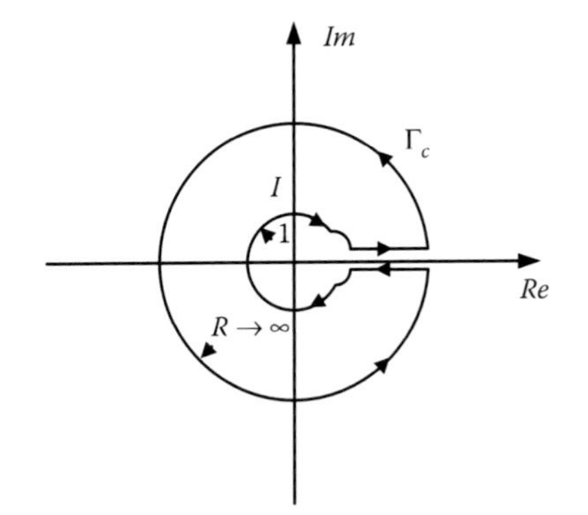

FIGURE 6.6
Nyquist path.

The gain margin is the amount of gain increase or decrease required to make the loop gain unity at the frequency where the phase angle is –180°.

The *phase margin* $\varphi_{m\arg}$ is the quantity arising from the expression (6.22) and it is the angle at which the diagram of $H(e^{j\omega T})$ should be rotated so as the point of $|H(e^{j\omega T})| = 1$ pass through the point $-1 + j0$ of the coordinates plane of $H(e^{j\omega T})$:

$$\varphi_{m\arg} = \pi + \arg(H(e^{j\omega_C T})) \tag{6.22}$$

where ω_C is the frequency where the amplitude $|H(e^{j\omega_C T})|$ equals to unity.

This stability measure is practically equal to the added phase delay, which is required before the system is turned to unstable.

A closed system is stable if $\varphi_{m\arg} > 0$.

6.7 Bode Stability Criterion (H.W. Bode—1930)

Using the bilinear transform $z = (1 + (T/2)w)/(1 - (T/2)w)$ the internal of the unit circle of the complex z-plane is depicted to the left w half-plane. In this way, we study the stability of a digital control system in w-domain, by utilizing the same methods applied to analog systems. Consider a control system with the loop transfer function:

$$G(z) = \frac{b_0 z^m + b_1 z^{m-1} + \cdots b_m}{z^n + a_1 z^{n-1} + \cdots + a_n}, \quad m \le n \tag{6.23}$$

Let the transform:

$$z = \frac{1 + (T/2)w}{1 - (T/2)w} \tag{6.24}$$

$$(6.23),(6.24) \Rightarrow G(w) = \frac{b_0 w^m + b_1 w^{m-1} + \cdots b_m}{w^n + a_1 w^{n-1} + \cdots + a_n} \tag{6.25}$$

Based on $G(w)$ and setting $w = jv$ (where v is the system angular frequency), the Bode diagram can be designed for $G(jv)$, while some useful outcomes

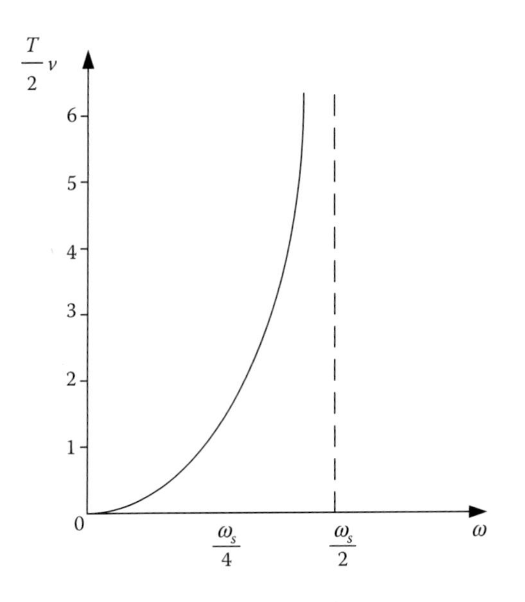

FIGURE 6.7
Relation between the analog and digital frequency.

regarding the stability of the closed-loop system can be extracted. Figure 6.7 illustrates the relation between w and $(T/2)v$.

6.8 Formula Table

The formula Tables 6.1 through 6.3 are discussed here.
where

$$b_k = \begin{vmatrix} a_0 & a_{n-k} \\ a_n & a_k \end{vmatrix}, \quad k = 0,1,2,\ldots,n-1$$

$$c_k = \begin{vmatrix} b_0 & b_{n-1-k} \\ b_{n-1} & b_k \end{vmatrix}, \quad k = 0,1,2,\ldots,n-2$$

$$d_k = \begin{vmatrix} c_0 & c_{n-2-k} \\ c_{n-2} & c_k \end{vmatrix}, \quad k = 0,1,2,\ldots,n-3$$

$$q_0 = \begin{vmatrix} p_0 & p_3 \\ p_3 & p_0 \end{vmatrix}, \quad q_2 = \begin{vmatrix} p_0 & p_1 \\ p_3 & p_2 \end{vmatrix}$$

TABLE 6.1

Algebraic Stability Criteria

1. Routh/Möbius $w = \dfrac{z+1}{z-1} \Rightarrow z = \dfrac{w+1}{w-1}$

2. Jury Table

Row	z^0	z^1	z^2	z^3	…	z^{n-1}	z^{n-1}	z^n
1	a_0	a_1	a_2	a_3	…	a_{n-2}	a_{n-1}	a_n
2	a_n	a_{n-1}	a_{n-2}	a_2	…	a_1	a_0	
3	b_0	b_1	b_2	b_3	…	b_{n-2}	b_{n-1}	b_n
4	b_n	b_{n-1}	b_{n-2}	b_{n-3}	…	b_2	b_1	b_0
$2n-5$	p_0	p_1	p_2	p_3	…			
$2n-4$	p_3	p_2	p_1	p_0	…			
$2n-3$	q_0	q_1	q_2					

TABLE 6.2

Steps of Root Locus Approximate Establishment for the Characteristic Equation of a Discrete System

α/α	Formulas	Comments
1	$G(z)H(z)$	Open-loop transfer function
2	$\lvert K\rvert \dfrac{\prod_{i=1}^{m}\lvert z+z_i\rvert}{\prod_{j=1}^{n}\lvert z+p_j\rvert}=1$ $-\infty < K < \infty$	Measure condition of the root locus points
3	$\sum_{i=1}^{m}\sphericalangle(z+z_i)-\sum_{j=1}^{n}\sphericalangle(z+p_j)=\begin{cases}(2\rho+1)\pi, & K>0\\ 2\rho\pi, & K<0\end{cases}$ $\rho=\pm1,\pm2,\ldots$	Phase condition of the root locus points
4	$l=\max(m,n)$	Number of branches of root locus
5	$\sphericalangle\varphi_\alpha=\dfrac{(2\rho+1)\pi}{n-m},\quad \begin{cases}\rho=0,1,\ldots,\lvert n-m\rvert-1\\ K\geq 0\end{cases}$	Angles between the asymptotic lines with the real axis for $K\geq 0$
6	$\sigma_\alpha=\dfrac{\sum_{i=1}^{n}p_i-\sum_{j=1}^{m}z_j}{n-m}$	Intersection point of the asymptotic lines and the real axis
7.	$\alpha\begin{cases}\dfrac{dK}{dz}=0\Rightarrow z_{b_i}\\ 1+G(z_b)H(z_b)=0 \text{ and } K\in R\end{cases}$	Finding the breaking points z_b (1st way)
8	$\sphericalangle\varphi_d=(2\rho+1)\pi-\left(\sum_{i=1}^{n}\varphi_{p_i}-\sum_{j=1}^{m}\varphi_{z_j}\right)$	Departure angles of root locus from complex poles or arrival angles in complex zeros

TABLE 6.3

Nyquist Criterion

Open-Loop Transfer Function $H(z)$	Closed-Loop Transfer Function $\frac{H(z)}{1+H(z)}$

- *It should hold*: $N = -P$ (Nyquist stability criterion)
  where
  N = the number of surroundings for the point $-1 + j0$ clockwise to $G(j\omega)H(j\omega)$.
  P = the number of poles of $G(s)H(s)$ outside the unit circle.
- *Gain margin*

$$k_g = \frac{1}{H(e^{j\omega_C T})}$$

ω_C = the critical frequency where the Nyquist diagram of $H(e^{j\omega T})$ intersects the axis Re$\{GH\}$, that is $\arg(H(e^{j\omega_C T}) = -\pi$

- *Phase margin*

$$\varphi_{marg} = \pi + \arg(H(e^{j\omega_C T})$$

Ω_C = the frequency where the amplitude $|H(e^{j\omega_C T})|$ equals to unity.

It should hold: $\begin{cases} k_g > 0 \\ \varphi_{marg} > 0 \end{cases}$ to preserve stability

Jury Stability Conditions

$$F(1) > 0$$

$$(-1)^n F(-1) > 0$$

$$|a_0| < |a_n|$$

$$|b_0| > |b_{n-1}|$$

$$|c_0| > |c_{n-2}|$$

$$\vdots$$

$$|r_0| > |r_2|$$

6.9 Solved Exercises

EXERCISE 6.1

Define the region of values for the parameter K so as the following systems are stable:

S1: $y(nT) - Ky(nT - T) + K^2y(nT - 2T) = x(nT)$
S2: $y(nT) - 2Ky(nT - T) + K^2y(nT - 2T) = x(nT)$

Solution

For the system S1, transform the difference equation in z-domain and calculate its transfer function.

$$Z(y(nT) - Ky(nT-T) + K^2y(nT-2T)) = Z(x(nT))$$
$$\Rightarrow Y(z) - Kz^{-1}Y(z) + K^2z^{-2}Y(z) = X(z)$$
$$H(z) = \frac{Y(z)}{X(z)} = \frac{1}{1 - Kz^{-1} + K^2z^{-2}} \tag{6.1.1}$$

Finding the poles of $H(z)$

$$1 - Kz^{-1} + K^2z^{-2} = 0 \Rightarrow z^2 - Kz + K^2 = 0 \tag{6.1.2}$$

$$\text{Yet: } \quad z^2 - Kz + K^2 = z^2 - Kz + (K/2)^2 + \frac{3}{4}K^2$$
$$\Rightarrow z^2 - Kz + K^2 = \left(\frac{z-K}{2}\right)^2 + \frac{3}{4}K^2 = 0$$
$$\Rightarrow \left(\frac{z-K}{2}\right)^2 = -\frac{3}{4}K^2 \Rightarrow z = \frac{K}{2} \pm j\sqrt{3}\frac{K}{2} \tag{6.1.3}$$

The modulus of z is given by

$$|z| = \sqrt{\left(\frac{K}{2}\right)^2 + \left(\sqrt{3}\frac{K}{2}\right)^2} = \sqrt{K^2} \tag{6.1.4}$$

To obtain stability, $|z| < 1$ should hold. Thus, solving the latter inequality we have that the region of values for the parameter K, to preserve stability, is $-1 < K < 1$.

For the system S2, transform the difference equation in z-domain and calculate its transfer function.

$$Z(y(nT) - 2Ky(nT-T) + K^2y(nT-2T)) = Z(x(nT))$$
$$Y(z) - 2Kz^{-1}Y(z) + K^2z^{-2}Y(z) = X(z)$$

$$H(z) = \frac{1}{1 - 2Kz^{-1} + K^2z^{-2}} \tag{6.1.5}$$

Finding the poles of $H(z)$

$$1 - 2Kz^{-1} + K^2z^{-2} = 0 \Rightarrow (z-K)^2 = 0 \Rightarrow z = K$$

The modulus of z is given by

$$|z| = \sqrt{K^2} \tag{6.1.6}$$

To obtain stability, $|z| < 1$ should hold. Thus, solving the latter inequality we have that the region of values for the parameter K, to preserve stability, is $-1 < K < 1$.

EXERCISE 6.2

The characteristic polynomial of a system is $\alpha(z) = z^2 + 0.7z + 0.1$. Evaluate the system stability. Repeat the procedure if the characteristic polynomial is $z^3 - 1.2z^2 - 1.375z - 0.25 = 0$.

Solution

a. Set $z = (w + 1)/(w - 1)$ into $\alpha(z)$ so as to transfer from z-domain to w-domain, having that

$$\alpha(w) = \left(\frac{w+1}{w-1}\right)^2 + 0.7\frac{w+1}{w-1} + 0.1 = \frac{(w+1)^2 + 0.7(w^2-1) + 0.1(w-1)^2}{(w-1)^2}$$

$$\Rightarrow \alpha(w) = \frac{1.8w^2 + 1.8w + 0.4}{(w-1)^2} \tag{6.2.1}$$

The numerator of $\alpha(w)$ is the characteristic polynomial where we apply the Routh criterion.

Routh table is presented as

W^2	1.8	0.4
W^1	1.8	0
W^0	0.4	

The coefficients of the first column have the same sign, hence the system is stable.

Analyzing $\alpha(z)$ in product terms, we get

$$\alpha(z) = (z + 0.5)(z + 0.2) \tag{6.2.2}$$

Its two roots are $p_1 = -0.5$ and $p_2 = -0.2$, which are placed inside the unit circle and, therefore, the system is stable by using the unit circle criterion.

b. Set $z = (w + 1)/(w - 1)$ into $\alpha(z)$ so as to transfer from z-domain to w-domain. The characteristic equation becomes:

$$-1.875w^3 + 3.875w^2 + 4.875w + 1.125 = 0 \tag{6.2.3}$$

The Routh table is presented as

W^3	−1.875	4.875
W^2	3.875	1.125
W^1	5.419	0
W^0	1125	

From the above table we see that there is a sign change in the first column, so given the three roots of the characteristic equation, one is on the right complex plane thus the closed system is unstable.

EXERCISE 6.3

The characteristic polynomial of a system is $\alpha(z) = z^3 - 1.3z^2 - 0.8z + 1$. Evaluate the system stability using the Jury stability criterion.

Solution

It holds that $\alpha(1) = -0.1$. Since $\alpha(1) < 0$, the necessary condition is not satisfied, hence the system is unstable.

Indeed, the roots of the characteristic equation $\alpha(z) = 0$ are

$$p_1 = -0.8841$$
$$p_2 = 1.3402$$
$$p_1 = 0.8440$$

Due to $p_2 = 1.3402$, which is outside the unit circle, the closed-loop system is unstable.

EXERCISE 6.4

The characteristic polynomial of a system is $\alpha(z) = z^2 + \alpha_1 z + \alpha_2$. Evaluate the system stability using the Jury stability criterion.

Solution

We formulate the Jury table

1	a_1	a_2	$B_2 = a_2$
a_2	a_1	1	
$1-a_2^{\,2}$	$a_1(1-a_2)$		
$a_1(1-a_2)$	$1-a_2^{\,2}$		$B_1 = \dfrac{a_1}{1+a_2}$
$1-a_2^{\,2} - \dfrac{a_1^{\,2}(1-a_2)}{(1+a_2)}$			

All roots of the characteristic polynomial are located inside the unit circle if

$$\begin{aligned}&1-\alpha_2^2>0\\&\frac{1-\alpha_2}{1+\alpha_2}\left((1+\alpha_2)^2-\alpha_1^2\right)>0\end{aligned} \tag{6.4.1}$$

Thereby, it suffices to show that

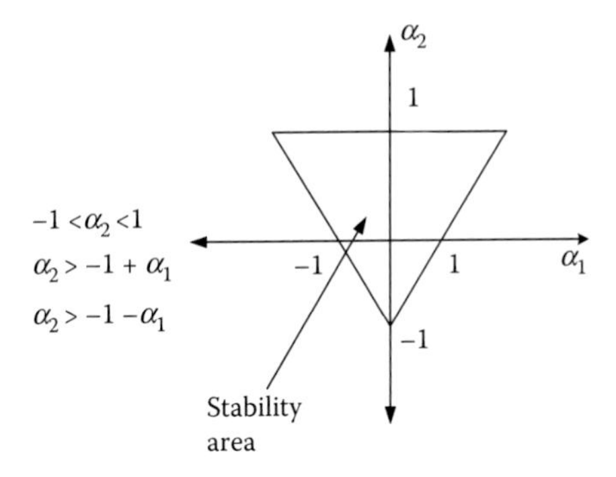

EXERCISE 6.5

The characteristic polynomial of a system is $F(z) = z^5 + 2.6z^4 - 0.56z^3 - 2.05z^2 + 0.0775z + 0.35 = 0$. Evaluate the system stability using the Jury stability criterion.

Solution

We formulate the Jury Table

Row	z^0	z^1	z^2	z^3	z^4	z^5
1	0.35	0.0775	−2.05	−0.56	2.6	1
2	1	2.6	−0.56	−2.05	0.0775	0.35
3	b_0	b_1	b_2	b_3	b_4	
4	b_4	b_3	b_2	b_1	b_0	
5	c_0	c_1	c_2	c_3		
6	c_3	c_2	c_1	c_0		
7	d_0	d_1	d_2			

The third row of the above table is calculated as

$$b_0 = \begin{vmatrix} a_0 & a_5 \\ a_5 & a_0 \end{vmatrix} = \begin{vmatrix} 0.35 & 1 \\ 1 & 0.35 \end{vmatrix} = -0.8775$$

$$b_1 = \begin{vmatrix} a_0 & a_4 \\ a_5 & a_1 \end{vmatrix} = \begin{vmatrix} 0.35 & 2.6 \\ 1 & 0.0775 \end{vmatrix} = -2.5728$$

$$b_2 = \begin{vmatrix} a_0 & a_3 \\ a_5 & a_2 \end{vmatrix} = \begin{vmatrix} 0.35 & -0.56 \\ 1 & -2.05 \end{vmatrix} = -0.1575$$

$$b_3 = \begin{vmatrix} a_0 & a_2 \\ a_5 & a_3 \end{vmatrix} = \begin{vmatrix} 0.35 & -2.05 \\ 1 & -0.56 \end{vmatrix} = 1.854$$

$$b_4 = \begin{vmatrix} a_0 & a_1 \\ a_5 & a_4 \end{vmatrix} = \begin{vmatrix} 0.35 & 0.0775 \\ 1 & 2.6 \end{vmatrix} = 0.8352$$

Similarly, calculate all the other terms of the table, which yields

Row	z^0	z^1	z^2	z^3	z^4	z^5
1	0.35	0.0775	−2.05	−0.56	2.6	1
2	1	2.6	−0.56	−2.05	0.0775	0.35
3	−0.8775	−2.5728	−0.1575	1.854	0.8352	
4	0.8352	1.854	−0.1575	−2.5728	−0.8775	
5	0.077	0.7143	0.2693	0.5151		
6	0.5151	0.2693	0.7143	0.077		
7	−0.2593	−0.0837	−0.3472			

The stability conditions for a fifth-order system are

$$F(1) > 0$$
$$(-1)^5 F(-1) > 0$$
$$|a_0| < a_5$$
$$|b_0| > |b_4|$$
$$|c_0| > |c_3|$$
$$|d_0| > |d_2|$$

$$F(1) = 1 + 2.6 - 0.56 - 2.05 + 0.0775 + 0.3514175$$

Thus, the condition is satisfied.

$$F(-1) = -1 + 2.6 + 0.56 - 2.05 - 0.0775 + 0.35 = 0.3825$$

The condition $(-1)^5 F(-1) > 0$ is not satisfied.
The condition $|a_0| < a_5$ is satisfied.
The condition $|b_0| > |b_4|$ is satisfied.
The condition $|c_0| > |c_3|$ is not satisfied.
The condition $|d_0| > |d_2|$ is satisfied.
The closed-loop system is unstable.

Verification using MATLAB®: The function Jury.m (https://www.mathworks.com/matlabcentral/fileexchange/13904-jury/content/jury.m) estimates the Jury Table and will be used in the following exercise.

```
function [J,C] = jury(coeff)

J = [coeff;flipdim(coeff,2)];
typ = class(coeff);
n = length(coeff)-1;

if strcmp(typ,'sym')
   for i=3:2:(2*n+1)
      try
         alph = J(i-1,1)/J(i-2,1);
      catch
         disp('Your polynomial seems to be critical')
         rethrow(lasterror);
         break;
      end
      newrow_1 = J(i-2,:)-alph*J(i-1,:);
      newrow = simplify(newrow_1);
      J = [J ; newrow ;
         [flipdim(newrow(1:end-(i-1)/2),2),
zeros(1,(i-1)/2)]
                     ];
   end
else
   for i=3:2:(2*n+1)
      try
         alph = J(i-1,1)/J(i-2,1);
      catch
         disp('Your polynomial seems to be critical')
         rethrow(lasterror);
         break;
      end
      newrow = J(i-2,:)-alph*J(i-1,:);
      J = [J ; newrow ;
         [flipdim(newrow(1:end-(i-1)/2),2),
zeros(1,(i-1)/2)]
                     ];
   end
end

J = J(1:end-1,:)
C = J(1:2:end,1)
```

```
% Provide the vector of coefficients for the characteristic
polynomial
coeff=[1 2.6 -0.5 -2.05 0.0775 0.35];
% With the command jury, formulate the Jury Table
jury(coeff)
ans =
     1.0000    2.6000   -0.5000   -2.0500    0.0775    0.3500
     0.3500    0.0775   -2.0500   -0.5000    2.6000    1.0000
     0.8775    2.5729    0.2175   -1.8750   -0.8325         0
    -0.8325   -1.8750    0.2175    2.5729    0.8775         0
     0.0877    0.7940    0.4238    0.5659         0         0
```

```
   0.5659    0.4238    0.7940    0.0877         0         0
  -3.5646   -1.9413   -4.7005         0         0         0
  -4.7005   -1.9413   -3.5646         0         0         0
   2.6337    0.6186         0         0         0         0
   0.6186    2.6337         0         0         0         0
   2.4884         0         0         0         0         0
```

From the coefficients of the first column, it is clear that they do not have the same sign, therefore the system is unstable.

```
% Find the roots of the characteristic polynomial
>> roots(coeff)
ans =
   -2.4708
    0.6812
    0.5106
   -0.8306
   -0.4904
```

Indeed, there is a pole at –2.4708, which is outside the unit circle, causing the instability of the closed system.

EXERCISE 6.6

Derive the region of values for K, such that the closed system shown in the following scheme is stable (a) using the Routh criterion and (b) using the Jury criterion.

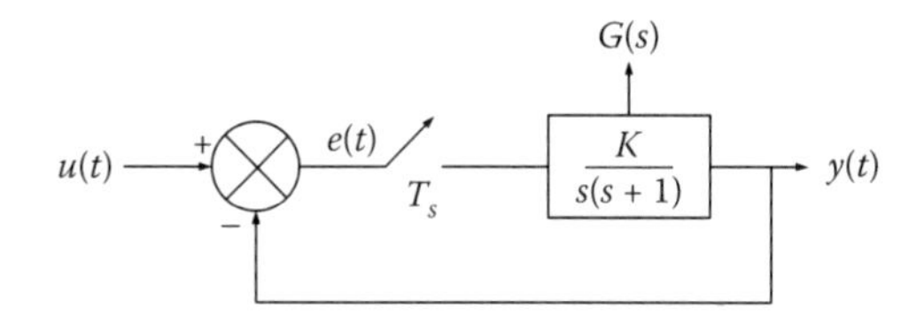

Solution

The transfer function of the given system is

$$G(s)=\frac{K}{s(s+1)}=K\left(\frac{1}{s}-\frac{1}{s+1}\right) \tag{6.6.1}$$

$$(6.6.1)\overset{ILT}{\Rightarrow} g(t)=K(1-e^{-t})\overset{t=kT}{\Rightarrow} g(kT)=K(1-e^{-kT}) \tag{6.6.2}$$

$$(6.6.2)\Rightarrow G(z)=z[g(kT)]=K\frac{z(1-e^{-T})}{z^2-(1+e^{-T})z+e^{-T}} \tag{6.6.3}$$

The poles of the closed system can be found by solving $1 + G(z) = 0$, that is, they are the roots of

$$P(z)=z^2-\left[e^{-T}(1+K)+1-K\right]z+e^{-T}=0 \tag{6.6.4}$$

a. Find the stability using the Routh criterion
In the expression (6.6.4), set $z = (w + 1)/(w - 1)$, thus

$$P(w) = \frac{w^2(K(1-e^{-T})) + w(2(1-e^{-T})) + 2(1+e^{-T}) - K(1-e^{-T})}{(w-1)^2} = 0 \tag{6.6.5}$$

Formulate the Routh table

w^2	$K(1 - e^{-T})$	$2(1 + e^{-T}) - K(1 - e^{-T})$
w^1	$2(1 - e^{-T})$	
w^0	$2(1 + e^{-T}) - K(1 - e^{-T})$	

To obtain stability, it should hold that

$$\begin{aligned} &K(1-e^{-T}) > 0 \\ &2(1-e^{-T}) > 0 \\ &2(1+e^{-T}) - K(1-e^{-T}) > 0 \end{aligned} \tag{6.6.6}$$

From

$$K(1-e^{-T}) > 0 \Rightarrow \boxed{K > 0}$$

From

$$2(1+e^{-T}) - K(1-e^{-T}) > 0 \Rightarrow \boxed{\frac{K}{2} < \frac{1+e^{-T}}{1-e^{-T}} = \cot\left(\frac{T}{2}\right)}$$

Consequently, we get

$$\boxed{0 < K < 2\frac{1+e^{-T}}{1-e^{-T}} = 2\cot\left(\frac{T}{2}\right)} \tag{6.6.7}$$

b. Find the stability using the Jury criterion
Formulate the Jury table

1	$-e^{-T}(1 + K) - 1 + K$	e^{-T}
e^{-T}	$-e^{-T}(1 + K) - 1 + K$	1
$1 - e^{-2T}$		$(e^{-T} - 1)[e^{-T}(1 + K) + 1 - K]$
$(e^{-T} - 1)[e^{-T}(1 + K) + 1 - K]$		$1 - e^{-2T}$
$1-e^{-2T} - \frac{(e^{-T}-1)^2}{1-e^{-2T}}(e^{-T}(1+K)+1-K)^2$		

According to the Jury criterion, the necessary stability conditions are

$$1>0:\text{True},\quad 1-e^{-2T}>0:\text{True}$$

$$1-e^{-2T}-\frac{(e^{-T}-1)^2}{1-e^{-2T}}(e^{-T}(1+K)+1-K)^2 \Rightarrow$$

$$(1-e^{-T})K^2-2K(1-e^{-2T})<0$$

In order to hold true the last inequality, K should be placed inside the range of roots of the polynomial at the left-hand side of the corresponding inequality, which is: 0 and $2((1+e^{-t})/(1-e^{-t}))=2\cot(T/2)$, that is, K should satisfy

$$\boxed{0<K<2\frac{1+e^{-T}}{1-e^{-T}}=2\cot\left(\frac{T}{2}\right)} \tag{6.6.8}$$

EXERCISE 6.7

Evaluate the system stability of the following transfer functions.

a. $H(z)=\dfrac{(z-0.5)}{(z+0.75)}$

b. $H(z)=\dfrac{(z^2+1)}{(z^2-0.25)}$

c. $H(z)=\dfrac{z(z-1)}{(z^2+0.5z-0.5)}$

d. $H(z)=\dfrac{(z-0.5)(z+0.5)}{(z^2+z+0.75)}$

Write the appropriate MATLAB command to design the step responses of the above systems.

Solution

a. $H(z)=\dfrac{(z-0.5)}{(z+0.75)}$

   The system is stable because the system pole $p=-0.75$ is inside the unit circle.

b. $H(z)=\dfrac{(z^2+1)}{(z^2-0.25)}$

   The system is stable because the system poles $p_1=-0.5$ and $p_2=0.5$ are inside the unit circle.

Subsequently, the diagram of poles and zeros is drawn for the two systems.

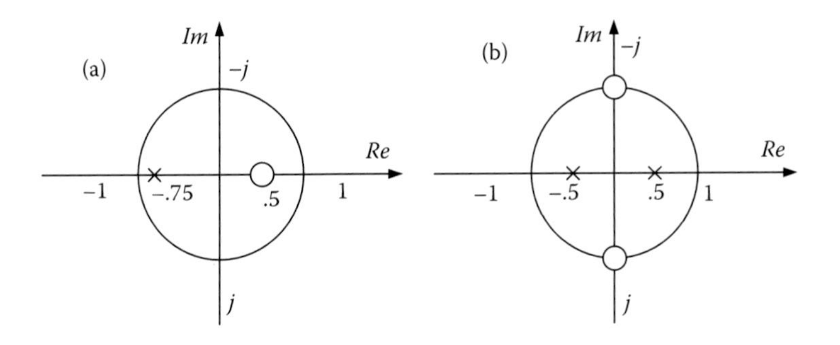

c. $H(z)=\dfrac{z(z-1)}{(z^2+0.5z-0.5)}=\dfrac{z(z-1)}{(z-1)(z-0.5)}$

The system is marginally stable because one of the system poles ($p=-1$) is on the circumference of the unit circle.

d. $H(z)=\dfrac{(z-0.5)(z+0.5)}{(z^2+z+0.75)}=\dfrac{(z-0.5)(z+0.5)}{(z+0.5+07j)(z+0.5-07j)}$

The system is stable because the system complex conjugate poles $p_1=0.86e^{j}126^0$ and $p_2=0.86e^{-j}126^0$ are inside the unit circle.

Subsequently, the diagram of poles and zeros is drawn for the two systems.

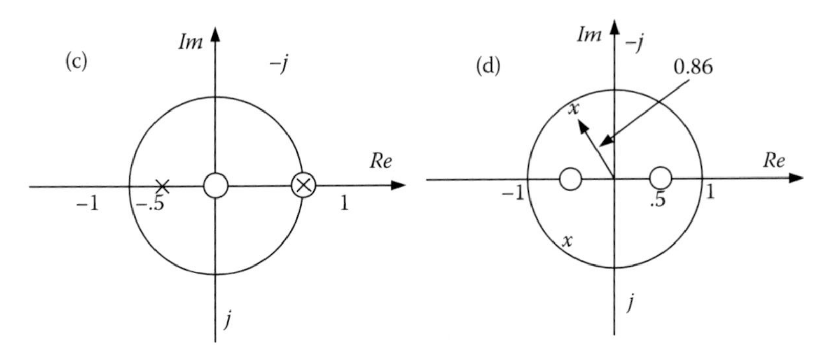

To design the corresponding step responses, the following MATLAB commands are used:

```
step([1, -0.5],[1, 0.75])    % for system (a).
step([1, 0, 1],[1, 0, -0.25])    % for system (b).
step([1, -1, 0],[1, 0.5, -0.5])    % for system (c).
step([1, -0.25],[1, 1, 0.75])    % for system (d).
```

EXERCISE 6.8

The following transfer functions are given:

$$\frac{y_1(z)}{u(z)} = \frac{0.2}{z-0.8}; \quad \frac{y_2(z)}{u(z)} = \frac{1.8}{z+0.8}; \quad \frac{y_3(z)}{u(z)} = \frac{2}{z+1};$$

$$\frac{y_4(z)}{u(z)} = \frac{2}{z-1}; \quad \frac{y_5(z)}{u(z)} = \frac{1.25}{(z-j0.5)(z+j0.5)}$$

For a step input, derive the first ten output values of each system. Evaluate the stability of these systems.

Solution

The difference equations of the given systems are

$$y_1[nT_s] = 0.2u[(n-1)T_s] + 0.8y_1[(n-1)T_s] \tag{6.8.1}$$

$$y_2[nT_s] = 1.8u[(n-1)T_s] - 0.8y_2[(n-1)T_s] \tag{6.8.2}$$

$$y_3[nT_s] = 2u[(n-1)T_s] - y_3[(n-1)T_s] \tag{6.8.3}$$

$$y_4[nT_s] = 2u[(n-1)T_s] + y_4[(n-1)T_s] \tag{6.8.4}$$

$$y_5[nT_s] = 1.25u[(n-2)T_s] - 0.25y_5[(n-2)T_s] \tag{6.8.5}$$

The first ten output values of each system, for a step input function, are provided in the following table.

From the values of this table, the following outcomes emerged:

- The pole of the pulse transfer function for the first system is real positive number and is inside the unit circle ($p = 0.8$). The response exponentially reaches the final value, that is, unity. The system is stable.
- The pole of the third system is $p = -1$. The response oscillates with fixed amplitude. The system is marginally stable.
- The pole of the fourth system is $p = +1$, the system acts (operates) as an integrator.
- The poles of pulse transfer function of the fifth system are imaginary conjugates and are located inside the unit circle. The response tends to unity with small oscillations. The system is stable.

n	y_1	y_2	y_3	y_4	y_5
0	0	0	0	0	0
1	0.2	1.8	2	2	0
2	0.36	0.36	0	4	1.25
3	0.488	0.512	2	6	1.25
4	0.5904	1.3904	0	8	0.9375
5	0.6723	0.6877	2	10	0.9375

Continued

6	0.7378	1.2499	0	12	1.0156
7	0.7902	0.8001	2	14	1.0256
8	0.8322	1.1599	0	16	0.9961
9	0.8658	0.8721	2	18	0.9961
10	0.8926	1.1023	0	20	1.001

EXERCISE 6.9

A system with the closed-loop transfer function $P(z) = K((1-e^{-T_s/2})/(z-e^{-T_s/2}))$ is given. Derive the sampling period if the parameter $K = 4$ for critical stability.

Solution

Since $K_{cr} = 4$, the transfer unction becomes:

$$P(z) = 4\frac{1-e^{-T_s/2}}{z-e^{-T_s/2}} \tag{6.9.1}$$

The characteristic equation of the system is presented as

$$1+P(z) = 1+4\cdot\frac{1-e^{-T_s/2}}{z-e^{-T_s/2}} = 0 \tag{6.9.2}$$

$$\Rightarrow z-e^{-T_s/2}+4(1-e^{-T_s/2}) = 0 \tag{6.9.3}$$

Solving with respect to z, we have

$$z = -4+5e^{-T_s/2} \tag{6.9.4}$$

For marginal stability, $z = -1$, hence from the expression (6.9.4) we get

$$5e^{-T_s/2} = 3 \Rightarrow e^{-T_s/2} = 3/5 \Rightarrow \frac{-T_s}{2} = \ln 0.6 = -0.5108 \Rightarrow T_s = 1.0217\,\text{s} \tag{6.9.5}$$

EXERCISE 6.10

For a closed-loop discrete system, the transfer function of the process is $P(z) = (1/z)$, while for the serial controller is $C(z) = Kz/(z-1)$. Derive the maximum value of parameter K such that the closed-loop system is stable. For $K = K_{max}/2$ calculate the system output for $k = 0$, 1, and 2. Assume a unit-step input.

Solution

The discrete open-loop transfer function is

$$L(z) = C(z)P(z) = \frac{Kz}{z-1}\cdot\frac{1}{z} = \frac{K}{z-1} \tag{6.10.1}$$

The characteristic equation of the closed system is expressed as

$$1+L(z)=0 \Rightarrow z-1+K=0 \tag{6.10.2}$$

In order for the closed-loop system to be stable, it suffices to hold $|z|<1$, which leads to the condition $K_{max}=2$.

For $K=K_{max}/2=1$ we have

$$L(z)=\frac{K}{z-1}=\frac{1}{z-1} \tag{6.10.3}$$

The total transfer function is given by

$$\frac{Y(z)}{R(z)}=\frac{L(z)}{1+L(z)}=\frac{1/(z-1)}{1+(1/(z-1))}=\frac{1}{z}=z^{-1} \tag{6.10.4}$$

For step input, output is

$$y[k]=1[k-1] \Rightarrow y[0]=0,\ y[1]=1,\ y[2]=1.$$

EXERCISE 6.11

The loop transfer function of a discrete system is given as $G(z)=(0.632Kz)/(z^2-1.368z+0.368)$. N Derive the region of values for K to obtain stability of the closed system.

Solution

The system characteristic equation is

$$1+G(z)=1+\frac{0.632Kz}{z^2-1.368z+0.368}=0 \tag{6.11.1}$$

In the expression (6.11.1), set $z=(w+1)/(w-1)$.

$$\begin{aligned}&1+\frac{0.632Kz}{z^2-1.368z+0.368}=0\\&\Rightarrow 0.632Kw+1.264w+(2.736-0.632K)=0\end{aligned} \tag{6.11.2}$$

Formulate the Routh table.

$$\begin{array}{cc}0.632K & 2.736-0.632K\\ 1.264 & \\ 2.736-0.632K & \end{array}$$

For stability, it should hold:

$$2.736-0.632K>0 \Rightarrow 0<K<4.33 \tag{6.11.3}$$

EXERCISE 6.12

The system of the following scheme is given.

1. For $G(s) = 1/(s(s+1))$, design the Nyquist diagram ($T = 1$ s)

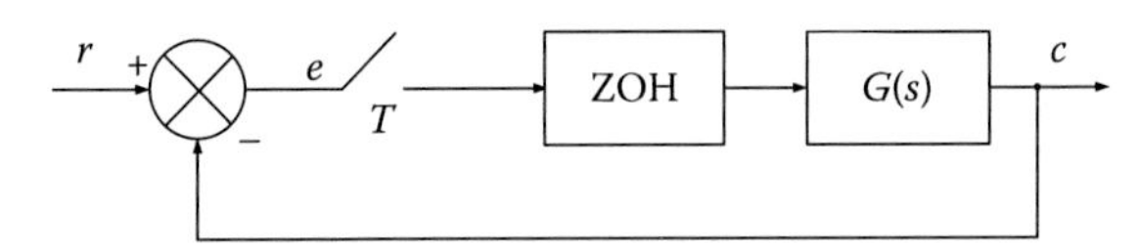

2. For $G(s) = (1/s^2)$, design the Bode diagram ($T = 1$ s).

Solution

We use MATLAB to design the requested diagrams.

```
% Design the Nyquist diagram
clear
clf
np=[0 0 10];
dp=[1 5 0];
[num den]=c2dm(np,dp,1,'z');
dnyquist(num,den,1);axis([-1.5 0 -20 20])
```

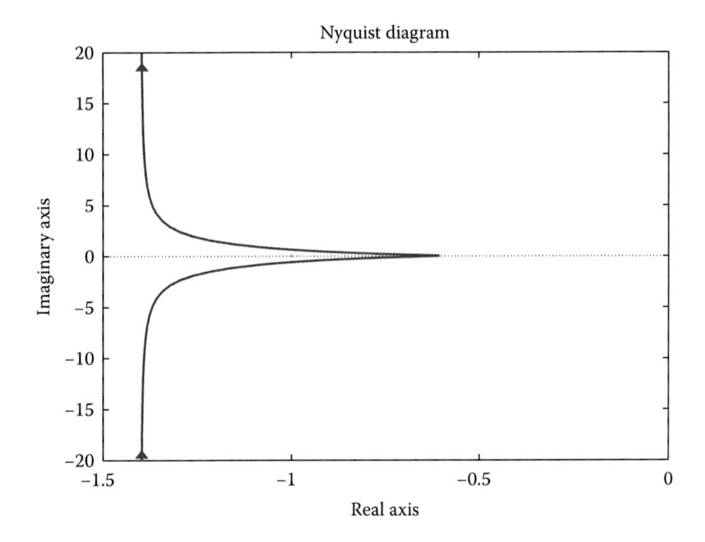


```
% Design the Bode diagram
w=[logspace(-1, .4) 1000];
np=[0 0 1];
dp=[1 0 0];
[a,b,c,d]=tf2ss(np,dp);
sysc=ss(a,b,c,d);
T=1;
sysd=c2d(sysc,T,'zoh');
[magph]= bode(sysd,w);
loglog(w,magph(1,:)) ; grid
```

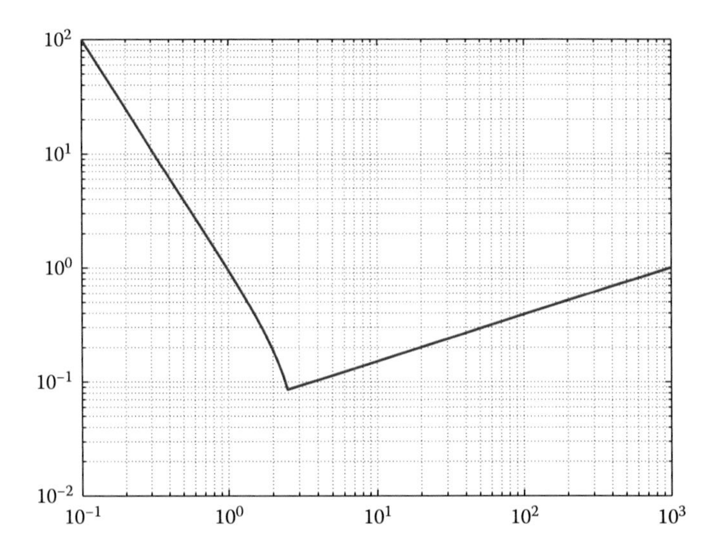

EXERCISE 6.13

Design the root locus of a system with the open-loop transfer function:

$$D(z)G(z) = \frac{K(z+0.995)}{(z-1)(z-0.905)}$$

Solution

The loop transfer function is

$$D(z)G(z) = \frac{K(z+0.995)}{(z-1)(z-0.905)} \tag{6.13.1}$$

The system characteristic equation is given by the expression (6.13.2). Solve with respect to K and equate the derivative (dK/dz) to zero, to obtain the breaking points. We have that

$$1 + D(z)G(z) = 0 \Rightarrow 1 + \frac{K(z+0.995)}{(z-1)(z-0.905)} = 0 \tag{6.13.2}$$

$$(6.13.2) \Rightarrow K = -\frac{(z-1)(z-0.905)}{z+0.995} = -\frac{z^2 - 1.905z + 0.905}{z+0.995} \tag{6.13.3}$$

$$\frac{dK}{dz} = 0 \Rightarrow \begin{array}{l} z^2 + 1.99z - 2.76 = 0 \\ z_1 = 0.954, z_2 = -2.934 \end{array} \tag{6.13.4}$$

Both roots of $(dK/dz) = 0$ are breaking points since:

$$K\,|_{z=0.954} = 0.001 > 0 \quad K\,|_{z=-2.934} = 2.03 > 0 \tag{6.13.5}$$

Calculate the intersections point of root locus with the unit circle

$$(6.13.2) \Rightarrow z^2 + (K - 1.905)z + 0.905 + 0.995K = 0 \tag{6.13.6}$$

It should hold

$$\left.\begin{matrix} A(1) > 0 \\ A(-1) > 0 \\ |A(0)| < 1 \end{matrix}\right\} \Rightarrow 0 < K < 0.095 \tag{6.13.7}$$

For

$$K_{cr} = 0.095 \Rightarrow \begin{matrix} z^2 - 1.81z + 1 = 0 \\ z_1 = 0.905 \pm j0.425 \end{matrix}$$

Design the root locus as

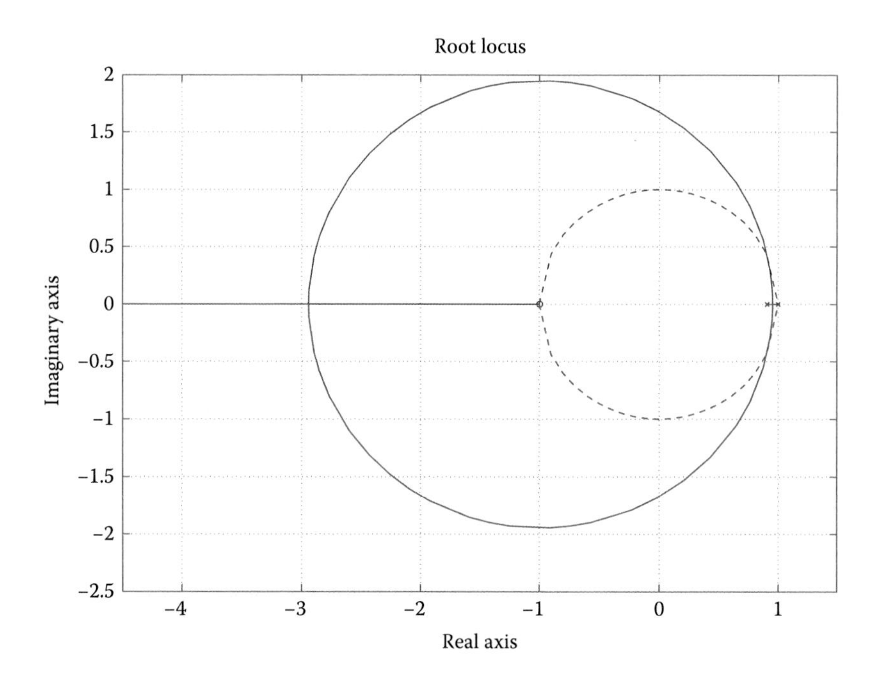

EXERCISE 6.14

Design the root locus for the system of the following scheme.

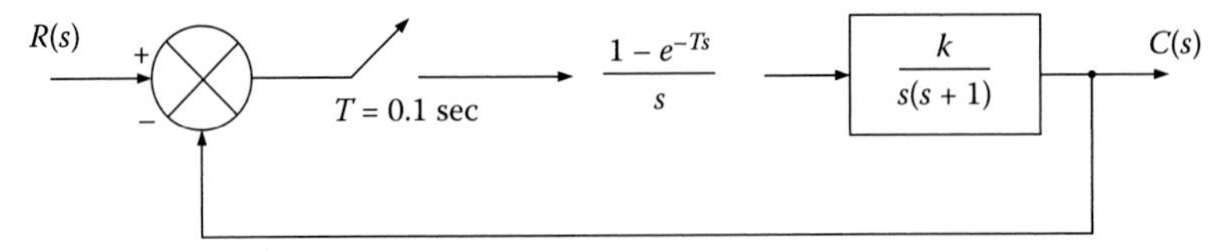

Solution

The open-loop transfer function is $G(z)$

$$G(z)=z\left[\frac{1-e^{-Ts}}{s}\frac{k}{s(s+1)}\right]=(1-z^{-1})z\left[\frac{k}{s^2(s+1)}\right]$$

$$\Rightarrow G(z)=\frac{z-1}{z}z\left[\frac{k}{s^2}-\frac{k}{s}+\frac{k}{s+1}\right]=\frac{z-1}{z}\left[\frac{kTz}{(z-1)^2}-\frac{kz}{z-1}+\frac{kz}{z-e^{-T}}\right] \quad (6.14.1)$$

$$\Rightarrow G(z)=k\frac{z(T-1+e^{-T})+1-e^{-T}-Te^{-T}}{(z-1)(z-e^{-T})}$$

For

$$T=0.1\text{s}\Rightarrow G(z)=k\frac{0.00484(z+0.9672)}{(z-1)(z-0.9048)} \quad (6.14.2)$$

The open-loop transfer function has two poles ($P1 = 1$, $P2 = 0.9048$) and a zero ($Z1 = -0.9672$).

Find the departure point from the horizontal axis

$$\frac{dk}{dz}=0 \quad (6.14.3)$$

where k is calculated from the characteristic equation $1 + G(z) = 0$.

$$\frac{dk}{dz}=\left(\frac{(z-1)(z-0.9048)}{0.00484(z-0.9672)}\right)'=0$$

$$\Rightarrow z^2+1.9344z-2.7471=0\begin{cases} z1=0.9516 \\ z2=-2.886 \end{cases}$$

Both roots of $(dk/dz) = 0$ are breaking points since there are positive values for the amplification parameter of the system. The critical value of K, K_{cr}, for which the system becomes stable, can be found by using the Jury criterion or the Routh criterion with the aid of bilinear Möbius transform, yielding that $K_{cr} = 165$. The design of root locus of the characteristic equation follows, where we observe that the digitalized system is unstable for $K \geq 165$, regardless of the fact that the original system is asymptotically stable.

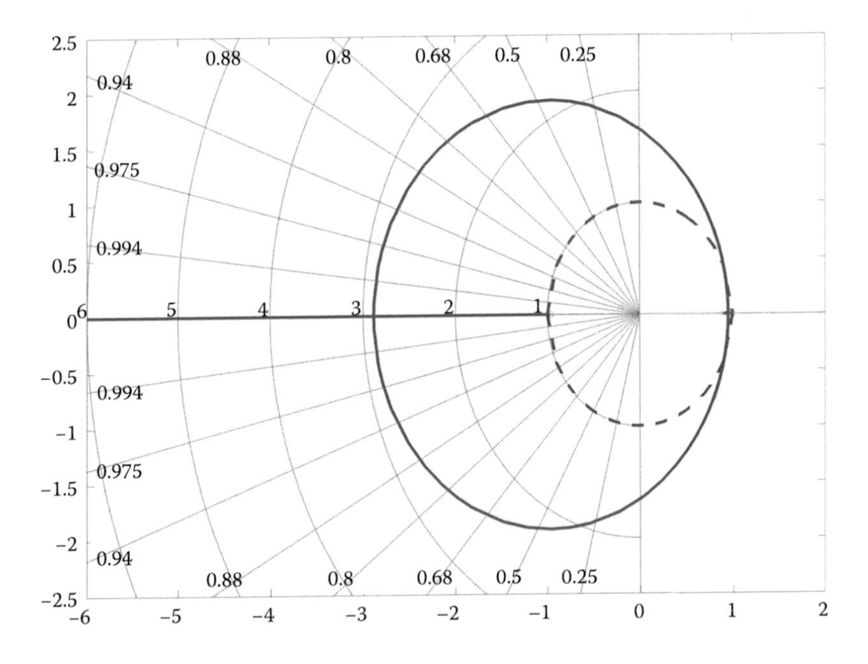

EXERCISE 6.15

Approximately design the root locus of the systems with open-loop transfer functions, as given below. Evaluate the stability of the closed-loop systems.

$$G_1(z) = \frac{z}{z-0.5}, \qquad G_2(z) = \frac{z}{(z-0.5)^2}, \qquad G_3(z) = \frac{z}{(z-0.1)(z-0.8)},$$

$$G_4(z) = \frac{z+0.6}{(z-0.1)(z-0.8)} \quad G_5(z) = \frac{z}{z^2-0.4z+0.6}, \quad G_6(z) = \frac{z(z-0.6)}{z^2-0.4z+0.6}$$

Solution

$$\boxed{G_1(z) = \frac{z}{z-0.5}}$$

For the first system, the transfer function has a zero at 0 and a pole at 0.5. In the following scheme, the root locus of the system characteristic equation is presented. Since the locus branch is located inside the unit circle, the system is stable.

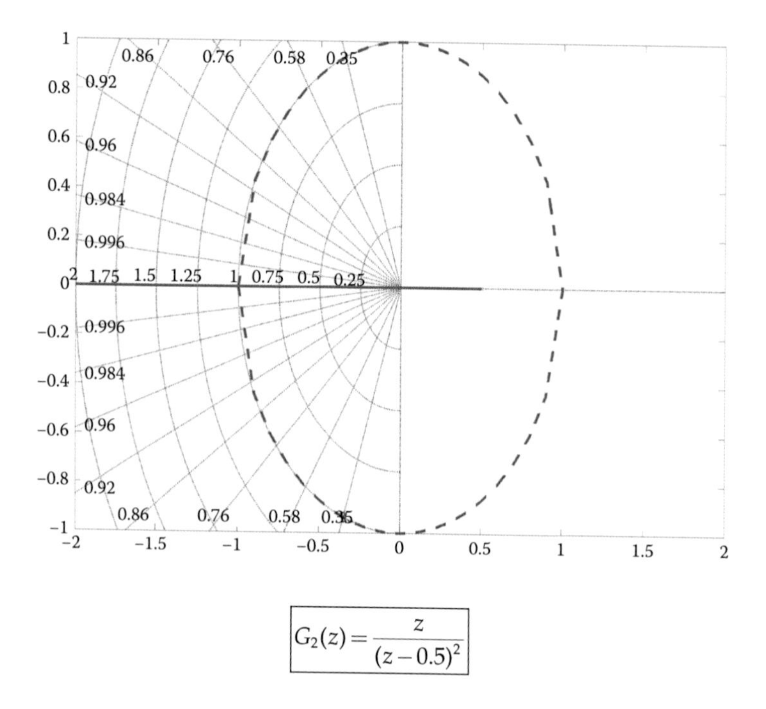

$$G_2(z) = \frac{z}{(z-0.5)^2}$$

For the second system, the transfer function has a zero at 0 and a double pole at 0.5. In the following scheme, the root locus of the system characteristic equation is presented. Since the two locus branches are located inside the unit circle, the closed-loop system is stable.

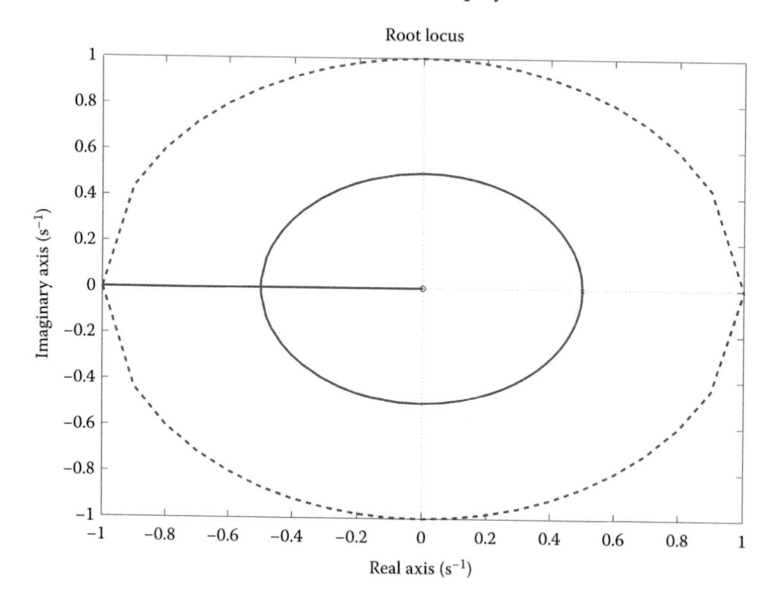

$$G_3(z) = \frac{z}{(z-0.1)(z-0.8)}$$

For the third system, the transfer function has a zero at 0 and two real poles at 0.1 and 0.8. In the following scheme, the root locus of the system characteristic equation is presented. Since the two locus branches are located inside the unit circle, the closed-loop system is stable.

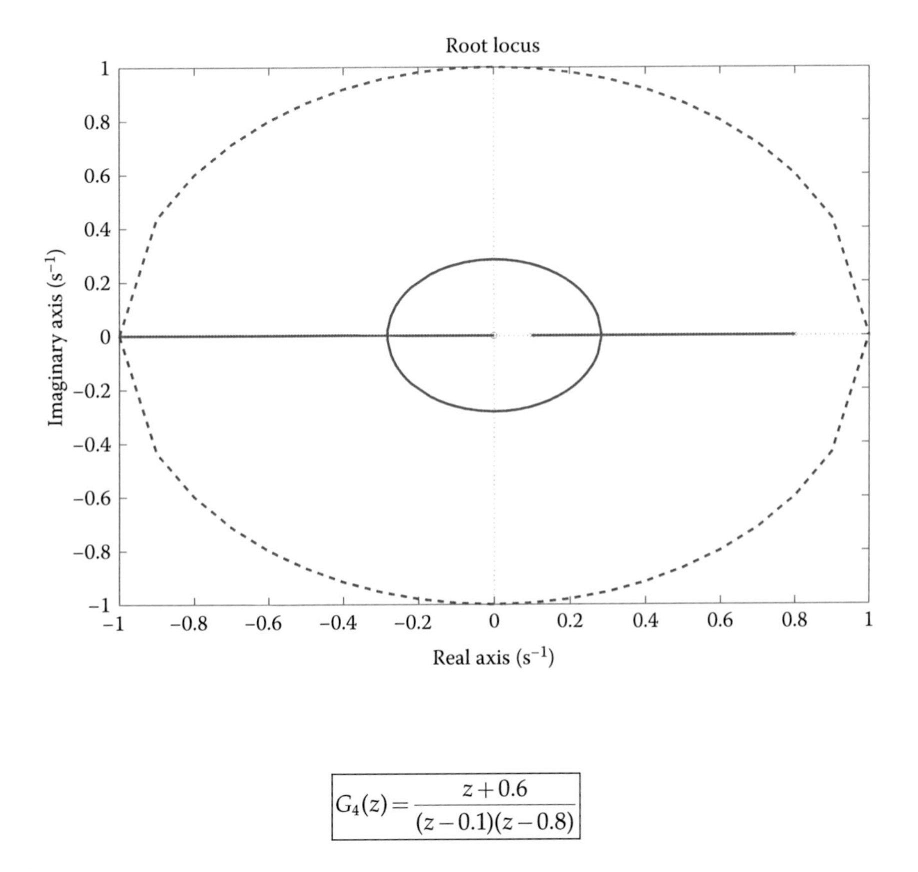

$$G_4(z) = \frac{z+0.6}{(z-0.1)(z-0.8)}$$

For the fourth system, the transfer function has a zero at -0.6 and two real poles at 0.1 and 0.8. In the following scheme, the root locus of the system characteristic equation is presented. Since the two locus branches are not located inside the unit circle, the closed-loop system is conditionally stable.

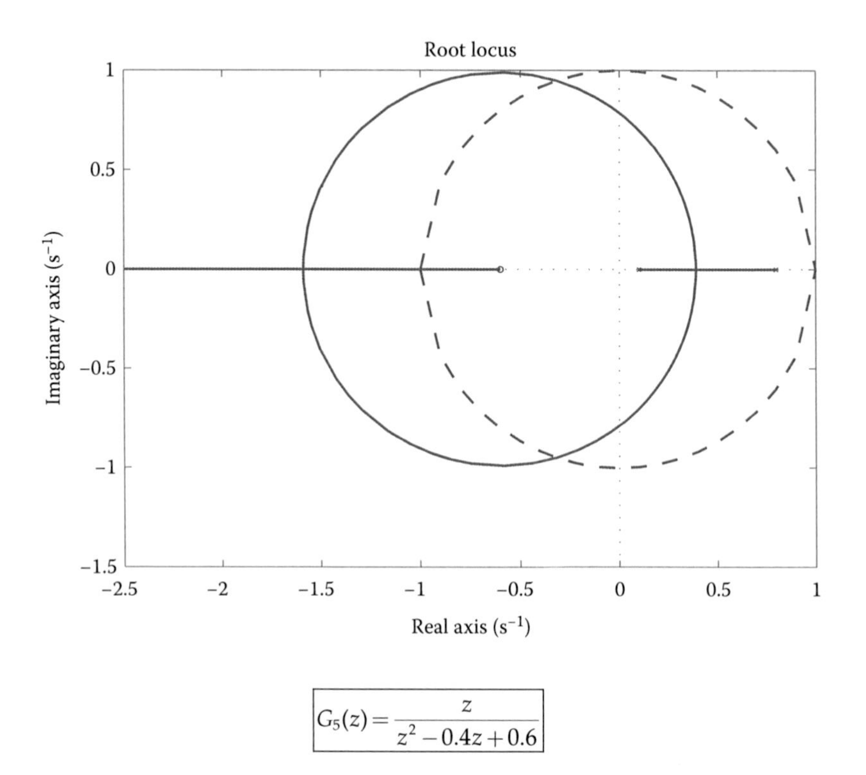

$$G_5(z) = \frac{z}{z^2 - 0.4z + 0.6}$$

For the fifth system, the transfer function has a zero at 0 and two complex conjugate poles at $0.2000 + 2.4413i$ and $0.2000 - 2.4413i$. In the following scheme, the root locus of the system characteristic equation is presented. Since the two locus branches are located inside the unit circle, the closed-loop system is stable.

The breaking point is computed as

$$1 + G_5(z) = 0 \Rightarrow 1 + \frac{Kz}{z^2 - 0.4z + 0.6} = 0$$

$$\Rightarrow K = -\frac{z^2 - 0.4z + 0.6}{z}$$

$$\frac{dK}{dz} = 0 \Rightarrow z^2 = 0.6$$

$$z = \pm 0.7746$$

$$z_1 = 0.7746 \Rightarrow K = -1.1492$$

$$z_2 = -0.7746 \Rightarrow K = 1.9492$$

Breaking point is the root $z_2 = -0.7746$ where it holds that $K = 1.9492 > 0$. The angle of departure from the complex pole is $0.2000 + 2.4413i$ is

$$\varphi_d = 180° - (q1 - f) = 180° - (90° - 75°) = 165°$$

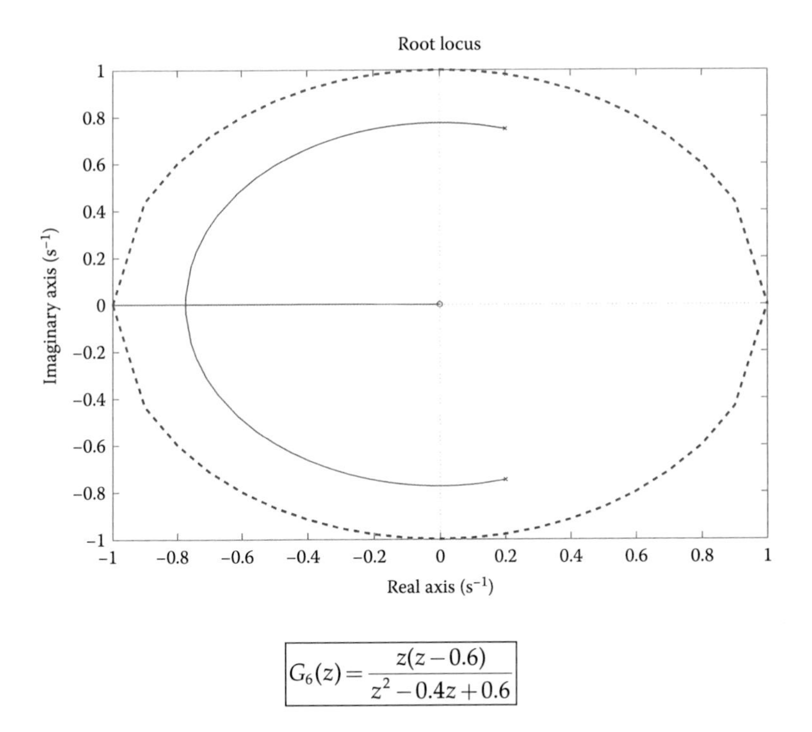

$$G_6(z) = \frac{z(z-0.6)}{z^2 - 0.4z + 0.6}$$

For the sixth system, the transfer function has two zeros at 0 and 0.6, and two complex conjugate poles at $0.2000 + 2.4413i$ and $0.2000 - 2.4413i$. In the following scheme, the root locus of the system characteristic equation is presented. Since the two locus branches are located inside the unit circle, the closed-loop system is stable.

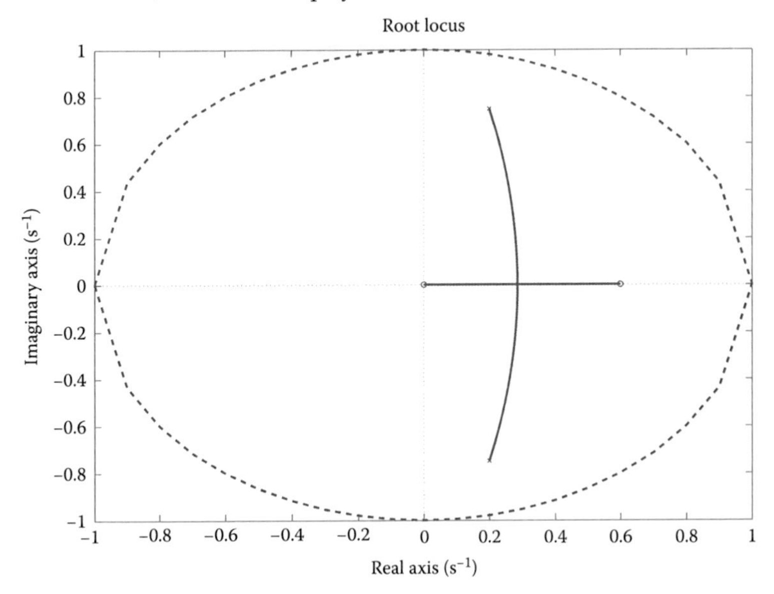

EXERCISE 6.16

Design the root locus of the characteristic equation for the system of the following scheme and provide some relevant outcomes regarding its stability. Assume that

$$G_p(s) = \frac{1}{s+1} \quad \text{and} \quad G_D(z) = \frac{K}{1-z^{-1}} = \frac{Kz}{z-1}$$

(T: 0.5, 1, and 2 s)

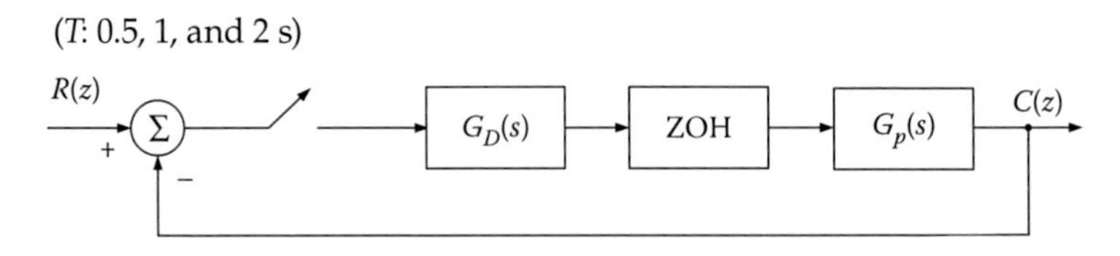

Solution

Calculate the transfer function of the given system using ZOH.

$$\begin{aligned} Z[G_h(s)G_p(s)] &= Z\left[\frac{1-e^{-Ts}}{s}\frac{1}{s+1}\right] = (1-z^{-1})Z\left[\frac{1}{s(s+1)}\right] \\ &= (1-z^{-1})Z\left[\frac{1}{s} - \frac{1}{s+1}\right] \\ &= \frac{z-1}{z}\left(\frac{z}{z-1} - \frac{z}{z-e^{-T}}\right) = \frac{1-e^{-T}}{z-e^{-T}} \end{aligned} \tag{6.16.1}$$

The open-loop transfer function is

$$G(z) = G_D(z)Z[G_h(s)G_p(s)] = \frac{Kz}{z-1}\frac{1-e^{-T}}{z-e^{-T}} \tag{6.16.2}$$

The characteristic equation of the system is given by

$$1 + \frac{Kz(1-e^{-T})}{(z-1)(z-e^{-T})} = 0 \tag{6.16.3}$$

1. For sampling period $T = 0.5$ s, we have

$$(6.16.2) \Rightarrow G(z) = \frac{0.3935Kz}{(z-1)(z-0.6065)} \tag{6.16.4}$$

   $G(z)$ has two poles at $z = 1$ and $z = 0.6065$, and a zero at $z = 0$. The finding process of the breaking points follows:

$$(6.16.3),(6.16.4) \Rightarrow K = -\frac{(z-1)(z-0.6065)}{0.3935z} \tag{6.16.5}$$

$$\frac{dK}{dz} = -\frac{z^2 - 0.6065}{0.3935z^2} = 0 \Rightarrow z^2 = 0.6065 \tag{6.16.6}$$

$$(6.16.6) \Rightarrow z = 0.7788 \quad \text{and} \quad z = -0.7788.$$

For $z = 0.7788 \Rightarrow K = 0.1244$ and for $z = -0.7788 \Rightarrow K = 8.041$. Since both values of the amplification parameter are positive, $z = 0.7788$ is the point of departure for the branches of the horizontal axis (break away point) and $z = -0.7788$ is the break-in point to the horizontal axis. In the following scheme, the root locus of the characteristic equation for $T = 0.5$ s is depicted.

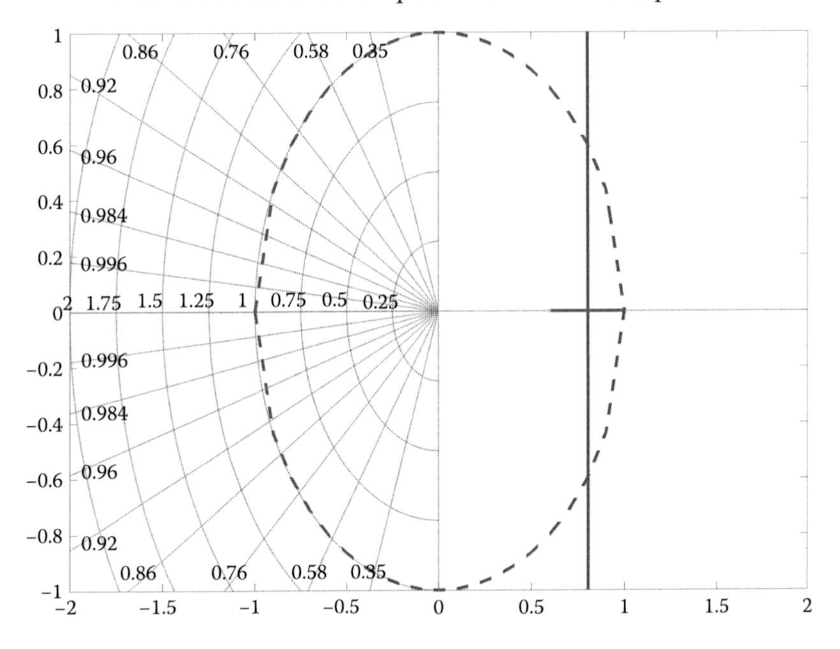

The marginal value of K parameter for stability is calculated as

$$\begin{aligned} A(z) &= 1 + \frac{0.3935Kz}{(z-1)(z-0.6065)} = 0 \\ &\Rightarrow A(z) = z^2 - 1.6065z + 0.6065 + 0.3935Kz = 0 \end{aligned} \tag{6.16.7}$$

It should hold

$$\left.\begin{aligned} |A(0)| &< 1 \Rightarrow 0.6065 < 1 \\ |A(1)| &> 0 \Rightarrow 0.3935K > 0 \Rightarrow K > 0 \\ |A(-1)| &> 0 \Rightarrow 1.6065 * 2 - 0.3935K > 0 \Rightarrow K > 8.1652 \end{aligned}\right\} \tag{6.16.8}$$

Consequently: $K_{cr} = 8.1652$

The poles of the closed-loop system for $K = 2$ are given by

$$z1 = 0.4098 + j0.6623 \text{ and } z2 = 0.4098 - j0.6623$$

These poles are marked with bullets in the root locus diagram.

2. For sampling period $T = 1$ s, we have

$$G(z) = \frac{0.6321Kz}{(z-1)(z-0.3679)} \tag{6.16.9}$$

In the following scheme, the root locus of the characteristic equation for $T = 1$ s is presented.

$G(z)$ has two poles at $z = 1$ and $z = 0.3679$, and a zero at $z = 0$. Using the process of finding the breaking points, the breakaway point is at $z = 0.6065$ and the break-in point at $z = -0.6065$. The corresponding gains are $K = 0.2449$ and $K = 4.083$, respectively. The marginal value of the parameter K for stability is 4.328. The poles of the characteristic equation which correspond to $K = 2$ are $z1 = 0.05185 + j0.6043$ and $z2 = 0.05185 - j0.6043$, which are presented in bullets at the root locus diagram.

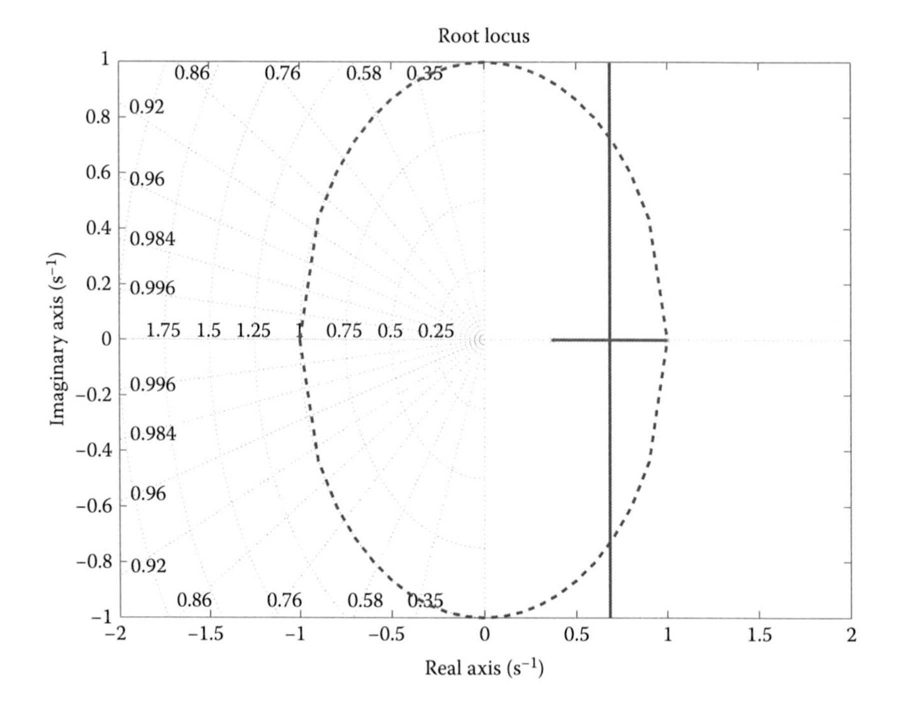

3. For sampling period $T = 2$ s, we have

$$G(z) = \frac{0.8647Kz}{(z-1)(z-0.1353)} \tag{6.16.10}$$

In the following scheme, the root locus of the characteristic equation for $T = 2$ s is presented.

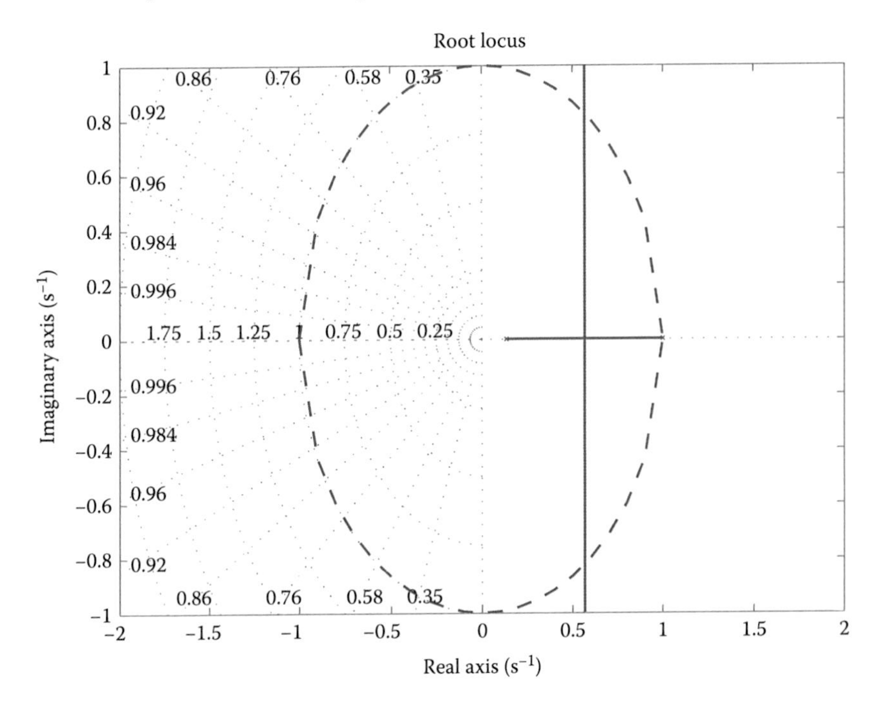

$G(z)$ has two poles at $z = 1$ and at $z = 0.1353$, and a zero at $z = 0$. Using the process of finding the breaking points, the breakaway point is at $z = 0.3678$ and the break-in point at $z = -0.3678$. The corresponding gains are $K = 0.4622$ and $K = 2.164$, respectively. The marginal value of the parameter K for stability is 2.626. The poles of the characteristic equation which correspond to $K = 2$ are $z1 = -0.2971 + j0.2169$ and $z2 = -0.2971 - j0.2169$, which are presented in bullets at the root locus diagram.

EXERCISE 6.17

Calculate the region of values for the parameter K ($K > 0$) for stability of the closed system using the Nyquist and Jury criteria for the open-loop transfer function: $G(z) = K/((z - 0.2)(z - 0.4))$ for $T = 1$ s.

Repeat the procedure if: $G(z) = K/(z(z - 0.2)(z - 0.4))$.

Solution

1.

$$G(z) = \frac{K}{(z-0.2)(z-0.4)}$$

Evaluation of stability using the Nyquist criterion: The frequency response of the open-loop system is provided by setting $z = e^{j\omega T} \overset{T=1}{=} e^{j\omega}$ *into*

$$G(z) = \frac{K}{(z-0.2)(z-0.4)}. \tag{6.17.1}$$

$$\begin{aligned} G(e^{j\omega}) &= \frac{K}{(e^{j\omega}-0.2)(e^{j\omega}-0.4)} \\ &= \frac{K}{(\cos\omega + j\sin\omega - 0.2)(\cos\omega + j\sin\omega - 0.4)} \end{aligned} \tag{6.17.2}$$

$$\begin{aligned} G(e^{j\omega}) &= \frac{K}{(\cos^2\omega - \sin^2\omega - 0.6\cos\omega + 0.08) + j(2\sin\omega\cos\omega - 0.6\sin\omega)} \\ &= \frac{K}{(\cos^2\omega - (1-\cos^2\omega) - 0.6\cos\omega + 0.08) + j(2\sin\omega\cos\omega - 0.6\sin\omega)} \\ &= \frac{K}{(2\cos^2\omega - 0.6\cos\omega - 0.92) + j(2\sin\omega\cos\omega - 0.6\sin\omega)} \end{aligned} \tag{6.17.3}$$

The points of interest for stability are the ones where the frequency response intersects the real axis. In this case, the imaginary part of the frequency response is zero.

$$\begin{aligned} 2\sin\omega\cos\omega - 0.6\sin\omega &\Rightarrow \sin\omega(2\cos\omega - 0.6) = 0 \\ &\Rightarrow \sin\omega = 0 \quad \text{and} \quad 2\cos\omega - 0.6 = 0 \\ &\Rightarrow \omega = 0 \quad \text{and} \quad \omega = \arccos(0.3) \end{aligned} \tag{6.17.4}$$

For positive values of *K*, the solution of $\omega = 0$ ($z = e^0 = 1$) is the starting point of frequency response; therefore is more interesting to examine the behavior of frequency response at $\omega = \arccos(0.3)$.

$$G(e^{j\omega}) = G(e^{j\arccos(0.3)}) = \frac{K}{2(0.3)^2 - 0.6\cdot 0.3 - 0.92} = \frac{K}{-0.92} \tag{6.17.5}$$

For the stability of the closed system, it should hold $G(e^{j\omega})$. Then, the Nyquist diagram lets the point (–1, 0) to the left.

$$\frac{K}{-0.92} > -1 \Rightarrow \boxed{K < 0.92} \tag{6.17.6}$$

Evaluation of stability using the Jury criterion: The closed-loop transfer function is

$$G_{cl}(z) = \frac{G(z)}{1+G(z)} = \frac{\dfrac{K}{(z-0.2)(z-0.4)}}{1+\dfrac{K}{(z-0.2)(z-0.4)}} = \frac{K}{z^2-0.6z+0.08+K}$$

$$\Rightarrow G_{cl}(z) = \frac{K}{z^2-0.6z+a} \tag{6.17.7}$$

where

$$a = 0.08 + K \tag{6.17.8}$$

Formulate the Jury Table

1	-0.6	a	$a_2 = a$
a	-0.6	1	
$1 - a^2$	$-0.6 + 0.6a$		
$a_1 = \dfrac{-0.6+0.6a}{1-a^2}$			
$-0.6 + 0.6a$	$1 - a^2$		
$1-a^2-\dfrac{[-0.6(1-a)]^2}{1-a^2}$			

The stability conditions are

$$\begin{aligned} &1 > 0 \\ &1-a^2 > 0 \\ &1-a^2-\frac{0.36(1-a)^2}{1-a^2} > 0 \end{aligned} \tag{6.17.9}$$

Examine the condition: $1 - a^2 > 0$.

$$\begin{aligned} &1-a^2 > 0 \Rightarrow a^2 < 1 \Rightarrow -1 < a < 1 \\ &\qquad \Rightarrow -1 < 0.08+K < 1 \Rightarrow -1.08 < K < 0.92 \end{aligned} \tag{6.17.10}$$

Examine the condition: $1-a^2-\dfrac{0.36(1-a)^2}{1-a^2} > 0$.

$$\begin{aligned} &1-a^2-\frac{0.36(1-a)^2}{1-a^2} > 0 \Rightarrow 1-a^2 > \frac{0.36(1-a)^2}{1-a^2} \Rightarrow (1-a^2)^2 > 0.36(1-a)^2 \\ &\qquad \Rightarrow -(1-a^2) < 0.6(1-a) < 1-a^2 \end{aligned} \tag{6.17.11}$$

Examine separately the two inequalities

$$-(1-a^2)<0.6(1-a)\Rightarrow -1+a^2<0.6(1-a)$$
$$\Rightarrow (0.08+K)^2+0.6(0.08+K)-1.6<0$$
$$\Rightarrow K^2+0.16K+0.0064+0.048+0.6K-1.6<0$$
$$\Rightarrow K^2+0.76K-1.5456<0\Rightarrow \boxed{-1.68<K<0.92} \tag{6.17.12}$$

$$0.6(1-a)<1-a^2\Rightarrow 0.6-0.6a<1-a^2$$
$$\Rightarrow (0.08+K)^2-0.6(0.08+K)-0.4<0 \tag{6.17.13}$$
$$\Rightarrow K^2+0.16K+0.0064-0.048-0.6K-0.4<0$$
$$\Rightarrow K^2-0.44K-0.4416<0\Rightarrow \boxed{-0.48<k<0.92}$$

Finally, the region of values for the amplification parameter K for stability of the closed-loop system is $\boxed{K<0.92}$.
Using MATLAB®, the root locus of the system characteristic equation is designed. By clicking onto the intersection points of the locus branches with the unit circle, we see that $K\approx 0{,}92$ (particularly, we write the commands: H = zpk([],[0.2 0.4],1,1); rlocus(H)).

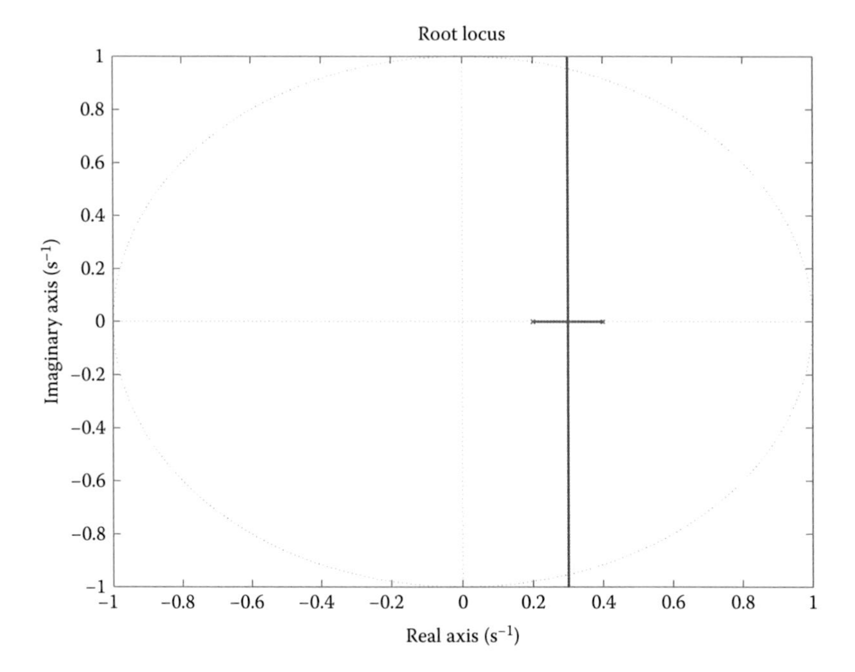

2.

$$G(z)=\frac{K}{z(z-0.2)(z-0.4)}$$

The closed-loop transfer function is

$$G_{cl}(z)=\frac{G(z)}{1+G(z)}=\frac{\frac{K}{(z(z-0.2)(z-0.4))}}{1+\frac{K}{(z(z-0.2)(z-0.4))}}=\frac{K}{z^3-0.6z^2+0.08z+K} \quad (6.17.14)$$

Formulate the Jury table

1	−0.6	0.08	K	$a_3=\frac{K}{1}=K$
K	0.08	−0.6	1	
$1-K^2$	$-0.6-0.08K$	$0.08+0.6K$		$a_2=\frac{0.08+0.6K}{1-K^2}$
$0.08-0.6K$	$-0.6-0.08K$	$1-K^2$		
$\frac{K^4-2.36K^2-0.096K+0.9936}{1-K^2}$				$\frac{0.08K^3+0.6064K^2+0.016K-0.24}{1-K^2}$
$\frac{0.08K^3+0.6064K^2+0.016K-0.24}{1-K^2}$				$\frac{K^4-2.36K^2-0.096K+0.9936}{1-K^2}$
				$a_1\frac{0.08K^3+0.6064K^2+0.016K-0.24}{K^4-2.36K^2-0.096K+0.9936}$

Due to the complexity of table terms, we will avoid evaluating the stability using the Jury criterion and we proceed to the Nyquist criterion.

With the aid of the following MATLAB commands, the Nyquist diagram is designed.

```
K = 1;
H = zpk([],[0 0.2 0.4],K,1);
nyquist(H)
```

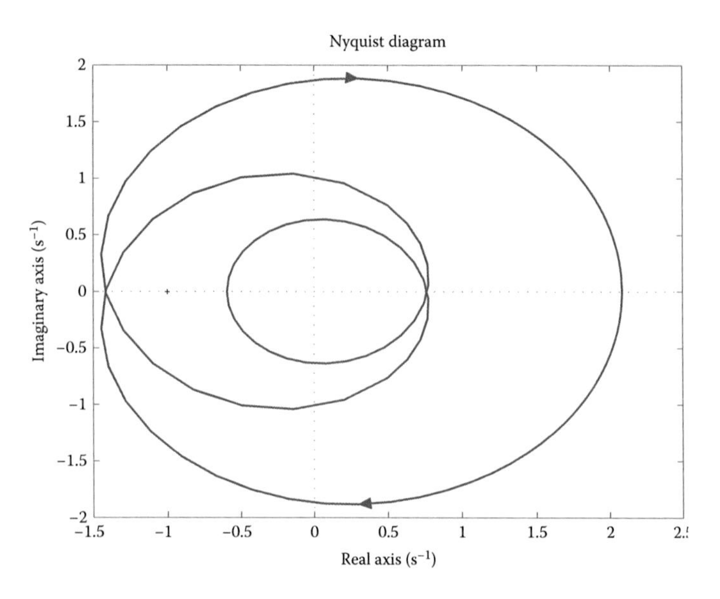

Observe that the Nyquist diagram surrounds the point $(-1, 0)$ as ω increases, therefore the closed-loop system for $K = 1$ is stable. In addition, the Nyquist curve intersects the horizontal axis at the point 1.42, thus the system is stable for $K < (1/1.42) = 0.70$. Using the root locus diagram of the system characteristic equation, we take the same value for K.

```
rlocus(H)
```

It is clear that the poles $z = 0.2$ and $z = 0.4$ are transferred outside the unit circle when the gain increases. The gain is 0.706.

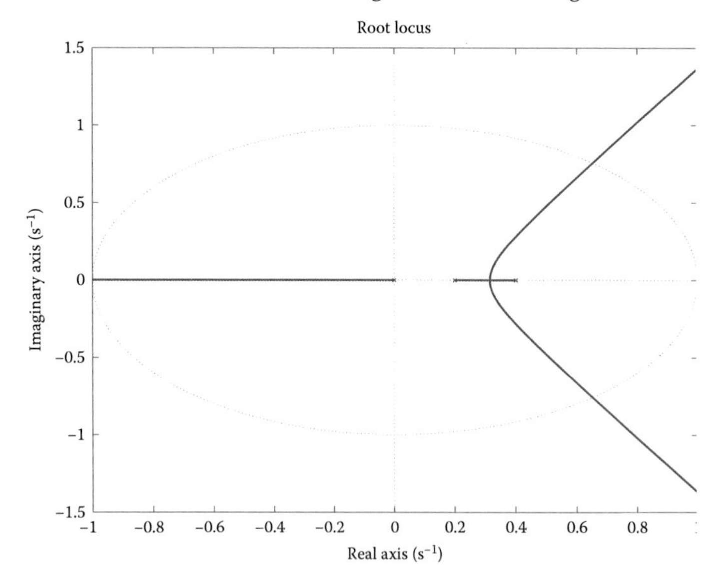

EXERCISE 6.18

A continuous-time system is given with the transfer function $G(s) = 2/((s + 1)(s + 2))$.

1. Design the Bode diagram of the discrete system that arises from a system sampling with period $T = 1$ and a ZOH circuit.
2. Using the transform $w = (2/T)\ (z - 1)/(z + 1)$, design the Bode diagrams of $G(s)$ and $\hat{G}(w) = G_d(z)\big|_{z=f(w)}$.
3. At the discrete system, a controller is implemented as in the scheme. Define the values of K so as the closed-loop system is stable, using the Jury criterion.
4. Design the root locus of the characteristic equation for the discretized system.

Solution

1. Calculate the transfer function $G_d(z)$ of the discrete-time system that arises from the sampling of the continuous-time system with period $T = 1$ s and ZOH circuit.

$$G_d(z) = (1 - z^{-1})Z\left\{L^{-1}\left(\frac{2}{s(s+1)(s+2)}\right)\right\}$$

$$= (1 - z^{-1})Z\left\{L^{-1}\left(\frac{1}{s} + \frac{-2}{s+1} + \frac{1}{s+2}\right)\right\} = (1 - z^{-1})Z\left\{(1 - 2e^{-t} + e^{-2t})u(t)\right\}$$

$$\Rightarrow G_d(z) = (1 - z^{-1})\left\{\frac{z}{z-1} + \frac{-2z}{z - e^{-T}} + \frac{z}{z - e^{-2T}}\right\} \tag{6.18.1}$$

For $T = 1$ s, the transfer function of the discretized system becomes

$$(6.18.1) \Rightarrow G_d(z) = (1 - z^{-1})\left\{\frac{z}{z-1} + \frac{-2z}{z - 0.368} + \frac{z}{z - 0.135}\right\}$$
$$\Rightarrow G_d(z) = \frac{0.4z + 0.148}{(z - 0.368)(z - 0.135)} \tag{6.18.2}$$

For $z = e^{j\omega T}$, the Bode diagram appears to the following scheme and MATLAB code, correspondingly.

```
for K=1:200
t(K)=0.1*(1.03)⊥K;
z(K)=exp(t(K)*i);
g(K)=(0.4*z(K)+0.148)/(z(K)-0.368)/(z(K)-0.135);
gr(K)=real(g(K));
gi(K)=imag(g(K));
bodem(K)=20*log(sqrt(gr(K)⊥2+gi(K)⊥2));
end
semilogx(t,bodem);axis equal
```

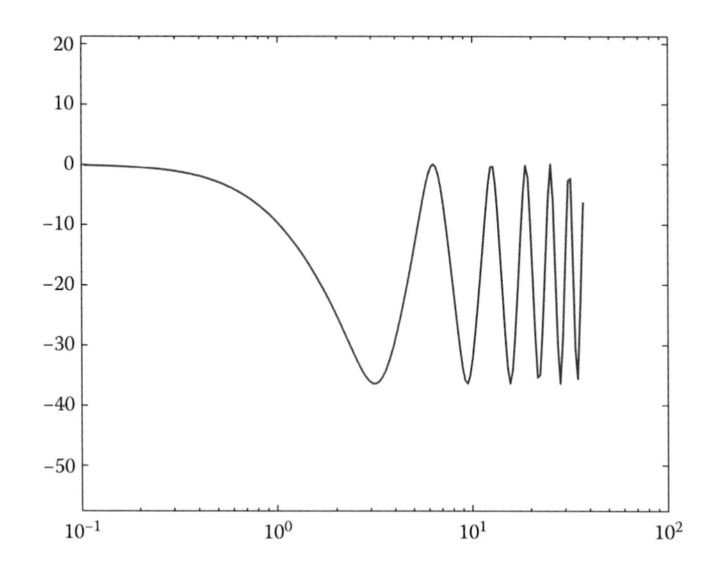

2. It holds that

$$z=\frac{1+(T/2)w}{1-(T/2)w}=\frac{2+w}{2-w} \tag{6.18.3}$$

Substituting in the expression (6.18.2), it stems that

$$\begin{aligned}\hat{G}(w)=G_d(z)\big|_{z=\frac{2+w}{2-w}}&=\frac{0.4z+0.148}{(z-0.368)(z-0.135)}\big|_{z=\frac{2+w}{2-w}}\\ \Rightarrow \hat{G}(w)&=\frac{(2-w)(0.252w+1.096)}{(1.368w+1.264)(1.135w+1.73)}\end{aligned} \tag{6.18.4}$$

In the following scheme, the Bode diagrams of $G(s)$ and $\hat{G}(w)=G_d(z)\big|_{z=\frac{2+w}{2-w}}$ are presented, using MATLAB. Observe that the Bode diagrams of the analog and discretized system are not matched (there is a frequency distortion).

```
for K=1:200
t(K)=0.1*(1.03)^K;
omega(K)=i*t(K);
z(K)=exp(t(K)*i);
g(K)=2/(omega(K)+1)/(omega(K)+2);
gr(K)=real(g(K));
gi(K)=imag(g(K));
bodem(K)=20*log(sqrt(gr(K)^2+gi(K)^2));
    gh(K)=(2-omega(K))*(0.252*omega(K)+1.096)/
(1.368*omega(K)+1.264)/(1.135*omega(K)+1.73);
ghr(K)=real(gh(K));
ghi(K)=imag(gh(K));
```

```
bodemh(K)=20*log(sqrt(ghr(K)^2+ghi(K)^2));
end
semilogx(t,bodem,t,bodemh,'b');axis equal
```

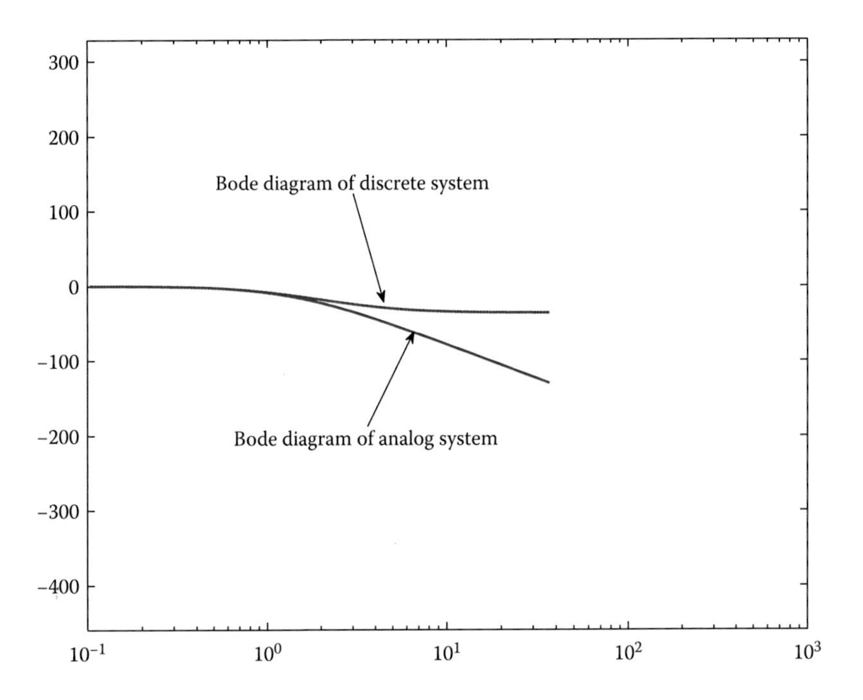

3. The characteristic equation of the closed-loop system is

$$\begin{aligned} a(z) &= (z-0.368)(z-0.135)+K(0.4z+0.148)=0 \\ \Rightarrow a(z) &= z^2+(0.4K-0.5)z+0.148K+0.05=0 \end{aligned} \tag{6.18.5}$$

Based on the Jury criterion, to obtain stability for the closed-loop system, the inequalities of the expressions (6.18.6) through (6.18.8) should hold.

$$a(1)=1+(0.4K-0.5)+0.148K+0.05=0.548K+0.55>0 \tag{6.18.6}$$

$$(-1)^2a(-1)=1+(-0.4K+0.5)+0.148K+0.05=-0.252K+1.55>0 \tag{6.18.7}$$

$$|0.148K+0.05|<1 \tag{6.18.8}$$

From the first inequality, we have $K > -1.004$.
From the second inequality, we have $K < 6.15$.
From the third inequality, we have $-1 < 0.148K + 0.05 < 1$ or equivalently $K > -7.09$ and $K < 6.419$.
The inequalities are jointly true for $-1.004 < K < 6.15$. (6.18.9)

4. The root locus of the compensated system is derived in the same way as in the case of the continuous-time system. It will be a circle with center at $z = -(0.148/0.4) = -0.37$ and parts of the real axis, as in the following scheme:

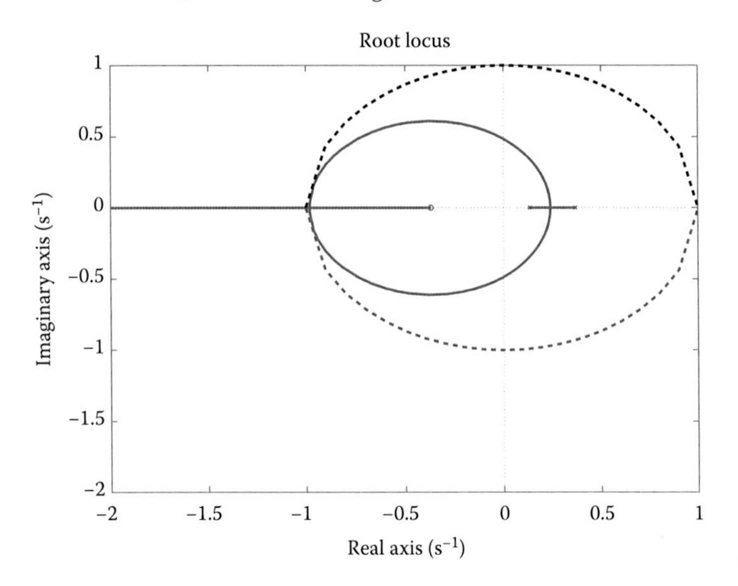

EXERCISE 6.19

For the sampled data system of the following scheme, it is required to

1. Calculate the region of values for the parameter K so as the closed-loop system to be stable using the Routh and Jury criteria.
2. Design the root locus of the system characteristic equation.
3. Design the Bode diagram.

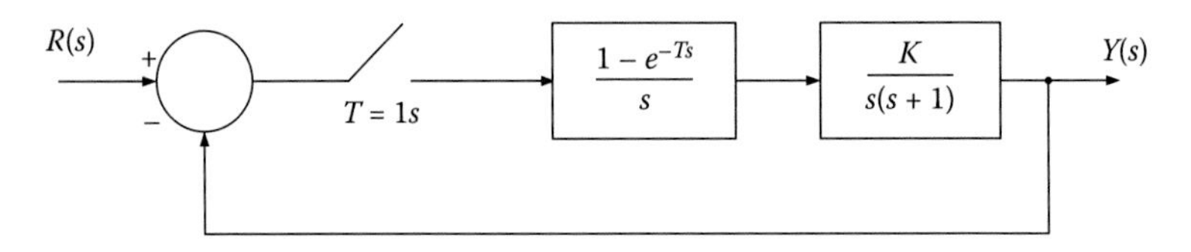

Solution

1. The open-loop transfer function is

$$G(s) = \frac{1-e^{-Ts}}{s}\frac{K}{s(s+1)} = \frac{K}{s^2(s+1)}(1-e^{-Ts}) \tag{6.19.1}$$

From the expression (16.19.1), using z-transform, the discrete open-loop transfer function arises.

$$G(z) = z\left[\frac{K}{s^2(s+1)}\right] * (1 - z^1) = K\frac{0.368z + 0.264}{(z-1)(z-0.368)} \tag{6.19.2}$$

Stability Evaluation Using the Routh Criterion

To find the region of values of K for stability of the closed system, we use the *Routh criterion* with the bilinear transform:

$$z = \frac{1 + (T/2)w}{1 - (T/2)w} \tag{6.19.3}$$

Hence, we transfer to w-domain. The transfer function $G(w)$ is

$$\begin{aligned} G(w) = G(z)\,|\, z = \frac{1 + (T/2)w}{1 - (T/2)w} &= \frac{1 + 0.5w}{1 - 0.5w} \\ &= \frac{-0.0381K(w-2)(w+12.14)}{w(w+0.924)} \end{aligned} \tag{6.19.4}$$

The characteristic equation is

$$\begin{aligned} &F(w) = 1 + G(w) = 0 \\ &\Rightarrow \frac{(1 - 0.0381K)w^2 + (0.924 - 0.386K)w + 0.924K}{w(w+0.924)} = 0 \\ &\Rightarrow (1 - 0.0381K)w^2 + (0.924 - 0.386K)w + 0.924K = 0 \end{aligned} \tag{6.19.5}$$

Formulate the Routh table

w^2	$1 - 0.0381K$	$0.924K$
w^1	$0.0924 - 0.386K$	0
w^0	$0.924K$	

For the stability of the closed system, the following inequalities should be satisfied:

$$\begin{aligned} &1 - 0.0381K > 0 \Rightarrow K > 26.2 \\ &0.924 - 0.386K > 0 \Rightarrow K < 2.39 \\ &K > 0 \end{aligned} \tag{6.19.6}$$

Hence, it should hold:

$$0 < K < 2.39 \tag{6.19.7}$$

Stability Evaluation Using the Jury Criterion

The characteristic equation of the system is

$$\begin{aligned} &1 + G(z) = 0 \Rightarrow 1 + K\frac{0.368z + 0.264}{(z-1)(z-0.368)} = 0 \\ &\Rightarrow F(z) = z^2 + (0.368K - 1.368)z + (0.368 + 0.264K) = 0 \end{aligned} \tag{6.19.8}$$

For the stability of the closed system, the following inequalities should be satisfied:

$$F(1)=1+0.368K-1.368+0.368+0.264\,K>0\Rightarrow K>0 \quad (6.19.9)$$

$$(-1)^2F(-1)=1-0.368K+1.368+0.368+0.264K>0\Rightarrow K<26.3 \quad (6.19.10)$$

$$|a_0|<|a_2|\Rightarrow|0.368+0.264K|<1\Rightarrow K<2.39 \quad (6.19.11)$$

z0	z1	z2
0.368 + 0.264K	0.368K − 1.368	1

Hence, it should hold: $0 < K < 2.39$ (6.19.12)

2. The loop transfer function is written as

$$G(z)=K\frac{0.368z+0.2104}{(z-1)(z-0.368)}=\frac{0.368K(z+0.717)}{(z-1)(z-0.368)} \quad (6.19.12)$$

The poles are $p_1 = 1, p_2 = 0.368$
The zeros are $z_1 = -0.717$
Asymptotes point:

$$\sigma=\frac{1+0.368-0.717}{2-1}=2.085 \quad (6.19.13)$$

Finding the breaking points

$$\begin{aligned}&1+G(z)=0\Rightarrow 1+K\frac{0.368z+0.264}{(z-1)(z-0.368)}=0\\&\Rightarrow K=-\frac{(z-1)(z-0.368)}{0.368z+0.264}\end{aligned} \quad (6.19.14)$$

$$\begin{aligned}&\frac{dK}{dz}=0\Rightarrow\frac{d}{dz}\left(\frac{(z-1)(z-0.368)}{0.368z+0.264}\right)=0\\&\Rightarrow z^2+1.434z-1.3489=0\\&z_1=-2.08,\quad z_2=0.648\end{aligned} \quad (6.19.15)$$

Apparently, both roots of the expression (6.19.15) are accepted as breaking points because in each one of them the gain K is positive, while they belong in the right-most regions, wherein the number of poles and zeros is odd.

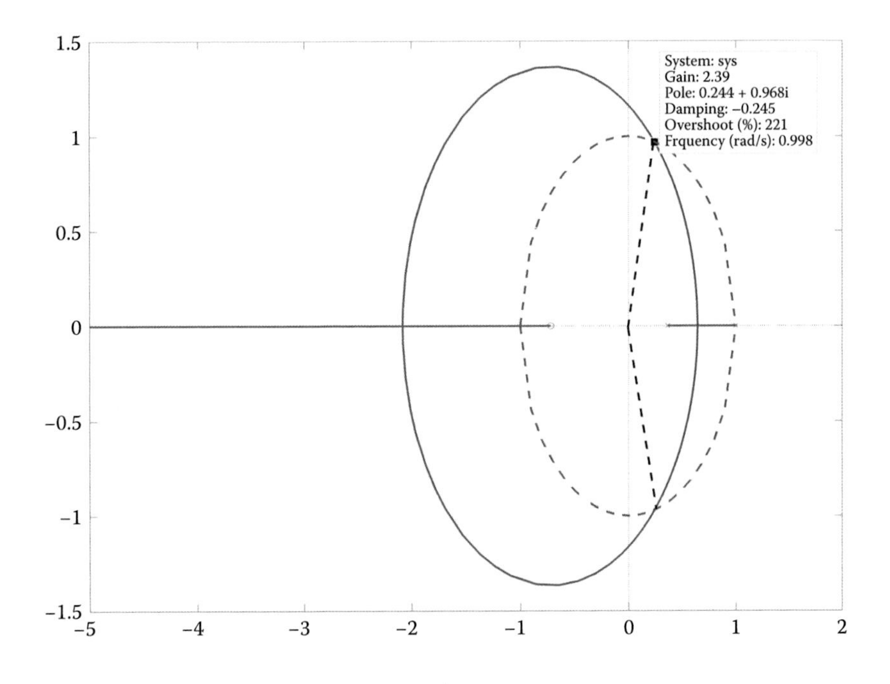

We know from the first part of the exercise that we have stability for $K = 2.39$. Indeed, at the intersection point of the branches of root locus with the unit circle, the value of K is 2.39.

Moreover, $1 + KL(z)\Big|_{K=2.39} = 0$, in $z = 1\angle \pm 75.8°$.

3. The loop transfer function is $G(z) = (0.368z + 0.264)/(z^2 - z + 0.632)$.
Set:

$$z = \frac{1 + 0.5w}{1 - 0.5w} \tag{6.19.16}$$

$$\Rightarrow G(w) = G(z)\Big|_{z=\frac{1+0.5w}{1-0.5w}} = \frac{-0.0381(w-2)(w+12.14)}{w(w+0.924)} \tag{6.19.17}$$

The frequency response arises by setting $w = j\omega_w$. We have that

$$G(j\omega_w) = \frac{(1-(j\omega_w/2))(1+(j\omega_w/12.14))}{j\omega_w(1+(j\omega_w/0.924))} \tag{6.19.18}$$

$$\left|G(j\omega_w)\right|_{\lim \omega_w \to \infty} = \frac{\left|\frac{\omega_w}{2}\right|\left|\frac{\omega_w}{12.14}\right|}{\left|\omega_w\right|\left|\frac{\omega_w}{0.924}\right|} = 0.0381$$

$$20\log(0.0381) = -28.4\,\text{dB}$$

The diagram of the digital system is subsequently designed, where the gain and phase margins for stability can be estimated.

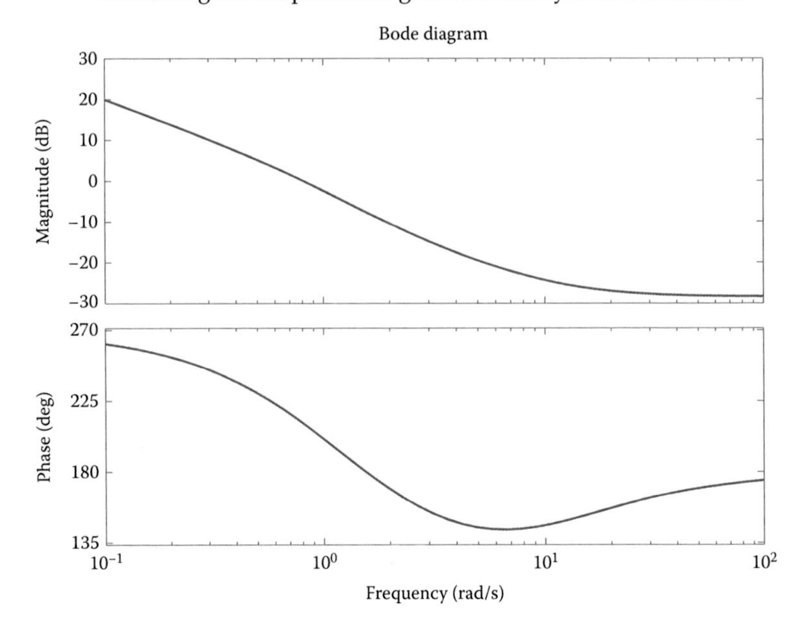

EXERCISE 6.20

For the system of the following scheme, calculate the region of values of K for stability. Design the response of the closed system for $K = K_{cr}$ and extract some relevant conclusions. The evaluation should be done for $T = 0.1$ s and $T = 1$ s

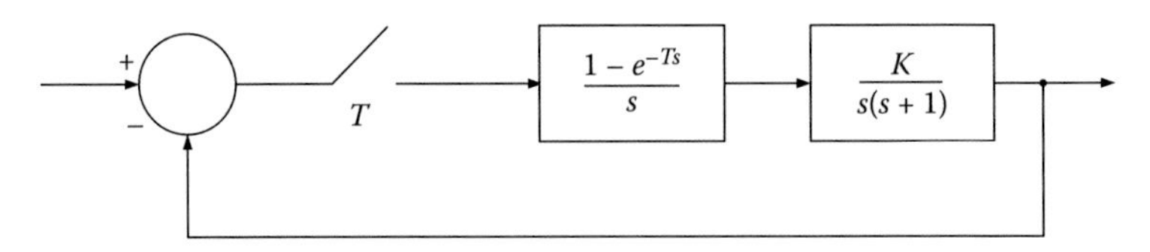

Solution

1. Derive the loop transfer function $G(z)$ using MATLAB for $T = 0.1$ s

```
num_s=[1];den_s=conv([1 0],[1 1]);
G_s=tf(num_s,den_s);
T=0.1;G_z=c2d(G_s,T,'zoh')
```

We get the transfer function

$$G(z) = \frac{0.004837\,z + 0.004679}{z^2 - 1.905z + 0.9048} = \frac{0.004837\,z + 0.004679}{(z-1)(z-0.905)} \quad (6.20.1)$$

The bilinear transform $z = (1+(T/2)w)\,/\,(1-(T/2)w)$ reflects the internal of the unit circle in the left w half plane. It is $z = (1 + 0.05w)/(1 - 0.05w)$ since $T = 0.1$, thus the loop transfer function becomes the corresponding one of $G(w)$.

$$G(w) = \frac{-0.00016w^2 - 0.1872w + 3.81}{3.81w^2 + 3.8w} \tag{6.20.2}$$

The characteristic equation is given by: $1 + KG(w)$=0, so

$$(3.81 - 0.00016K)w^2 + (3.8 - 0.1872K)w + 3.81K = 0 \tag{6.20.3}$$

The Routh Table will be of a form

w^2	$3.81 - 0.00016K$	$3.81K$
w^1	$3.80 - 0.1872K$	
w^0	$3.81K$	

where the region of values for the parameter K for stability is $0 < K < 20.3$. Obviously, for $K = K_{cr} = 20.3$, the closed system will be marginally stable, which is numerically verified as follows:

```
K=20.34;
Hcl_z=K*G_z/(1+K*G_z);
pole(Hcl_z)
ans =
   0.9032 + 0.4292i
   0.9032 - 0.4292i
1.0000
   0.9048
step(Hcl_z)
```

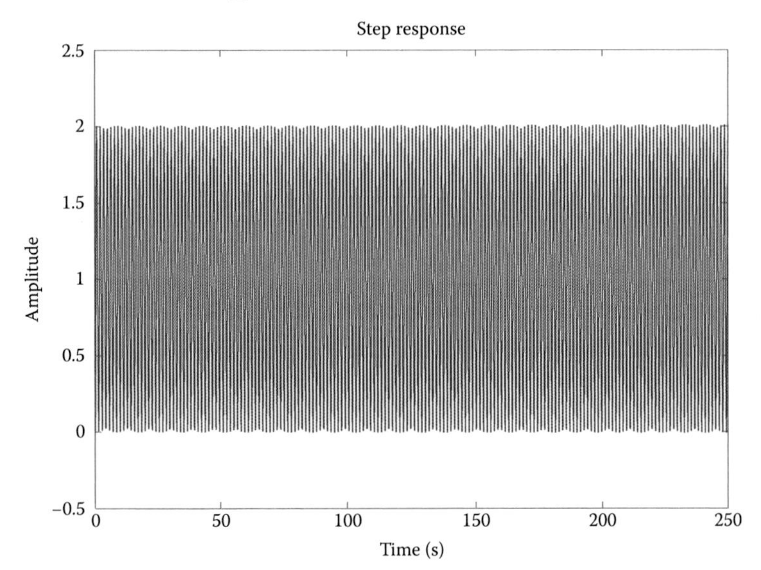

Observe that the system, for step input, oscillates with fixed amplitude that does not fade out.

2. Derive the loop transfer function $G(z)$ using MATLAB for $T = 1$ s

```
num_s=[1];den_s=conv([1 0],[1 1]);
G_s=tf(num_s,den_s);
T=1;G_z=c2d(G_s,T,'zoh')
```

The transfer function arises as

$$G(z) = \frac{0.36788\,(z + 0.7183)}{(z-1)(z-0.3679)} \tag{6.20.4}$$

Apply the bilinear transform to the given loop transfer function to obtain $G(w)$. Write the new characteristic equation as

$$\begin{aligned} 1 + KG(w) = 0 \Rightarrow (1 - 0.03788K)w^2 \\ + (0.9242 - 0.3864K)w + 0.9242K = 0 \end{aligned} \tag{6.20.5}$$

The Routh table will be of a form

w^2	1–0.03788K
w^1	0.9242–0.3864K
w^0	0.9242K

where the region of values for the parameter K for stability is $0 < K < 2.39$. Apparently, for $K = K_{cr} = 2.29$, the closed system will be marginally stable. Observe that for higher sampling period, the region of values of the parameter K is drastically reduced to preserve stability, as expected.

EXERCISE 6.21

The system of the following scheme is given. Using LabVIEW, derive its total transfer function and design the poles-zeros diagram. Is the system stable?

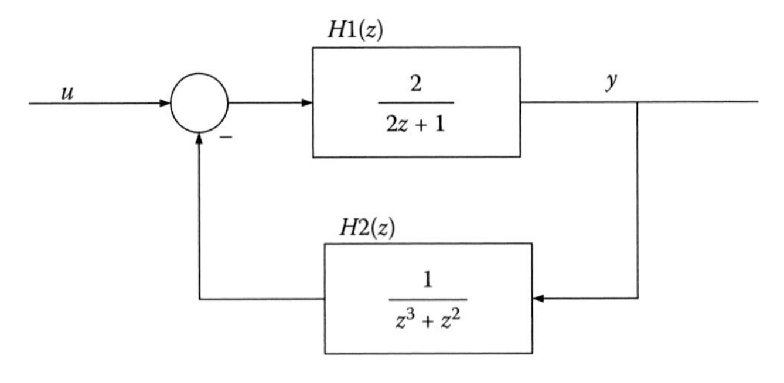

Solution

In the following schemes, the front panel and block diagram of the designed vi are presented.

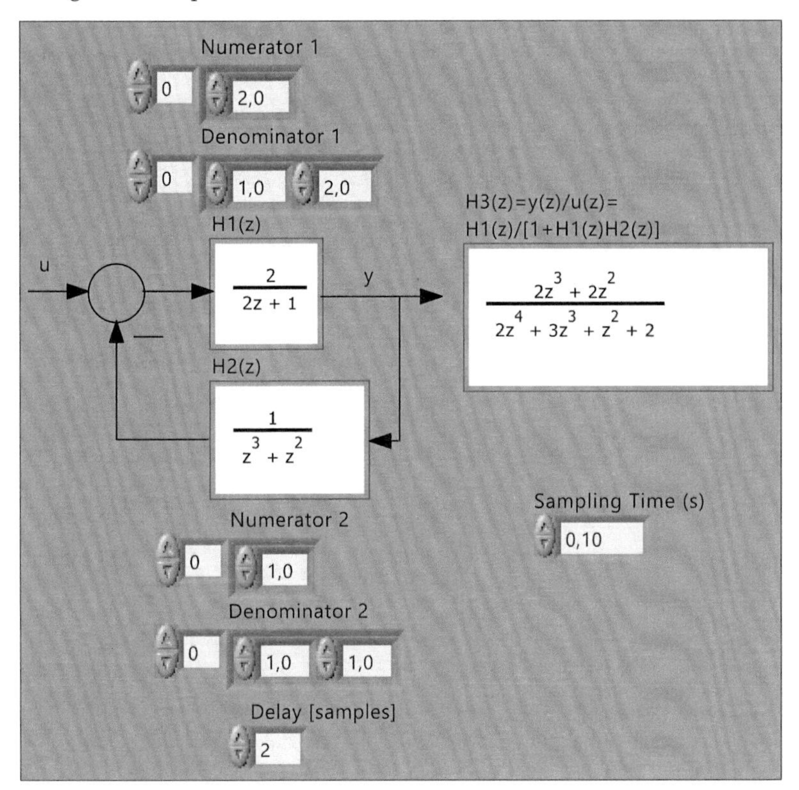

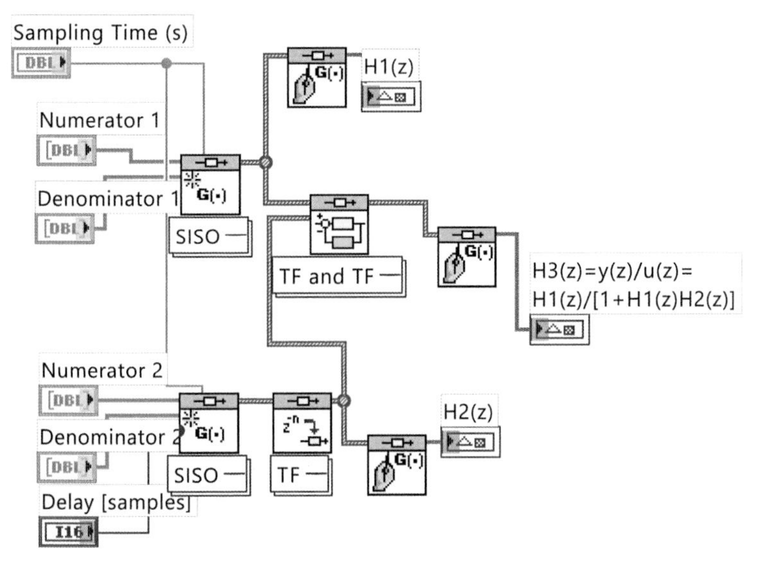

The resultant transfer function is

$$H_3(z) = \frac{H_1(z)}{1 + H_1(z)H_2(z)} = \frac{2z^3 + 2z^2}{2z^4 + 3z^3 + z^2 + 2} \tag{6.21.1}$$

In the following schemes, the front panel and block diagram of the poles-zeros diagram is presented. It appears that complex conjugate poles exist outside the unit circle, so the system is unstable.

Indeed, the poles of the closed-loop system are

$-1.0708 + 0.8247i$
$-1.0708 - 0.8247i$
$0.3208 + 0.6667i$
$0.3208 - 0.6667i$

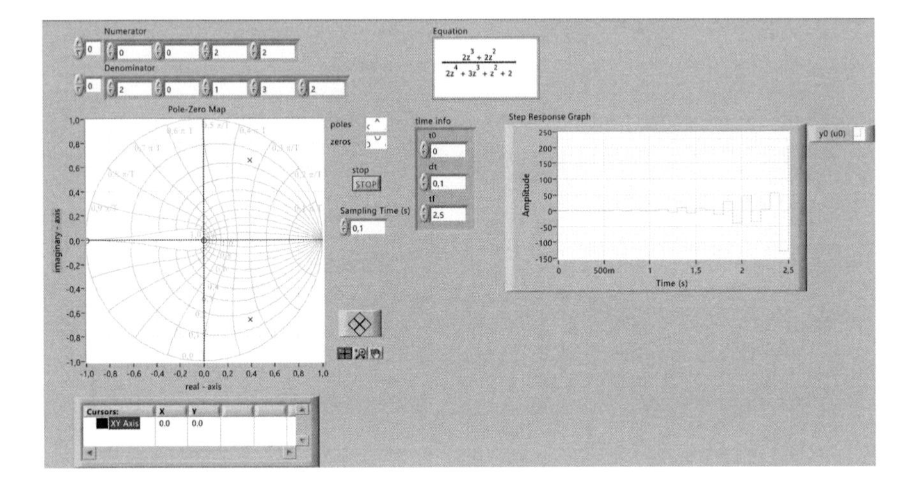

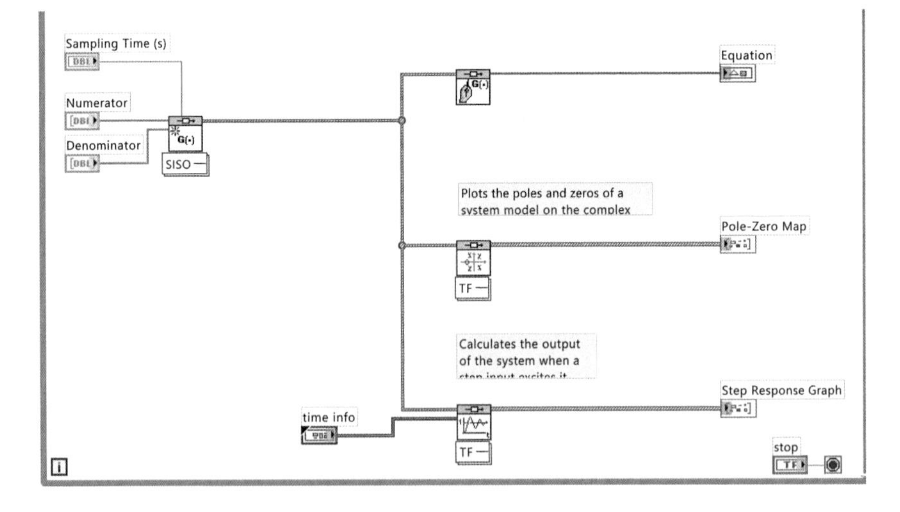

EXERCISE 6.22

Let the following loop transfer function: $G(s) = 2Ks/((s+1)^2(s+2))$. Discretize it using the sample equivalence method (or sampled inverse Laplace transform) and design the root locus diagram of the system. Is the given closed-loop system stable for $K = 1$?

Solution

Insert the transfer function to the program Comprehensive Control—CC with the command

```
CC>G=enter(1,1,2,0, 2,1,1,2,2,1,2,1)
```

Using the command Convert, the transfer function $G(s)$ is being discretized. Select the sample equivalence method for sampling period $T = 1$ s. Writting the command

```
CC>gd=convert(G,7,1)
```

the following transfer function in z-domain arises as

$$gd(z) = \frac{0.1944z(z-1.248)}{(z-0.3679)^2(z-0.1353)} \tag{6.22.1}$$

To design the root locus of the characteristic equation, write

```
CC>rootlocus(gd)
```

From the root locus diagram, the marginal value of the parameter K arises, which is approximately 7.2. Hence, for $K = 1$, the closed system is stable.

$$s = 1 + 0.002445j(\text{Mag} = 1, \text{Zeta} = -1)$$
$$\text{gain} = 7.189 + 0.1294j(\text{Mag} = 7.19, \text{Phase} = 1.031\,\text{deg})$$

Theoretical proof

The open-loop transfer function, which is provided in the expression (6.22.1), has three poles (a double one and a simple one) and two zeros

$$\begin{aligned} p_1 = p_2 &= 0.368, \quad p_3 = -0135 \\ z_1 &= 1.24 \end{aligned} \tag{6.22.2}$$

Thus, the root locus will have three branches. In the real axis, a locus is placed between z_1 and p_1 and between p_2 and p_3, while $z_1 \to -\infty$.

Asymptotes are

$$\Phi_\alpha = \frac{(2*\mu+1)*180^\circ}{n_p - n_z} = (2*\mu+1)*180^\circ \xrightarrow{\mu=0} \Phi_\alpha = 180^\circ \tag{6.22.3}$$

Intersection point of asymptotes with the real axis

$$\sigma_{\alpha} = \frac{\Sigma p - \Sigma z}{n_p - n_z} = 0.871 - 1.248 \Rightarrow \sigma_{\alpha} = -0.377 \tag{6.22.4}$$

Break away points of the real axis

$$\frac{2}{|\sigma_b| - 0.368} + \frac{1}{|\sigma_b| - 0.135} = \frac{1}{|\sigma_b| - 1.248} \Rightarrow \sigma_b = \begin{cases} -0.150 \\ +0.198 \end{cases} \tag{6.22.5}$$

The latter two values are accepted as break away points of the branches from the real axis since there is an odd number of poles and zeros to the right-most side of each of them. Also, for both σ_b values, the system gain is positive (for $\sigma_b = -0.15$, there is almost a gain of 1.88 and for $\sigma_b = 0.198$, there is almost a gain of 0.0447).

The system characteristic equation with unit feedback $H(z) = 1$ (closed system) is given by (assuming that $K = 1$)

$$p(z) = 0 \Rightarrow 1 + gd(z) = 0 \Rightarrow z^3 - 0.677z^2 - 0.008z - 0.018 = 0 \tag{6.22.6}$$

Apply the Jury test for the latter equation

z_0	z_1	z_2	z_3
−0.018	−0.008	−0.677	1
1	−0.677	−0.008	−0.018
−0.999	0.677	0.020	

$$b_k = \begin{vmatrix} a_0 & a_{n-k} \\ a_n & a_k \end{vmatrix}, \quad n = 3 \Rightarrow b_0 = -0.999, \quad b_1 = 0.677, \quad b_2 = 0.020$$

Obviously: $a_0 = -0.018, \quad a_1 = -0.008, \quad a_2 = -0.677, \quad a_3 = 1$

The following inequalities should hold

$$\begin{aligned} & p(1) = 0.29 > 0 \\ & (-1)^n * p(-1) = 1.68 > 0 \\ & |a_0| < |a_3| \\ & |b_0| > |b_2| \end{aligned} \tag{6.22.7}$$

All the above equations hold true, thereby the closed-loop system of unit feedback is stable for $K = 1$.

7

Time and Harmonic Response Analysis Steady-State Errors

7.1 Time Response

The time response of a system denotes the behavior of the system over time for a given input. The time response of a control system consists of two parts:

1. The *transient response* and
2. The *steady-state response*

It holds that

$$y(k) = y_t(k) + y_{ss}(k) \tag{7.1}$$

where $y_t(k)$ = the transient response
and $y_{ss}(k)$ = the steady-state response

By *transient* we mean the response of the system directly after its excitation and before stabilization of its corresponding output.

The term *steady state* denotes the remaining part of the response after the attenuation of the transitional part, holding that

$$y_{ss}(k) = \lim_{k \to \infty} y(k) \tag{7.2}$$

The time response of a discrete system can be calculated in two ways

1. If the system is being described by a transfer function, first $Y(z)$ is calculated and then $y(k)$ is calculated via the inverse z-transform.
2. If the system is being described by state space equations, first the solution of state vector $x(k)$ is derived and then $y(k)$ is calculated.

The *design specifications* for a control system include, among others, various parameters of the corresponding time response with respect to a given input function along with the required precision that should be preserved during the steady state. The specifications, determined in accordance to the required operation measures, represent an indicator of the system *quality*. The system time response denotes a feature of the most interest. In the case when a system is stable, its time response, given a particular input signal, provides valuable information regarding the general system performance.

Generally, certain *typical input signals* are selected so as to correlate the system response to a given signal and its operational behavior under canonical conditions.

The most common *input signals* are

- Unit step function
- Ramp function
- Dirac function
- Parabola function
- Sinusoid function, etc.

The time response of a closed-loop control system can be described as a function of the location of poles of the transfer function in the complex plane. The information obtained from the knowledge of the relative location of poles of a system practically corresponds to a graphical method for determining its behavior. The *poles* of the closed-loop transfer function $G_{cl}(z)$ determine the form of the corresponding response, while the *zeros* of $G_{cl}(z)$ determine the fixed terms of the corresponding functions. Specifically, moving a zero closer to a pole, the influence rate of the function corresponding to this pole in the system response is decreased.

The time response of a system can be described as a function of two factors

- The velocity of the response, as expressed by the rise time and peak time.
- The matching level between the actual and the desired system response, as it is expressed by the percentage of overshoot and settling time.

In principle, the above factors contradict to each other and therefore some relative compromises must be performed. In practical control systems, the transient response manifests damped oscillations before reaching the steady state.

7.1.1 Impulse Time Response of First-Order Systems

Consider *the first-order system* with a single pole at $z = a$, as presented in Figure 7.1.

In Figure 7.2, the time responses for $|a| < 1$ are presented, where the system is stable; and for $|a| > 1$ in Figure 7.3, where the system is unstable.

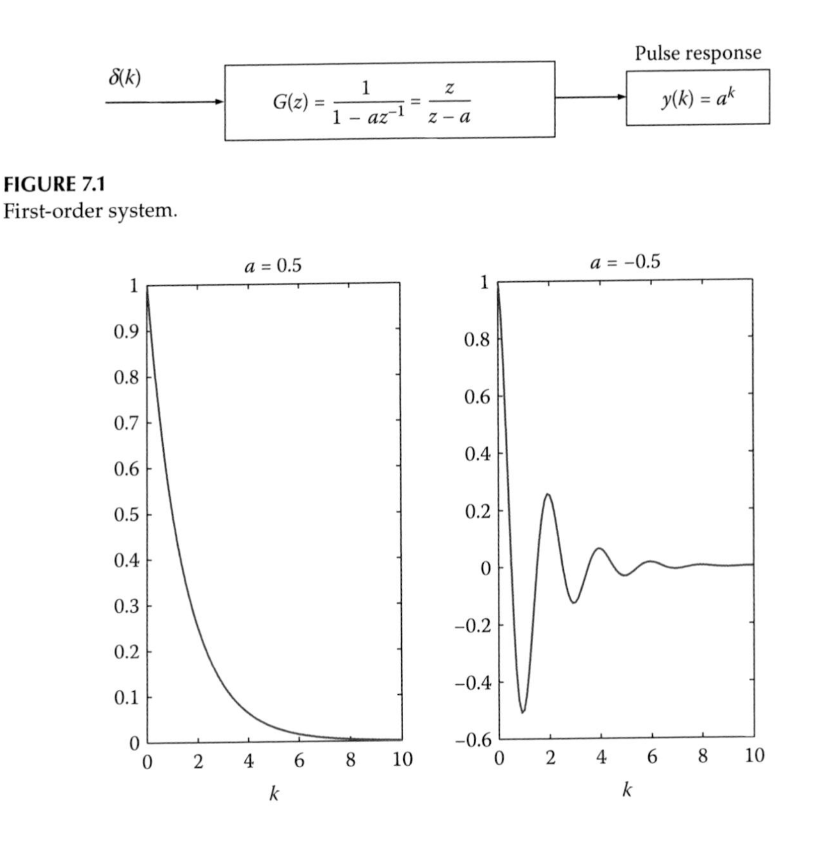

FIGURE 7.1
First-order system.

FIGURE 7.2
Time responses for the step response $y(k) = a^k$ when $|a| < 1$.

7.1.2 Impulse Time Response of Second-Order Systems

Consider *the second-order system* with a pair of complex conjugate poles at $z = re^{\pm j\theta}$

$$G(z) = \frac{Nz^{-1}}{(1 - re^{j\theta}z^{-1})(1 - re^{-j\theta}z^{-1})} \tag{7.3}$$

The impulse response of the second-order system is expressed as

$$y(k) = 2r^k(a\cos(k\theta) - \beta\sin(k\theta)) \tag{7.4}$$

N is the number of samples per oscillation of a sinusoidal signal

$$N = \frac{2\pi}{\theta}\Big|_{rad} = \frac{360}{\theta}\Big|_{\deg} \tag{7.5}$$

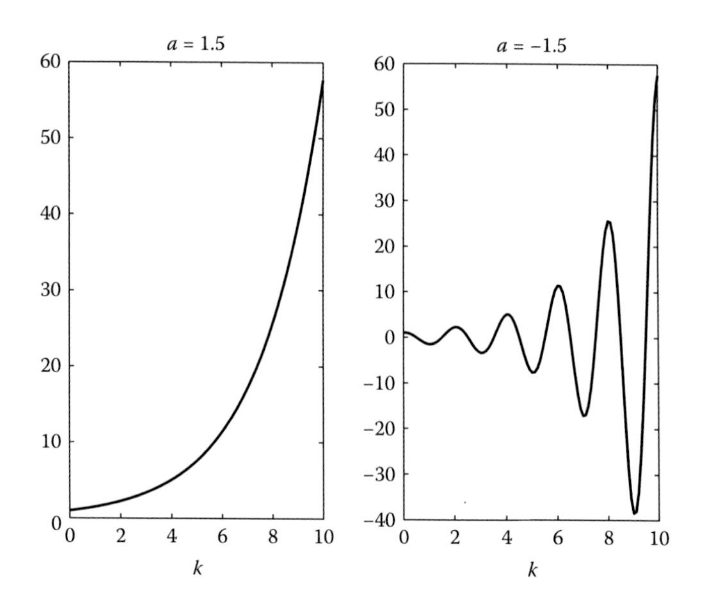

FIGURE 7.3
Time responses for the step response $y(k) = a^k$ when $|a| > 1$.

7.1.3 Step Response of First-Order System

Consider a first-order system with a single pole at $z = p$

$$G(z) = \frac{b}{z - p} \tag{7.6}$$

The step response is given by

$$y(k) = \frac{b}{1 - p}(1 - p^k) \tag{7.7}$$

From the expression (7.7), it can be seen that for $p = 1$ the system acts as an integrator, for $0 < p < 1$ the output exponentially follows the input, for $p = 0$ the system acts as a delay component, for $p = -1$ the step response results to oscillations of fixed amplitude (which do not fade out), and for $-1 < p < 0$ the step response oscillates tending to a fixed value.

7.1.4 General Form for the Step Response of Discrete Time System

Consider a system with the transfer function (where $n > m$)

$$G_{CL}(z) = \frac{p(z)}{q(z)} = \frac{b_m z^m + b^{m-1} z_{m-1} + \cdots + b_1 z + b_0}{a_n z^n + a_{n-1} z^{n-1} + \cdots + a_1 z + a_0} \tag{7.8}$$

$$= \frac{b_m}{a_n} \times \frac{(z-z_1)(z-z_2)\cdots(z-z_j)\cdots(z-z_m)}{(z-p_1)(z-p_2)\cdots(z-p_i)\cdots(z-p_n)}$$

$$= \frac{b_m}{a_n} \times \frac{\prod_{j=1}^{m}(z-z_j)}{\prod_{i=1}^{n}(z-p_i)} \tag{7.9}$$

For distinguished real poles, the step response yields as

$$Y(z) = \frac{p(z)}{q(z)}\frac{z}{z-1} = \frac{Az}{z-1} + \sum_{i=1}^{n}\frac{B_i z}{z-p_i} \tag{7.10}$$

$$\text{where } A = \left.\frac{p(z)}{q(z)}\right|_{z=1} \qquad B_i = \left.\frac{p(z)(z-p_i)}{q(z)(z-1)}\right|_{z=p_i} \tag{7.11}$$

1. *p_i is a real positive pole*

   The transient response, in this case, is given by

$$Y(z) = \sum_{i=1}^{n}\frac{B_i z}{z-p_i} \Rightarrow y(kT) = \sum_{i=1}^{n} B_i p_i^k \quad \left(Z[a^k] = \frac{1}{1-az^{-1}}\right) \tag{7.12}$$

   Let $\alpha = (1\backslash T)\ln p_i$. Then, the transient part of the system response is $B_i e^{\alpha kT}$. For $|p_i| < 1 \Rightarrow a < 0$. The transient response is a exponentilly decreasing curve. The smaller the value of $|p_i|$ and higher the value of $|a|$, the faster the transient response gets.

2. *p_i is a real negative pole*

   The transient response, in this case, is given by

$$Y(z) = \sum_{i=1}^{n}\frac{B_i z}{z-p_i} \Rightarrow y(kT) = \sum_{i=1}^{n} B_i p_i^k \tag{7.13}$$

$$\text{where } B_i p_i^k = \begin{cases} B_i |p_i|^k & k \quad \text{is even} \\ -B_i |p_i|^k & k \quad \text{is odd} \end{cases} \tag{7.14}$$

The negative poles correspond to high-frequency oscillations with a frequency $\omega_s/2$.

3. *p_i equals to zero*

   The transient response is presented as

$$Y(z) = \sum_{i=1}^{n} \frac{B_i z}{z - p_i} \Rightarrow y_i(kT) = B_i \delta(k) \tag{7.15}$$

   When the pole p_i equals to zero, the transient response is faster and is called *deadbeat* control.

4. *Case of multiple poles with multiplication factor m*

   The transient response is presented as

$$\begin{aligned} Y(z) &= \frac{p(z)}{q(z)} \times \frac{1}{1 - z^{-1}} \\ &= \frac{p(1)}{q(1)} + \frac{z a_1 \Phi_1(z)}{(z-p)^m} + \frac{z a_2 \Phi_2(z)}{(z-p)^{m-1}} + \cdots \Rightarrow \end{aligned}$$

$$y(kT) = a_1 k^{m-1} p^{k-1} + a_2 k^{m-2} p^{k-1} + \cdots \tag{7.16}$$

   For $|p| < 1, p \neq 0$ it holds: $\lim\limits_{k \to \infty} \frac{a_i k^{m-i}}{(1/p)^{k-1}} = 0,\ i = 1,2,3,\ldots,$ for $0 < p < 1$ the response exponentially decreases, for $-1 < p < 0$ decreases with high-frequency oscillations at a frequency $\omega_s/2$.

5. *Complex conjugate poles*

$$\begin{aligned} &a_i, a_{i+1} = |a_i| e^{\pm j\phi_i} \\ &p_i = |p_i| e^{j\theta_i}, p_{i+1} = |p_i| e^{-j\theta_i} \end{aligned}$$

   where

$$|a_i| = \left| \frac{p(z)(z-p_i)}{q(z)(z-1)} \right|_{z=p_i} \qquad \phi_i = \arg \left| \frac{p(z)(z-p_i)}{q(z)(z-1)} \right|_{z=p_i} \tag{7.17}$$

   The transient response is presented as

$$\begin{aligned} y_i(kT) + y_{i+1}(kT) &= a_i |p_i|^k + a_{i+1} |p_{i+1}|^k \\ &= |a_i| e^{j\phi_i} \cdot |p_i|^k e^{jk\theta_i} + |a_i| e^{-j\phi_i} \cdot |p_i|^k e^{-jk\theta_i} \\ &= 2|a_i| \cdot |p_i|^k \cos(k\theta_i + \phi_i) \quad 0 < \theta_i < \pi \end{aligned} \tag{7.18}$$

The transient response performs decreasing oscillations having a periodic form. The higher the value of θ_i (ωT: $0 \rightarrow \pi/2 \rightarrow \pi$), the more intense the occurred oscillations (oscillation frequency $0 \rightarrow \omega s/4 \rightarrow \omega s/2$).

7.1.5 Correlation between Analog and Discrete Time Response

Consider the block diagram of the *second-order system*, as shown in Figure 7.4.

The transfer function is of a form

$$G(s) = \frac{Y(s)}{X(s)} = \frac{\omega_n{}^2}{s^2 + 2J\omega_n s + \omega_n{}^2} \tag{7.19}$$

The constant J is called *damping ratio* of the system, the constant ω_n is called *undamped natural frequency* and the constant $\omega_d = \omega_n\sqrt{1-J^2}$ is called *damped natural frequency*.

In control system applications, only the case $0 < J < 1$ is of practical interest since it corresponds to stable systems. In such a case, the characteristic equation has two complex conjugate poles

$$s_{1,2} = -J\omega_n \pm j\omega_n\sqrt{1-J^2} = -J\omega_n \pm j\omega_d \tag{7.20}$$

In Figure 7.5, the two complex planes, s and z, and their joint correlation are illustrated according to the relation $z = e^{Ts}$.

- *Relation between z and J, ω_n, ω_d*

$$z = e^{Ts} = e^{\sigma T} e^{j\omega T} = e^{\sigma T} e^{j2\pi \frac{f}{f_s}} \tag{7.21}$$

$$\begin{aligned} z = e^{Ts} &= \exp(-J\omega_n T + j\omega_d T) \\ &= \exp\left(-\frac{2\pi J}{\sqrt{1-J^2}}\frac{\omega_d}{\omega_s} + j2\pi\frac{\omega_d}{\omega_s}\right) \Rightarrow \end{aligned}$$

$$\left.\begin{aligned} \Rightarrow |z| &= \exp\left(-\frac{2\pi J}{\sqrt{1-J^2}}\frac{\omega_d}{\omega_s}\right) \\ \angle z &= 2\pi\frac{\omega_d}{\omega_s} \end{aligned}\right\} \tag{7.22}$$

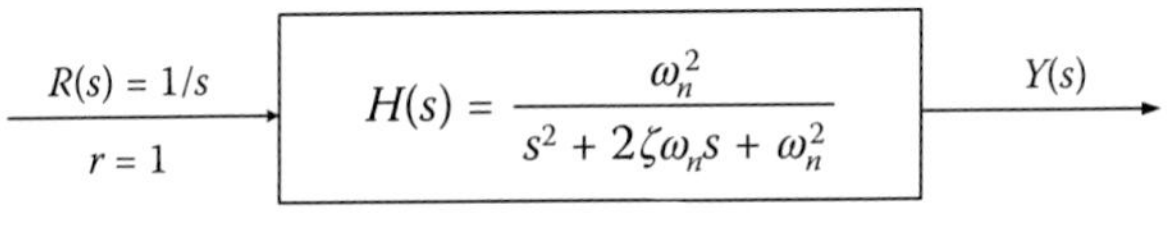

$$R(s) = 1/s,\ r = 1 \longrightarrow \boxed{H(s) = \frac{\omega_n^2}{s^2 + 2\zeta\omega_n s + \omega_n^2}} \longrightarrow Y(s)$$

FIGURE 7.4
Continuous-time second-order system.

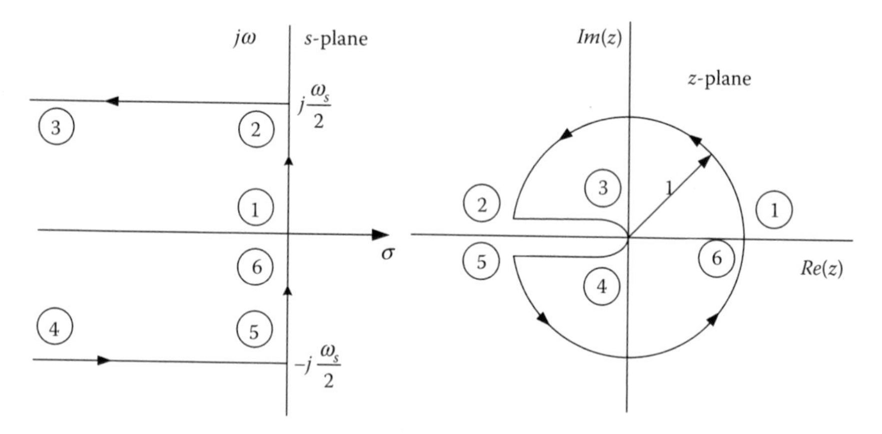

FIGURE 7.5
Joint correlation of planes s and z, according to the relation $z = e^{Ts}$.

- *Percentage of overshoot*

$$POT = \frac{y_{\max} - y_{ss}}{y_{ss}} 100\% \tag{7.23}$$

where $y_{\max}$ and y_{ss}, denote the maximum value and the steady-state value, respectively, of the system response $y(k)$.

- *Settling time*

$$t_s = k_s T \tag{7.24}$$

where k_s satisfies the condition

$$\begin{gathered} |y(k) - y_{ss}| \le \frac{\varepsilon y_{ss}}{100}, \quad \forall k \ge k_s \Leftrightarrow \\ \left(1 - \frac{\varepsilon}{100}\right) y_{ss} \le y(k) \le \left(1 + \frac{\varepsilon}{100}\right) y_{ss}, \quad \forall k \ge k_s \end{gathered} \tag{7.25}$$

Typically, it holds that

$$t_s = \frac{3}{J\omega_n} \tag{7.26}$$

- *Dominant poles*, that is, the poles that are closer to the circumference of the unit circle.

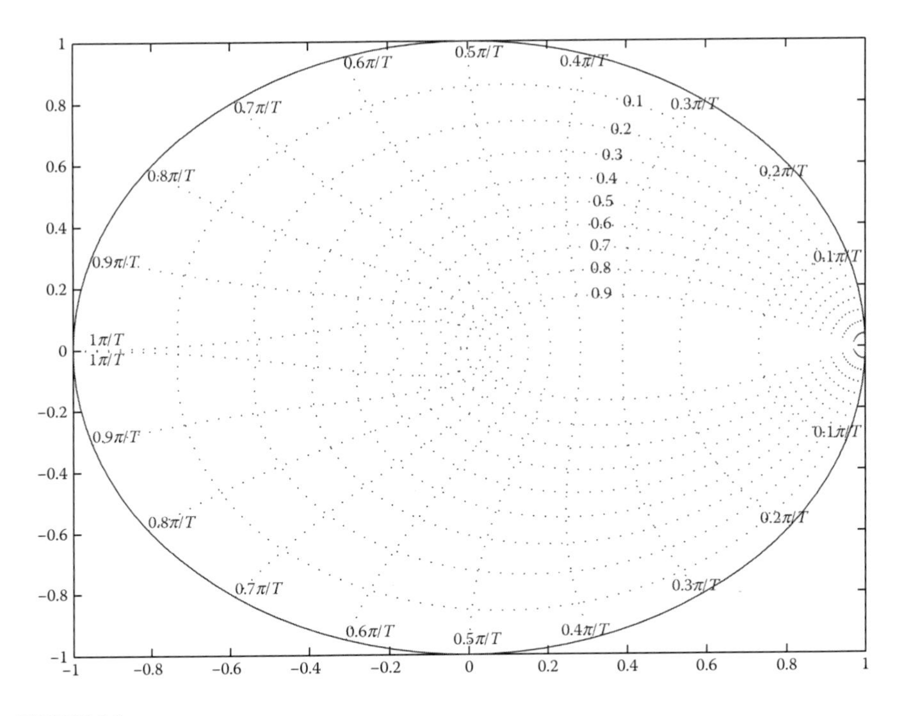

FIGURE 7.6
Illustration of the fixed damping coefficient lines (in logarithmic spiral) and the natural oscillation frequency lines (in radial curves) from *s*- to *z*-plane.

$$z_{1,2\surd} = re^{j\varphi} \Rightarrow$$
$$J = \frac{-\ln r}{\sqrt{(\ln r)^2 + \varphi^2}} \tag{7.27}$$

$$\omega_n = \frac{1}{T}\sqrt{(\ln r)^2 + \varphi^2} \tag{7.28}$$

In Figure 7.6, the fixed damping coefficient lines (in logarithmic spiral) and the natural oscillation frequency lines (in radial curves) are illustrated from *s*- to *z*-plane. This scheme has been developed via the MATLAB® command **`zgrid`**.

7.2 Steady-State Errors

An important factor related to the operation of control systems corresponds to the *error in the steady state* (steady-state error—$e_{ss}(kT)$), which appears to the system output after the transient response period.

The *steady state* is quite important since the design of an automatic control system is designed, among others, to maintain a predetermined steady state for the output $y_{ss}(kT)$, which is usually the input function $u(kT)$. In other words, the system is designed so as to hold $y_{ss}(kT) = u_{ss}(kT)$, when the system is stimulated by $u(kT)$. Otherwise, we have a static error or steady-state error.

Note that the steady-state error of closed-loop stable system is usually much smaller than its open-loop counterpart.

From the theory of analog control systems, it is known that the steady-state error depends on the input function and the system features.

The system features are studied with the aid of three error factors, namely, position, velocity, and acceleration (Kp, Kv, and Kα). It will be shown, subsequently, that these factors can be utilized in digital control systems as well.

To study the steady-state error, the block diagram of Figure 7.7 is be used, assuming that the system is stable. Otherwise, the error would increase without any bound, reflecting a system self-destruction.

It holds that

$$e(t) = r(t) - b(t) \tag{7.29}$$

To make the analysis feasible, the signal $e^*(t)$ is used, thus the steady-state error, during sampling, is

$$e_{ss}^* = \lim_{t \to \infty} e^*(t) = \lim_{k \to \infty} e(kT) \tag{7.30}$$

Applying the z-transform and the final value theorem, we get

$$e_{ss}^* = \lim_{t \to \infty} e^*(t) = \lim_{z \to 1} (z-1)E(z) \tag{7.31}$$

For the system of the scheme, we have that

$$E(z) = \frac{R(z)}{1 + z[GH(s)]} \tag{7.32}$$

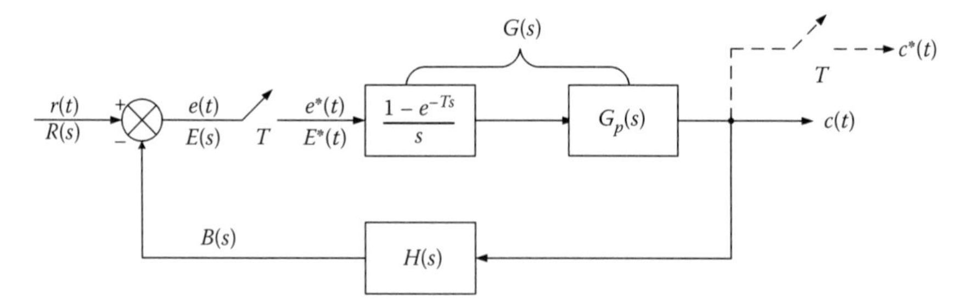

FIGURE 7.7
Sampled data system.

$$\overset{(7.31),(7.32)}{\Rightarrow} \quad e_{ss}^{*} = \lim_{t\to\infty} e^{*}(t) = \lim_{z\to 1}(z-1)\frac{R(z)}{1+z[G(s)H(s)]} \tag{7.33}$$

From the expression (7.33), it is explicitly indicated that the steady-state error depends on the input signal *r*(*t*) and the system features.

Consider the open-loop transfer function in the general form

$$GH(z) = \frac{k\prod^{m}(z-z_i)}{(z-1)^N \prod^{P}(z-z_j)}, \quad z_j \neq 1, z_i \neq 1 \tag{7.34}$$

N plays a crucial role since it influences the system features and defines the system as a *zero-*, *first-* or *second*-order, when $N = 0$, 1, and 2.

Define K_{dc} as

$$K_{dc} = \left.\frac{k\prod^{m}(z-z_i)}{\prod^{P}(z-z_j)}\right|_{z=1} \tag{7.35}$$

where K_{dc} is the gain of the open system for zero frequency, when the pole at $z = 1$ has been removed.

Next, we study the impact of input function for *three different signal types*. The most common ones are the step, ramp and parabolic input functions.

1. *Step input*

   The z-transform for a step input of range A is

$$R(z) = \frac{Az}{z-1} \tag{7.36}$$

From:

$$\overset{(7.33),(7.36)}{\Rightarrow} \quad e_{ss}^{*} = \lim_{z\to 1}\frac{A}{1+\lim_{z\to 1} GH(z)} = \frac{A}{1+k_p} \tag{7.37}$$

where k_p is the position error constant:

$$k_p = \lim_{z\to 1}(GH(z)) \tag{7.38}$$

We can easily show, from the expressions (7.34), (7.35), and (7.39), that $k_p = k_{dc}$ for $N = 0$ and $k_p = \infty$ for $N \geq 1$.

$$GH(z) = z[GH(s)] = (1 - z^{-1})z\left[\frac{GpH(s)}{s}\right] \tag{7.39}$$

Consequently, for $N = 0$, the position error is

$$e_{ss}^{*} = \frac{A}{1 + k_p} = \frac{A}{1 + k_{dc}} \tag{7.40}$$

while for $N \geq 1$ the position error is zero ($e_{ss} = 0$).

2. *Ramp input*

   The z-transform for a ramp input of range A is

$$R(z) = \frac{ATz}{(z-1)^2} \tag{7.41}$$

$$\begin{aligned} &\overset{(7.33),(7.41)}{\Rightarrow} e_{ss}^{*} = \lim_{z \to 1} \frac{AT}{(z-1)(1+GH(z))} \\ &\Rightarrow e_{ss}^{*} = \frac{A}{\lim_{z \to 1}((z-1)/T)GH(z)} = \frac{A}{k_v} \end{aligned} \tag{7.42}$$

   where k_v is the velocity error constant.

$$k_v = \lim_{z \to 1}\left(\frac{z-1}{T}\right)GH(z) \tag{7.43}$$

   For $N = 0 \Rightarrow k_v = 0$ and $e_{ss} = \infty$
   For $N \geq 1 \Rightarrow k_v = \infty$ and $e_{ss} = 0$
   For $N = 1 \Rightarrow k_v = (k_{dc}/T)$ so: for $N = 1$ the velocity error is

$$e_{ss}^{*} = \frac{A}{k_v} = \frac{AT}{k_{dc}} \tag{7.44}$$

3. *Parabolic input*

   The z-transform for a parabolic input of range A is

$$R(z) = \frac{AT^2 z(z+1)}{2(z-1)^3} \tag{7.45}$$

TABLE 7.1

Steady-State Errors

Error Type (Number of Integrators of $G(z)$)	Error Constants	Steady-State Errors $e_{ss}^* = \lim_{t\to\infty} e^*(t) = \lim_{z\to 1}(z-1)\frac{R(z)}{1+z[G(s)H(s)]}$		
		Position	**Velocity**	**Acceleration**
0	Position: k_p Velocity: 0 Acceleration: 0	$\frac{A}{1+k_p} = \frac{A}{1+k_{dc}}$	∞	∞
1	Position: ∞ Velocity: k_v Acceleration: 0	0	$\frac{A}{k_v} = \frac{AT}{k_{vcd}}$	∞
2	Position: ∞ Velocity: ∞ Acceleration: k_a	0	0	$\frac{A}{k_a} = \frac{AT^2}{k_{dc}}$

$$\begin{aligned} &\overset{(7.33),(7.45)}{\Rightarrow} \quad e_{ss}^* = \frac{AT^2}{2}\lim_{z\to 1}\frac{z+1}{(z-1)^2(1+GH(z))} \\ &\Rightarrow e_{ss}^* = \frac{A}{\lim\limits_{z\to 1}((z-1)^2/T^2)GH(z)} = \frac{A}{k_a} \end{aligned} \tag{7.46}$$

where k_a is the acceleration error constant.

$$k_a = \lim_{z\to 1}\left\{\frac{(z-1)^2}{T}GH(z)\right\} \tag{7.47}$$

For $N \leq 1 \Rightarrow k_a = 0$ and $e_{ss} = \infty$

For $N = 2$

$$k_a = \frac{k_{dc}}{T^2} \Rightarrow e_{ss}^* = \frac{A}{k_a} = \frac{AT^2}{k_{dc}} \tag{7.48}$$

Overall, the steady-state errors and the corresponding error factors are presented in Table 7.1, corresponding to the appropriate discrete system type.

7.3 Harmonic Response of Discrete Systems

The term *harmonic response* or *frequency response* refers to the time response of the steady state for a linear time invariant system, whose input is a sinusoidal signal of fixed amplitude and variable frequency. For a sinusoidal input signal, the time response of a linear system is also a sinusoidal signal, whose frequency is identical to the input signal, while their corresponding amplitudes and phases are different.

The frequency response of a discrete system arises from its transfer function, by replacing the complex variable z with the imaginary variable $e^{j\Omega}$, where Ω is the digital frequency in the range $(0, \pi)$.

The resulting function $H(e^{j\omega}) = H(e^{j\omega T_s})$ is a complex function of a real variable having certain *magnitude* and *phase*. The magnitude and phase diagrams versus the frequency provide valuable information during the analysis and design process of a control system.

In the case when the input of a linear and discrete-time invariant system with impulse response $h(n)$ is a sinusoidal signal, that is, being of the form of $x(n) = Ae^{j\omega n}$, the system response is presented as

$$y(n) = h(n) * x(n0) = \sum_{k=-\infty}^{\infty} h(k)x(n-k) = \sum_{k=-\infty}^{\infty} h(k)e^{j\omega(n-k)}$$

$$= e^{j\omega n} \sum_{k=-\infty}^{\infty} h(k0e)^{-j\omega k} \Rightarrow y(n) = e^{j\omega n} H(e^{j\omega}) = e^{j\omega n} \left|H(e^{j\omega})\right| \angle H(e^{j\omega}) = \tag{7.49}$$

$$y(n) = Ae^{j\omega n} H\left(e^{j\omega}\right)$$

The following outcomes are provided:

- The output is a sinusoidal signal with circular frequency equal to the circular frequency of input function.
- The amplitude of output signal is $A|H(e^{j\Omega})|$, where $|H(e^{j\Omega})|$ is the complex magnitude of the z-transform for $z = e^{j\Omega}$.
- The output $y(n)$ equals to the input $x(n) = e^{j\Omega n}$ multiplied with a weighted function $H(e^{j\Omega})$.

It is noteworthy that the above conditions apply only when the function $H(e^{j\Omega})$ does not tend to infinity, that is, the region of convergence of the discrete-time system includes the unit circle.

7.4 Formula Tables

The formula Tables 7.2 through 7.6 are discussed here.

TABLE 7.2

Time Response of Discrete System

Impulse time response of first-order system $G(z)=\frac{z}{z-a}$	$y(k)=a^k$
Step time response of first-order system $G(z)=\frac{b}{z-p}$	$y(k)=\frac{b}{1-p}(1-p^k)$
Impulse time response of second-order system $G(z)=\frac{Nz^{-1}}{(1-re^{j\theta}z^{-1})(1-re^{-j\theta}z^{-1})}$ $N=\frac{2\pi}{\theta}\vert_{rad}=\frac{360}{\theta}\vert_{\deg}$	$y(k)=2r^k(a\cos(k\theta)-\beta\sin(k\theta))$
General form for step response of discrete system $G_{CL}(z)=\frac{p(z)}{q(z)}=\frac{b_m z^m+b^{m-1}z_{m-1}+\cdots+b_1 z+b_0}{a_n z^n+a_{n-1}z^{n-1}+\cdots+a_1 z+a_0}$ $=\frac{b_m}{a_n}\times\frac{\prod_{j=1}^{m}(z-z_j)}{\prod_{i=1}^{n}(z-p_i)}$	$Y(z)=\frac{p(z)}{q(z)}\frac{z}{z-1}=\frac{Az}{z-1}+\sum_{i=1}^{n}\frac{B_i z}{z-p_i}$ where $A=\left.\frac{p(z)}{q(z)}\right\vert_{z=1}$ $B_i=\left.\frac{p(z)(z-p_i)}{q(z)(z-1)}\right\vert$ $y(k)=IZT(Y(z))$

TABLE 7.3

Correlation of Time Response between Analog and Discrete Control System

Relation between z and J, ω_n, ω_d	$\lvert z\rvert=\exp\left(-\frac{2\pi J}{\sqrt{1-J^2}}\frac{\omega_d}{\omega_s}\right)$ $\angle z=2\pi\frac{\omega_d}{\omega_s}$
Percentage of overshoot	$POT=\frac{y_{\max}-y_{ss}}{y_{ss}}100\%$
Settling time	$t_s=k_s T$ $\left(1-\frac{\varepsilon}{100}\right)y_{ss}\le y(k)\le\left(1+\frac{\varepsilon}{100}\right)y_{ss}$ $\forall k\ge k_s$
Dominant poles	$z_{1,2\surd}=re^{j\varphi}\Rightarrow$ $J=\frac{-\ln r}{\sqrt{(\ln r)^2+\varphi^2}}$ $\omega_n=\frac{1}{T}\sqrt{(\ln r)^2+\varphi^2}$

TABLE 7.4

Steady-State Errors

Error Type (Number of Integrators of $G(z)$)	Error Constants	Steady-State Error $e_{ss}^{*} = \lim_{t\to\infty} e^{*}(t) = \lim_{z\to 1}(z-1)\frac{R(z)}{1+z[G(s)H(s)]}$		
		Position	**Velocity**	**Acceleration**
0	Position: k_p Velocity: 0 Acceleration: 0	$\frac{A}{1+k_p} = \frac{A}{1+k_{dc}}$	∞	∞
1	Position: ∞ Velocity: k_v Acceleration: 0	0	$\frac{A}{k_v} = \frac{AT}{k_{vcd}}$	∞
2	Position: ∞ Velocity: ∞ Acceleration: k_a	0	0	$\frac{A}{k_a} = \frac{AT^2}{k_{dc}}$

TABLE 7.5

Error Constants for Various Forms of Closed-Loop Digital Control Systems

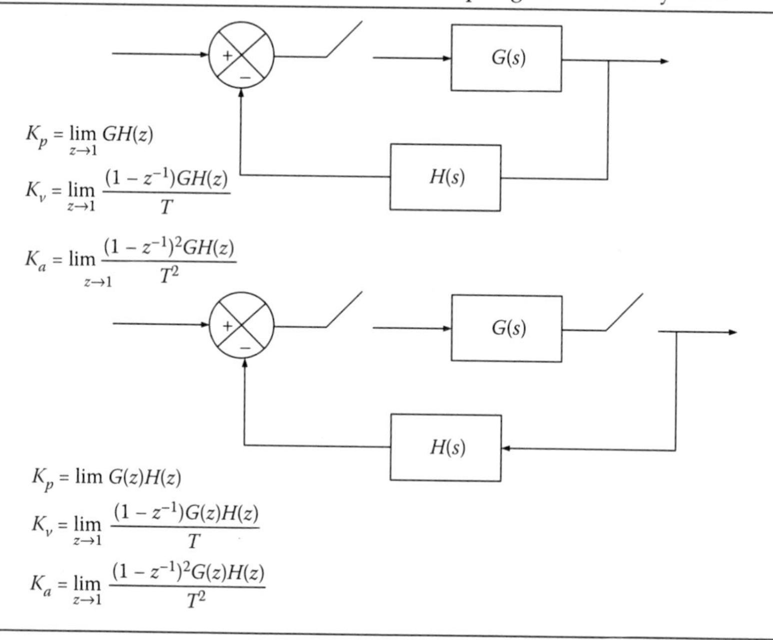

7.5 Solved Exercises

EXERCISE 7.1

Consider the mathematical model of a discrete system, described by $y[n] = 0.3y[n-1] + 0.7x[n]$ and $n \geq 0$

TABLE 7.6

Error Constants for Various Forms of Closed-Loop Digital Control Systems

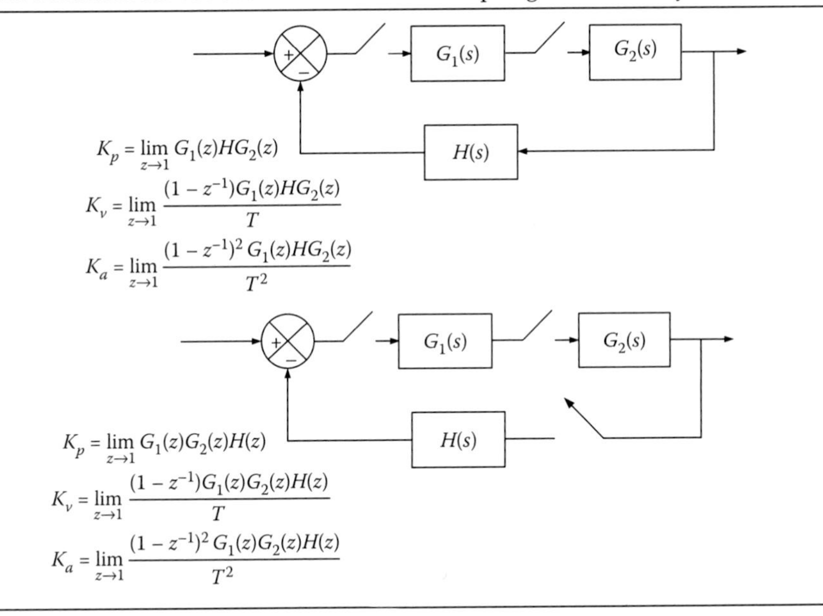

$$K_p = \lim_{z\to 1} G_1(z)HG_2(z)$$

$$K_v = \lim_{z\to 1} \frac{(1-z^{-1})G_1(z)HG_2(z)}{T}$$

$$K_a = \lim_{z\to 1} \frac{(1-z^{-1})^2 G_1(z)HG_2(z)}{T^2}$$

$$K_p = \lim_{z\to 1} G_1(z)G_2(z)H(z)$$

$$K_v = \lim_{z\to 1} \frac{(1-z^{-1})G_1(z)G_2(z)H(z)}{T}$$

$$K_a = \lim_{z\to 1} \frac{(1-z^{-1})^2 G_1(z)G_2(z)H(z)}{T^2}$$

a. Derive the frequency response $H(e^{j\Omega})$.
b. Derive the response for input function: $x[n] = \sin(0,3\pi n)$ and $n \geq 0$

Solution

a. To solve the problem, z-transform is applied to the given difference equation so as to derive its transfer function.

$$y[n] = 0.3y[n-1] + 0.7x[n] \Rightarrow Y(z) = 0.3\,z^{-1}Y(z) + 0.7X(z) \Rightarrow \\ (1-0.3z^{-1})Y(z) = 0.7X(z) \tag{7.1.1}$$

The system transfer function is

$$H(z) = \frac{Y(z)}{X(z)} = \frac{0.7}{1-0.3z^{-1}} = \frac{0.7z}{z-0.3} \tag{7.1.2}$$

The frequency response is obtained by setting: $z = e^{j\Omega}$
Hence, we have

$$H(e^{j\Omega}) = H(z)\Big|_{z=e^{j\Omega}} = \frac{0.7e^{j\Omega}}{e^{j\Omega}-0.3} \tag{7.1.3}$$

Yet: $e^{j\Omega} = \cos\Omega + j\sin\Omega$
Thus,

$$H(e^{j\Omega}) = \frac{0.7e^{j\Omega}}{\cos\Omega - 0.3 + j\sin\Omega} \quad (7.1.4)$$

The magnitude of $H(e^{j\Omega})$ is

$$\left|H(e^{j\Omega})\right| = \left|\frac{0.7e^{j\omega}}{\cos\omega - 0.3 + j\sin\omega}\right| = \frac{0.7}{\sqrt{(\cos\omega - 0.3)^2 + \sin^2\omega}} \quad (7.1.5)$$

The corresponding phase of $H(e^{j\Omega})$ is

$$\varphi(\Omega) = Arg\frac{0.7e^{j\Omega}}{\cos\Omega - 0.3 + j\sin\Omega} = \Omega - \tan^{-1}\frac{\sin\Omega}{\cos\Omega - 0.3} \quad (7.1.6)$$

b. For input $x(n]) = \sin(0.3\pi n)$, when $n \geq 0$, we have: $\Omega = 0.3\pi = 0.94$ rad

The magnitude of frequency response for $\Omega = 0.3\pi$ is

$$\left|H(e^{j\Omega})\right| = 0.815 \quad (7.1.7)$$

The corresponding phase for $\Omega = 0.3\pi$ is

$$\phi(\Omega) = 0.3\pi - \tan^{-1}(2.81) = -0.29 \text{ rad} \quad (7.1.8)$$

Therefore, the system response is

$$y(n) = e^{j0.94n} \cdot 0.815e^{-j0.29} = 0.815 \cdot e^{j(0.94n - 0.29)}$$

or

$$y(n) = 0.815\sin(0.94n - 0.29) \quad (7.1.9)$$

EXERCISE 7.2

For the discrete systems, described by the following transfer functions, derive and design the discrete frequency responses.

$$H_1(z) = \frac{z+1}{z}, \quad H_2(z) = \frac{1+z+z^2}{3z^1},$$

$$H_3(z) = \frac{z^2 - z + 1}{z^2}, \quad H_4(z) = \frac{z}{z-a}$$

$$H_5(z) = \frac{5(1+0.25z)}{(1-0.5z)(1-0.1z)}, \quad H_6(z) = \frac{1}{z+3}$$

Solution

1.

$$\boxed{H_1(z) = \frac{z+1}{z}}$$

The frequency response arises by setting: $z = e^{j\Omega}$.

$$H(\Omega) = [H(z)]_{z=e^{j\Omega}} = 1 + e^{-j\Omega} = e^{-j\Omega/2}(e^{j\Omega/2} + e^{-j\Omega/2}) = 2e^{-j\Omega/2} \cdot \cos\left(\frac{\Omega}{2}\right) \tag{7.2.1}$$

Amplitude:

$$|H(\Omega)| = \left|2e^{(-j\Omega/2)} \cdot \cos\left(\frac{\Omega}{2}\right)\right|, \quad |\Omega| \leq \pi$$

and since $|e^{(-j\Omega/2)}| = 1$, we get

$$|H(\Omega)| = \left|2 \cdot \cos\left(\frac{\Omega}{2}\right)\right|, \quad |\Omega| \leq \pi \tag{7.2.2}$$

Phase:

$$\theta(\Omega) = \angle \mathrm{H}(\Omega) = \angle 2e^{\frac{-j\Omega}{2}} \cdot \cos\left(\frac{\Omega}{2}\right) = -\frac{\Omega}{2}, \quad |\Omega| \leq \pi, \tag{7.2.3}$$

2.

$$\boxed{H_2(z) = \frac{1+z+z^2}{3z^1}}$$

The frequency response arises by setting: $z = e^{j\Omega}$

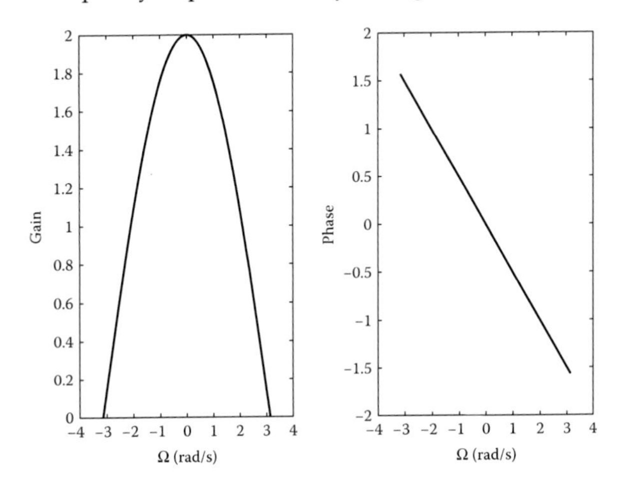

$$H(\Omega) = [H(z)]_{z=e^{j\Omega}} = \frac{1}{3}(e^{-j\Omega} + 1 + e^{j\Omega}) = \frac{1}{3}(1 + 2\cos\Omega) \tag{7.2.4}$$

Amplitude:

$$|H(\Omega)| = |1 + 2\cos(\Omega)|, \; |\Omega| \leq \pi \tag{7.2.5}$$

3.

$$\boxed{H_3(z) = \frac{z^2 - z + 1}{z^2}}$$

$$H(\Omega) = [H(z)]_{z=e^{j\Omega}} = 1 - e^{-j\Omega} + e^{-2j\Omega} = e^{-j\Omega}(2\cos\Omega - 1)$$

Amplitude:

$$|H(\Omega)| = |2\cos(\Omega) - 1|, \quad |\Omega| \leq \pi \tag{7.2.6}$$

Phase:

$$\angle H(\Omega) = -\Omega \tag{7.2.7}$$

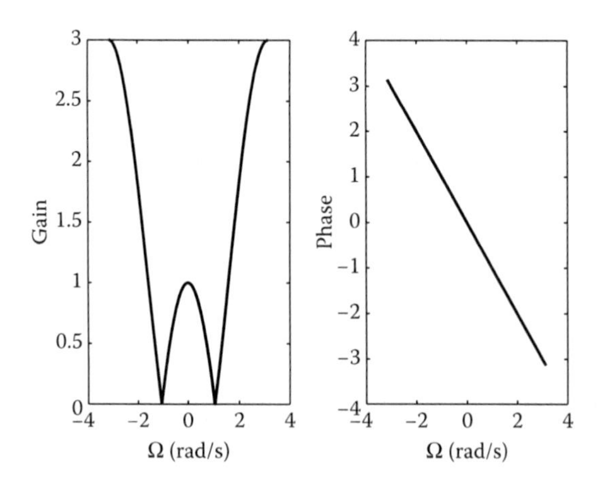

4.

$$\boxed{H_4(z) = \frac{z}{z-a}}$$

$$H(\Omega) = [H(z)]_{z=e^{j\Omega}} = \frac{1}{1-ae^{-j\Omega}} \tag{7.2.8}$$

Amplitude:

$$|H(\Omega)| = \left|\frac{1}{1-a\cos\Omega + ja\sin\Omega}\right| = \left|\frac{(1-a\cos\Omega) - ja\sin\Omega}{(1-a\cos\Omega)^2 + (\alpha\sin\Omega)^2}\right|$$
$$= \frac{\sqrt{(1-a\cos\Omega)^2 + (\alpha\sin\Omega)^2}}{(1-a\cos\Omega)^2 + (\alpha\sin\Omega)^2} = \frac{1}{\sqrt{(1-a\cos\Omega)^2 + (\alpha\sin\Omega)^2}} |H(\Omega)|$$
$$= \frac{1}{\sqrt{1-2a\cos\Omega + a^2}} \tag{7.2.9}$$

Phase:

$$\angle H(\Omega) = \tan^{-1}\left(\frac{-a\sin\Omega}{1-a\cos\Omega}\right) = -\tan^{-1}\left(\frac{a\sin\Omega}{1-a\cos\Omega}\right) \tag{7.2.10}$$

To design the corresponding diagrams, a has to obtain a certain value ($a = 0.6$).

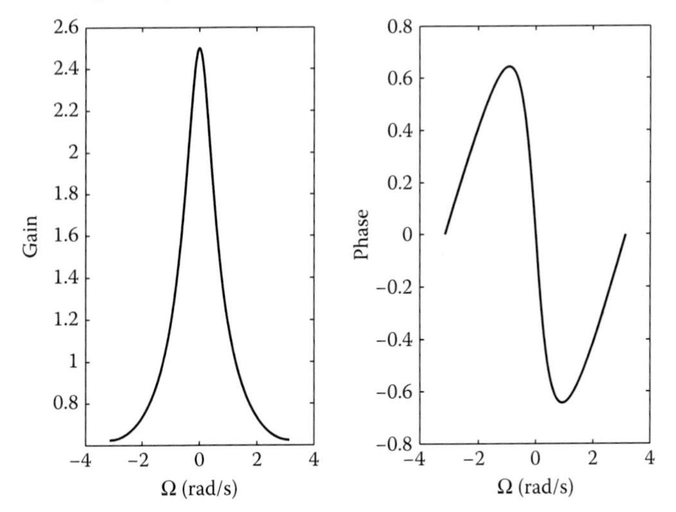

5.

$$\boxed{H_5(z) = \frac{5(1+0.25z)}{(1-0.5z)(1-0.1z)}}$$

$$H_5(z) = \frac{z+5}{(1-0,5z)(1-0,2z)} = \frac{20z+100}{(z-2)(z-10)} \tag{7.2.11}$$

Set: $z = e^{j\omega} \Rightarrow$

$$H_5(e^{j\Omega}) = 20\frac{5+e^{j\Omega}}{(e^{j\Omega}-2)(e^{j\Omega}-10)} = \left|H_5(e^{j\Omega})\right|\angle\phi(\Omega) \tag{7.2.12}$$

The amplitude of the response equals to

$$\begin{aligned}\left|H_5(e^{j\Omega})\right| &= 20\frac{\left|5+e^{j\Omega}\right|}{\left|e^{j\Omega}-2\right|\left|e^{j\Omega}-10\right|} \\ &= 20\frac{\sqrt{(5+\cos\Omega)^2+\sin^2\Omega}}{\sqrt{(\cos\Omega-2)^2+\sin^2\Omega}\sqrt{(\cos\Omega-10)^2+\sin^2\Omega}}\end{aligned} \tag{7.2.13}$$

The phase of the response is

$$\varphi(\Omega) = \tan^{-1}\frac{\sin\Omega}{\cos\Omega+5} - \tan^{-1}\frac{\sin\Omega}{\cos\Omega-2} - \tan^{-1}\frac{\sin\Omega}{\cos\Omega-10} \tag{7.2.14}$$

6.

$$H_6(z) = \frac{1}{z+3} \tag{7.2.15}$$

Set: $z = e^{j\omega} \Rightarrow$

$$H_6(e^{j\Omega}) = 20\frac{1}{e^{j\Omega}+3} = \left|H_6(e^{j\Omega})\right|\angle\phi(\Omega) \tag{7.2.16}$$

The amplitude of the response equals to

$$\left|H_6(e^{j\Omega})\right| = \frac{1}{\left|e^{j\Omega}+3\right|} = \frac{1}{\left|\cos\Omega + j\sin\Omega + 3\right|} = \frac{1}{\sqrt{(\cos\Omega+3)^2+\sin^2\Omega}} \tag{7.2.17}$$

The phase of the response is

$$\varphi(\Omega) = -\tan^{-1}\frac{\sin\Omega}{\cos\Omega+3} \tag{7.2.18}$$

In the following scheme, the magnitude of the discrete frequency response is illustrated for $-10\pi \leq \Omega \leq 10\pi$.

Ω	-10π	$-15\pi/2$	-5π	$-5\pi/2$	0	$5\pi/2$	5π	$15\pi/2$	10π
$\lvert H_6(e^{j\Omega})\rvert$	0.5	0.4168	0.316	0.265	0.25	0.265	0.3162	0.4168	0.5

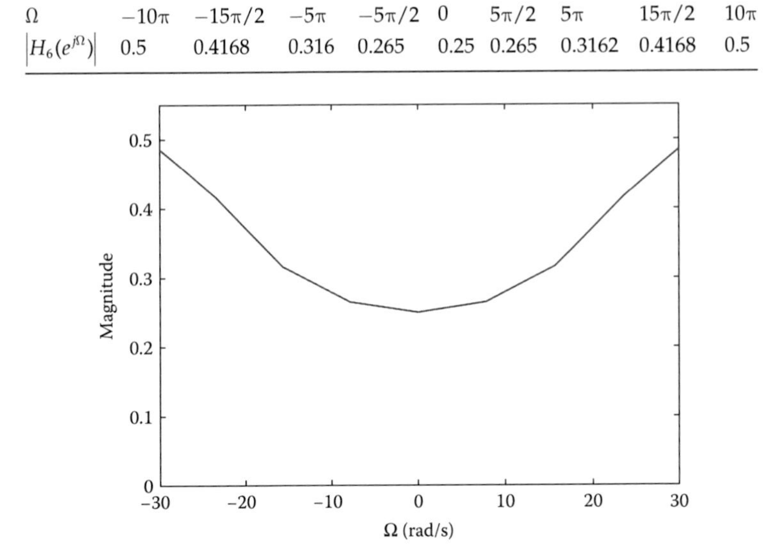

EXERCISE 7.3

Derive and design the frequency response of a ZOH circuit. In addition, calculate the amplitude and phase of the frequency response of a FOH circuit.

Solution

The transfer function of ZOH is given by

$$G_{h_0}(s) = L[1(t) - 1(t-T)] = \frac{1-e^{-Ts}}{s} \tag{7.3.1}$$

The frequency response of ZOH is obtained by setting $z = e^{Tj\omega}$ in the latter expression, yielding

$$\begin{aligned} G_{h0}(j\omega) &= \frac{1-e^{-Tj\omega}}{j\omega} \\ &= \frac{2e^{-(1/2)Tj\omega}(e^{(1/2)Tj\omega} - e^{-(1/2)Tj\omega})}{2j\omega} = T\frac{\sin(\omega T/2)}{\omega T/2}e^{-(1/2)Tj\omega} \end{aligned} \tag{7.3.2}$$

The amplitude of the frequency response is

$$\lvert G(j\omega)\rvert = \left\lvert T\frac{\sin(\omega T/2)}{\omega T/2}\right\rvert \tag{7.3.3}$$

The corresponding phase is

$$\begin{aligned}\angle G_{h0}(j\omega) &= \angle T\frac{\sin(\omega T/2)}{\omega T/2}e^{-(1/2)Tj\omega} \\ &= \angle \sin\frac{\omega T}{2} + \angle e^{-(1/2)Tj\omega} \\ &= \angle \sin\frac{\omega T}{2} - \frac{\omega T}{2}\end{aligned} \tag{7.3.4}$$

The transfer function of FOH is presented as

$$G_{foh}(s) = \frac{Ts+1}{s}\left[\frac{1-e^{-Ts}}{s}\right]^2 \tag{7.3.5}$$

The frequency response of FOH is obtained by setting $z = e^{Tj\omega}$ in the latter expression, yielding

$$G_{foh}(j\omega) = \frac{1+j\omega T}{T}\left[\frac{1-e^{-j\omega T}}{j\omega}\right]^2 \tag{7.3.6}$$

The amplitude and phase of the frequency response are, respectively, expressed as

$$|G_{foh}(j\omega)| = T\sqrt{1+\frac{4\pi^2\omega^2}{\omega_s^2}}\left[\frac{\sin(\pi\ \omega/\omega_s)}{\pi\ \omega/\omega_s}\right]^2 \tag{7.3.7}$$

$$= Arg(G_{foh}(j\omega)) = \tan^{-1}\left(\frac{2\pi\ \omega}{\omega_s}\right) - \frac{2\pi\ \omega}{\omega_s} \tag{7.3.8}$$

In the following schemes, the amplitude and phase of ZOH are illustrated. Observe that there is a low-pass filter with features depending on the sampling period.

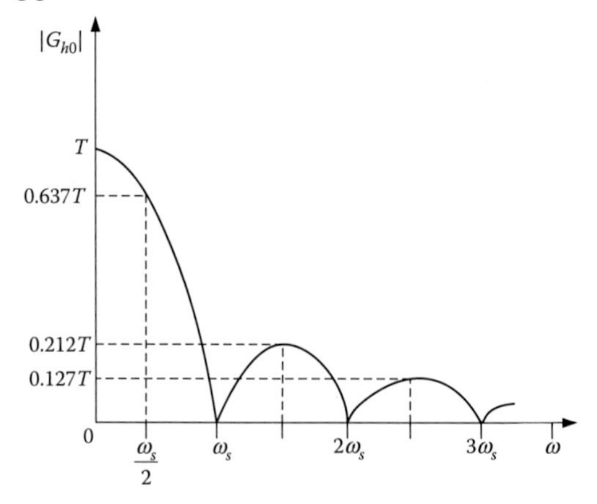

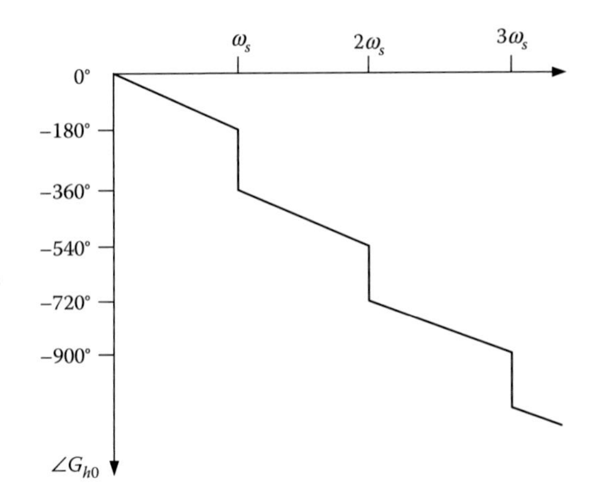

EXERCISE 7.4

Consider a discrete control system with the closed-loop transfer function $H(z) = (Y(z)/U(z)) = (0.813/(z - 0.8187))$.

Derive and design its step response.

Solution

The difference equation resulting from the given transfer function is

$$\begin{aligned} H(z) &= \frac{Y(z)}{U(z)} = \frac{0.813}{z-0.8187} \Rightarrow Y(z)(z-0.8187) = 0.813U(z) \\ &\Rightarrow zY(z) - 0.8187Y(z) = 0.813U(z) \\ &\Rightarrow Y(z) - 0.8187z^{-1}Y(z) = 0.813z^{-1}U(z) \\ &\Rightarrow y(k) = 0.8187y(k-1) + 0.813u(k) \end{aligned} \tag{7.4.1}$$

From the above difference equation, assuming the unit step function as an input signal, the following table arises.

k	$u(k)$	$y(k)$
0	1	0
1	1	0.1813
2	1	0.3297
3	1	0.4513
4	1	0.5507
5	1	0.6322
6	1	0.6989
7	1	0.7535
.	.	.
.	.	.
.	.	.
∞	1	1.0000

Based on this table, observe that the step response reaches to the 63% of the final value for $k = 5$ or $t = 0.5$ s.

These values are verified by calculating the step response via the Laplace transform:

$$\begin{aligned} &H(z) = \frac{Y(z)}{U(z)} = \frac{0.813}{z - 0.8187} \Rightarrow Y(z) = \frac{0.813}{z - 0.8187} U(z) \\ &\Rightarrow Y_u(z) = \frac{0.813}{z - 0.8187} \frac{z}{z - 1} \end{aligned} \tag{7.4.2}$$

$$\begin{aligned} &(7.4.2) \Rightarrow Y_u(z) = \frac{0.813z}{(z-1)(z-0.8187)} = \frac{1}{z-1} - \frac{0.8187}{z-0.8187} \\ &\Rightarrow \; Y_u(z) = z^{-1}\frac{z}{z-1} - 0.8187z^{-1}\frac{z}{z-0.8187} \\ &\Rightarrow \; y_u(k) = u(k-1) - 0.8187^k \end{aligned} \tag{7.4.3}$$

Design the step response using MATLAB as

```
u = ones(size(0:10));
num = [0 0.1813];
den = [1 -0.8187];
y = dlsim(num,den,u)
y =
0
0.1813
0.3297
0.4513
0.5507
0.6322
0.6989
0.7535
0.7982
0.8348
0.8647
plot(0:10,y,'o',0:10,y);
```

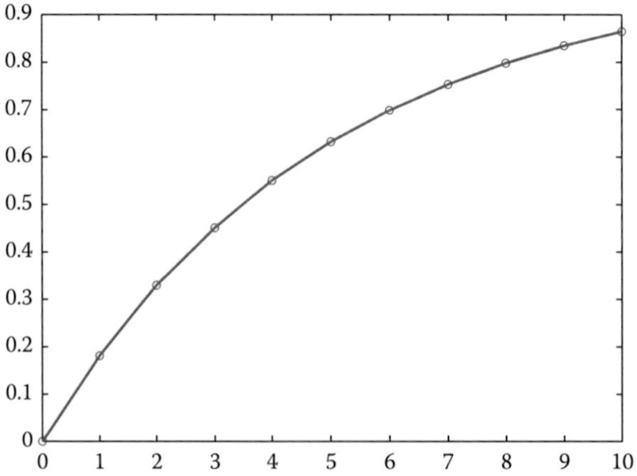

EXERCISE 7.5

Consider a discrete control system with the closed-loop transfer function $G_{CL}(z) = (C(z)/(X(z) = (2z - 1)/(z^2)$. Derive and design the step response and ramp response.

Solution

Step response:

$$X(z) = Z(u(k)) = \frac{z}{z-1}$$

The output in z-domain is computed by

$$C(z) = G_{CL}(z)R(z) = \frac{2z-1}{z^2} \cdot \frac{z}{z-1} = \frac{2z-1}{z^2 - z} \tag{7.5.1}$$

$$\Rightarrow C(z) = 2z^{-1} + z^{-2} + z^{-3} + z^{-4} + \cdots \tag{7.5.2}$$

Below, the desired system step response is provided

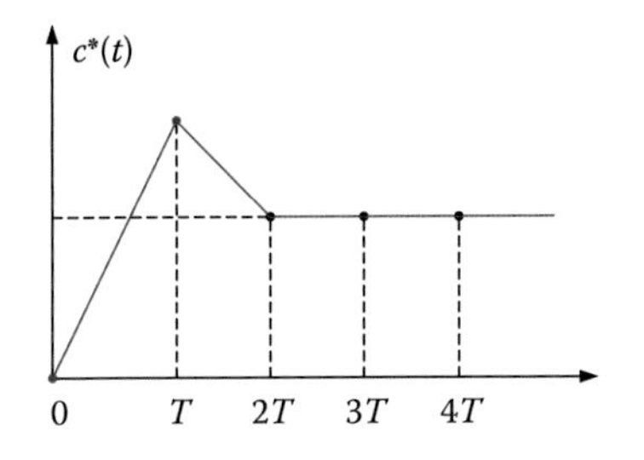

Ramp response:

$$X(z) = Z(r(k)) = \frac{Tz}{(z-1)}$$

The output in z-domain is computed by

$$C(z) = G_{CL}(z)R(z) = \frac{2z-1}{z^2} \cdot \frac{Tz}{(z-1)^2} \tag{7.5.3}$$

$$\Rightarrow C(z) = \frac{T(2z-1)}{z^3 - 2z^2 + z} = 2Tz^{-2} + 3Tz^{-3} + 4Tz^{-4} + \cdots \tag{7.5.4}$$

Below, the desired system step response is provided

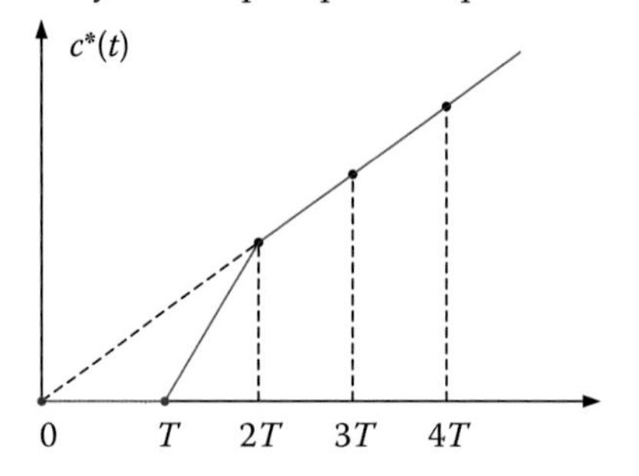

EXERCISE 7.6

Consider a discrete control system with the transfer function: $G(z) = (z + 0.5)/(z^2 - 0.6z + 0.3)$. Calculate the steady-state error for step and impulse input function.

Solution

The poles of transfer function $G(z)$ are complex conjugate, such that

$$p_1 = 0.3000 + 0.4583i, \quad p_2 = 0.3000 - 0.4583i \tag{7.6.1}$$

Since both the poles are located inside the unit circle, the final value theorem is invoked.

The system step response is

$$Y_u(z) = \frac{z+0.5}{z^2-0.6z+0.3}\frac{z}{z-1} \tag{7.6.2}$$

Applying the final value theorem, the final value of the step response is derived as

$$\begin{aligned} y_{ss,u} &= \lim_{z\to 1}(z-1) \\ Y_u(z) &= \lim_{z\to 1}(z-1)\frac{z+0.5}{z^2-0.6z+0.3}\cdot\frac{z}{z-1} = 2.14 \end{aligned} \tag{7.6.3}$$

Thus, the final value of the step response in steady state for the given system is 2.14. This reflects that the steady-state error is in the order of 114%.

The above theoretical results are verified using MATLAB as

```
numDz=[1 0.5];
denDz=[1 -0.6 0.3];
T = 05;
sys = tf(numDz,denDz,T);
[x,t]=step(sys,5);
stairs(t,x)
```

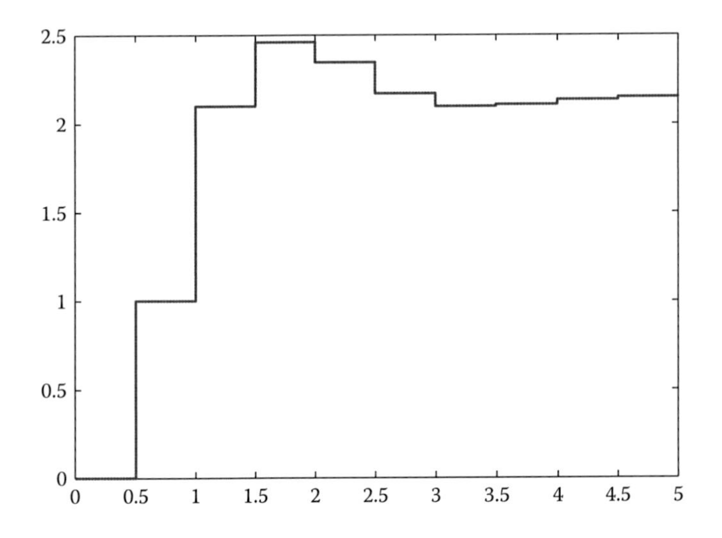

The steady-state error is 2.14, as expected.
The system impulse response is

$$Y_\delta(z) = \frac{z+0.5}{z^2-0.6z+0.3}\cdot 1 \tag{7.6.4}$$

Applying the final value theorem, the final value of the impulse response is derived as

$$\begin{aligned} y_{ss,\delta} &= \lim_{z\to 1}(z-1) \\ Y_\delta(z) &= \lim_{z\to 1}(z-1)\frac{z+0.5}{z^2-0.6z+0.3}\cdot 1 = 0 \end{aligned} \tag{7.6.5}$$

Thus, the final value of the impulse response in steady state for the given system is zero. This reflects that the steady-state error is in the order of 0%.

EXERCISE 7.7

For the system of the following scheme, derive the position and velocity errors.

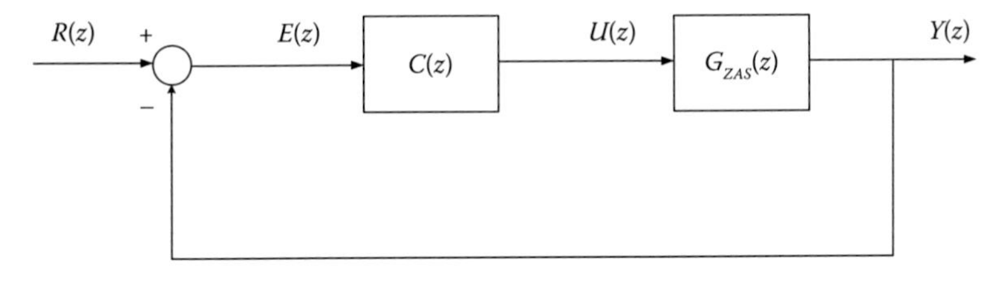

$$G_{ZAS}(z) = \frac{K(z+a)}{(z-1)(z-b)}$$

$$C(z) = \frac{K_c(z-b)}{z-c}$$

For $0 < a, b, c < 1$

Solution

The error transfer function of the given system is

$$\frac{E(z)}{R(z)} = \frac{1}{1+G_{ZAS}(z)G(z)} \tag{7.7.1}$$

For the position error, we have

$$e(\infty) = \frac{1}{1+K_p} \tag{7.7.2}$$

The position error constant is calculated by

$$K_p = \lim_{z\to1} G_{ZAS}(z)G(z) = \lim_{z\to1} \frac{K(z+a)K_c(z-b)}{(z-1)(z-b)(z-c)} \tag{7.7.3}$$

Hence

$$K_p = \frac{K(1+a)K_c(1-b)}{(1-1)(1-b)(1-c)} = \infty \tag{7.7.4}$$

Consequently, the position error is

$$e(\infty) = \frac{1}{1+K_p} = 0 \tag{7.7.5}$$

For the velocity error, we have

$$e(\infty) = \frac{1}{K_v} \tag{7.7.6}$$

The velocity error constant is calculated by

$$K_v = \frac{1}{T}\lim_{z\to1}(z-1)G_{ZAS}(z)G(z) = \frac{1}{T}\lim_{z\to1}(z-1)\frac{K(z+a)K_c(z-b)}{(z-1)(z-b)(z-c)}$$

Hence

$$K_v = \frac{1}{\tau}\frac{K(1+a)K_c(1-b)}{(1-b)(1-c)} = \frac{KK_c(1+a)}{\tau(1-c)} \tag{7.7.7}$$

Consequently, the velocity error is

$$e(\infty) = \frac{\tau(1-c)}{KK_c(1+a)} \tag{7.7.8}$$

EXERCISE 7.8

Consider a discrete control system with the transfer function $G(s) = 10/(s(s+1))$, $T = 1$ s. Derive the steady-state error for different kinds of input.

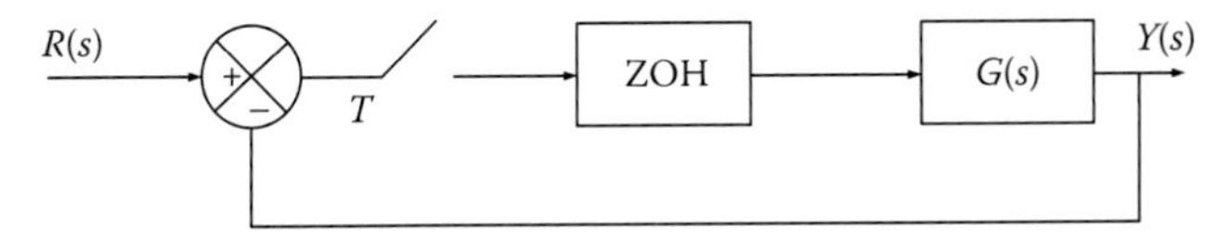

Solution

Calculate the open-loop transfer function $G(s)$ with the aid of ZOH

$$G(s) = \frac{10e^{-Ts}}{s^2(s+1)} = 10e^{-Ts}\left(\frac{1}{s^2} - \frac{1}{s} + \frac{1}{s+1}\right) \tag{7.8.1}$$

The z-transform of $G(s)$ is

$$\begin{aligned} G(z) &= 10(1-z^{-1})\left(\frac{Tz}{(z-1)^2} - \frac{z}{z-1} + \frac{z}{z-e^{-T}}\right) \\ &= 10\left(\frac{T}{z-1} - 1 + \frac{z-1}{z-e^{-T}}\right) \end{aligned} \tag{7.8.2}$$

For step input, the steady-state error equals to

$$K_p = \lim_{z\to 1} G(z) = \infty, \quad e_{ss,p} = \frac{1}{1+K_p} = 0 \tag{7.8.3}$$

For ramp input, the steady-state error equals to

$$K_v = \frac{1}{T}\lim_{z\to 1}(z-1)G(z) = 10, \quad e_{ss,v} = \frac{1}{K_v} = 0.1 \tag{7.8.4}$$

For parabolic input, the steady-state error equals to

$$K_a = \frac{1}{T^2}\lim_{z\to 1}(z-1)^2 G(z) = 0, \quad e_{ss,a} = \frac{1}{K_a} = \infty \tag{7.8.5}$$

EXERCISE 7.9

Consider a discrete control system with the transfer function $G(s) = K/(s(s+5))$, $T = 1$ s.

Derive the steady-state error for $r(t) = 1 + t$.

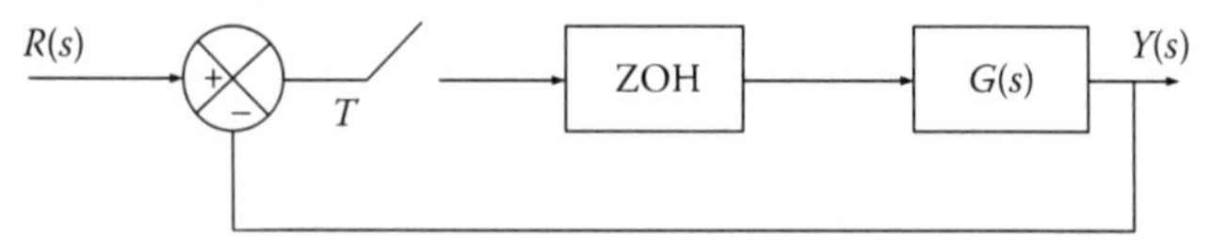

Solution

Calculate the open-loop transfer function $G(s)$ with the aid of ZOH

$$\begin{aligned} G(z) &= Z\left[\frac{1-e^{-Ts}}{s}\cdot\frac{K}{s(s+5)}\right] = (1-e^{-Ts})Z\left[\frac{K}{s^2(s+5)}\right] \\ &= (1-e^{-Ts})Z\left[\frac{K/5}{s^2} + \frac{-K/5}{s} + \frac{K/25}{s+5}\right] \\ &= (1-z^{-1})\left(\frac{KTz/5}{(z-1)^2} - \frac{Kz/5}{z-1} + \frac{Kz/25}{z-e^{-5T}}\right)\Bigg|_{T=1} \\ \Rightarrow G(z) &\approx -\frac{K}{5}\cdot\frac{z^2 - 2.2067z + 0.2135}{(z-1)(z-0.0067)} \end{aligned} \tag{7.9.1}$$

Since the system input is $r(t) = 1 + t$, the steady-state error is the sum of the position error and velocity error, that is

$$e_{ss} = \frac{1}{1+K_p^*} + \frac{T}{K_v^*} \tag{7.9.2}$$

where

$$K_p = \lim_{z\to 1} G(z) = \lim_{z\to 1}\frac{K}{5}\cdot\frac{z^2 - 2.2067z + 0.2135}{(z-1)(z-0.0067)} = \infty \tag{7.9.3}$$

$$K_v^* = \lim_{z\to 1}(z-1)G(z) = \lim_{z\to 1} -\frac{K}{5}\cdot\frac{z^2 - 2.2067z + 0.2135}{(z-0.0067)} \approx 0.2K \tag{7.9.4}$$

Finally, the requested steady-state error is given by

$$e_{ss} = \frac{1}{1+K_p^*} + \frac{T}{K_v^*} = 0 + \frac{T}{0.2K}\bigg|_{T=1} = \frac{5}{K} \tag{7.9.5}$$

EXERCISE 7.10

Consider a discrete control system with the transfer function: $G(s) = 1/(s+1)$.

a. Derive the difference equation.
b. Simulate the given system.
c. Derive the step response for $T = 1$ s

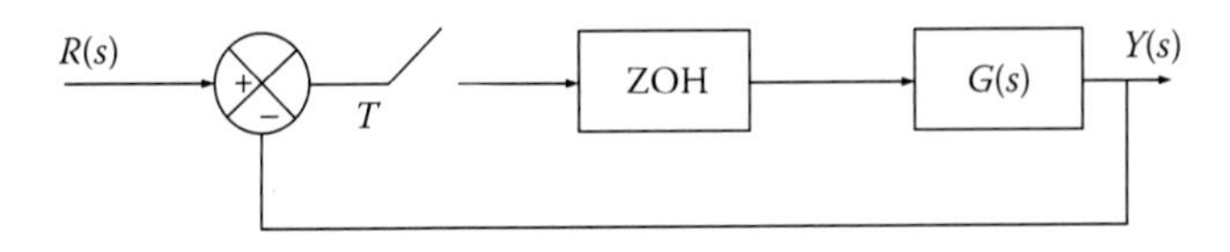

Solution

a. The system transfer function with ZOH is

$$G(z) = \frac{(1-e^{-T})z^{-1}}{1-e^{-T}z^{-1}} \tag{7.10.1}$$

Setting $A = 1 - e^{-T}$ and $B = e^{-T}$, then the latter function becomes

$$G(z) = \frac{(1-e^{-T})z^{-1}}{1-e^{-T}z^{-1}} = \frac{Az^{-1}}{1-Bz^{-1}} = \frac{Y(z)}{X(z)} \tag{7.10.2}$$

The difference equation is derived by

$$\begin{aligned} &Y(z)(1-Bz^{-1}) = AX(z) \Rightarrow Y(z) - BY(z)z^{-1} = AX(z)z^{-1} \\ &\Rightarrow y(k) - By(k-1) = Ax(k-1) \\ &\Rightarrow y(k) = Ax(k-1) + By(k-1) \end{aligned} \tag{7.10.3}$$

b. The system parameters and its input are $A = 1 - e^{-T}$, $B = e^{-T}$, $x(k) = 1$

The initial conditions are $x(k-1) = 0$, $y(k) = y(k-1) = 0$, $k = 0$

Simulation program

```
While k<100 do
  y(k)=Ax(k−1)+By(k−1); Calculation of output
  x(k−1)=x(k); y(k−1)=y(k); x(k)=1; k=k+1; Data update
  print k, x(k), y(k); Print input and output
End
```

c. For $T = 1$ s, we have $A = 0.6321$ and $B = 0.3679$.

For step input, the response is

$$y(k) = 0.6321x(k-1) + 0.3679y(k-1) \tag{7.10.4}$$

We proceed by formulating the table of values; output signal follows its corresponding input.

k	0	1	2	3	4	5
$x(k)$	1	1	1	1	1	1
$y(k)$	0	0.6321	1	1	1	1

EXERCISE 7.11

For a sampled data closed-loop system, the z-transform of the error signal is given as $E(z) = z^{-1} + 0.6\,z^{-2} + 0.2z^{-3}$. Derive the system output $y(k)$ for $k = 0, 1, 2, 3, 4, 5$ when the input is a ramp function.

Solution

We know that the error is defined as

$$e(k) = r(k) - y(k) \tag{7.11.1}$$

Using the z-transform, we get

$$E(z) = z^{-1} + 0.6z^{-2} + 0.2z^{-3} \tag{7.11.2}$$

In the discrete-time domain, the error is

$$\begin{aligned} e(k) = Z^{-1}\{E(z)\} &= 0\cdot\delta(k) + 1\cdot\delta(k-1) + 0.6\cdot\delta(k-2) \\ &+ 0.2\cdot\delta(k-3) + 0\cdot\delta(k-4) + 0\cdot\delta(k-5) \end{aligned} \tag{7.11.3}$$

In the discrete-time domain, the system input is

$$\begin{aligned} r(k) = k\cdot 1(k) &= 0\cdot\delta(k) + 1\cdot\delta(k-1) + 2\cdot\delta(k-2) \\ &+ 3\cdot\delta(k-3) + 4\cdot\delta(k-4) + 5\cdot\delta(k-5) + \cdots \end{aligned} \tag{7.11.4}$$

Hence, the ramp response is presented as

$$\begin{aligned} y(k) = r(k) - e(k) &= 0\cdot\delta(k) + 0\cdot\delta(k-1) + 1.4\cdot\delta(k-2) \\ &+ 2.8\cdot\delta(k-3) + 4\cdot\delta(k-4) + 5\cdot\delta(k-5) \end{aligned} \tag{7.11.5}$$

Thus, the ramp output of the system is

$$y(0) = 0, \quad y(1) = 0, \quad y(2) = 1.4, \quad y(3) = 2.8, \quad y(4) = 4, \quad y(5) = 5$$

Therefore,

$$y(k) = k \quad \text{and} \quad k \geq 4 \tag{7.11.6}$$

EXERCISE 7.12

Consider the transfer function of an analog system, given as $G(s) = K/(s + b)$.

a. Derive the transfer function in z-domain with or without ZOH.
b. Derive the final value of the step response for both cases.
c. For $K = b = 1$, design the closed-loop step response using LabVIEW.

Solution

a. Applying the inverse Laplace transform at the given transfer function, it yields

$$g(t) = ILT[G(s)] = ILT\left[\frac{K}{s+b}\right] = Ke^{-bt} \tag{7.12.1}$$

So, the discretized transfer function is

$$G(z) = K\frac{z}{z - e^{-bT_0}} = \frac{a_0}{1 + b_1 z^{-1}} \tag{7.12.2}$$

$$\textit{where}\; a_0 = K \;\textit{and}\; b_1 = -e^{-bT_0} \tag{7.12.3}$$

The corresponding transfer function with ZOH is presented as

$$\begin{aligned} G_{zoh}(z) &= \frac{z-1}{z} Z\left[\frac{G(s)}{s}\right] = \frac{z-1}{z} Z\left[\frac{K}{s(s+b)}\right] \\ &= \frac{K}{b}\frac{z-1}{z} Z\left[\frac{1}{s} - \frac{1}{s+b}\right] \\ &= \frac{K}{b}\frac{z-1}{z}\left[\frac{z}{z-1} - \frac{z}{z-e^{-bT_0}}\right] \\ &= \frac{K}{b}\frac{1-e^{-bT_0}}{z-e^{-bT_0}} = \frac{a_1 z^{-1}}{1+b_1 z^{-1}} \end{aligned} \tag{7.12.4}$$

where

$$a_1 = K(1 - e^{-bT_0})/b \quad \text{and} \quad b_1 = -e^{-bT_0} \tag{7.12.5}$$

The ZOH circuit introduces a delay term along with the parameter $(1 - e^{-bT_0})/b$ in the numerator.

b. The fixed value of the step response without ZOH is

$$
\begin{aligned}
y(\infty) &= \lim_{z\to 1}\frac{z-1}{z} \\
Y(z) &= \lim_{z\to 1}\frac{z-1}{z}K\frac{z}{z-e^{-bT_0}}\frac{z}{z-1} = \frac{K}{1-e^{-bT_0}}
\end{aligned}
\tag{7.12.6}
$$

The fixed value of the step response with ZOH is

$$
\begin{aligned}
y(\infty) &= \lim_{z\to 1}\frac{z-1}{z}Y(z) = \lim_{z\to 1}\frac{z-1}{z}G(z)U(z) \\
&= \lim_{z\to 1}\frac{z-1}{z}\frac{K}{b}\frac{1-e^{-bT_0}}{z-e^{-bT_0}}\frac{z}{z-1} = \frac{K}{b}
\end{aligned}
\tag{7.12.7}
$$

The fixed value of the step response for the analog system is

$$
y(\infty) = \lim_{s\to 0} sY(s) = \lim_{z\to 1} s\frac{K}{s+b}\frac{1}{s} = \frac{K}{b}
\tag{7.12.8}
$$

c. The front panel and block diagram of the vi implemented to design the requested step function are presented as follows:

It holds that: $G(s) = 1/(s+1)$ and the resultant pulse transfer function with ZOH is $G(z) = (0.0952)/(z - 0.9048)$ for $T = 0.1$ s.

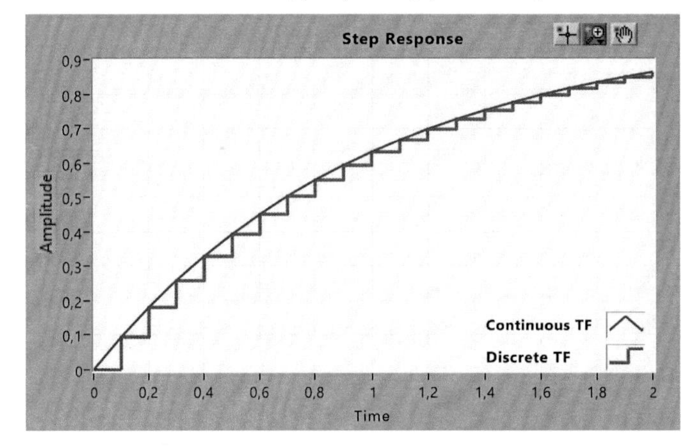

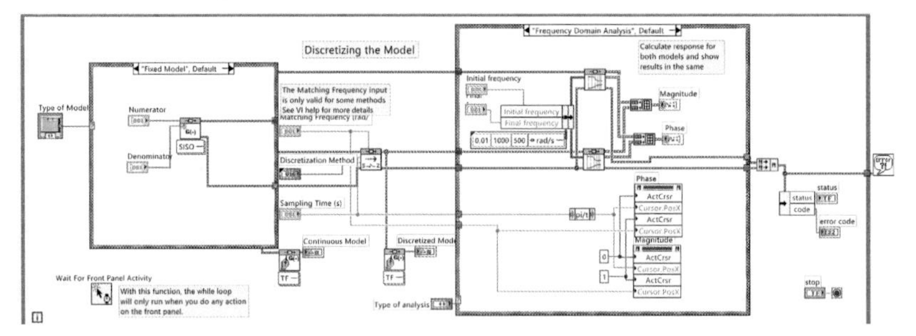

Typically, using the above vi, any analog transfer function discretized with ZOH can be given, along with the desired sampling period, while the corresponding discretized open- and closed-loop transfer function can be calculated, and the system step response can be designed.

EXERCISE 7.13

Consider a discrete control system with the transfer function $G(s) = 2e^{-s}/(s + 1)$. Derive the transfer function $G(z)$ when $fs = 5$ Hz and calculate the first seven samples of the system step response.

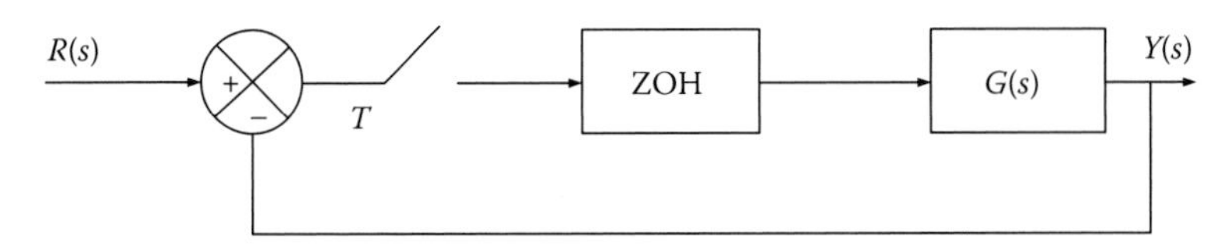

Solution

Calculate with ZOH the transfer function $1/(s + 1)$.

$$\mathbb{Z}OH\left(\frac{1}{s+1}\right) => \left(\frac{1-e^{-T}}{z-e^{-T}}\right) \xrightarrow{T=0.2} \frac{0.1813}{z-0.8187} \simeq \frac{0.18}{z-0.82} \quad (7.13.1)$$

Similarly, Calculate with ZOH the transfer function $2e^{-s}/(s + 1)$.

$$G(z) = \mathbb{Z}OH\left(\frac{2e^{-s}}{s+1}\right) = \frac{0.36z^{-5}}{z-0.82} \quad (7.13.2)$$

The closed-loop transfer function is derived by

$$\begin{aligned}\frac{Y(z)}{R(z)} &= \frac{(0.36z^{-5}/z-0.82)}{1+(0.36z^{-5}/z-0.82)} = \frac{0.36z^{-5}}{z-0.82+0.36z^{-5}} \\ &= \frac{0.36z^{-6}}{1-0.82z^{-1}+0.36z^{-6}}\end{aligned} \quad (7.13.3)$$

From the latter expression, the difference equation of the system arises.

$$y(k) = 0.82y(k-1) - 0.36y(k-6) + 0.36r(k-6) \quad (7.13.4)$$

The first seven samples of the step response are $y(0{:}5) = 0$, $y(6) = 0.36$, $y(7) = 0.82(0.36) + 0.36 = 0.655$.

EXERCISE 7.14

Consider a discrete control system with the transfer function $G(s) = 10/(s+2)(s+3)$.

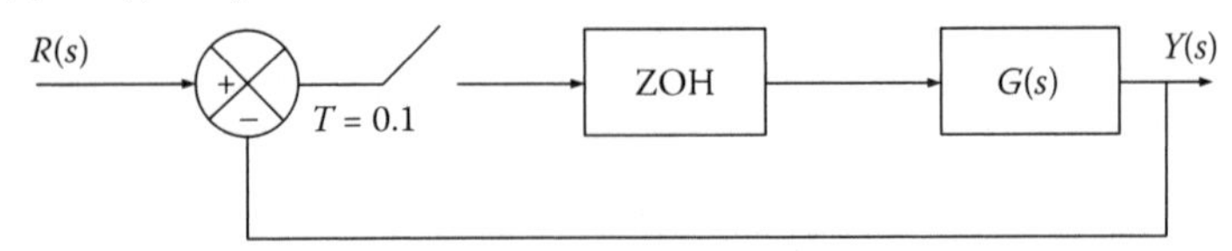

1. Derive the closed-loop transfer function of the system.
2. Derive the system step response.
3. Calculate the parameters of the system time response: percentage of overshoot, settling time, and steady-state error.

Solution

1. The closed-loop transfer function is presented as

$$G_{cl}(z) = \frac{G(z)}{1+G(z)} \tag{7.14.1}$$

where

$$G(z) = (1-z^{-1})Z\left(\frac{G(s)}{s}\right) = (1-z^{-1})Z\left(\frac{10}{s(s+2)(s+3)}\right) =$$

$$= 10(1-z^{-1})\frac{z(Az+B)}{(z-1)(z-e^{-0.2})(z-e^{-0.3})} \tag{7.14.2}$$

$$\Rightarrow G(z) = \frac{0.042z+0.036}{(z-0.819)(z-0.741)}$$

$$(7.14.1),(7.14.2) \Rightarrow G_{cl}(z) = \frac{(0.042z+0.036)/(z-0.819)(z-0.741)}{1+(0.042z+0.036)/(z-0.819)(z-0.741)}$$

$$\Rightarrow G_{cl}(z) = \frac{0.042z+0.036}{z^2-1.518z+0.643} \tag{7.14.3}$$

2. Derivation of the system step response.

$$Y(z) = G_{cl}(z)R(z) = \frac{0.042z+0.036}{z^2-1.518z+0.643}R(z)$$

$$= \frac{0.042z^{-1}+0.036z^{-2}}{1-1.518z^{-1}+0.643z^{-2}}R(z) \Rightarrow$$

$$Y(z)(1-1.518z^{-1}+0.643z^{-2}) = (0.042z^{-1}+0.036z^{-2})R(z) \tag{7.14.4}$$

From the expression (7.14.4), the difference equation of the system is derived as

$$y(k) = 1.518y(k-1) - 0.643y(k-2) + 0.042r(k-1) \\ = 0.036r(k-2) \tag{7.14.5}$$

For step input: $r(k) = u(k) = 1, \ \forall k \geq 0$ and initial conditions: $y(-1) = y(-2) = 0$, the expression (7.14.5) is iteratively solved as

$$y(k) = \{0, 0.0420, 0.1418, 0.2662, 0.3909, 0.5003, 0.5860, 0.6459, \\ 0.6817, 0.6975, 0.6985, 0.6898, 0.6760, 0.6606, 0.6461, 0.6341, 0.6251, \ldots\}$$

3. Steady-state response

$$y_{ss} = \lim_{z\to 1}(1 - z^{-1})Y(z) = \lim_{z\to 1}(1 - z^{-1})G_{cl}(z)R(z) \\ = \lim_{z\to 1}(1 - z^{-1})\frac{0.042z + 0.036}{z^2 - 1.518z + 0.643}\frac{1}{1 - z^{-1}} = 0.624 \tag{7.14.6}$$

The maximum value of step response is

$$y_{\max} = 0.6985 \tag{7.14.7}$$

Percentage of overshoot is

$$POT = \frac{y_{\max} - y_{ss}}{y_{ss}}100\% = \frac{0.6985 - 0.624}{0.624}100\% = 11.94\% \tag{7.14.8}$$

Settling time with a constraint 5%

$$t_s = k_s T \tag{7.14.9}$$

k_s should follow the condition of maintaining 5% steady-state error; thereby

$$k_s \geq 14 \tag{7.14.10}$$

$$(7.14.9), (7.14.10) \Rightarrow t_s = 14 \cdot 0.1 = 1.4\text{s} \tag{7.14.11}$$

Steady-state error

$$e_{ss} = r_{ss} - y_{ss} = 1 - 0.624 = 0.376 \tag{7.14.12}$$

NOTE: POT and t_s can be computed by the dominant poles of the systems.

The closed system poles are the roots of the equation

$$z^2 - 1.518z + 0.643 = 0 \Rightarrow \tag{7.14.13}$$

$$z_{1,2} = 0.7590 \pm j0.2587 = 0.8019e^{j0.3285} \tag{7.14.14}$$

It holds that

$$J = \frac{-\ln r}{\sqrt{(\ln r)^2 + \varphi^2}} = \frac{-\ln 0.8019}{\sqrt{(\ln 0.8019)^2 + 0.3285^2}} = 0.5579 \tag{7.14.15}$$

$$\omega_n = \frac{1}{T}\sqrt{(\ln r)^2 + \varphi^2} = \frac{1}{0.1}\sqrt{(\ln 0.8019)^2 + 0.3285^2} = 0.3958 \tag{7.14.16}$$

Thus

$$\begin{aligned} POT &= \exp\left(-\frac{J\pi}{\sqrt{1-J^2}}\right)100\% \\ &= \exp\left(-\frac{0.5579 \cdot 3.14}{\sqrt{1-0.5579^2}}\right)100\% = 12.11\% \end{aligned} \tag{7.14.17}$$

$$t_s = \frac{3}{J\omega_n} = \frac{3}{0.5579 \cdot 0.3958} = 1.36\,\text{sec} \tag{7.14.18}$$

The values of the expressions (7.14.17) and (7.14.18) are very close to the ones of the expressions (7.14.8) and (7.14.11).

EXERCISE 7.15

Consider the discrete system with pulse transfer function: $H(z) = (0.9z - 0.8)/(z^2 - 1.3z + 0.4)$. Design the system response for input $u(t) = \sin(\pi t)$ using MATLAB for $T = 0.2$ s and verify theoretically the results.

Solution

The MATLAB code is

```
T=0.2; % Sampling period
w=pi; % Frequency of Sinusoidal signal in rad/s
t = 0:T:6; % Create vector from time samples
u = sin(w*t); % Define input
num = [0 0 1]; % Numerator of transfer function
den = [1 -0.5 0.5]; % Denominator of transfer function
G = tf(num,den,T); % Define the discrete LTI system
y = lsim(G,u,t); % Calculate the time response
```

```
plot(t,u,'k-o',t,y,'r--*'); % Design the input and output
legend('Input','Output'); % Title of diagram
```

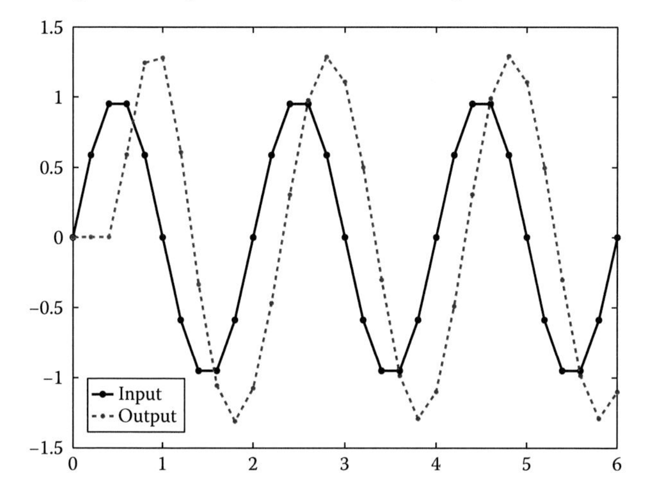

From the input and output waveforms, we conclude that the output is a sinusoidal signal with circular frequency equal to the circular frequency of the input, yet with different amplitude and phase.

The input signal amplitude is obviously 1; the corresponding output signal amplitude is approximately 1.3. the phase difference is $\varphi = \omega t_\varphi \simeq -0.3\pi = -0.94\ rad$.

The above values can be verified in two ways.

1. Setting $\omega T = \pi T$ in the amplitude and phase of the system transfer function:

$$H(z) = \frac{0.9z - 0.8}{z^2 - 1.3z + 0.4} \tag{7.15.1}$$

$$H(e^{j\omega T}) = \frac{0.9\, e^{j\omega T} - 0.8}{e^{2j\omega T} - 1.3e^{j\omega T} + 0.4} \tag{7.15.2}$$

$$\begin{aligned}
\left|H(e^{j\omega T})\right| &= \left|\frac{0.9\, e^{j\omega T} - 0.8}{e^{2j\omega T} - 1.3e^{j\omega T} + 0.4}\right| \\
&= \left|\frac{0.9(\cos\omega T + j\sin\omega T) - 0.8}{\cos 2\omega T + j\sin 2\omega T - 1.3(\cos\omega T + j\sin\omega T) + 0.4}\right| \\
&= \left|\frac{0.9\cos\omega T - 0.8 + j\sin\omega T}{\cos 2\omega T - 1.3\cos\omega T + 0.4 + j(\sin 2\omega T - 1.3\sin\omega T)}\right| \\
&= \frac{\sqrt{(0.9\cos\omega T - 0.8)^2 + \sin^2\omega T}}{\sqrt{(\cos 2\omega T - 1.3\cos\omega T + 0.4)^2 + (\sin 2\omega T - 1.3\sin\omega T)^2}} \\
&(7.15.3) \Rightarrow \left|H(e^{j0.2\pi})\right| \simeq 1.3
\end{aligned} \tag{7.15.3}$$

$$Arg(H(e^{j\omega T})) = Arg(0.9\,e^{j\omega T} - 0.8) - Arg(e^{2j\omega T} - 1.3e^{j\omega T} + 0.4)$$
$$= \tan^{-1}\frac{\sin\omega T}{0.9\cos\omega T - 0.8} - \tan^{-1}\frac{\sin 2\omega T - 1.3\sin\omega T}{\cos 2\omega T - 1.3\cos\omega T + 0.4}$$
$$(7.15.4) \Rightarrow Arg(H(e^{j0.2\pi})) \simeq -0.94 \quad (7.15.4)$$

2. Setting in the pulse transfer function

$$z = e^{j\omega T} = e^{j\pi T} = \cos(\pi T) + j\sin(\pi T) = 0.809 + j0.5878 \quad (7.15.5)$$

and then calculate the amplitude and phase as

$$|H(0.809 + j0.5878)| = |0.6793 - j1.1036| \simeq 1.3$$

$$Arg(H(0.809 + j0.5878)) \simeq -0.94$$

EXERCISE 7.16

Consider the discrete system with pulse transfer function: $H(z) = (0.9z - 0.8)/(z^2 - 1.3z + 0.4)$ Design the ramp response using MATLAB.

Solution

We calculate the ramp response in three different ways.

a. From the corresponding expression for ramp response we have

$$y(k) = u + 2 - (6 + (2/3)) * (0.8)^k + (4 + (2/3)) * (0.5)^k; \quad (7.16.1)$$

b. Using the command lsim and
c. Calculating the IZT of the product

$$H(z)R(z) = \frac{0.9\,z - 0.8}{z^2 - 1.3z + 0.4} \cdot \frac{z}{(z-1)^2} \quad (7.16.2)$$

```
% Define H(z)
H = tf([0.9 -0.8],conv([1 -0.8],[1 -0.5]),1);
% Define input function
k = [0:20];
u = k;
% Calculate ramp response with three different ways
y1 = u + 2 - (6+(2/3))*(0.8).^k + (4+(2/3))*(0.5).^k ;
y2 = lsim(H,u,k);
Y = tf([0.9 -0.8 0],conv(conv([1 -0.8],[1 -0.5]),
[1 -2 1]),1);
```

```
y3 = impulse(Y,k);
% Plot the results
plot(k,y1,'-',k,y1,'x',k,y2,'-',k,y2,'+',k,y3,'-',k,y3,'s')
xlabel('time k')
ylabel('y(k)')
grid
legend({'','formula','','lsim','','impulse'},'Location',
'SouthEast')
```

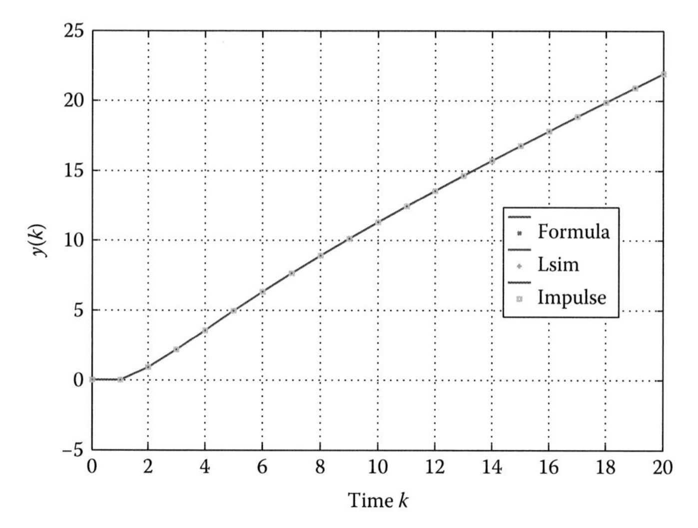

Observe that there is a sharp matching of the response over the three ways.

EXERCISE 7.17

Consider the system with transfer function: $G(s) = 1/(s^2)$ which is controlled by a controller with the transfer function: $G_D(s) = (70\ s + 140)/(s + 10)$.

Discretize the system with the ZOH method for sampling period $T = 0.05$ s and $T = 0.025$ s. Compare the step response of the closed-loop analog and discretized system.

Solution

The MATLAB code is

```
% Define the given system
numG=1;
denG=[1 0 0];
% Define the controller
Ko=70;
a=2; % D(s) zero
b=10; % D(s) pole
numD=Ko*[1 a];
denD=[1 b];
% Define the closed-loop transfer function
```

```
num=conv(numG,numD);
den=conv(denG,denD);
[numcl,dencl]=feedback(num,den,1,1);
% Define the step response
tf=1;
t=0:.01:tf;
yc=step(numcl,dencl,t);
subplot(2,1,1)
plot(t,yc,'-'),grid
axis([0 1 0 1.5])
hold on
% Discretize the system for T=0.05 s.
Ws= 20; % Hz
T=1/Ws; T=0.05
[numGd,denGd]=c2dm(numG,denG,T,'zoh');
% Define the digital controller
numDd=Ko*[1 -(1-a*T)];
denDd=[1 -(1-b*T)];
>> printsys(numDd,denDd,'z')
% Define the closed-loop transfer function
numd=conv(numGd,numDd);
dend=conv(denGd,denDd);
[numcld,dencld]=feedback(numd,dend,1,1);
% Define the step response
N=tf*Ws;
yd=dstep(numcld,dencld,N);
td=0:T:(N-1)*T;
plot(td,yd,'*')
plot(td,yd,'-')
ylabel('output y')
title('Continuous and digital response using zoh method')
text(.25,.1,'*-----*-----* digital control')
text(.25,.3,'------------- analog control')
text(.35,.6,' (a) 20 Hz')
hold off
% Repeat the process for T=0.025 s.
Ws= 40; % Hz
T=1/Ws; %T=0.025
[numGd,denGd]=c2dm(numG,denG,T,'zoh');
numDd=Ko*[1 -(1-a*T)];
denDd=[1 -(1-b*T)];
numd=conv(numGd,numDd);
dend=conv(denGd,denDd);
[numcld,dencld]=feedback(numd,dend,1,1);
N=tf*Ws;
subplot(2,1,2)
plot(t,yc,'-'),grid
hold on
yd=dstep(numcld,dencld,N);
td=0:T:(N-1)*T;
plot(td,yd,'*')
plot(td,yd,'-')
xlabel('time (sec)')
ylabel('output y')
text(.25,.1,'*-----*-----* digital control')
text(.25,.3,'------------- analog control')
```

```
text(.35,.6,' (a) 40 Hz')
hold off
```

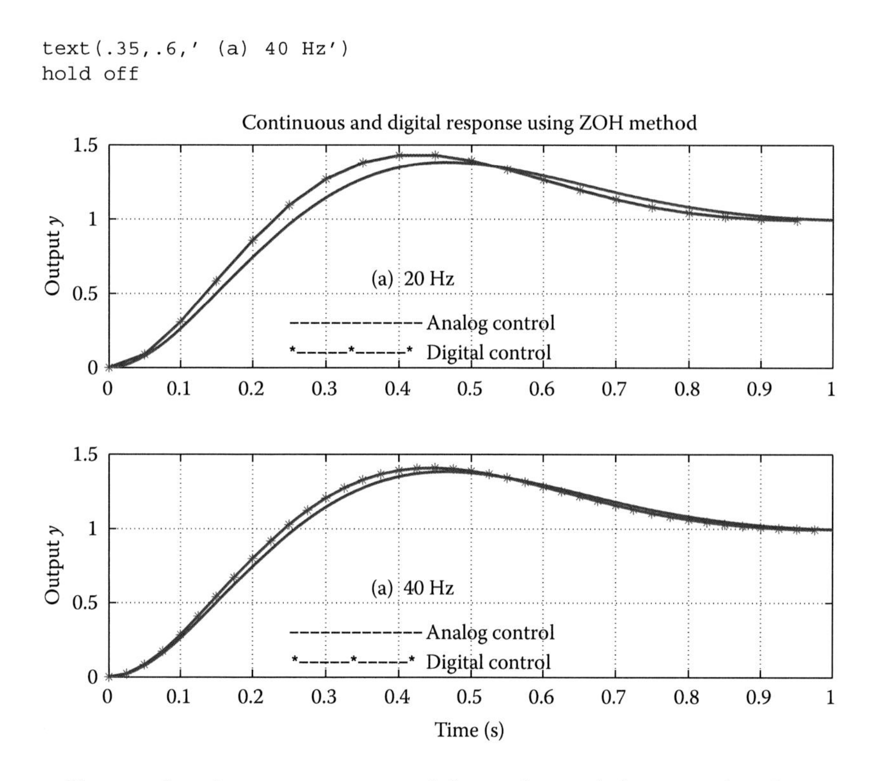

Observe that the step responses of the analog and discretized with ZOH system more closely match each other as the sampling time decreases.

EXERCISE 7.18

Illustrate the poles and zeros and design the step response of a system with transfer function: $G(z) = (1/(z^2 - 0.3z + 0.5)$ using the software platforms MATLAB and LabVIEW.

Solution

Illustration of poles and zeros

a. Using MATLAB, utilize the following program:

```
numDz = 1;
denDz = [1 -0.3 0.5];
sys = tf(numDz,denDz,1/20)
pzmap(sys)
axis([-1 1 -1 1])
zgrid on
```

We used the command **`pzmap`** to illustrate the poles and zeros in the complex z-plane. The poles at the denominator are complex conjugate: $p_1 = 0.1500 + 0.6910i$ and $p_2 = 0.1500 - 0.6910i$, while they are placed inside the unit circle, thus the system is stable. The sampling period is $T = 0.05$ s.

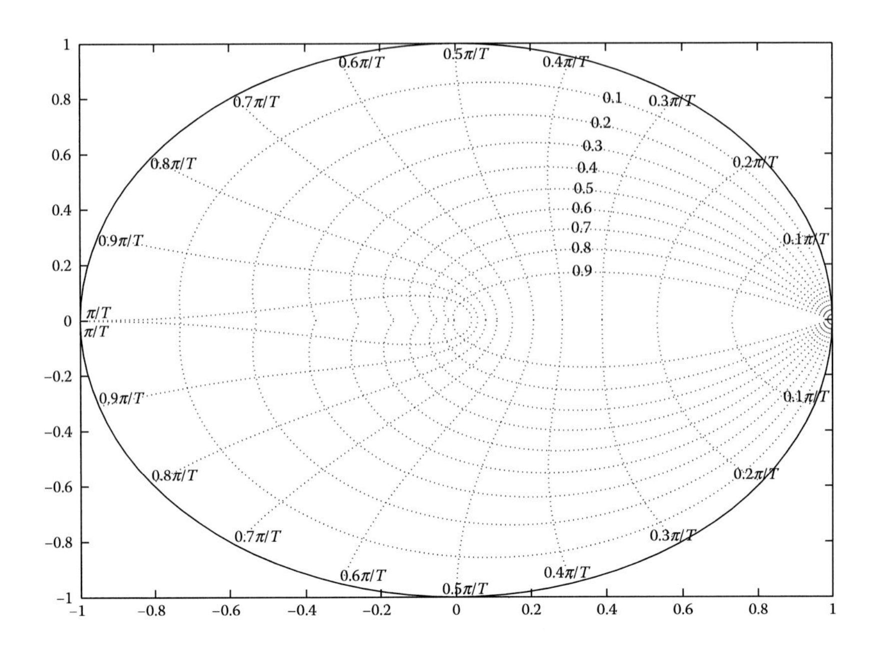

b. Using *LabVIEW* and *LabVIEW Control Design and Simulation Module*, discrete transfer functions can be developed and digital control systems can be efficiently simulated.

The poles and zeros of the transfer function can be designed with the aid of *CD Pole–Zero Map VI*.

The block diagram and the poles-zeros diagram are respectively presented below.

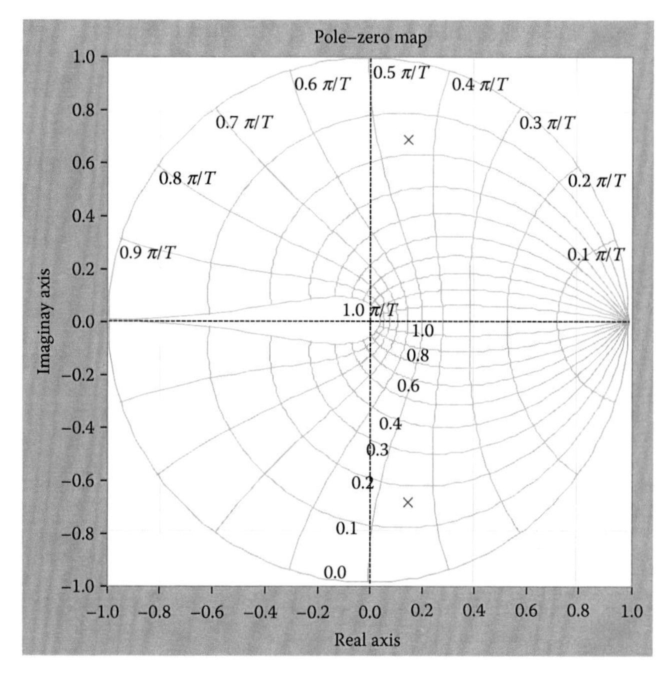

Step response:

a. Using **MATLAB**, we have the following program that illystrates the step response

```
numDz = 1;
denDz = [1 -0.3 0.5];
sys = tf(numDz,denDz,1/20);
step(sys,2.5);
```

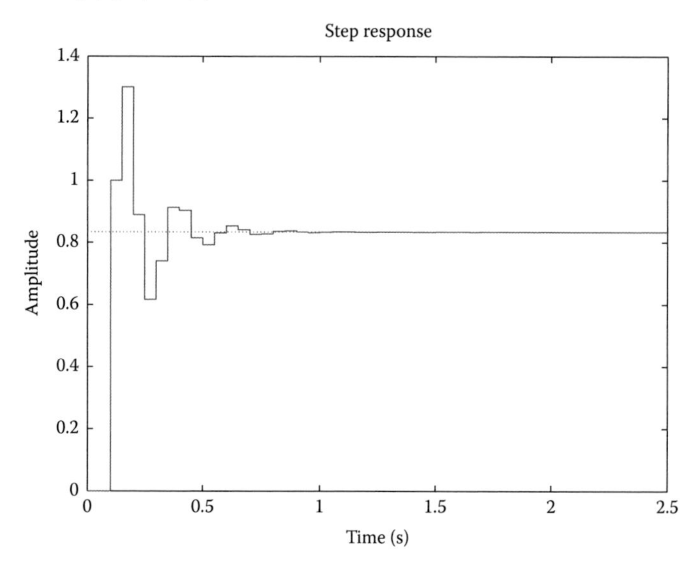

b. Using *LabVIEW* and *LabVIEW Control Design Module,* we add the *CD Step Response VI* in the block diagram.

The block diagram and the step response are illustrated below.

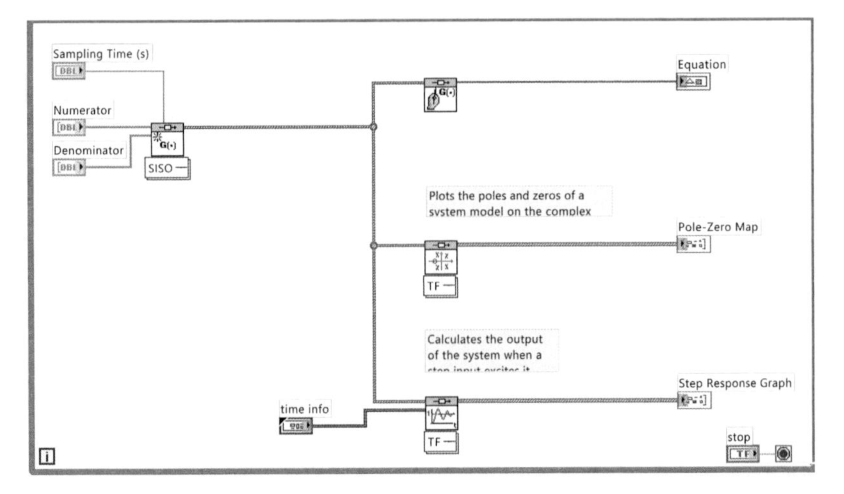

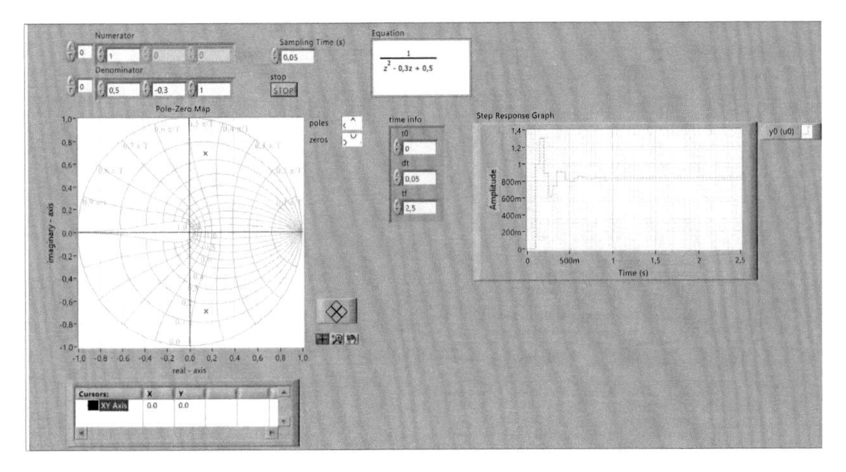

EXERCISE 7.19

Evaluate a second-order system with pulse transfer function $H(z) = (z^2 - 0.3\,z - 0.1)/(z^2 - 0.55z + 0.595)$ and a system with two stages connected in series with first-order transfer functions: $H_1(z) = (z + 0.2)/(z - 0.7)$, $H_2(z) = (z - 0.5)/(z - 0.85)$ for a rectangular input signal of frequency 0.2 Hz.

Solution

Using the *LabVIEW Simulation Module,* we analyze the system model by presenting its block diagram as follows:

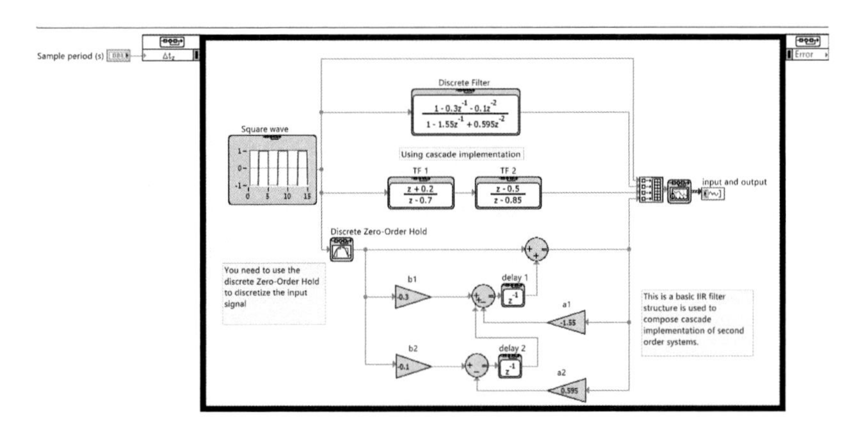

In this simulation scenario, the second-order discrete transfer function $H(z) = (z^2 - 0.3z - 0.1)/(z^2 - 0.55z + 0.595)$ and its equivalent are implemented, while it holds that $H_1(z)H_2(z) = H(z)$. The simulation goal is the system evaluation for a rectangular input signal with frequency 0.2 Hz. The corresponding front panel for $T = 0.01$ s is given below.

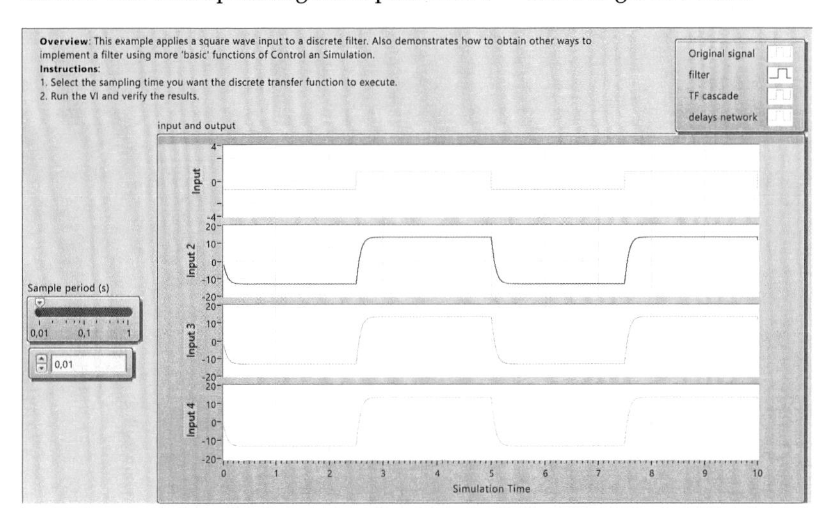

8

Compensation of Digital Control Systems

8.1 Introduction

System compensation corresponds to the system design such that the given requirements or specifications are satisfied. Typically, an additional system is introduced in the appropriate connection with the system to be controlled in order to compensate its behavior. These systems are called *compensators* or *controllers* and depend on the system structure.

The most common compensation units are

- Cascade compensation, where the compensator is placed in a series to the system to be controlled, which is the most usual compensation form in control systems.
- Feedback compensation, where the compensator is placed for feedback.
- Mixed compensation, where the compensator is placed in a series and for feedback with the system to be controlled.

The compensator is a device which automatically produces an input signal. In the case when the compensator utilizes an output measurement, then it is called *feedback control*. The algorithm that is installed in the controller is called *feedback law*.

The goal of the control systems design is the determination of the appropriate input signal so as the controlled process achieves the desired performance. This performance accordingly varies with the features and the exact operation of the process.

The selection of the appropriate compensator depends on

- The structure of the system to be controlled (plant)
- The design specifications

Typically, the *design specifications* are summarized as follows:

1. The system output should be bounded for every bounded input signal and for every initial value of the input and output (bounded response).

2. The system output should follow a desired response (command following).
3. The system poles should be placed in desired positions.
4. In transition state, the rise time should be small.
5. In steady state, the error should be as small as possible.
6. System robustness should be preserved, that is, the closed-loop system should not be susceptible to alterations of the process conditions and to errors onto the process model.

Under typical compensation applications, all the specifications cannot be simultaneously achieved, because some are conflicting. Yet, a balanced tradeoff between two important objectives should be delivered; namely, performance and robustness.

A compensated system exhibits satisfactory efficiency when it performs a fast and smooth response to changes and disturbances of the desired value with little or no oscillation, and it is robust if it provides satisfactory performance for a wide range of process conditions and the expected error-case present in the model process. The robustness can be achieved by selecting a conservative calibration of the modulator, but this option tends to result in low efficiency levels. Thereby, the conservative regulatory controller actions sacrifice a part of good performance to achieve the desired robustness.

For the control systems design in the z-domain or s-domain, a common practice is to neutralize undesirable poles and zeros of the controlled process with corresponding zeros or poles inserted in the controller. This method should be applied with caution, especially in the case when implemented near the perimeter of the circle ($z = 1$), which may cause instability due to inaccuracies that always exist during the neutralization.

The prevailing methods of digital control systems design are divided into *indirect methods* (indirect design methods) and *direct methods* (direct design methods).

8.2 Indirect Design Methods

Consider the analog control system in Figure 8.1 with $G(s)$ and $C(s)$ indicating the transfer function of the system to be controlled and the analog controller, respectively.

The objective is the design of a discrete-time controller with transfer function $C(z)$ such that the closed-system performance (Figure 8.2) approaches as close as possible the performance of the analog system. First, the continuous-time controller is designed and then the discrete-time controller is calculated

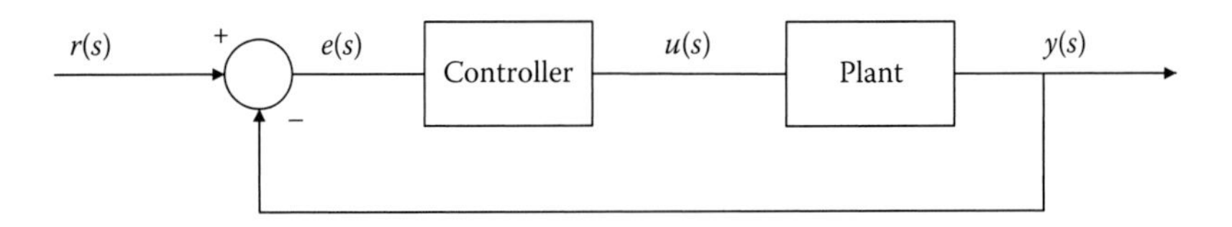

FIGURE 8.1
Continuous-time control system.

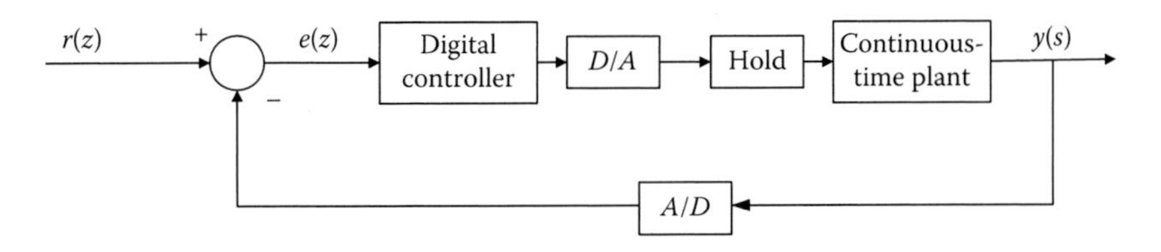

FIGURE 8.2
Digital control system.

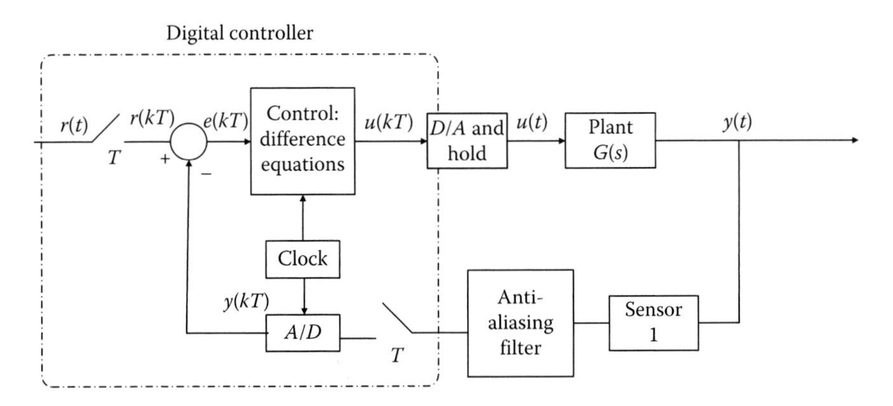

FIGURE 8.3
Digital control system with a pre-antialiasing filter.

using one of the available discretization techniques (as thoroughly analyzed in Chapter 3).

The disadvantages of this method are the largest phase delay of the continuous-time system and the dependence of the dynamic behavior of the system by the sampling period.

The advantages of indirect technique is the use of control engineering experience in continuous-time controllers and the facilitation of the discrete-time controller design in the case of the analog controller's existence.

This method is implemented in the following steps:

1. Selection of sampling period and design of the pre-antialiasing filter to cutoff the high frequencies (Figure 8.3).
2. Design or discretization of the analog controller.

3. Control of the discrete system performance.
4. System optimization by using the subsequent approaches, in the case when the specifications are not satisfied:
   - Select a more suitable discretization method.
   - Increase the sampling frequency.
   - Modification of the analog controller.

The indirect design technique has an implementation difficulty due to the applied holding system. In the simplest case, the holding system is zero order, so

$$G_h(s) = \frac{1-e^{-Ts}}{s} \tag{8.1}$$

After the approximation of the term e^{-Ts}, the approximated expression of $G_h(s)$ arises, which is given by the expression (8.3) and is often used in practice:

$$e^{-Ts} = \frac{e^{-Ts/2}}{e^{Ts/2}} = \frac{1-(Ts/2)+((Ts)^2/8)-\cdots}{1+(Ts/2)+((Ts)^2/8)+\cdots} \simeq \frac{1-(Ts/2)}{1+(Ts/2)} \tag{8.2}$$

$$(8.1),(8.2) \Rightarrow G_h(s) = \frac{1}{s}\left(1-\frac{1-(Ts/2)}{1+(Ts/2)}\right) = \frac{T}{(Ts/2)+1} \tag{8.3}$$

8.2.1 Selection of Sampling Rate

The *sampling frequency* is a critical parameter for the control system and determines the stability of the closed system. The selection of the optimal sampling rate for a digital control system is actually a matter of compromise. Generally, the smaller the sampling period, the closer is the digital system performance to its analog counterpart. In the asymptotic limit where the period approaches zero, the behavior of the discrete system coincides with the analog one.

The performance of a digital controller improves as the sampling frequency increases, while, however, increasing the implementation cost. On the other hand, a smaller sampling rate means more execution time of the control algorithm, so it becomes possible to use slower (and hence more cost-efficient) CPUs. Based on the above, we conclude that the ideal sampling rate is the largest possible that will simultaneously meet all the performance and cost requirements.

An important criterion in the selection of sampling frequency is the avoidance of the frequency aliasing effect. The Nyquist frequency determines the theoretical minimum bound on the sampling rate. However, in practical control systems, it is almost always necessary to choose a sampling rate much higher than the one indicated by the Nyquist criterion. Accordingly, the sampling frequency should be at least twice the largest system frequency, namely, $\omega_{sp} > 2\omega_{\max}$ where ω_s is the sampling frequency and $\omega_{\max}$ is the maximum system frequency.

The above condition is not sufficient for proper operation of the control system. It is a common practice to select sampling rates that are at least five times greater of $\omega_{\max}$ for first-order systems, while at least 10 times greater of $\omega_{\max}$ is required for second- and higher-order systems. This arises from the necessity to maintain a distance that is as small as possible between successive samples and to make the system capable of monitoring any possible alteration of the signal. This reduces the deviation of samples from the actual signal values.

Also, the D/A converter, which is always between the discrete control system and the analog physical system, holds the control signal during a sampling period. As already remarked, this retention (hold) introduces a delay in the control signal by approximately $T/2$, as shown in Figure 8.4.

The so-called *dead time*, which is introduced, does not affect the amplitude of frequency response; however, it causes the phase margin increase by $\Delta\varphi = -\omega_c T/2$ (rad). Therefore, the $T/2$ delay negatively affects the control loop stability.

The sampling time should generally be a flexible parameter. A physical system that is difficult to be controlled, the high precision achievement, or the use of a differential term, all require a faster sampling rate. On the other hand, if the controlled system is simple or if a differential term is not used,

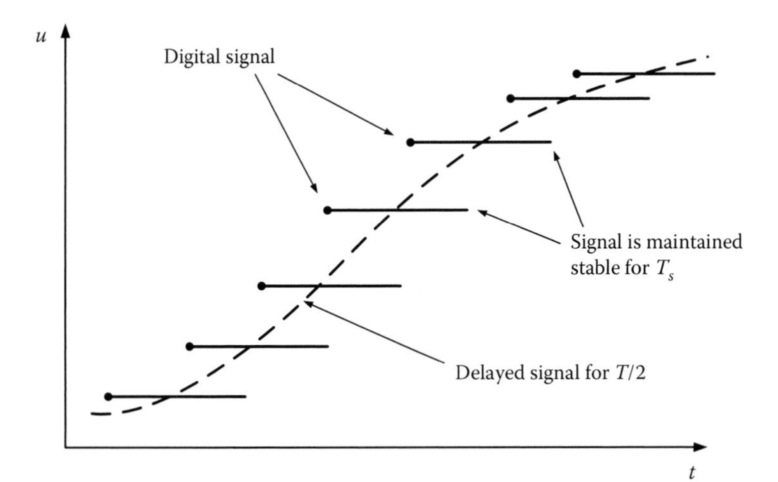

FIGURE 8.4
Control system delay by $T/2$ due to the hold process.

then we can reduce the sampling time, yet no more than 1/5 of the desired recovery time.

It should also be understood that a part of information may not be recovered due to sampling. This is the reason why we perform sampling at a rate higher than the theoretical one.

8.3 Direct Design Methods

Using these methods, the digital controller is directly designed without first having designed the standard analog controller. First, a mathematical discrete-time model of the continuous-time control system is determined and then the design in z-domain is implemented; thereby, the discrete-time controller directly arises.

There are *three direct design methods*:

1. Design via analytical methods.
2. Root locus design in z-domain.
3. Design in the frequency domain (w-plane design).

8.3.1 Design via Analytical Methods

Consider the system in Figure 8.5.

Assume that

$$HP(z) = \frac{B(z)}{A(z)}, \quad C(z) = \frac{S(z)}{V(z)}, \quad F(z) = \frac{T(z)}{V(z)} \tag{8.4}$$

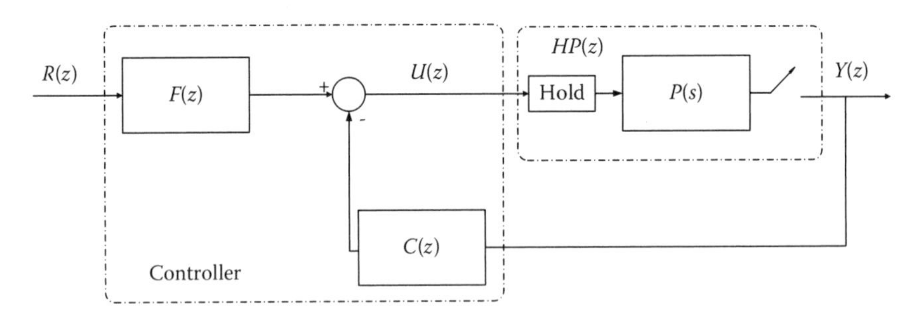

FIGURE 8.5
Digital control system with a controller.

The analytical design methods define the polynomials $V(z)$, $S(z)$, and $T(z)$ so as to successfully achieve a predetermined closed-loop transfer function of the form

$$\frac{Y(z)}{R(z)} = \frac{B(z)T(z)}{A(z)V(z)+B(z)S(z)} \tag{8.5}$$

In the simplest form, it suffices that $T(z) = S(z)$ so the closed-loop system is a typical unit-feedback system, the controller is selected to satisfy the steady-state error condition, and the poles of characteristic polynomial coincide with the poles of the desired polynomial.

The following steps are followed during the design process:

1. Selection of the desired characteristic polynomial.
2. Selection of the appropriate $m \geq 0$ and polynomials' definition

$$V(z) = z^m + u_1 z^{m-1} + \cdots + u_m \tag{8.6}$$

$$S(z) = s_0 z^m + s_1 z^{m-1} + \cdots + s_m \tag{8.7}$$

3. Selection of the coefficients u_i and s_i to satisfy the error specifications in steady state and to preserve a matching of the closed-loop characteristic polynomial with the desired one.

The selection of parameter m is based on the subsequent assumptions:

- Let $n = \deg(A(z))$ where $A(z)$ is a monic polynomial (i.e., its leading coefficient equals to 1) and the system transfer function with ZOH $HP(z)$ has an order ≥ 1. Then, the polynomial $(B(z)T(z))/(A(z)V(z)+B(z)S(z))$ is monic with order $n + m$.
- The parameters s_i, $i = 0 \ldots m$ and u_i, $i = 0 \ldots m$ are $2m + 1$.
- The design parameters should satisfy $n + m$ conditions (due to the zeros of the characteristic polynomial) and p conditions to satisfy the steady-state error specifications.
- The design problem has a unique solution if $2m+1 = n+m+p = \Rightarrow m = n+p-1$.

8.3.2 Design Using the Root Locus Method of the Characteristic Equation

The root locus method represents a direct method for defining the transfer function of the digital controller.

Consider a closed sampled-data system with the closed-loop transfer function $G_{cl}(z) = G(z)/1+G(z)F(z)$ with $F(z)$ denoting the transfer function of feedback branch and characteristic equation $1 + G(z)F(z) = 0$.

Let the open-loop transfer function $G(z)F(z)$ be in the special form

$$G(z)F(z) = K\frac{\prod_{i=1}^{m}(s+z_i)}{\prod_{i=1}^{n}(s+p_i)} \tag{8.8}$$

The most appropriate parameter value K is selected according to the root locus of $1 + G(z)F(z) = 0$ in order for the closed system specifications to be satisfied. Note that the discrete-time controller is an amplifier with gain K.

The steps that should be followed during the design are

1. Definition of sampling period T.
2. Calculation of the equivalent discrete transfer function of the controlled system (which is connected in series to the hold circuit).
3. Root locus design of a controller $C(z)$ so as to satisfy the required specifications.

8.3.3 Design in the Frequency Domain

The design techniques of controllers, based on the response of a continuous-time system in the frequency domain, can be extended to discrete-time systems according to the expression $z = e^{sT}$. To maintain the simplicity of the logarithmic curves (Bode diagrams) and the discrete-time systems, the bilinear transformation of the expression (8.9) is utilized, where the unit circle in z-domain reflects the left w half plane.

$$z = \frac{1+\frac{Tw}{2}}{1-\frac{Tw}{2}} \quad \text{or} \quad w = \frac{2}{T}\left[\frac{z-1}{z+1}\right] \tag{8.9}$$

In the case when the product ωT (where $\omega = 2\pi f = 2\pi/T$ denotes the circular frequency of the analog system) is small, the curves of harmonic response in s and w domains are identical.

For high ωT values, the frequency scales ω and Ω (where Ω is the frequency in w-domain) are distorted because of the relation $\Omega = (2/T)\tan(\omega T/2)$.

Consequently, discrete-time controllers can be designed with the aid of Bode diagrams.

The steps that should be followed during the design are

1. Definition of sampling period T.
2. Calculation of the equivalent discrete transfer function of the controlled system (which is connected in series to the hold circuit).

3. Transform of the discrete transfer function into the frequency domain via the expression (8.9).
    a. Design of the controller using analog methods into the frequency domain $C(w)$.
    b. Transform of $C(w)$ to $C(z)$.
4. Analysis of the closed system dynamic behavior.

8.4 PID Digital Controller

The *three-term controller* is actually a controller-cascade compensator that is placed in the direct branch of the closed system and regulates the signal that drives the system taking into account the deviation (error) of the input from output. The acronyms of the PID controller is from the initials of *proportional, integral,* and *derivative,* corresponding to the analog, integrated, and differential operation, performed by the controller. The block diagram of an analog PID controller has the form of Figure 8.6.

It holds that

$$u(t) = K_p\left(e(t) + \frac{1}{T_i}\int_0^t e(\tau)\,d\tau + T_d\frac{de(t)}{dt}\right) \tag{8.10}$$

The transfer function of the analog PID controller is given by

$$C(s) = \frac{U(s)}{E(s)} = K_p\left(1 + \frac{1}{T_i s} + T_d s\right) = K_p + \frac{K_i}{s} + K_d s$$
$$\text{with}\quad K_i = \frac{K_p}{T_i},\ K_d = K_p T_d \tag{8.11}$$

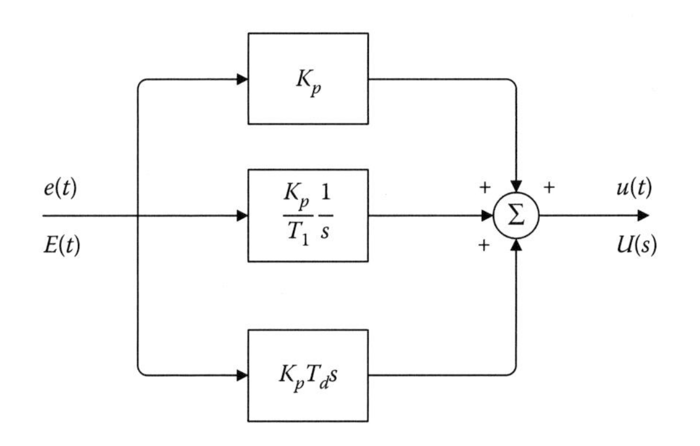

FIGURE 8.6
Block diagram of an analog PID controller.

A *digital controller* measures the controlled variable in certain time periods, which are separated to each other by a time fraction called the sampling time. Each sample (or measure) of the controlled variable is transformed into a binary number to enter into a digital computer or a microcomputer. This computer removes each sample of the measured variable to calculate a set of error samples. The block diagram of a digital PID controller has the form of Figure 8.7.

After the calculation of each error sample, a digital PID controller follows a process, the so-called *PID algorithm*, to calculate the controller's output, which is based on the error samples. The PID algorithm has two versions: (a) the position version and (b) the velocity incremental version.

- Implementation of digital controllers using the position algorithm

  Assume that $t = kT$, $k = 0, 1, 2, \ldots$ hence: $e(t) = e(kT)$

  Let

$$\left.\begin{aligned} &\int_0^t e(\tau)d\tau = \sum_{j=1}^{k} e(jT)T = T\sum_{j=1}^{k} e(jT) \\ &\frac{de(t)}{dt} \approx \frac{e(kT)-e(kT-T)}{T} \end{aligned}\right\} \tag{8.12}$$

  The algorithm of the digital PID controller is given by the expression (8.13), it is called *position (or absolute) algorithm* and calculates the valve position based on the error signals.

$$u(k) = K_p\left(e(k)+\frac{T}{T_i}\sum_{j=1}^{k} e(j)+\frac{T_d}{T}\{e(k)-e(k-1)\}\right) \tag{8.13}$$

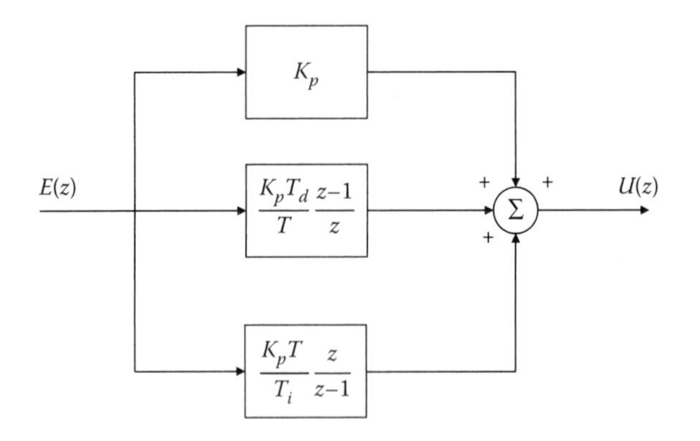

FIGURE 8.7
Block diagram of a digital PID controller.

- Implementation of digital controllers using the incremental algorithm

  From the expression (8.13), we have that

$$u(k-1)=K_p\left(e(k-1)+\frac{T}{T_i}\sum_{j=1}^{k-1}e(j)+\frac{T_d}{T}\{e(k-1)-e(k-2)\}\right) \tag{8.14}$$

But: $\Delta u(k) = u(k) - u(k-1)$ thus:

$$\Delta u(k)=K_p\left(e(k)-e(k-1)+\frac{T}{T_i}e(k)+\frac{T_d}{T}\{e(k)-2e(k-1)+e(k-2)\}\right) \tag{8.15}$$

or

$$u(k)=u(k-1)+K_p\left(e(k)-e(k-1)+\frac{T}{T_i}e(k)+\frac{T_d}{T}\{e(k)-2e(k-1)+e(k-2)\}\right) \tag{8.16}$$

The algorithm of the digital PID controller is given by the expression (8.15) or (8.16) and is called the *incremental algorithm*. The incremental algorithm fits very well in increasing output devices, such as stepping motors. The position algorithm is more physical and has the advantage that the controller "remembers" the position of the valve. The output of the controller, which is implemented by the incremental algorithm, represents the control signal increments.

The advantage of this algorithm is that the system is less affected if there is a computer error or if there is a switch between the manual and automatic operation. Also, the system is protected from jamming due to iterations (i.e., reset windup) because it does not include the error sum series that leads to a saturation of the control signal.

$$(8.16)\overset{ZT}{\Rightarrow} U(z)=z^{-1}U(z)+K_p(E(z)-z^{-1}E(z)+\frac{T}{T_i}E(z)+ \\ +\frac{T_d}{T}\{E(z)-2z^{-1}E(z)+z^{-2}E(z)\}) \tag{8.17}$$

The transfer function of *the PID controller* is presented as

$$\frac{U(z)}{E(z)}=\frac{1}{1-z^{-1}}K_p\left(1-z^{-1}+\frac{T}{T_i}+\frac{T_d}{T}(1-z^{-1})^2\right) \\ \Rightarrow \frac{U(z)}{E(z)}=K_p\left(1+\frac{T}{T_i(1-z^{-1})}+\frac{T_d}{T}(1-z^{-1})\right) \tag{8.18}$$

Equivalently, it holds that

$$\frac{U(z)}{E(z)} = \left.\frac{U(s)}{E(s)}\right|_{s=\frac{1-z^{-1}}{T}} = K_p\left.\left(1+\frac{1}{T_i s}+T_d s\right)\right|_{s=\frac{1-z^{-1}}{T}}$$

$$= K_p\left(1+\frac{T}{T_i(1-z^{-1})}+T_d\frac{1-z^{-1}}{T}\right) \tag{8.19}$$

- For $T_i = \infty, T_d = 0, C(z) = K_p \Rightarrow$ Analog control (P control)
- For $T_d = 0, C(z) = \dfrac{K_p(1+(T/T_i)) - K_p z^{-1}}{1-z^{-1}} \Rightarrow$ Analog—Integrated control (PI control)
- For $T_i = \infty, C(z) = K_p\left(1+\dfrac{T_d}{T}\right) - K_p\dfrac{T_d}{T}z^{-1} \Rightarrow$ Analog—Differential control (PD control).

NOTE: The integral term of a controller causes the output to vary as long as the error is nonzero. Often, the problem cannot be rapidly reduced to zero, resulting in a price rise of the integral term, and the saturation of regulatory action. This condition is also referred as *integral windup*. In this case, even if the error is zero, the regulatory action remains saturated. A PI controller requires special protective operations to meet the integral windup.

8.4.1 Digital PID Controller Tuning

Each *tuning method* should be aimed at compromise among many and often mutually conflicting requirements. The most common requirements are as follows:

- Accelerated depreciation of the effects of load disturbances
- Accelerated depreciation of noise effects in measurements
- Low sensitivity to variations of process parameters
- Short system response time

1. *Transition response method*. The method is characterized by two parameters related to the response delay and the maximum speed of the step response. Specifically, the process input is set equal to the continuous-time step signal with a relatively small amplitude, yet sufficient to enable the identification of the step response features.
   - To find the two parameters *Z–N*, the point where the maximum step response rate (i.e., the slope) is marked. At this point, the tangent that intersects the axes of amplitude and time is drawn.
   - The intersections define the parameters a and L, respectively. The intersection on the time axis determines the value of the dead time L, while the maximum rate is equal to the ratio α/L.

Figure 8.8 shows the *Z–N* parameters of the first tuning method, called Ziegler–Nichols method. Table 8.1 lists the parameters of the three-term controller as functions of the *Z–N* parameters, where K_{pD}, T_{iD}, and T_{dD} are the corresponding parameter values of the discrete-time controller (digital values).

2. *Stability limit method.* First, a *P* (proportional) controller is utilized. Then, the gain *K* of the the closed-loop system increases until it reaches to a critical point (sensitivity limit). The controller's parameters are computed by the expressions of Table 8.2, where T_0 is the oscillation period in the critical frequency and $K_{\kappa\rho}$ denotes the gain at the critical frequency.

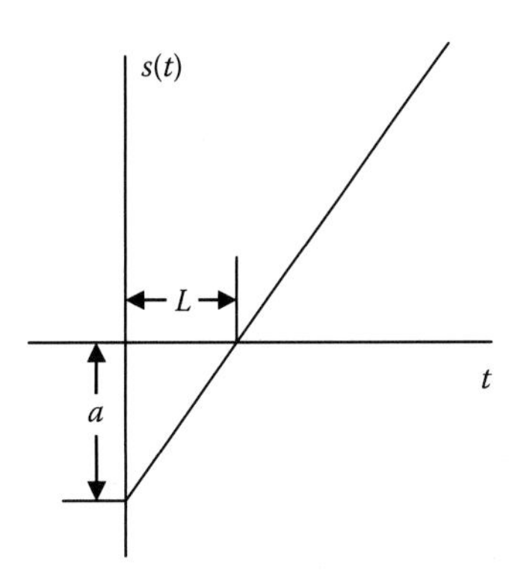

FIGURE 8.8
Illustration of the *Z–N* parameters of the first tuning method, the Ziegler–Nichols method.

TABLE 8.1
Controller Parameters for the Transition Response Method

	K_{pD}	T_{iD}	T_{dD}
P	$1/a$	–	–
PI	$0.9/a$	$3L$	–
PID	$1.2/a$	$2L$	$0.5L$

TABLE 8.2
Controller Parameters for the Stability Limit Method

	K_{pD}	T_{iD}	T_{dD}
P	$0.5K_{\kappa\rho}$	–	–
PI	$0.45K_{\kappa\rho}$	$T_o/1.2$	–
PID	$0.6K_{\kappa\rho}$	$T_o/2$	$T_o/8$

8.5 Deadbeat Digital Controller

A system is called *deadbeat* when all the poles of the closed system transfer function are placed at zero. Then, the system provides a remarkable identity: Its pulse response is zero after n pulses, where n denotes the system's degree.

The design of digital control systems to be deadbeat is accomplished by the appropriate choice of controller transfer function so as to cancel all the poles and zeros of the controlled process transfer function and to add a pole at the position $z = 1$.

This method of *pole cancellation* is applied either in the frequency domain (Bode diagrams) or in the time domain (root locus) and aims to

- Replace slow poles with faster ones to increase the system response
- Replace the dominant pole with a slow one to increase the accuracy of the system steady state
- Replace a pair of complex poles with different pair of complex poles to modify the transient response

The *deadbeat control* has been implemented in various applications, such as current control in AC voltage inverters, rectifiers, active filters, UPS, AC-DC converters, and electric induction motor torque control. Although the excellent dynamic behavior offered by the deadbeat control is a key advantage, errors in the model design or unforeseen variations of some parameters may lead to malfunctioning or even instability.

The design requirement is that the closed-loop transfer function should be of the form

$$G_{cl}(z) = z^{-k}, \quad k \geq 1 \tag{8.20}$$

where k stands for the system delay, expressed as an integer, multiple of the sampling period.

The transfer function of the deadbeat digital controller is provided by the expression (8.21) with $HP(z) = ZT(G_{zoh}(s)G_p(s))$ the system transfer function using ZOH.

$$C(z) = \frac{1}{HP(z)} \cdot \frac{z^{-k}}{1 - z^{-k}} \tag{8.21}$$

8.6 Phase Lead/Lag Digital Compensators

The *phase lead compensation* has approximately the same behavior as the *PD* controller and is applied to control systems that have satisfactory features in steady state but their transient response is not satisfactory and needs improvement.

The *phase lag compensation* has approximately the same behavior as the *PI* controller and is applied to control systems that have satisfactory features in transient response but the steady state does not show satisfactory features and requires improvement.

Consider an analog phase lead compensator with the transfer function

$$C(s) = \frac{s+a}{s+b} \tag{8.22}$$

The case when $b > a > 0$ results to a phase lead compensator and $a > b > 0$ results to a phase lag compensator.

Utilizing the *z*-transform of *C*(*s*), we get the transfer function of the digital phase lead/lag compensator.

$$\begin{aligned} C(z) &= z\left(\frac{s+a}{s+b}\right) = z\left(1+\frac{a-b}{s+b}\right) \\ &= z(\delta(t)+(a-b)e^{-bt}) \\ \Rightarrow C(z) &= (1+\mathrm{a}-\mathrm{b})\frac{z-e^{-bT}/(1+a-b)}{z-e^{-bT}} \end{aligned} \tag{8.23}$$

Hence, *C*(*z*) has a single pole and zero, correspondingly, which are of the form of

$$C(z) = K\frac{z-z_0}{z-z_p} \tag{8.24}$$

For the phase lead compensators, it holds that: $z_0 > z_p$.

8.7 Formula Table

The formula Tables 8.3 through 8.5 are discussed here.

TABLE 8.3

Digital PID Controllers

Controller Type	Transfer Function
P	$G_c(z) = \frac{U(z)}{E(z)} = K_p$
I	$G_c(z) = \frac{U(z)}{E(z)} = \frac{K_pT}{T_i}\frac{z}{z-1}$
D	$G_c(z) = \frac{U(z)}{E(z)} = \frac{K_pT_d}{T}\frac{z-1}{z}$
PID	$\frac{U(z)}{E(z)} = K_p(1+\frac{T}{T_i(1-z^{-1})}+\frac{T_d}{T}(1-z^{-1}))$

TABLE 8.4

Effect of PID Controllers on Time Response of Closed Control Systems

Controller Type	Rise Time	Overshoot	Recovery Time	Steady Error
P	Reduction	Increase	Slight change	Reduction
I	Reduction	Increase	Increase	Cancellation
D	Slight change	Reduction	Reduction	Slight change

TABLE 8.5

Tuning of PID Controllers

	K_{pD}	T_{iD}	T_{dD}
Transient Response Method			
P	1/a	–	–
PI	0.9/a	3L	–
PID	1.2/a	2L	0.5L
Stability Limit Method			
P	$0.5K_{\kappa\rho}$	–	–
PI	$0.45K_{\kappa\rho}$	$T_o/1.2$	–
PID	$0.6K_{\kappa\rho}$	$T_o/2$	$T_o/8$

8.8 Solved Exercises

EXERCISE 8.1

For the system of the following scheme with $G(s) = 1/s(s+1)$:

a. Derive the transfer function of the digital controller $D(z)$ when the desired transfer function of the closed system is

$$G_{cl}(z) = \frac{z+1}{z^2 - 1.14z + 0.403}$$

b. Derive and design the discrete-time response $y(kT)$ of the closed system when the input is a unit step signal.

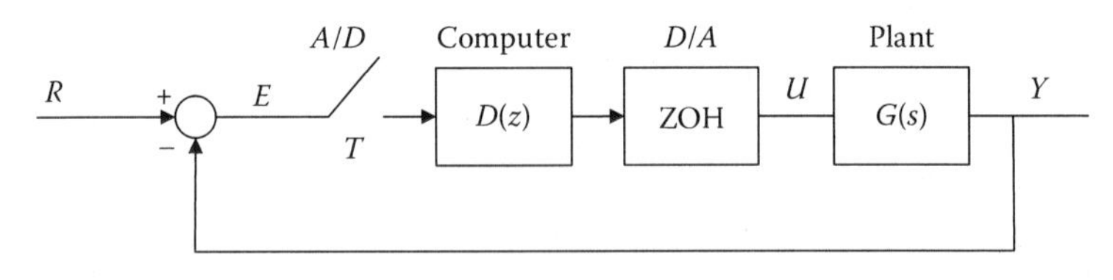

Let: $T = 0.1$ s.

Solution

a. The transfer function of the controlled system using ZOH is calculated as follows:

$$G(z) = z[G_H(s)G(s)] = \frac{z-1}{z} z\left[\frac{G_p(s)}{s}\right] = \frac{z-1}{z} z\left[\frac{1}{s^2(s+1)}\right] \quad (8.1.1)$$

$$\text{Yet}: \quad \frac{1}{s^2(s+1)} = \frac{c_1}{s+1} + \frac{c_{21}}{s} + \frac{c_{22}}{s^2} \quad (8.1.2)$$

Calculate c_1 using the Heaviside formula for simple poles.

$$c_1 = \lim_{s\to -1}\left(\frac{1}{s^2(s+1)}(s+1)\right) = \lim_{s\to -1}\left(\frac{1}{s^2}\right) = 1$$

Calculate c_{22} and c_{22} using the Heaviside formula for multiple poles with multiplication factor $n = 2$.

$$c_{22} = \lim_{s\to 0}\left(\frac{1}{s^2(s+1)}s^2\right) = \lim_{s\to -1}\left(\frac{1}{(s+1)}\right) = 1$$

$$c_{21} = \lim_{s\to 0}\frac{d}{ds}\left(\frac{1}{s^2(s+1)}s^2\right) = \lim_{s\to -1}\frac{d}{ds}\left(\frac{1}{(s+1)}\right) = -1$$

Consequently, substituting, we get

$$\frac{1}{s^2(s+1)} = \frac{1}{s+1} + \frac{-1}{s} + \frac{1}{s^2} \quad (8.1.3)$$

Next, calculate $G(z)$

$$G(z) = \frac{z-1}{z} z\left[\frac{1}{s^2(s+1)}\right] = \frac{z-1}{z} z\left[\frac{1}{s+1} + \frac{-1}{s} + \frac{1}{s^2}\right] \Rightarrow$$

$$G(z) = \frac{z-1}{z}\left[\frac{z}{z-e^{-T}} + \frac{-z}{z-1} + \frac{Tz}{(z-1)^2}\right] \Rightarrow$$

$$G(z) = \frac{z-1}{z-e^{-T}} - 1 + \frac{T}{z-1} \Rightarrow$$

$$G(z) = \frac{z(T-1+e^{-T})+1-e^{-T}-Te^{-T}}{(z-1)(z-e^{-T})} \tag{8.1.4}$$

For sampling period $T = 0.1$ s, we have

$$G(z) \simeq \frac{0.00484(z+0.9672)}{(z-1)(z-0.9048)} \tag{8.1.5}$$

The above analysis is verified by the following MATLAB® code example

```
n=[0 0 1];
d=[1 1 0];
[a,b]=c2dm(n,d,0.1,'zoh')
a = 0 0.0048 0.0047
b =1.0000 -1.9048 0.9048
printsys(a,b,'z')
```

The closed system transfer function is

$$G_{cl}(z) = \frac{Y(z)}{R(z)} = \frac{D(z)G(z)}{1+D(z)G(z)} \tag{8.1.6}$$

The transfer function of the digital controller is computed from the solution of

$$G_{cl}(z) = \frac{z+1}{z^2-1.14z+0.403} = \frac{D(z)\dfrac{0.00484(z+0.9672)}{(z-1)(z-0.9048)}}{1+D(z)\dfrac{0.00484(z+0.9672)}{(z-1)(z-0.9048)}} \tag{8.1.7}$$

$$\overset{(8.1.7)}{\Rightarrow} D(z) = \frac{20.66(z+1)(z-1)(z-0.9048)}{(z+0.9672)(z+0.25)(z-2.39)} \tag{8.1.8}$$

b. The closed system step response is calculated by the inverse z-transform of

$$Y(z) = G_{cl}(z)\frac{z}{z-1} = \frac{z(z+1)}{(z-1)(z^2-1.14z+0.403)} \tag{8.1.9}$$

$$(8.1.9) \overset{IZT}{\Rightarrow} y(kT) = (7.6 + 11.1e^{-0,454kT}\cos(0,456kT + 113.18°))u(kT) \tag{8.1.10}$$

EXERCISE 8.2

For the system of the following scheme with $G(s) = 1/(s+1)$ and $T = 0.1$ s, derive the controller $D(z)$ that satisfies the steady-state error requirement <0.01, when a unit ramp function is applied.

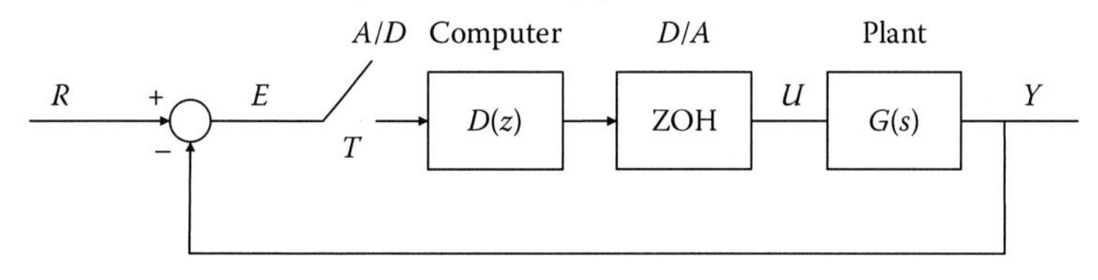

Solution

The transfer function of the controlled system with ZOH, which is the open-loop transfer function due to the unit feedback, is calculated as

$$G(z) = z\left[G_H(s)\frac{1}{s+1}\right] = \frac{z-1}{z}\left[\frac{1}{s(s+1)}\right] = \frac{z-1}{z}\left(\frac{z}{z-1} - \frac{z}{z-e^{-T}}\right)$$

$$= \frac{1-e^{-T}}{z-e^{-T}} = \frac{0.0952}{z-0.9048} \tag{8.2.1}$$

The discrete system is zero type, so the steady-state error, for a ramp input function (velocity error), is infinite.

A PI controller is used (because the integration operation optimizes the steady-state error), such that the denominator of the open-loop transfer function has the coefficient $(z-1)$:

$$D(z) = \frac{K_i z}{z-1} + K_p \tag{8.2.2}$$

The velocity error constant is

$$K_v = \lim_{z\to 1}\left(\frac{z-1}{T}D(z)G(z)\right) = \lim_{z\to 1}\left(\frac{z-1}{T}\left(\frac{(K_i+K_p)z-K_p}{z-1}\frac{0.0952}{z-0.9048}\right)\right)$$

$$\Rightarrow K_v = \frac{K_i}{T} = \frac{K_i}{0.1} \tag{8.2.3}$$

But, the steady-state error is

$$e_{ss} = \frac{A}{K_v} = \frac{1}{K_v} = \frac{0.1}{K_i} \le 0.01 \Rightarrow K_i \ge 10 \tag{8.2.4}$$

The above analysis holds only if the system is stable. Therefore, the $(K_p)_{cr}$ should be estimated, using the Jury criterion or Routh criterion with Möbius transform.

The system characteristic equation is

$$1 + D(z)G(z) = 0 \Rightarrow 1 + \frac{z(10+K_p) - K_p}{z-1}\frac{0.0952}{z-0.9048} = 0$$
$$\Rightarrow z^2 - (0.953 - 0.0952K_p)z + 0.905 - 0.0952K_p = 0 \tag{8.2.5}$$

To implement the Routh criterion, the bilinear transform $z=1+w/1-w$ is enforced, hence, we have

$$0.952w^2 + (0.19 + 0.1904K_p)w + 2.858 - 0.1904K_p = 0 \tag{8.2.6}$$

The Routh table is formed:

w^2	0.952	$2.858-0.1904K_p = 0$
w^1	$0.19 + 0.1904K_p$	0
w^0	$2.858 - 0.1904K_p$	0

To present stability, it suffices that

$$2.858 - 0.1904K_p > 0 \Rightarrow K_p \leq 15 \tag{8.2.7}$$

EXERCISE 8.3

For the system of the following scheme with $G(s) = e^{-2s}/(1+0.5s)$ and $T = 0.5$ s

a. Derive the digital PI controller using the pole cancellation technique. Define the controller gain so as to satisfy that the cut-off frequency of the open-loop system is approximately 0.2.
b. Provide the transfer function and the difference equation of the controller.
c. Define the first 15 values of the step response of the closed system.

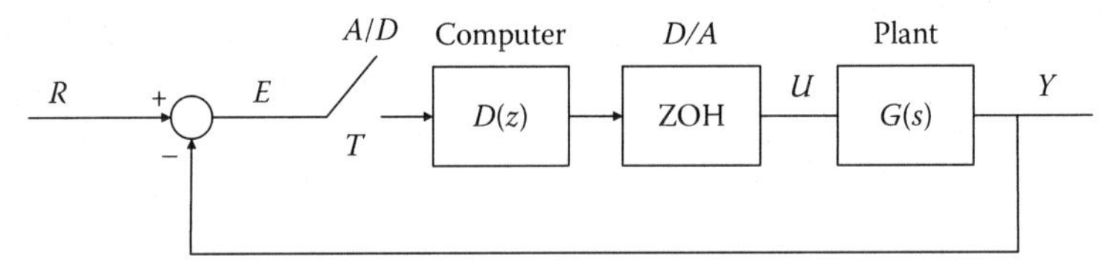

Solution

a. The system transfer function with ZOH is

$$G(z) = \frac{(1-e^{-1})}{z-e^{-1}}z^{-4} = \frac{0.6321}{z-0.3679}z^{-4} \tag{8.3.1}$$

The transfer function of the digital PI controller, by eliminating the pole at 0.3679, is

$$D(z) = K\frac{z-0.3679}{z-1} \tag{8.3.2}$$

The discrete open-loop transfer function is presented as

$$W_o(z) = D(z)G(z) = \frac{0.6321K}{z-1}z^{-4} \tag{8.3.3}$$

In the frequency domain, for low-range frequencies ($\omega < 1/T_s = 2$), the approximate function $W_o(j\omega)$ is

$$W_o(j\omega) \approx \frac{0.6321K}{0.5j\omega}e^{-2.25j\omega} \tag{8.3.4}$$

An additional dead time $T_s/2$ (arising from sampling) has been included in the calculation.

In the cut-off frequency, the absolute value $|W_o(j\omega_c)|$ is 1, thus

$$|W_o(j\omega_c)| = \frac{0.6321K}{0.5\omega_c} = 1 \Rightarrow K = \frac{0.5\cdot 0.2}{0.6321} = 0.1582 \tag{8.3.5}$$

b. The transfer function of the PI controller is

$$D(z) = \frac{U(z)}{E(z)} = 0.1582\frac{z-0.3679}{z-1} = \frac{0.1582-0.0582z^{-1}}{1-z^{-1}} \tag{8.3.6}$$

The difference equation is given by

$$u[nT] = u[(n-1)T] + 0.1582e[nT] - 0.0582e[(n-1)T] \tag{8.3.7}$$

c. The step response of the closed system is expressed as

$$\frac{Y(z)}{R(z)} = \frac{W_o(z)}{1+W_o(z)} = \frac{0.1582\cdot 0.6321z^{-4}}{z-1+0.1582\cdot 0.6321z^{-4}} = \frac{0.1z^{-5}}{1-z^{-1}+0.1z^{-5}} \tag{8.3.8}$$

The corresponding difference equation is

$$y[nT_s] = 0.1r[(n-5)T_s] + y[(n-1)T_s] - 0.1y[(n-5)T_s] \tag{8.3.9}$$

The first 15 values of the step response are

$$y[nT_s] = \{0,0,0,0,0,0.1,0.2,0.3,0.4,0.5,0.59,0.67,0.74,0.8,0.85\}$$

EXERCISE 8.4

For the system of the following scheme with $G(s)=e^{-2s}/(1+0.5s)$ and $T=0.5$ s, define the transfer function of the digital controller in order for the closed-loop system to operate as a deadbeat controller. Assume a step input signal.

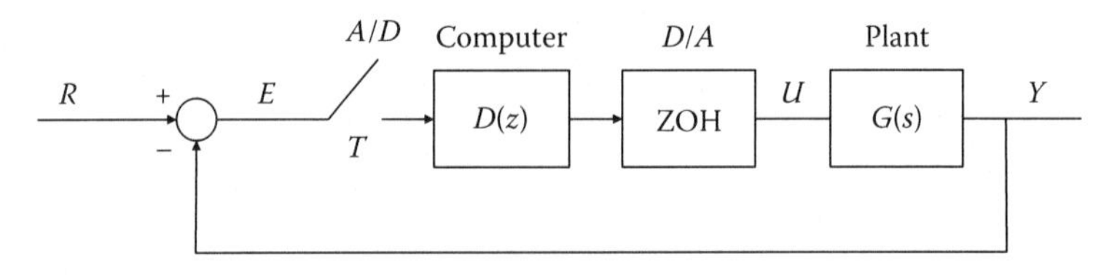

Solution

The system transfer function with ZOH is

$$G(z)=\frac{(1-e^{-1})}{z-e^{-1}}z^{-4}=\frac{0.6321}{z-0.3679}z^{-4} \tag{8.4.1}$$

The system output should reach the value 1 after $d=5$ (dead time + 1) samples for a step input.

The closed-loop transfer function, using the controller $C(z)$ is

$$\frac{D(z)G(z)}{1+D(z)G(z)}=z^{-d} \tag{8.4.2}$$

$$(8.4.2)\Rightarrow D(z)=\frac{z^{-d}}{G(z)(1-z^{-d})} \tag{8.4.3}$$

Substituting the value of $G(z)$ and d in the expression (8.4.3), we have that

$$D(z)=1.582\frac{1-0.3679z^{-1}}{1-z^{-5}} \tag{8.4.4}$$

For the samples $k=0,1,2,\ldots,8$, the output signal values are: 0, 0, 0, 0, 0, 1, 1, 1, 1,…

The transfer function $U(z)/R(z)$ is given by

$$\frac{U(z)}{R(z)}=\frac{D(z)}{1+D(z)G(z)}=1.582(1-0.3679z^{-1})=1.582-0.582z^{-1} \tag{8.4.5}$$

The values of the controlled signal for the first eight samples are: 1.582, 1, 1, 1, 1, 1, 1, 1, … Therefore, the controller operates as a deadbeat controller.

EXERCISE 8.5

For the system of the following scheme with $G(s) = e^{-0.1s}/(s^2 + 5s + 6)$ and $T = 0.05$ s

a. Define the digital PI controller using the pole cancellation technique, which should satisfy that the initial value of the controlled signal is $u[0] = 10$, assuming a unit step input function.
b. Calculate the controlled signal values for $k = 0$, 1, and 2.

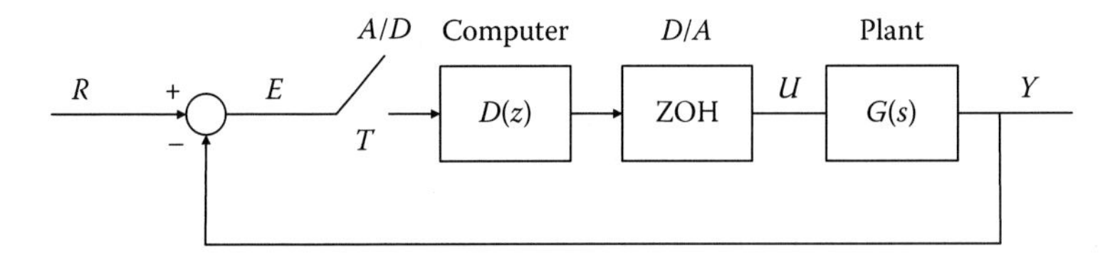

Solution

a. It holds that

$$\begin{aligned} z &= e^{sT} = e^{0.05s} \\ e^{-0.1s} &= e^{-0.05s\cdot 2} = z^{-2} \end{aligned} \tag{8.5.1}$$

Hence, $G(s)$ can be written as

$$G(s) = \frac{e^{-0.1s}}{s^2 + 5s + 6} = \frac{z^{-2}}{s^2 + 5s + 6} \tag{8.5.2}$$

The system transfer function with ZOH is

$$\begin{aligned} G(z) &= z^{-2}(1 - z^{-1})Z\left\{\frac{1}{s(s^2 + 5s + 6)}\right\} = z^{-2}(1 - z^{-1})Z\left\{\frac{1}{s(s+2)} - \frac{1}{s(s+3)}\right\} \\ &= z^{-2}\cdot\frac{1}{2}\cdot\frac{1 - e^{-T_s/0.5}}{z - e^{-T_s/0.5}} - z^{-2}\cdot\frac{1}{3}\cdot\frac{1 - e^{-T_s/0.33}}{z - e^{-T_s/0.33}} = z^{-2}\left(\frac{0.0476}{z - 0.9048} - \frac{0.0464}{z - 0.8607}\right) \\ &= 0.0012\cdot\frac{z^{-2}(z + 0.92)}{(z - 0.9048)(z - 0.8607)} \end{aligned} \tag{8.5.3}$$

The transfer function of the digital controller follows the form of

$$\begin{aligned} C(z) = \frac{U(z)}{E(z)} &= K\cdot\frac{z - 0.9048}{z - 1}\cdot\frac{z - 0.8607}{z} \\ &= K\cdot\frac{z^2 - 1.7655z + 0.7788}{z(z - 1)} \end{aligned} \tag{8.5.4}$$

or, equivalently, in the discrete-time domain

$$u[k] = u[k-1] + Ke[k] - 1.7655Ke[k-1] + 0.7788Ke[k-2] \quad (8.5.5)$$

To satisfy the condition $u[0] = 10$, $K = 10$ should hold.

b. The time delay of the process results to: $e[0] = e[1] = e[2] = 1$, so

$$u[0] = 10$$

$$u[1] = 10 + 10 - 17.655 = 2.345$$

$$u[2] = 2.345 + 10 - 17.655 + 7.788 = 2.478$$

EXERCISE 8.6

For the system of the following scheme with $G(s) = K/1 + 2s$ and $D(z) = 1$, define the sampling period assuming that the limit value of the K parameter to preserve stability of the closed system is equal to 4.

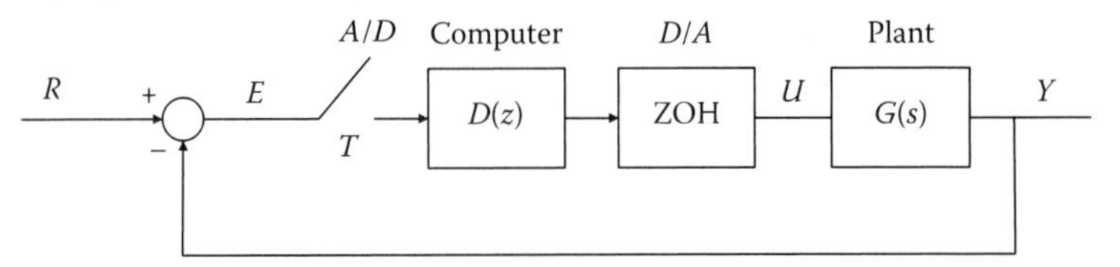

Solution

The open-loop system transfer function is

$$G(z) = (1 - z^{-1})Z\left\{\frac{K}{1+2s}\right\} = K \cdot \frac{1 - e^{-T/2}}{z - e^{-T/2}} \quad (8.6.1)$$

Setting the given value of K, we get

$$G(z) = 4 \cdot \frac{1 - e^{-T/2}}{z - e^{-T/2}} \quad (8.6.2)$$

The characteristic equation of the closed-loop system is

$$1 + G(z) = 1 + 4 \cdot \frac{1 - e^{-T/2}}{z - e^{-T/2}} = 0 \Rightarrow$$

$$z - e^{-T/2} + 4(1 - e^{-T/2}) = 0 \quad (8.6.3)$$

$$(8.6.3) \Rightarrow z = -4 + 5e^{-T/2} \quad (8.6.4)$$

To preserve marginal stability, it is required that $z=-1$, hence, substituting into the expression (8.6.4), the desired sampling period arises

$$5e^{-T/2}=3\Rightarrow e^{-T/2}=\frac{3}{5}\Rightarrow\frac{-T}{2}=\ln 0.6=-0.5108\Rightarrow T=1.0217\,\text{s}$$

EXERCISE 8.7

For the system of the following scheme with $G(z)=10(z+0.5)/(z^2-1.5z+0.5)$, design a cascade phase lead controller with the transfer function $D(z)=k(z-z_1/z-p_1)$ such that its output $y(k)$ follows the unit step input function in a minimum time without reaching the overshoot time.

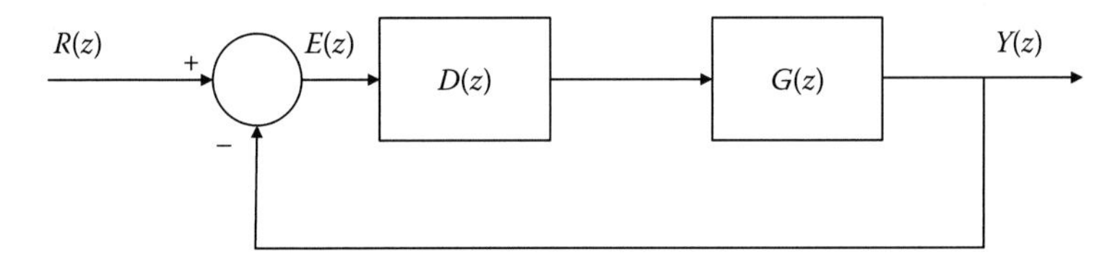

Solution

Design the controller applying pole–zero cancellation, with an additional pole at $z=1$, having

$$D(z)=\frac{1}{G(z)\cdot(z-1)} \tag{8.7.1}$$

$$G(z)=\frac{10(z+0.5)}{z^2-1.5z+0.5} \tag{8.7.2}$$

$$(8.7.1),(8.7.2)\Rightarrow D(z)=\frac{1}{\dfrac{10(z+0.5)(z-1)}{z^2-1.5z+0.5}}=\frac{0.1(z-0.5)}{z+0.5} \tag{8.7.3}$$

Observe that the transfer function of the digital controller follows the requested form (i.e., phase lead)

$$D(z)=\frac{0.1(z-0.5)}{z+0.5}=k\frac{z-z_1}{z-p_1}\Rightarrow(k=0.1,\ z_1=0.5,\ p_1=-0.5) \tag{8.7.4}$$

The closed-loop transfer function is

$$G_{cl}(z)=\frac{D(z)\cdot G(z)}{1+D(z)\cdot G(z)}=\frac{(1/(z-1))}{1+(1/(z-1))}=z^{-1} \tag{8.7.5}$$

The closed system output is presented as

$$Y(z) = R(z) \cdot G_{cl}(z)$$

$$= \frac{z}{z-1} \cdot \frac{1}{z} = \frac{1}{z-1} = z^{-1} + z^{-2} + z^{-3} + z^{-4} + \cdots \quad (8.7.6)$$

From the derived output signal, we conclude that the system tends to the steady state in a given time pulse without reaching overshoot. The system operates as a deadbeat controller.

EXERCISE 8.8

Design a digital controller for the analog system of type 0 so as to satisfy the specifications: zero position error, attenuation factor $J = 0.7$ and recovery time $T_s \leq 1$ s.

$$G(s) = \frac{1}{(s+1)(s+10)}$$

The controller's design should be implemented (a) with the indirect and (b) with the direct method.

Solution

a. *Indirect method*
   First we design the analog controller. To meet the zero position error requirement, we choose the PI controller. The simplest design can be made by the pole–zero cancellation technique at the pole in –1, that is, the transfer function of the controller will be in the form of

$$D(s) = K\frac{s+1}{s} \quad (8.8.1)$$

The open-loop transfer function is

$$D(s)G(s) = \frac{K}{s(s+10)} \quad (8.8.2)$$

The closed system characteristic equation is

$$1 + D(s)G(s) = 0 \Rightarrow s(s+10) + K = 0 \quad (8.8.3)$$

The characteristic polynomial $s(s + 10) + K = s^2 + 10s + K$ is of a form of $s^2 + 2J\omega_n s + \omega_n^2$; thus, the value of K parameter is computed

$$s^2 + 10s + K \equiv s^2 + 2J\omega_n s + \omega_n{}^2 \Rightarrow$$

$$J\omega_n = 5 \Rightarrow \omega_n = \frac{5}{J} = \frac{5}{0.7} = 7.142 \text{ rad/s} \quad (8.8.4)$$

$$\omega_n{}^2 = K \Rightarrow K = 51.02 \quad (8.8.5)$$

The recovery time is

$$T_s = \frac{4}{J\omega_n} = \frac{4}{5} = 0.8 \text{ s} \quad (8.8.6)$$

Consequently, the transfer function of the analog controller is expressed as

$$D(s) = 51.02\frac{s+1}{s} \quad (8.8.7)$$

Set the sampling period for: $\omega_n \simeq 7.142$ rad/s. Let $T = 0.02$ s (usually, the sampling period should be less than $2\pi/40\omega_n$).

Calculate the transfer function of the analog system with ZOH as

$$G(z) = (1 - z^{-1})Z\left(\frac{1}{s(s+1)(s+10)}\right)$$
$$= 1.8604 \cdot 10^{-4} \frac{z + 0.9293}{(z - 0.8187)(z - 0.9802)} \quad (8.8.8)$$

Using the bilinear transform, we derive the digital controller from its analog counterpart as follows:

$$D(z) = D(s)\big|_{s=\frac{2}{T}\frac{z-1}{z+1}} \Rightarrow$$
$$D(z) = 1.01K\frac{z - 0.9802}{z - 1} \quad (8.8.9)$$

Results of the indirect design:

For $J = 0.7$, using the MATLAB command *rlocus*, we find that the gain of the open-loop transfer function is approximately 46.7 and the physical oscillation frequency is 6.85 rad/s.

The recovery time is greater than 0.83 s the corresponding recovery time of the analog design, yet it follows the specifications.

$$T_s = \frac{4}{J\omega_n} = \frac{4}{0.7 \cdot 6.85} = 0.83 \text{ s} \quad (8.8.10)$$

The gain is

$$K \simeq 1.01 \cdot 46.7 = 47.2 < 51.02 \tag{8.8.11}$$

b. *Direct design*
Set a sampling period $T = 0.02$ s and calculate the transfer function $G(z)$ with ZOH

$$G(z) = (1 - z^{-1})Z\left(\frac{G(s)}{s}\right) \tag{8.8.12}$$

From the expression (8.8.12), and after some straightforward manipulations, we get

$$G(z) = 0.186 \cdot 10^{-3} \frac{z + 0.9293}{(z - 0.8187)(z - 0.9802)} \tag{8.8.13}$$

A PI controller of type 1 is used (because the integration optimizes the steady-state error) in order for the denominator of the open-loop transfer function to have the parameter $(z-1)$

$$D(z) = \frac{K_i z}{z - 1} + K_p \tag{8.8.14}$$

It follows that the digital controller has a pole at $z = 1$ and a zero at $z = 0.98$, that is

$$D(z) = \frac{z - 0.98}{z - 1} \tag{8.8.15}$$

The above results closely approach the ones obtained using the indirect design method.

EXERCISE 8.9

Design a digital controller for a discrete system with the transfer function $G(z) = 1/(z-1)(z-0.5)$, $T = 0.1$ s.

Consider the case where the oscillation frequency with attenuation is $\omega_d = 5$ rad/s, the attenuation factor is $J = 0.7$ and $J\omega_n = 2$ rad/s.

Solution

The open-loop transfer function for the system with the analog controller is

$$D(z)G(z) = \frac{K}{(z-1)(z-0.5)} \tag{8.9.1}$$

The characteristic equation of the closed system is given by

$$1+D(z)G(z)=0\Rightarrow 1+\frac{K}{(z-1)(z-0.5)}$$
$$\Rightarrow z^2-1.5z+K+0.5=0 \tag{8.9.2}$$

Matching the characteristic polynomial with the typical polynomial of a second-order system, we have that

$$z^2-1.5z+K+0.5\equiv z^2-2(\cos\omega_d T)e^{-J\omega_n T}z+e^{-2J\omega_n T} \tag{8.9.3}$$

Equating the coefficients, we get

$$1.5=2(\cos\omega_d T)e^{-J\omega_n T} \tag{8.9.4}$$

$$K+0.5=e^{-2J\omega_n T} \tag{8.9.5}$$

Design

1. $\omega_d=5$ rad/s
   From the expression (8.9.4), it follows that

$$J\omega_n=\frac{1}{T}\ln\left(\frac{1.5}{2\cos(\omega_d T)}\right)=10\ln\left(\frac{1.5}{2\cos(0.5)}\right)=1.571 \tag{8.9.6}$$

   The physical oscillation frequency with attenuation is given as

$$\omega_d=\omega_n\sqrt{(1-J^2)} \tag{8.9.7}$$

$$(8.9.7)\Rightarrow \omega_d{}^2=\omega_n{}^2(1-J^2)\Rightarrow\frac{\omega_d{}^2}{(J\omega_n)^2}=\frac{1-J^2}{J^2}=\frac{25}{(1.571)^2}$$
$$\Rightarrow J=0.3,\quad \omega_n=5.24\text{ rad / s} \tag{8.9.8}$$

   From the expression (8.9.5), we have

$$K+0.5=e^{-2J\omega_n T}\Rightarrow K=e^{-2J\omega_n T}-0.5=e^{-2\cdot 1.571\cdot 0.1}-0.5\Rightarrow K=0.23 \tag{8.9.9}$$

2. $J\omega_n=2$ rad/s

$$\omega_d=\frac{1}{T}\cos^{-1}\left(\frac{1.5e^{-J\omega_n T}}{2}\right)=10\cos^{-1}\left(0.75e^{-0.2}\right)=4.127\text{ rad/s} \tag{8.9.10}$$

$$\omega_d^{\ 2} = \omega_n^{\ 2}(1 - J^2) \Rightarrow \frac{\omega_d^{\ 2}}{(J\omega_n)^2} = \frac{1 - J^2}{J^2} = \frac{(4.127)^2}{2^2} \Rightarrow \tag{8.9.11}$$
$$J = 0.436, \quad \omega_n = 4.586 \text{ rad / s}$$

$$K + 0.5 = e^{-2J\omega_n T} \Rightarrow K = e^{-2J\omega_n T} - 0.5 = e^{-2.2 \cdot 0.1} - 0.5 \Rightarrow K = 0.17 \tag{8.9.12}$$

3. $J = 0.7$

$$(8.9.4) \Rightarrow 1.5 = 2(\cos\omega_d T)e^{-J\omega_n T} = 2\cos(0.0714\omega_n)e^{-0.07\omega_n} \tag{8.9.13}$$

Solve the expression (8.9.13), numerically, using the trial and error method.

$$\omega_n = 3.63 \text{ rad/s} \tag{8.9.14}$$

$$K + 0.5 = e^{-2J\omega_n T} \Rightarrow K = e^{-2J\omega_n T} - 0.5 = e^{-0.14 \cdot 3.63} - 0.5 \Rightarrow K = 0.10 \tag{8.9.15}$$

EXERCISE 8.10

Consider the system of the following scheme with $G_p(s) = 1/(s+1)$

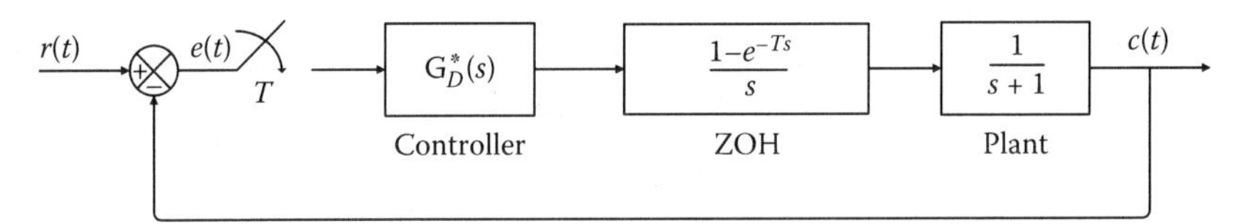

Design the integrated controller for $T = 0.5$ s and $K = 2$ (controller's gain), so as to satisfy a velocity error <0.3 s. How does the sampling period value affect the limit value for stability of the amplification factor?

Solution

The transfer function of the system to be controlled with ZOH is computed as

$$G_h G_p(z) = (1 - z^{-1})Z\left(\frac{G_p(s)}{s}\right) = (1 - z^{-1})Z\left(\frac{1}{s} - \frac{1}{s+1}\right)$$
$$= \frac{z-1}{z}\left[\frac{z}{z-1} - \frac{z}{z - e^{-T}}\right] = \frac{1 - e^{-T}}{z - e^{-T}} \tag{8.10.1}$$

The transfer function of the integrated controller is of the form of

$$G_D(z) = \frac{Kz}{z-1} \tag{8.10.2}$$

Design the root locus of the system characteristic equation to graphically capture the limit value of the K parameter and to provide stability of the closed system.

The open-loop transfer function is presented as

$$G(z) = G_D(z) \cdot G_h G_p(z) = \frac{Kz}{z-1} \cdot \frac{1-e^{-T}}{z-e^{-T}} \tag{8.10.3}$$

The characteristic equation of the closed system is obtained by $1 + G(z) = 0$

$$1 + G(z) = 0 \Rightarrow 1 + \frac{Kz}{z-1} \cdot \frac{1-e^{-T}}{z-e^{-T}} = 0 \tag{8.10.4}$$

for $T = 0.5$ s, the open-loop transfer function is given by the following expression and has two poles at $z = 1$ and $z = 0.605$, and one zero at $z = 0$.

$$(8.10.3) \Rightarrow G(z) = \frac{0.3935Kz}{(z-1)(z-0.6065)} \tag{8.10.5}$$

The breaking points can be derived by solving

$$\begin{aligned} &\frac{dK}{dz} = 0 \Rightarrow \frac{d}{dz}\left(-\frac{(z-1)(z-0.6065)}{0.3935z}\right) = 0 \Rightarrow z^2 - 0.6065 = 0 \\ &\Rightarrow z^2 = 0.6065 \Rightarrow z_1 = 0.7788 \quad \text{and} \quad z_2 = -0.7788 \end{aligned} \tag{8.10.6}$$

The limit value of K can be derived by using the magnitude criterion

$$\left|\frac{0.3935z}{(z-1)(z-0.6065)}\right| = \frac{1}{K} \tag{8.10.7}$$

The critical gain arises for $z = -1$, so

$$\left|\frac{-0.3935}{(-2)(-1.6065)}\right| = \frac{1}{K} \Rightarrow K_{cr} = 9.165 \tag{8.10.8}$$

The following scheme illustrates the root locus of the system characteristic equation for $T = 0.5$ s. The closed-loop system is stable for

$$0 < K < 8.165 \tag{8.10.9}$$

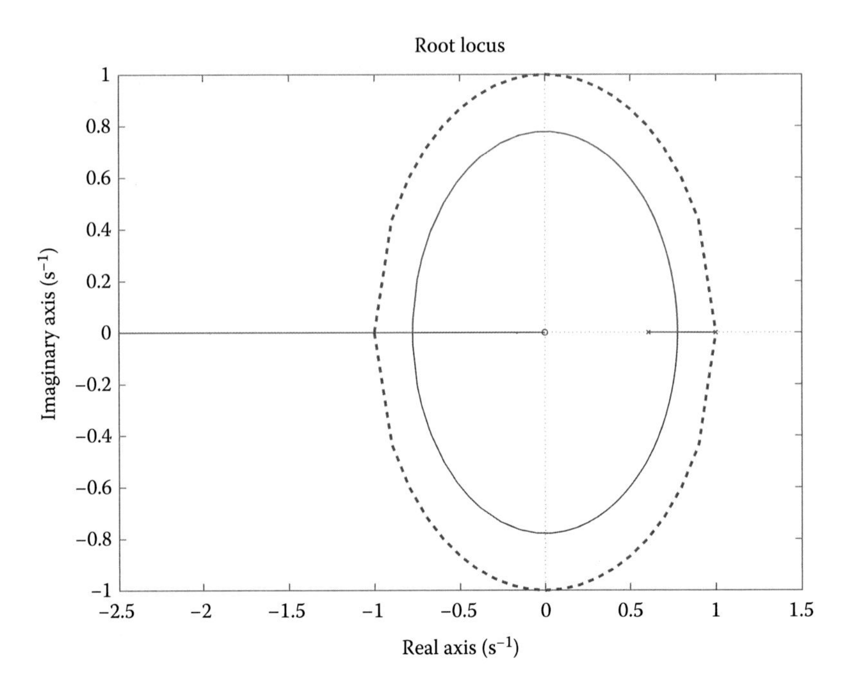

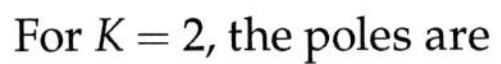

For $K = 2$, the poles are

$$z_{1,2} = 0.4098 \pm j0.6623 \tag{8.10.10}$$

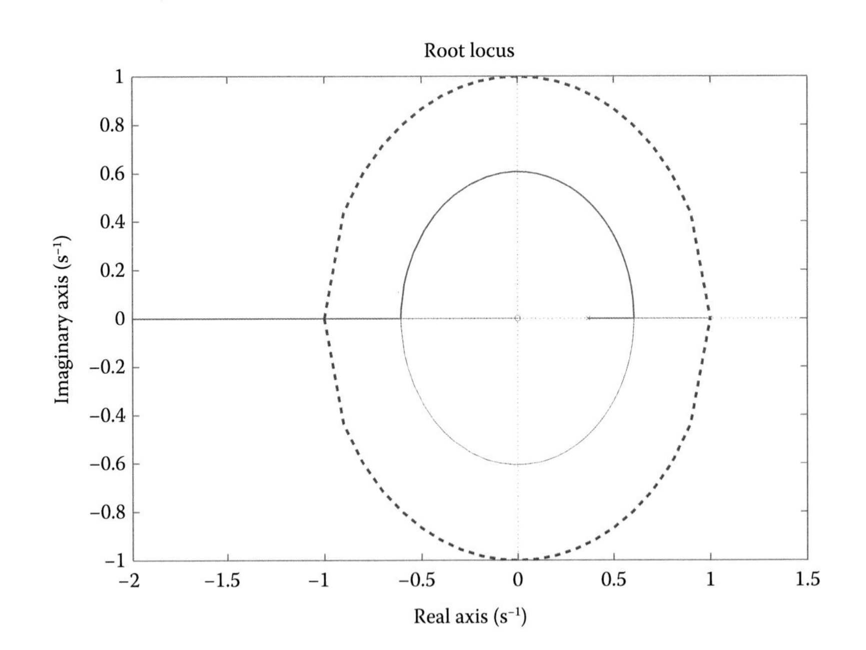

For $T = 0.5$ s and $K = 2$, the open-loop transfer function becomes

$$G(z) = \frac{0.787z}{(z-1)(z-0.6065)} \tag{8.10.11}$$

The velocity error constant is expressed as

$$K_V = \lim \frac{(1-z^{-1})G(z)}{T} = 4 \tag{8.10.12}$$

$$e_{ss,v} = \frac{1}{K_v} = \frac{1}{4} = 0.25 \tag{8.10.13}$$

The velocity error is equal to $0.25 < 0.3$. Hence, the integrated controller for $K = 2$, as shown in the expression (8.10.12), is acceptable.

For $T = 1$ s, the open-loop transfer function is

$$G(z) = \frac{0.6321Kz}{(z-1)(z-0.3679)} \tag{8.10.14}$$

The breaking points are

$$z^2 = 0.3679 \Rightarrow z_1 = 0.605 \quad \text{and} \quad z_2 = -0.605 \tag{8.10.15}$$

The critical gain for $z = -1$ is

$$K_{cr} = 4.328 \tag{8.10.16}$$

The following scheme illustrates the root locus of the system characteristic equation for $T = 1$ s. The closed-loop system is stable for

$$0 < K < 4.328 \tag{8.10.17}$$

For $K = 2$, the poles are

$$z_{1,2} = 0.05185 \pm j0.6043 \tag{8.10.18}$$

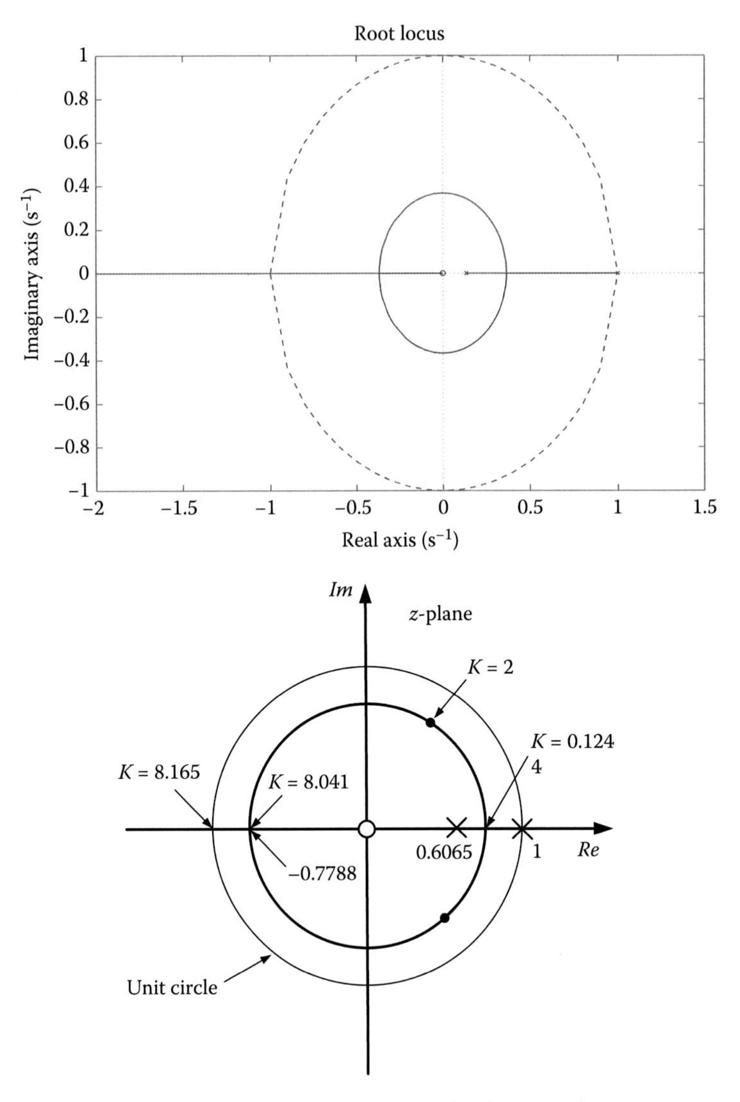

For $T = 1$ s and $K = 2$, the open-loop transfer function becomes

$$G(z) = \frac{1.2642z}{(z-1)(z-0.3679)} \tag{8.10.19}$$

The velocity error constant is given by

$$\begin{aligned} K_v &= \lim_{z\to 1} \frac{(1-z^{-1})G(z)}{T} \\ &= \lim_{z\to 1} \left[\frac{z-1}{z} \frac{1.2642z}{(z-1)(z-0.3679)} \right] = 2 \end{aligned} \tag{8.10.20}$$

$$e_{ss,v} = \frac{1}{K_v} = 0.5 \tag{8.10.21}$$

For $T = 2$ s, the open-loop transfer function becomes

$$G(z) = \frac{0.8647Kz}{(z-1)(z-0.1353)} \tag{8.10.22}$$

The breaking points are

$$z_1 = 0.3678 \quad \text{and} \quad z_2 = -0.3678 \tag{8.10.23}$$

The closed-loop system is stable for

$$0 < K < 2.626 \tag{8.10.24}$$

For $T = 2$ s and $K = 2$, the open-loop transfer function is

$$G(z) = \frac{1.7294z}{(z-1)(z-0.1353)} \tag{8.10.25}$$

The velocity error constant is given by

$$K_v = \lim_{z \to 1} \frac{(1-z^{-1})G(z)}{T} = 1 \tag{8.10.26}$$

$$e_{ss,v} = \frac{1}{K_v} = 1 \tag{8.10.27}$$

For $K = 2$, the poles are

$$z_{1,2} = -0.2971 \pm j0.2169 \tag{8.10.28}$$

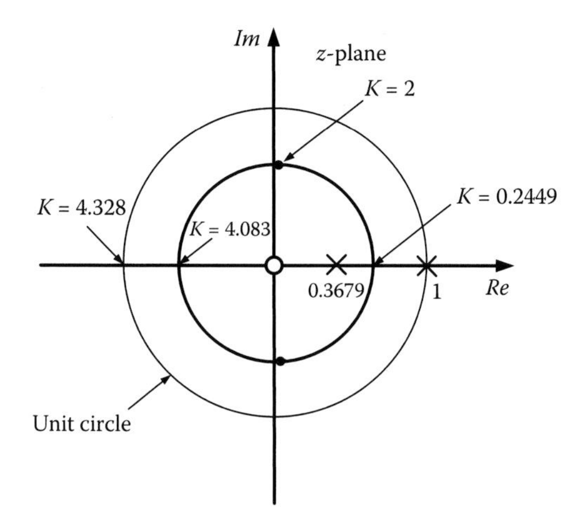

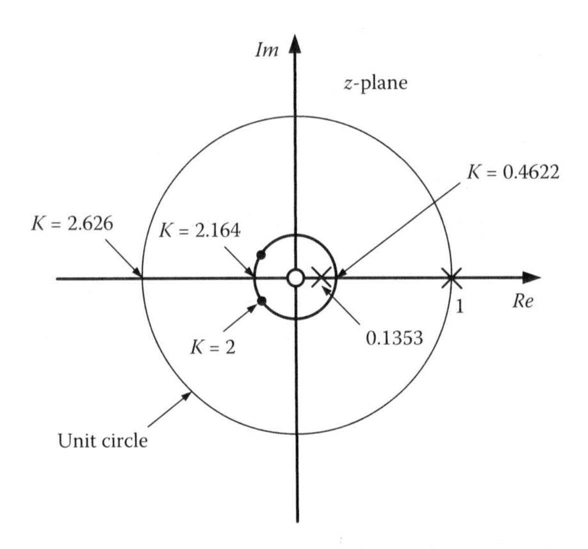

Overall, it is clear that an increase of sampling period negatively affects the relative stability of the closed system.

The attenuation factors for the closed-loop poles, which correspond to sampling periods $T = 0.5$, 1, and 2 have approximate values 0.24, 0.32, and 0.37, respectively.

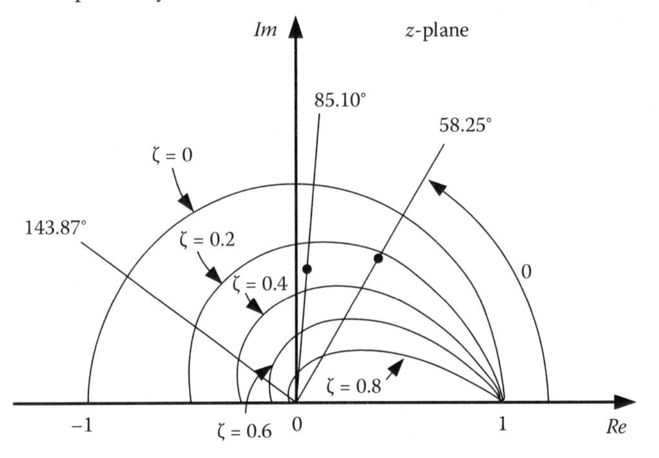

For $T = 0.5$ s and $K = 2$, the step response is computed as

$$c(k) = Z^{-1}(C(z)) = Z^{-1}\left(\frac{G(z)}{1+G(z)}\frac{1}{1-z^{-1}}\right) \tag{8.10.29}$$

where

$$\frac{C(z)}{R(z)} = \frac{G(z)}{1+G(z)} = \frac{Kz(1-e^{-T})}{(z-1)(z-e^{-T})+Kz(1-e^{-T})} \Rightarrow$$

$$C(z)=\frac{0.7870z^{-1}}{1-0.8195z^{-1}+0.6065z^{-2}}\frac{1}{1-z^{-1}} \tag{8.10.30}$$

In the following diagram, the step response of the closed system is approximately designed:

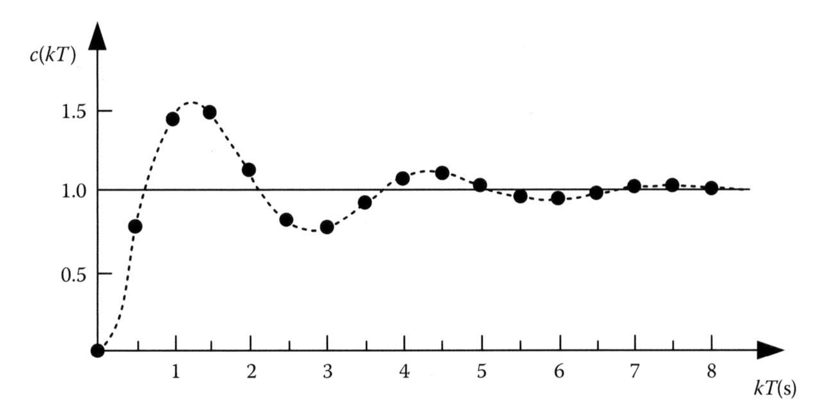

For $T = 1$ s and $K = 2$, the step response is calculated as

$$c(k)=Z^{-1}(C(z))=Z^{-1}\left(\frac{1.2642z^{-1}}{1-0.1037z^{-1}+0.3679z^{-2}}\frac{1}{1-z^{-1}}\right) \tag{8.10.31}$$

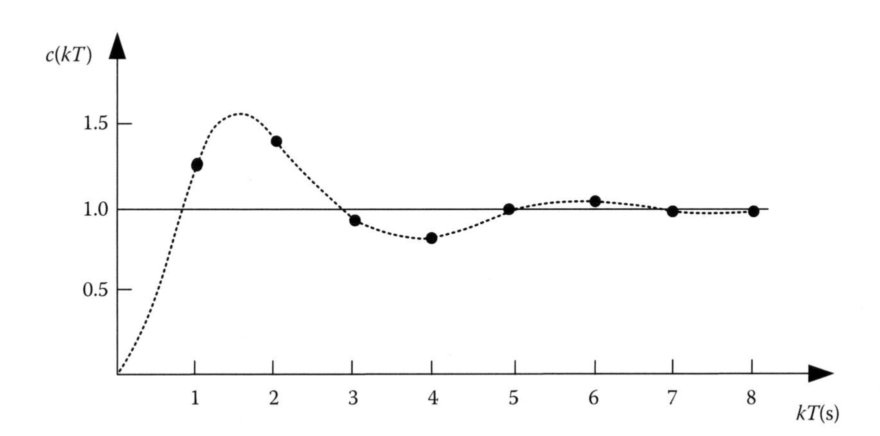

For $T = 2$ s and $K = 2$, the step response is calculated as

$$c(k)=Z^{-1}(C(z))=Z^{-1}\left(\frac{1.7294z^{-1}}{1-0.5941z^{-1}+0.1353z^{-2}}\frac{1}{1-z^{-1}}\right) \tag{8.10.32}$$

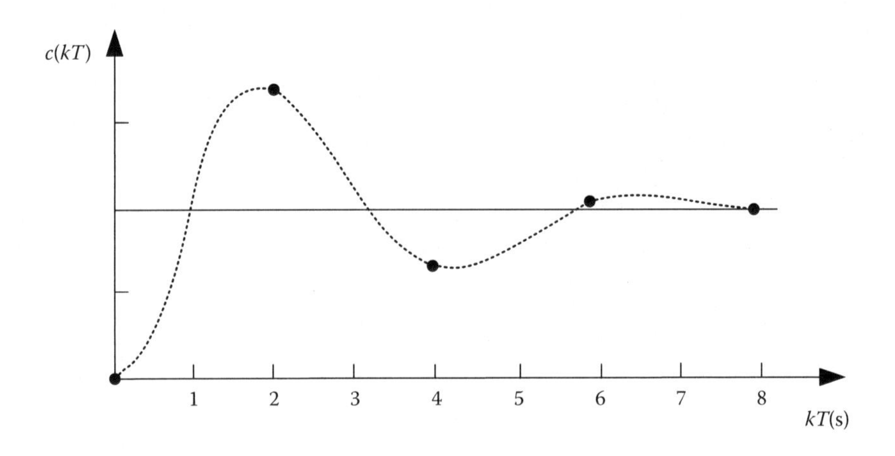

Observe that an increase of the sampling period also increases the system overshoot and recovery time.

EXERCISE 8.11

Consider the system to be controlled with the transfer function $G(s)=1/s(s+1)$, where a digital controller is applied, obtained from the discretization of an analog controller with the transfer function $D(s)=70((s+2)/(s+10))$ using

a. Bilinear transform (Tustin approximation)
b. FOH circuit
c. ZOH circuit

For sampling periods: $T=0.03$ s and $T=0.10$ s

Solution

```
% Calculate the loop transfer function
Gnum = [0 0 1] ;
Gden = [1 1 0] ;
Dnumc = 70*[1 2];
Ddenc = [1 10];
[nol,dol] = series(Dnumc,Ddenc,Gnum,Gden);
% Calculate the closed-loop transfer function of the system
[ncl,dcl] = cloop(nol,dol,-1);
% Discretize the analog controller for T=0.03s.
T = 0.03 ;
[Dnum1t,Dden1t] = c2dm(Dnumc,Ddenc,T,'tustin') ;
[Dnum1f,Dden1f] = c2dm(Dnumc,Ddenc,T,'foh') ;
[Dnum1z,Dden1z] = c2dm(Dnumc,Ddenc,T,'zoh') ;
% Set the step responses using the command loopstep
[y1t,u1t,t1t] = cl_stepzoh(Dnum1t,Dden1t,Gnum,Gden,T,ceil
(1.8/T)) ;
tc1t = t1t;
yc1t = step(ncl,dcl,tc1t);
```

```
[y1z,u1z,t1z]=cl_stepzoh(Dnum1z,Dden1z,Gnum,Gden,T,ceil(1.
8/T)) ;
tc1z = t1z;
yc1z = step(ncl,dcl,tc1z);
[y1f,u1f,t1f] = cl_stepzoh (Dnum1f,Dden1f,Gnum,Gden,T,ceil(
1.8/T)) ;
tc1f = t1f;
yc1f = step(ncl,dcl,tc1f);
% Discretize the analog controller for T=0.1s.
T = 0.1 ;
[Dnum3t,Dden3t] = c2dm(Dnumc,Ddenc,T,'tustin') ;
[Dnum3f,Dden3f] = c2dm(Dnumc,Ddenc,T,'foh') ;
[Dnum3z,Dden3z] = c2dm(Dnumc,Ddenc,T,'zoh') ;
[y3t,u3t,t3t] = cl_stepzoh (Dnum3t,Dden3t,Gnum,Gden,T,ceil(
1.8/T)) ;
tc3t = t3t;
yc3t = step(ncl,dcl,tc3t);
[y3z,u3z,t3z] = cl_stepzoh (Dnum3z,Dden3z,Gnum,Gden,T,ceil(
1.8/T)) ;
tc3z = t3z;
yc3z = step(ncl,dcl,tc3z);
[y3f,u3f,t3f] = cl_stepzoh (Dnum3f,Dden3f,Gnum,Gden,T,ceil(
1.8/T)) ;
tc3f = t3f;
yc3f = step(ncl,dcl,tc3f);
plot(t1t,y1t,'-',t1t,u1t/50,'-',tc1t,yc1t,'--','
Linewidth',3)
axis([0 2 -1 2])
grid
plot(t3t,y3t,'-',t3t,u3t/50,'-',tc3t,yc3t,'--',
'Linewidth',3)
axis([0 2 -1 2])
grid
```

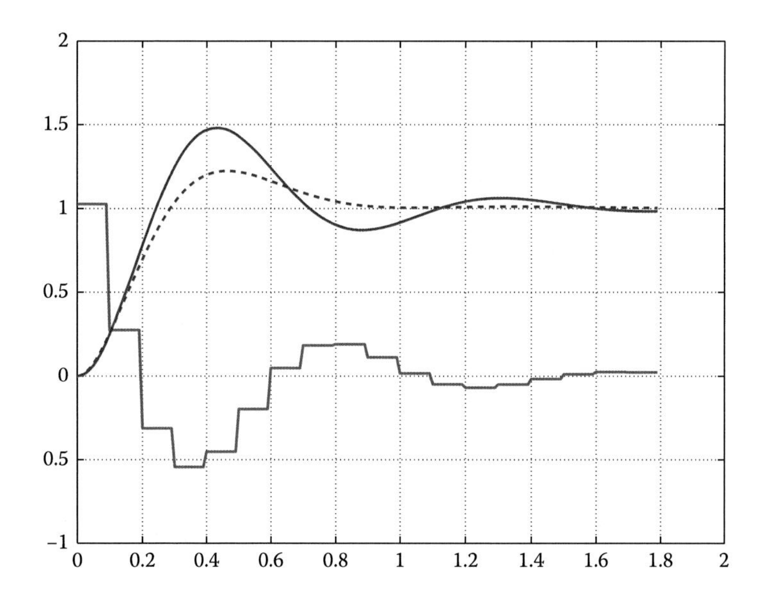

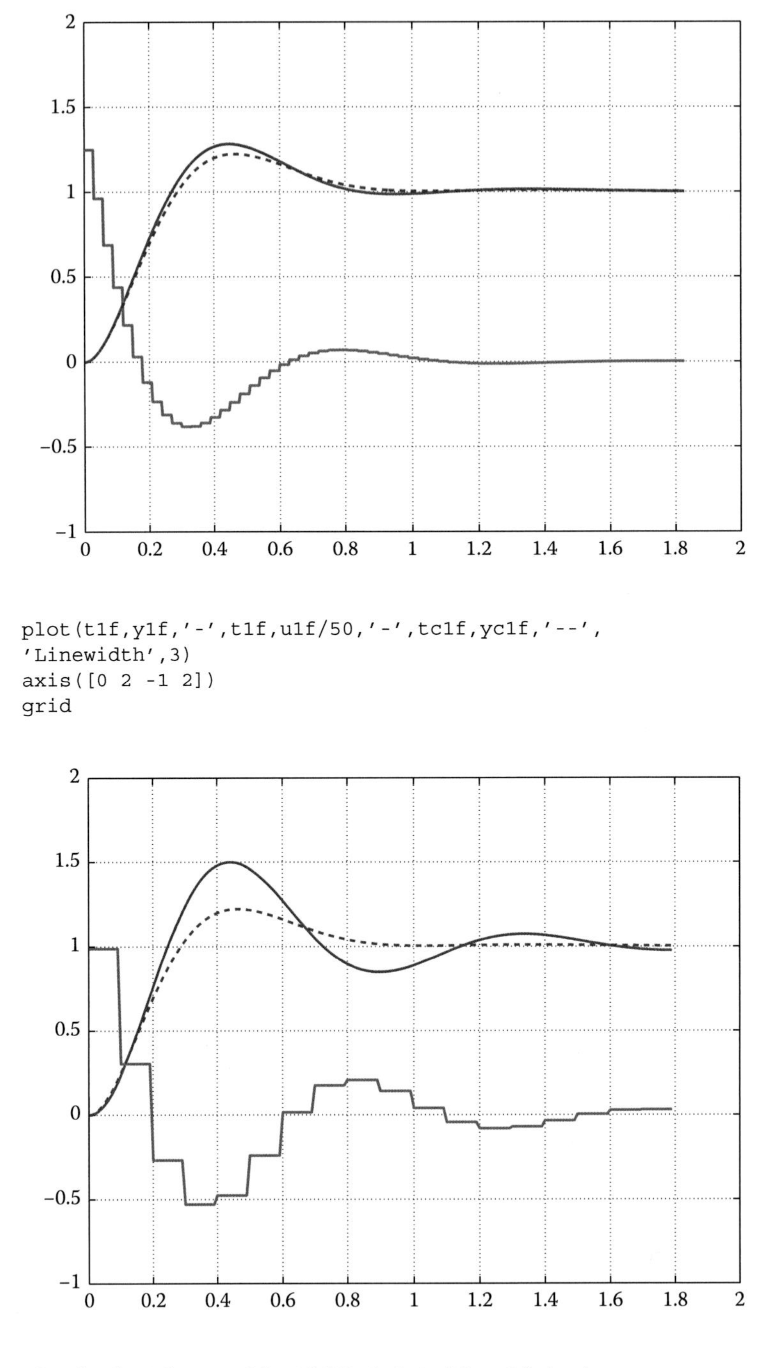


```
plot(t1f,y1f,'-',t1f,u1f/50,'-',tc1f,yc1f,'--',
'Linewidth',3)
axis([0 2 -1 2])
grid
```

```
plot(t3f,y3f,'-',t3f,u3f/50,'-',tc3f,yc3f,'--',
'Linewidth',3)
axis([0 2 -1 2]), grid
```

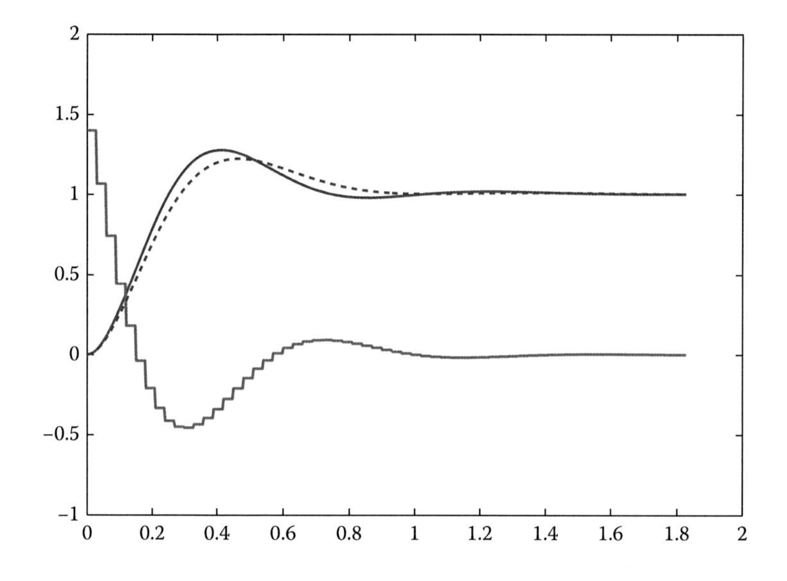


```
plot(t1z,y1z,'-',t1z,u1z/50,'-',tc1z,yc1z,'--',
'Linewidth',3)
axis([0 2 -1 2])
grid
plot(t3z,y3z,'-',t3z,u3z/50,'-',tc3z,yc3z,'--',
'Linewidth',3)
axis([0 2 -1 2]),grid
```

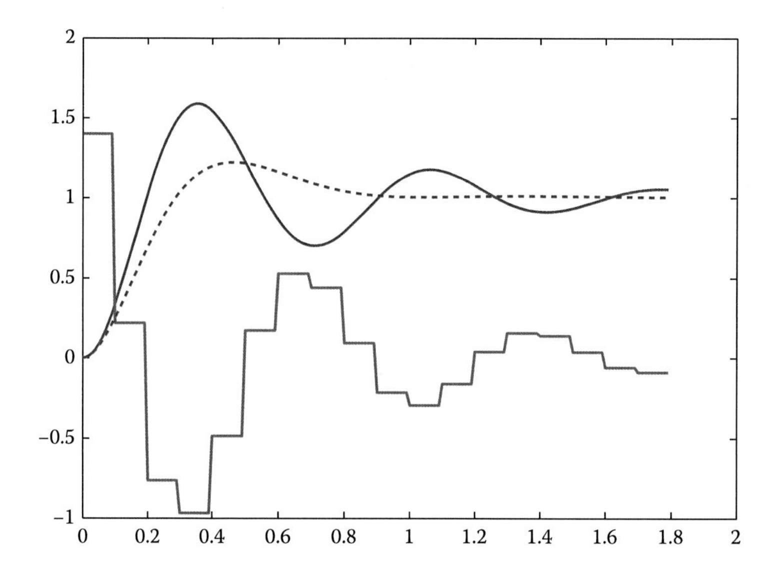


```
plot(t1t,y1t,'-',t1f,y1f,'-',t1z,y1z,'-',tc3z,yc3z,'--',
'Linewidth',3)
```

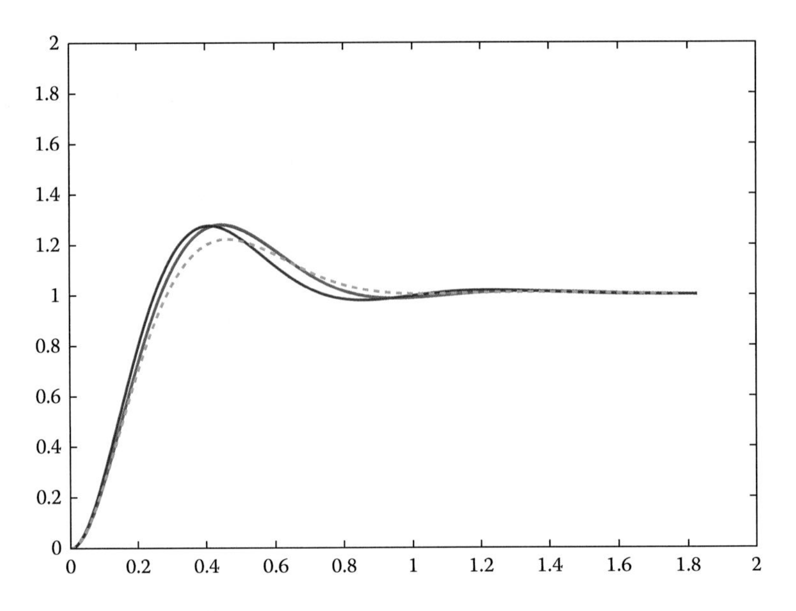

```
axis([0 2 -1 2])
legend({'bilinear','FOH','ZOH','Cts.-time'}),grid
title('Step response with T = 0.1',...
  'FontSize',LabelFont,'FontWeight',FontWeight)
xlabel('Time (sec.)','FontSize',LabelFont,'FontWeight',
FontWeight)
set(gca,'FontWeight','bold','FontSize',AxesFont)
plot(t3t,y3t,'-',t3f,y3f,'-',t3z,y3z,'-',tc3z,yc3z,'--',
'Linewidth',3)
```

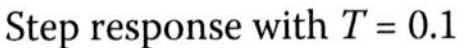

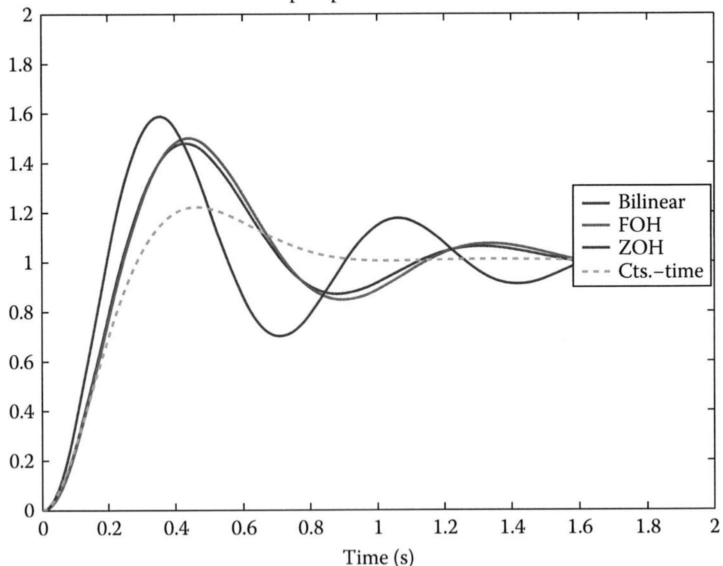

The closed-loop step responses of the analog system approximately coincide with the corresponding responses of the discretized system for a small sampling period.

Moreover, for higher sampling period, we can obtain better results if we discretize the analog controller using the FOH and Tustin methods.

EXERCISE 8.12

Consider the continuous-time system with $G_P(s) = 0.03/(s^2 + 0.4s)$, which is controlled by a relevant controller described by the transfer function $G_D(s) = (30s + 15)/(s + 1.5)$.

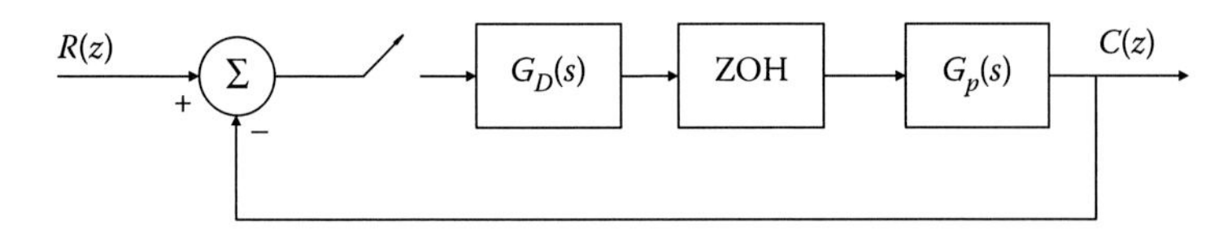

Design the digital controller that satisfies an overshoot of the order of 10% and a maximum recovery time of 10 s.

Solution

```
% Define the specifications
overshoot_limit = 10; % percent
settling_limit = 10; % in seconds
% Continuous-time system
G_c = tf(0.03,[1 0.4 0]);
% Controller
D_c = tf(30*[1 0.5],[1 1.5]);
% Calculate the phase margin
[Gm,Pm,Wcg,Wcp] = margin(series(D_c,G_c));
crossover_frequency = Wcp
phase_margin = Pm
crossover_frequency = 0.6027
phase_margin = 62.0016
% Design the Bode diagram
bode(series(D_c,G_c))
```

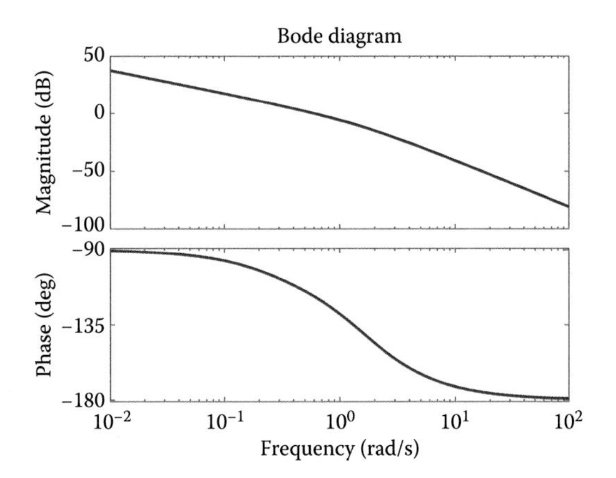

```
% Calculate the closed-loop step response
H = feedback(series(D_c,G_c),1);
t = [0:0.001:15];
y_c = step(H,t);
% Derive the values of overshoot and recovery time
overshoot_c = (max(y_c)-1)*100
has_settled = (y_c>0.99).*(y_c<1.01);
N = length(y_c);
ts_index = 0;
for n = 1:N
  if has_settled(n)==0,
   ts_index = n+1;
  end
end
t_s = t(ts_index)
overshoot_c = 7.9561
t_s = 8.4000
% Design the time response
figure
plot(t,y_c,'-',t,1.01*ones(size(t)),'--',t,0.99*ones
 (size(t)),'--',… t,(1+overshoot_limit/100)*ones(siz
 e(t)),'--')
xlabel('time t (sec)'),ylabel('step response')
title(['Continous-time control, overshoot = ',
num2str(overshoot_c),… ', t_s = ',num2str(t_s)]);
grid
```

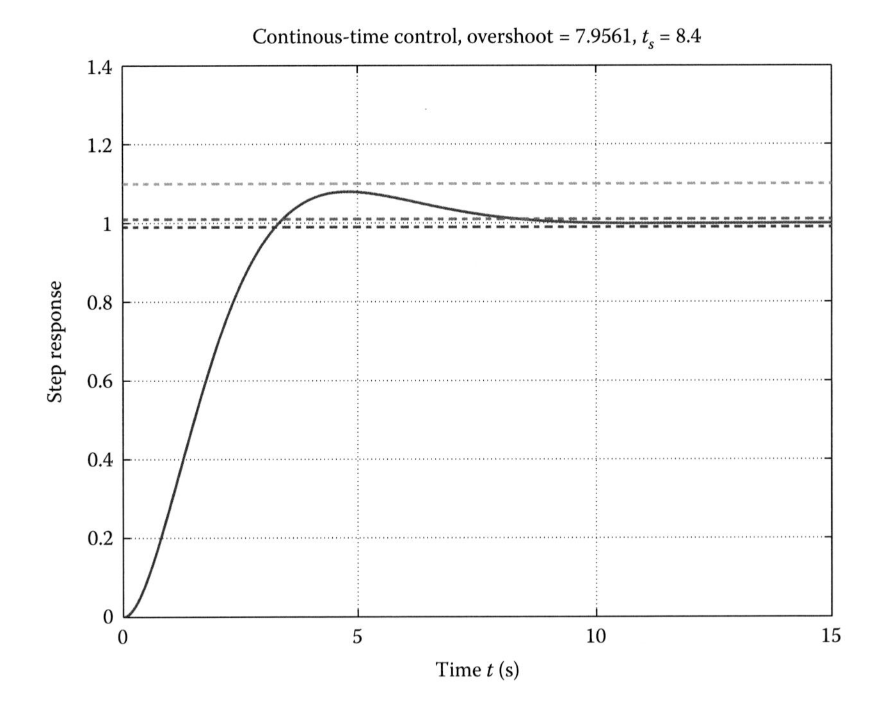

```
% Calculate the amplitude and phase of the controller's
frequency response at the crossover frequency.
D_c_at_wc = freqresp(D_c,Wcp)
D_c_at_wc = 12.7802 + 6.9191i
D_cross_over_mag = abs(D_c_at_wc)
D_cross_over_mag = 14.5330
D_cross_over_phase = angle(D_c_at_wc)
D_cross_over_phase = 0.4962
% Set a sampling period T=0.5s.
% Set the data of the system transfer function
[Gnum,Gden]=tfdata(G_c);
Gnum = Gnum{:};
Gden = Gden{:};
% Set a sampling period T=0.5s.
T = 0.5;
% Discretize the controller using Tustin method
D = c2d(D_c,T,'tustin');
[Dnum,Dden]=tfdata(D);
Dnum = Dnum{:};
Dden = Dden{:};
% Use of cl_stepzoh to set the step response
[y,u,t] = cl_stepzoh (Dnum,Dden,Gnum,Gden,T,round(15/T));
% Overshoot control, where an overshoot percent of 14.59%
arises
max(y)
ans = 1.1459
% Time recovery control, which is equal to 7.45s.
has_settled = (y>0.99).*(y<1.01);
N = length(y);
ts_index = 0;
  for n = 1:N
   if has_settled(n)==0,
    ts_index = n+1;
   end
end
t_s = t(ts_index)
t_s = 7.4500
% Repeat the process using the methods prewarp - zoh - foh
- matched and observe that the required specifications are
not satisfied.
D = c2d(D_c,T,'prewarp',0.6);
[Dnum,Dden]=tfdata(D);
Dnum = Dnum{:};
Dden = Dden{:};
[y,u,t] = cl_stepzoh (Dnum,Dden,Gnum,Gden,T,round(15/T));
D = c2d(D_c,T,'zoh');
[Dnum,Dden]=tfdata(D);
Dnum = Dnum{:};
Dden = Dden{:};
[y,u,t] = cl_stepzoh (Dnum,Dden,Gnum,Gden,T,round(15/T));
D = c2d(D_c,T,'foh');
[Dnum,Dden]=tfdata(D);
Dnum = Dnum{:};
Dden = Dden{:};
[y,u,t] = cl_stepzoh (Dnum,Dden,Gnum,Gden,T,round(15/T));
D = c2d(D_c,T,'matched');
```

```
[Dnum,Dden]=tfdata(D);
Dnum = Dnum{:};
Dden = Dden{:};
[y,u,t] = cl_stepzoh (Dnum,Dden,Gnum,Gden,T,round(15/T));
% Set a sampling period T=0.18s., where it follows that all
the system specifications are now satisfied.
% Design the step response for the controller's
discretization using the matched method.
T = 0.18;
D = c2d(D_c,T,'matched');
[Dnum,Dden]=tfdata(D);
Dnum = Dnum{:};
Dden = Dden{:};
[y,u,t] = cl_stepzoh (Dnum,Dden,Gnum,Gden,T,round(15/T));
plot(t,y,'-',t,1.01*ones(size(t)),'--',
 t,0.99*ones(size(t)),… '--',t,(1+overshoot_limit/100)*ones
 (size(t)),'--')
xlabel('time t (sec)'),ylabel('step response')
```

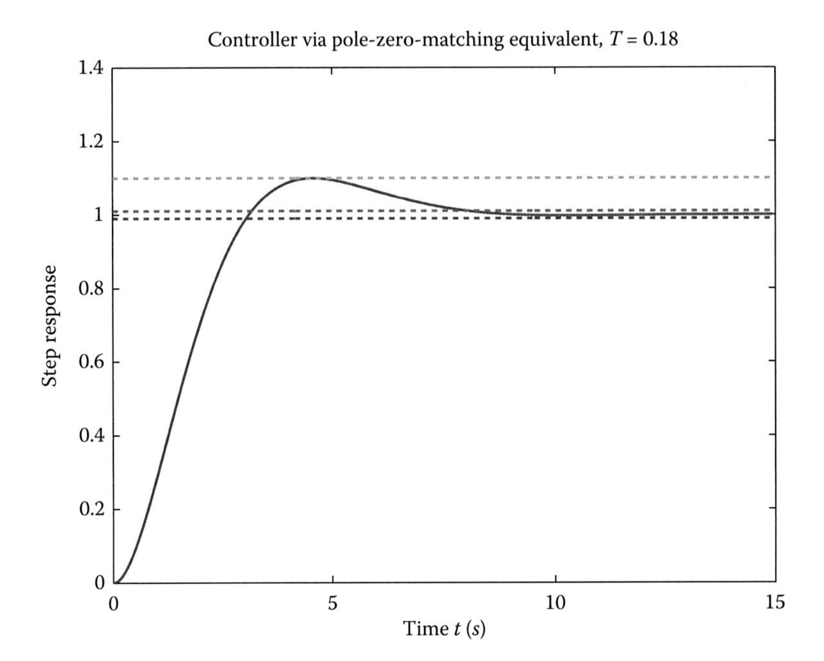

The transfer function of the digital controller, in the case when the specifications are satisfied, is $D(z) = (27.49z - 25.13)/(z - 0.7634)$.

EXERCISE 8.13

Consider the system of the following scheme with $G_P(s) = 10/((s+1)(s+2))$. Design a digital controller in order to preserve a zero steady-state error, for a step input signal and for a minimum velocity error constant equal to 5.

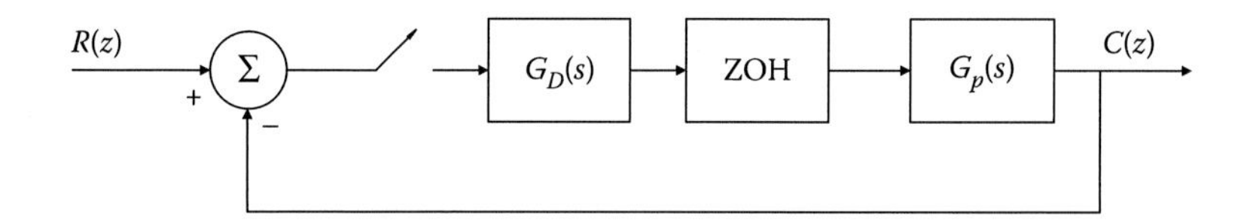

Solution

The discretized system transfer function with ZOH, for $T = 0.1$ s, is

$$G(z) = (1 - z^{-1})Z\left[\frac{10}{s(s+1)(s+2)}\right] \simeq \frac{0.04528(z+0.9048)}{(z-0.9048)(z-0.8187)} \tag{8.13.1}$$

```
% MATLAB code to compute G(z).
s=tf('s')
Gp=10/((s+1)*(s+2));
GhGp=c2d(Gp,0.1,'zoh')

Transfer function
0.04528 z + 0.04097
----------------------
z^2 - 1.724 z + 0.7408
Sampling time: 0.1

% Design of root locus for the system without the controller
rlocus(GhGp)
```

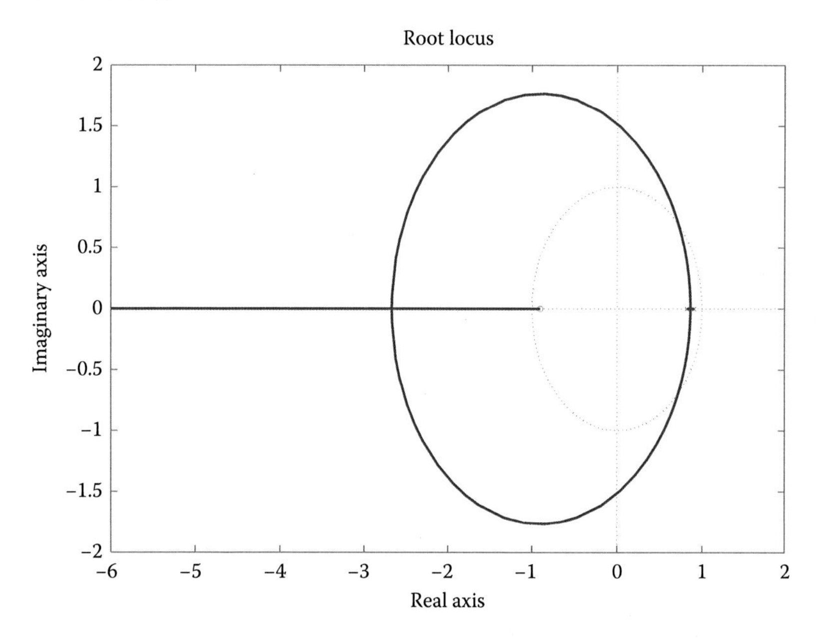


```
% Design of poles-zeros diagram for the system without
controller
pzplot(GhGp)
```

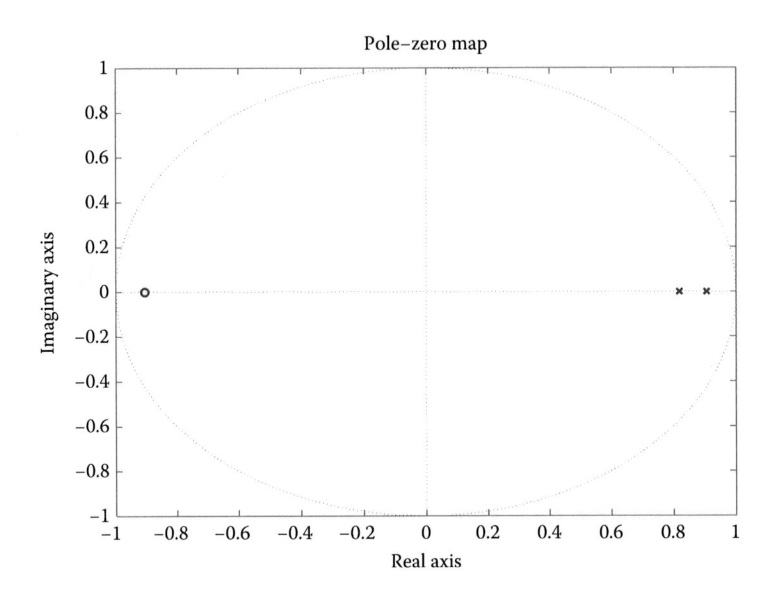

One of the design requirements specifies that the closed system should preserve a zero-steady-state error for step input. Hence, a PI controller follows, by using the backward rectangular integration method, as

$$G_D(z) = K_p + \frac{K_i T}{z-1} = \frac{K_p z - (K_p - K_i T)}{z-1} \tag{8.13.2}$$

The *Ki* parameter can be calculated using

$$k_v = \frac{1}{T}\lim_{z\to 1}(z-1)G_D(z)G(z) = 5K_i \geq 5 \tag{8.13.3}$$

The latter condition is satisfied for $Ki \geq 1$.

For $Ki = 1$, the characteristic equation becomes

$$\begin{aligned}&(z-1)(z-0.9048)(z-0.8187)+0.004528(z+0.9048)\\&+0.04528Kp(z-1)(z+0.9048)=0\end{aligned}$$

or

$$1+\frac{0.04528Kp(z-1)(z+0.9048)}{z3-2.724z2+2.469z-0.7367}=0 \tag{8.13.4}$$

```
% Design of root locus of the compensated system with Kp as
a variable.
z=tf('z',0.1);
Gcomp=0.04528*(z-1)*(z+0.9048)/(z^3 - 2.724*z^2 + 2.469*z
- 0.7367);
zero(Gcomp);
pole(Gcomp);
rlocus(Gcomp)
```

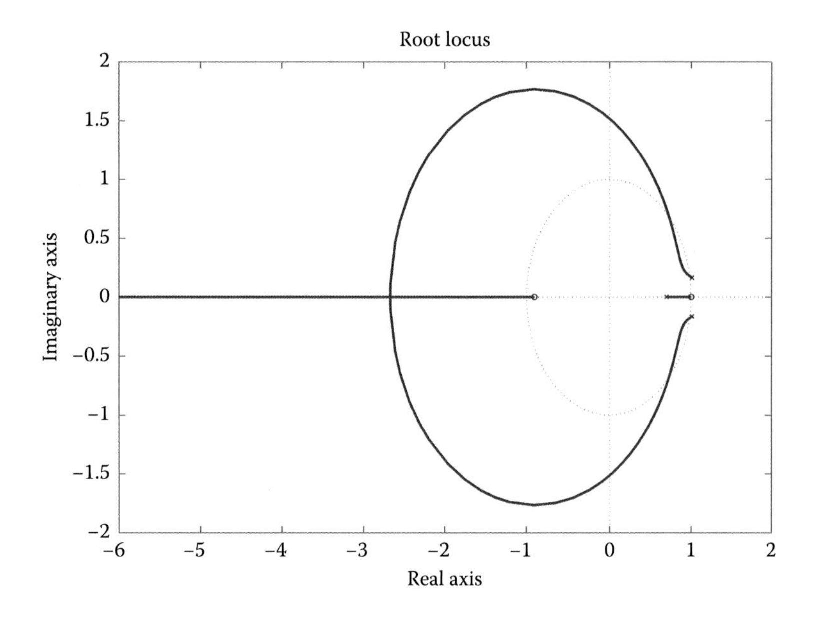

The system zeros are in 1 and −0.9048, while the system poles are placed in 1.0114 ± 0.1663i and 0.7013.

```
% Redesign the root locus
rlocus(Gcomp);axis([0.6 1.2 -0.8 0.8])
```

Obviously, the system is stable for quite a narrow neighborhood of Kp, particularly for $0.242 < Kp < 6.41$. For $K_p = 1$, the overshoot is approximately 45.5%, that is, it is extremely high for any practical system.

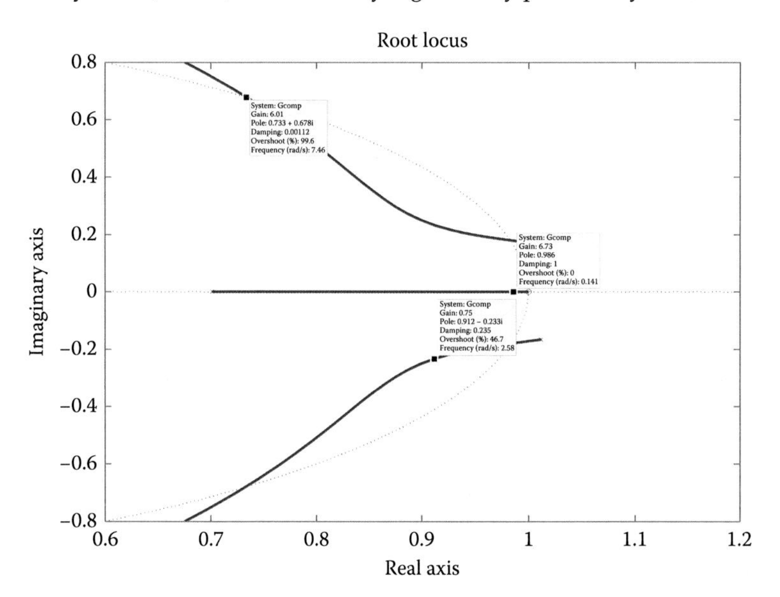

To optimize the relative stability, the differential operation of the controller is required. Consider the digital controller with the transfer function

$$G_d(z) = \frac{(K_pT + K_d)z^2 + (K_iT^2 - K_pT - 2K_d)z + K_d}{Tz(z-1)} \tag{8.13.5}$$

To satisfy the velocity error constant, $Ki \geq 1$ should hold. Assuming a 15% overshoot (which corresponds to J≈ 0.5 and recovery time 2 s, i.e., $\omega_n \approx 4$), the desired dominant poles are computed as

$$s_{1,2} = -J\omega_n \pm j\omega_n\sqrt{1-J^2} = -2 \pm j3.46 \tag{8.13.6}$$

Thus, the closed system poles are

$$z_{1,2} = e^{T(-2\pm j3.46)} \simeq 0.77 \pm 0.28j \tag{8.13.7}$$

The poles–zeros diagram of the closed system, including the poles of the controller, is illustrated as

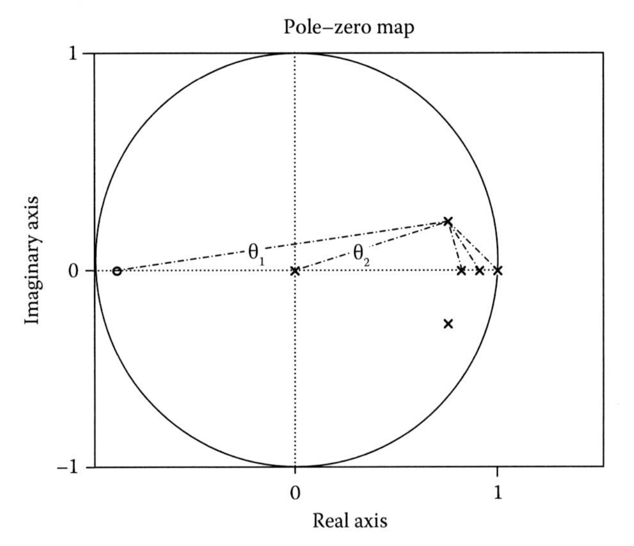

From the phase condition $argG(s)H(s) = \pm 180°(2k + 1)$, which defines the s values of the characteristic equation's roots, we get

$$\theta_1 - \theta_2 - \theta_3 - \theta_4 - \theta_5 = 9.5° - 20° - 99.9° - 115.7° - 129.4° = -355.5°$$

where the angles θ1 and θ2 arise as

$$\theta_1 = \tan^{-1}\frac{0.28}{0.77+0.9048} = 9.5°, \quad \theta_2 = \tan^{-1}\frac{0.28}{0.77} = 20°$$

Thereby, the two zeros of the controller should provide a phase equal to 355.5°–180° = 175.5°.

Let's place the two zeros at the same point; so, the phase of each zero would be 175.5°/2=87.75° and, apparently, will be placed left to the desired pole.

Hence

$$\tan^{-1}\frac{0.28}{0.77-z_x}=87.75^\circ\Rightarrow\frac{0.28}{0.77-z_x}=\tan(87.75^\circ)=25.45$$

$$\Rightarrow 0.77-z_x=\frac{0.28}{25.45}=0.011\Rightarrow z_x=0.77-0.011=0.759$$

Using the above technique, the digital controller has been defined, whose transfer function is

$$G_D(z)=K\frac{(z-0.759)^2}{z(z-1)} \tag{8.13.8}$$

In the following scheme, the root locus of the compensated system (including the digital controller) is illustrated, where the value of the closed system pole corresponds to $K = 4.33$.

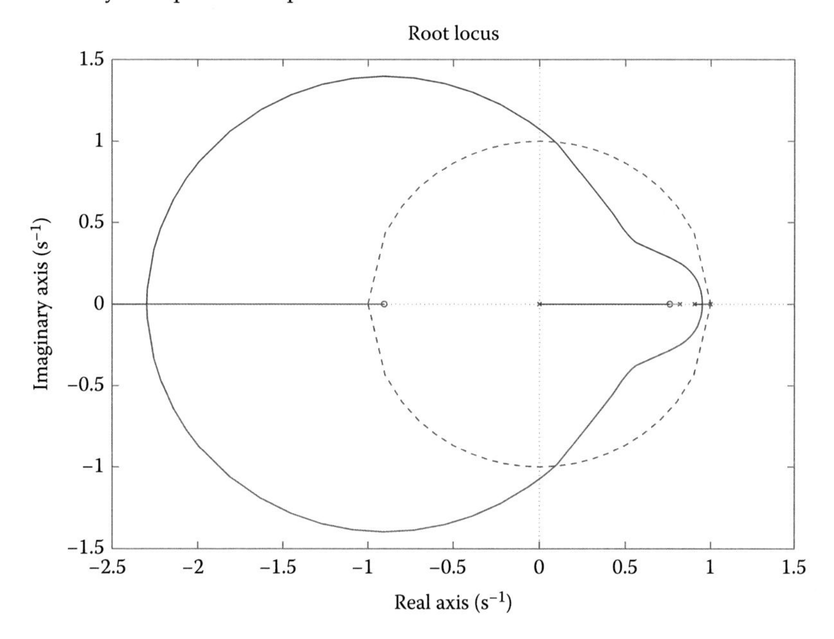

Thereby, the transfer function of the digital controller is

$$G_D(z)=4.33\frac{(z-0.759)^2}{z(z-1)} \tag{8.13.9}$$

From the resultant controller

$$G_d(z) = \frac{(KpT + Kd)z^2 + (KiT^2 - KpT - 2Kd)z + Kd}{Tz(z-1)}$$

the following expressions emerge

$$Kd/T = 0.5761 * 4.33 \Rightarrow Kd = 0.2495$$

$$KiT - Kp - 2Kd/T = -1.518 * 4.33 \Rightarrow Ki = 2.521 \geq 1$$

$$Kp + Kd/T = 4.33 \Rightarrow Kp = 1.835$$

EXERCISE 8.14

Consider the transfer function of a LEAD or LAG controller: $D(s) = (U(s)/E(s)) = K((s+a)/(s+b))$. Discretize it to a digital controller and formulate the resultant difference equation using the Simulink® software platform.

Solution

$$\begin{aligned} D(s) &= \frac{U(s)}{E(s)} = K\frac{s+a}{s+b} \\ &\Rightarrow (s+b)U(s) = K(s+a)E(s) \end{aligned} \tag{8.14.1}$$

Applying the inverse Laplace transform, we have that

$$\dot{u}(t) + bu(t) = K\dot{e}(t) + Kae(t) \tag{8.14.2}$$

Using the backward rectangular discretization method, the latter expression becomes

$$\frac{u(k) - u(k-1)}{T} + bu(k) = K\frac{e(k) - e(k-1)}{T} + Kae(k) \tag{8.14.3}$$

$$\begin{aligned} (8.14.3) &\Rightarrow u(k) - u(k-1) + bTu(k) = Ke(k) - Ke(k-1) + KTae(k) \\ &\Rightarrow u(k) = \frac{u(k-1) + K((1+aT)e(k) - e(k-1))}{1+bT} \end{aligned} \tag{8.14.4}$$

The latter difference equation can be easily illustrated in Simulink via the following diagram.

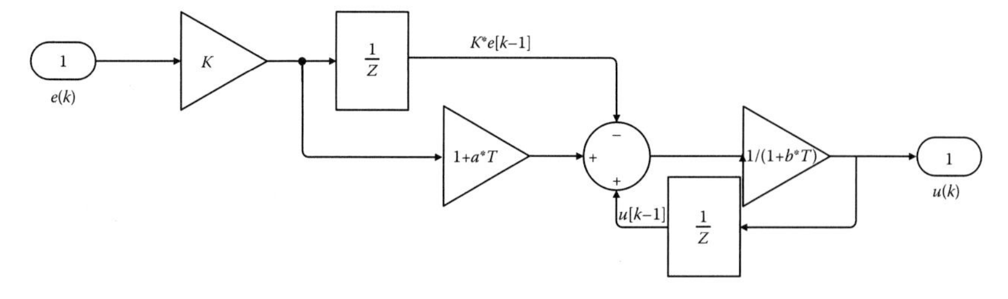

EXERCISE 8.15

Design a digital controller by replacing the analog controller $D(s) = 70(s+2)/(s+10)$ for the analog system $G(s) = 1/s(s+1)$ using a sampling frequency 20 Hz and 40 Hz.

Solution

The solution is provided using MATLAB.

```
clear
% Definition of the system to be controlled (plant)
numG=1;
denG=[1 1 0];
% Analog compensator
Ko=70;
a=2;
b=10;
numD=Ko*[1 a];
denD=[1 b];
% Calculation of the loop transfer function
num=conv(numG,numD);
den=conv(denG,denD);
% Calculation of the closed-loop transfer function
[numcl,dencl]=feedback(num,den,1,1);
% Design the step response of the analog system
tf=1;
t=0:.01:tf;
yc=step(numcl,dencl,t);
subplot(2,1,1)
plot(t,yc,'-'),grid
axis([0 1 0 1.5])
hold on
% Sampling frequency 20Hz
Ws= 20; % Hz
T=1/Ws;
% Discretization of the analog system with ZOH
[numGd,denGd]=c2dm(numG,denG,T,'zoh');
% Definition of the digital controller
numDd=Ko*[1 -(1-a*T)];
```

```
denDd=[1 -(1-b*T)];
% Calculation of the loop transfer function of the
discretized system
numd=conv(numGd,numDd);
dend=conv(denGd,denDd);
% Calculation of the closed-loop transfer function of the
discretized system
[numcld,dencld]=feedback(numd,dend,1,1);
% Design at the same diagram
N=tf*Ws;
yd=dstep(numcld,dencld,N);
td=0:T:(N-1)*T;
plot(td,yd,'*')
plot(td,yd,'-')
ylabel('output y')
title('Continuous and digital response using Eulers
method')
text(.25,.1,'*-----*-----* digital control')
text(.25,.3,'------------- analog control')
text(.35,.6,' (a) 20 Hz')
hold off
% Sampling frequency 40Hz
Ws= 40; % Hz
T=1/Ws;
% Discretization of the analog system with ZOH
[numGd,denGd]=c2dm(numG,denG,T,'zoh');
% Definition of the digital controller
numDd=Ko*[1 -(1-a*T)];
denDd=[1 -(1-b*T)];
% Calculation of the loop transfer function of the
discretized system
numd=conv(numGd,numDd);
dend=conv(denGd,denDd);
% Calculation of the closed-loop transfer function of the
discretized system
[numcld,dencld]=feedback(numd,dend,1,1);
% Design at the same diagram
N=tf*Ws;
subplot(2,1,2)
plot(t,yc,'-'),grid
hold on
yd=dstep(numcld,dencld,N);
td=0:T:(N-1)*T;
plot(td,yd,'*')
plot(td,yd,'-')
xlabel('time (sec)')
ylabel('output y')
text(.25,.1,'*-----*-----* digital control')
text(.25,.3,'------------- analog control')
text(.35,.6,' (b) 40 Hz')
hold off
```

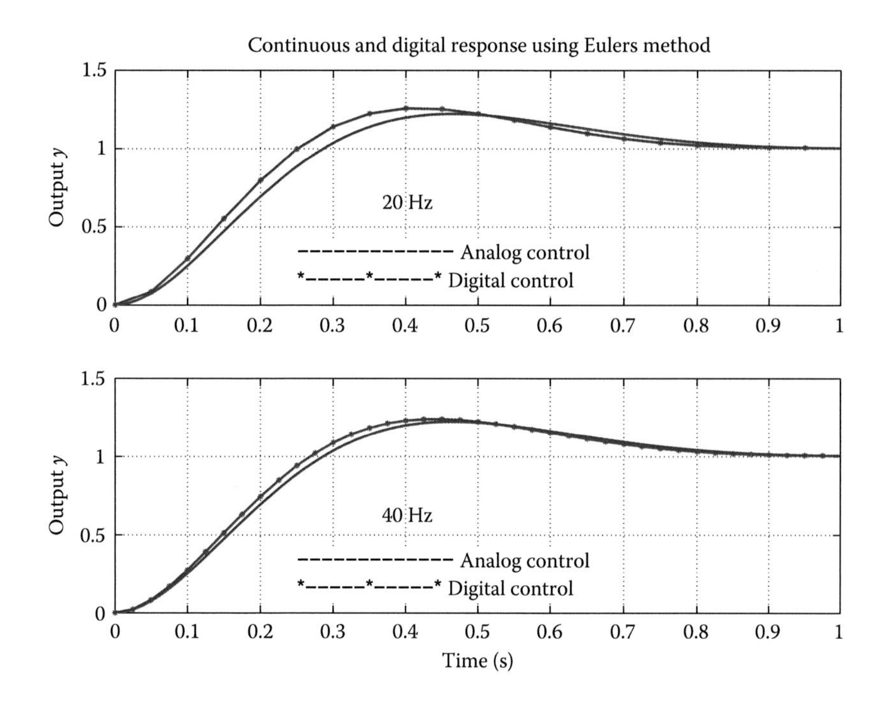

EXERCISE 8.16

Consider a second-order system with the transfer function

$$G(s) = \frac{Y(s)}{U(s)} = \frac{10000}{s^2 + 20s + 10000}$$

Design the digital controller that satisfies an overshoot percent <10%, rise time <0.03 s, and increases the system type over 1.

Solution

$$G(s) = \frac{Y(s)}{U(s)} = \frac{10000}{s^2 + 20s + 10000} = \frac{100^2}{s^2 + 2(0.1)100s + 100^2} \tag{8.16.1}$$

The analog system has an attenuation factor $J = 0.1$ and physical oscillation frequency $\omega_n = 100$ rad/s ($f_n \simeq 16$ Hz).

Design the system step response using MATLAB

```
zeta = 0.1;
wn = 100;
num = [0 0 wn^2];
den = [1 2*zeta*wn wn^2];
step(num,den);
```

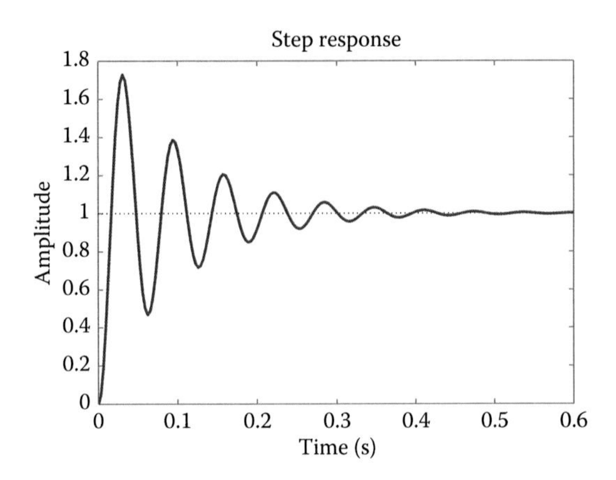

The attenuation factor will be computed from the overshoot requirement. Also, the physical oscillation frequency will be computed by the rise time requirement.

$$PO < 10\% \Rightarrow J > 0.54 \tag{8.16.2}$$

$$\text{tr} < 0.03\ \text{s} \Rightarrow \omega_n > \frac{1.8}{0.03} = 60\ \text{rad/s} \tag{8.16.3}$$

Set the sampling frequency $f_s = 100$ Hz so as to be 6 to 20 times higher than the maximum system frequency. Hence, since the system physical frequency is 16 Hz, set $f_s = 100$ Hz, which corresponds to a sampling period equal to $T = 0.01$ s.

The regions of complex z-plane which correspond to the limits J and Ω_n are illustrated into the following diagram:

```
T = 0.01;
zeta = 0.54;
wn = 60;
zgrid(zeta,wn*T,'new');
grid;
axis('square');
axis('equal');
```

A block diagram of the digital control system is presented as follows:

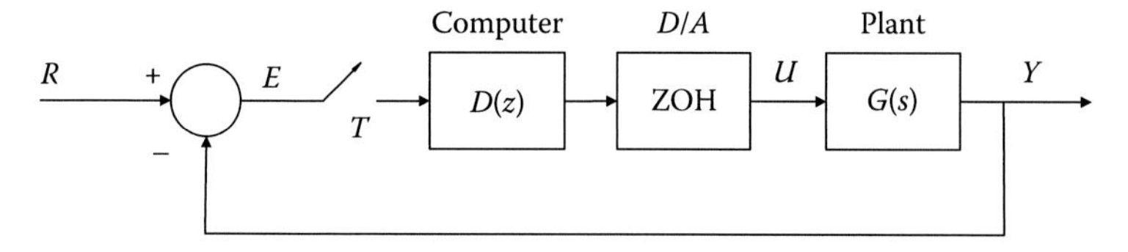

or equivalently

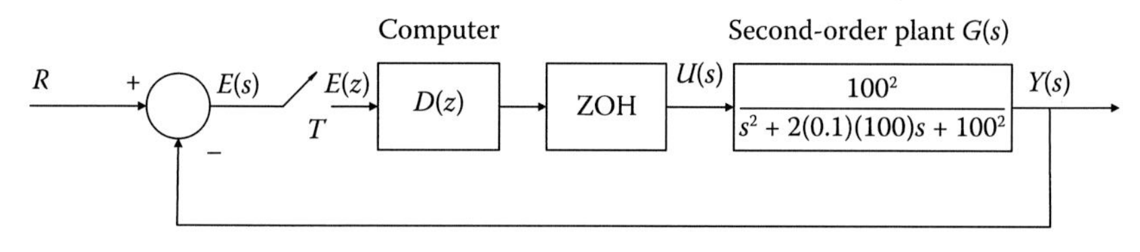

Discretize the analog system with ZOH and compute its pulse transfer function

```
T = 0.01;
wn_plant = 100;
zeta_plant = 0.1;
num_gc = [0 0 wn_plant^2];
den_gc = [1 2*zeta_plant*wn_plant wn_plant^2];
Gc = tf(num_gc,den_gc); % Continuous plant G(s)
Gd = c2d(Gc,T) % Discrete plant G(z)
```

$$G(z) = \frac{0.431\,z + 0.4023}{z^2 - 0.9854z + 0.8187} = \frac{0.431\,z^{-1} + 0.4023z^{-2}}{1 - 0.9854z^{-1} + 0.8187z^{-2}} \tag{8.16.4}$$

The latter can be written in a poles–zero form as

$$G(z) = \frac{0.431\,(z + 0.9334)}{z^2 - (0.4927 \pm j0.7589)} \tag{8.16.5}$$

The digital controller $D(z)$ should add a pole at $z = 1$ to make the system be from a type 0 to type 1. The obvious technique to do so is to add zeros so as to cancel the undesired dynamic system features and to add another pole to prevent the case where the poles be fewer than zeros. A digital controller of the following form is selected

$$D(z) = K\frac{z - 0.4927 \pm j0.7589}{(z-1)(z-p)} \tag{8.16.6}$$

The closed-loop transfer function is

$$\frac{Y(z)}{R(z)} = \frac{G(z)D(z)}{1 + G(z)D(z)} \tag{8.16.7}$$

The characteristic equation is

$$1 + G(z)D(z) = 0 \overset{(8.16.5),(8.16.6)}{\Rightarrow}$$
$$1 + \frac{0.4310K(z - 0.4927 \pm j0.7589)(z + 0.9334)}{(z - 0.4927 \pm j0.7589)(z - 1)(z - p)} = 0$$
$$\Rightarrow 1 + \frac{0.4310K(z + 0.9334)}{(z - 1)(z - p)} = 0 \tag{8.16.8}$$

Design the root locus of the characteristic equation for the compensated system as

```
num_rl = 0.4310*[1 0.9334];
den_rl = conv([1 -1],[1 0]);
zgrid('new');
rlocus(num_rl,den_rl);
```

The command **zgrid** designs a z-plane with straight lines for the attenuation factor ($J = 0.1, 0.2, 0.3, \ldots 0.8, 0.9$) and straight lines for the physical oscillation frequency. These lines facilitate the selection of the desired poles' position and is achieved with the aid of the command **rlocfind**.

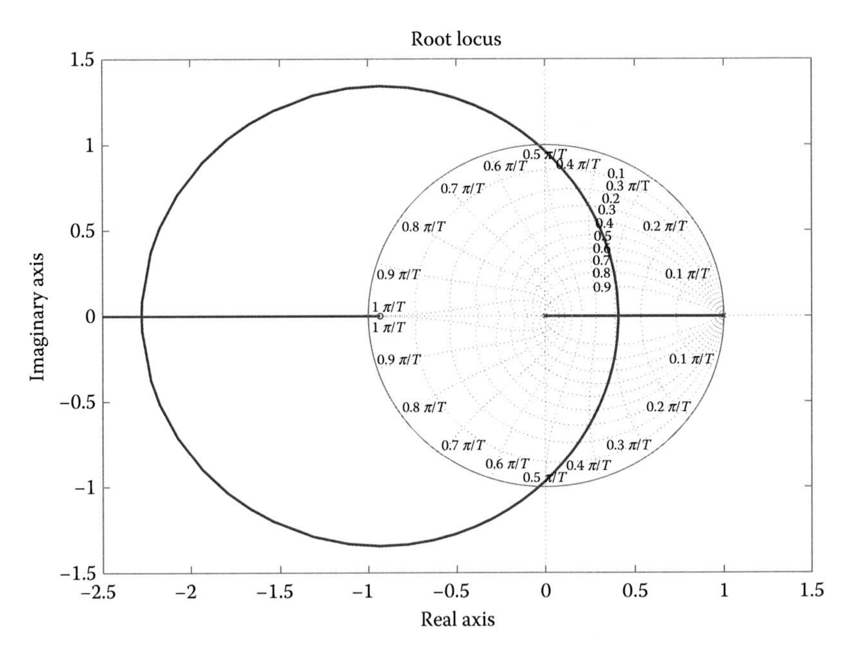


```
K = rlocfind(num_rl,den_rl);
Select a point in the graphics window
selected_point = 0.3707+0.3245i
K
K=0.6022
```

The designed digital controller has the transfer function

$$G(z) = \frac{0.6022\,(z - 0.4927 \pm j0.7589)}{z(z-1)} \tag{8.16.9}$$

The latter function results from MATLAB as

```
zeros_dz = [0.4927+j*0.7589 0.4927-j*0.7589]';
poles_dz = [0 1]'
K_dz = 0.6022;
[num_dz,den_dz] = zp2tf(zeros_dz,poles_dz,K_dz)
Dd = tf(num_dz,den_dz,T)

Transfer function
0.6022 z^2 - 0.5934 z + 0.493
-----------------------------
           z^2 - z
Sampling time: 0.01
```

So, the transfer function of the digital compensator is

$$G(z) = \frac{0.6022 - 0.5934z^{-1} + 0.4930z^{-2}}{1 - z^{-1}} \tag{8.16.10}$$

The closed system step response with the controller is calculated as

```
forward = series(Dd,Gd);
sys_cl = feedback(forward,1);
k = 0:10;
t = k*T;
y = step(sys_cl,t);
plot(t,y,'o',t,y,'-');
grid;
xlabel('Time(seconds)');
ylabel('Response y(t)');
title('Unit Step Response');
```

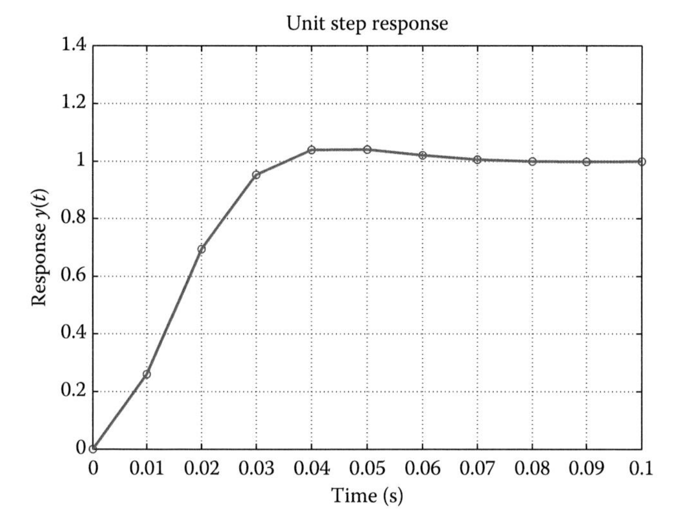

Observe that all the desired specifications are satisfied.
This system can be modeled by the following block diagram

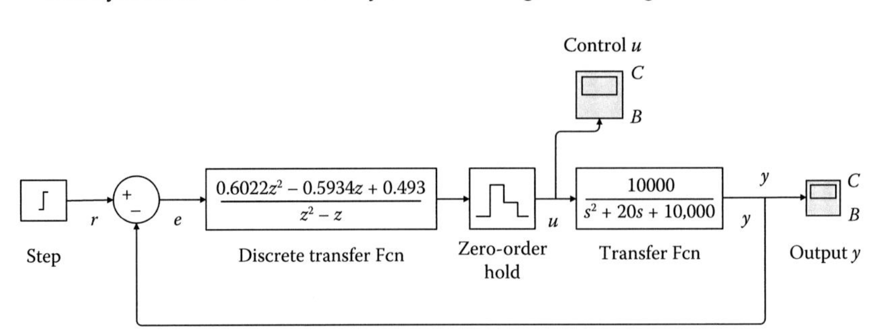

EXERCISE 8.17

Consider the analog system to be controlled:

$$W_c(s) = \frac{e^{-s}}{(1+10s)(1+5s)} = \frac{numPs}{denPs}e^{-s}$$

Design a digital controller so as to satisfy phase margin $\cong 60°$, small recovery time, the system response should not oscillate, and the position error should be zero. The sampling period is $T = 1$ s.

Solution

The MATLAB code is

```
Ts=1;
% Define the system to be controlled without time delay
numPs=1;
denPs=conv([10 1],[5 1]);
% Discretize the system with ZOH
[numPz,denPz]=c2dm(numPs,denPs,Ts,'zoh');
% Convert to a form of poles-zeros-gain (without delay)
[zerosd,polesd,gaind]=tf2zp(numPz,denPz);
% Add the time delay into the process: Z{e^-s} = z^-1 by
multiplying the denominator with z.
denPz=[denPz, 0];
```

The pulse transfer function with time delay is

$$W_p(z) = \frac{numPz}{denPz}\cdot\frac{1}{z} = 0.0091\frac{(z+0.9048)}{(z-0.9048)(z-0.8187)z} \tag{8.17.1}$$

Design the discrete controller so as to cancel the poles and zeros

$$W_c(z) = k_c\frac{z-0.9048}{z-1}\frac{z-0.8187}{z} \tag{8.17.2}$$

Assume that $k_c = 1$ and check the Bode diagram for discrete time

```
% Define the digital controller
kc=1;
numCz=conv([1, -0.9048],[1, -0.8187])
denCz=[1 -1 0];
% Calculate the loop transfer function L(z)=C(z)P(z)
[numLz,denLz]=series(numCz,denCz,numPz,denPz);
% Use of minreal.m to cancel the poles-zeros
[numLz,denLz]=minreal(numLz,denLz,0.01);
w=logspace(-1, 0, 100);
[Magd,Phased]=dbode(numLz,denLz,Ts,w);
M=[ Magd Phased w']
M =
 0.0693 -118.0289 0.2477
 0.0677 -118.6883 0.2535
 0.0661 -119.3633 0.2595
 0.0646 -120.0541 0.2656
 0.0631 -120.7612 0.2719
 0.0616 -121.4850 0.2783
 0.0602 -122.2257 0.2848
```

To satisfy the phase margin requirement PM = 60°, it is obvious from the above table that a phase shift of 119.3633° is achieved for an amplitude 0.0661. Consequently, a gain 1/0.0661 = 15.13 determines the appropriate gain k_c for the digital controller.

```
kc=1/0.0661
% Results verification by designing the Bode diagram
numLz=kc*numLz;
[Magd,Phased]=dbode(numLz,denLz,Ts,w);
[GM,PM,wpi,wc]=margin(Magd,Phased,w)
margin(Magd,Phased,w)
 GM = 3.22
 PM = 60.64
 wpi = 0.796
 wc = 0.26
```

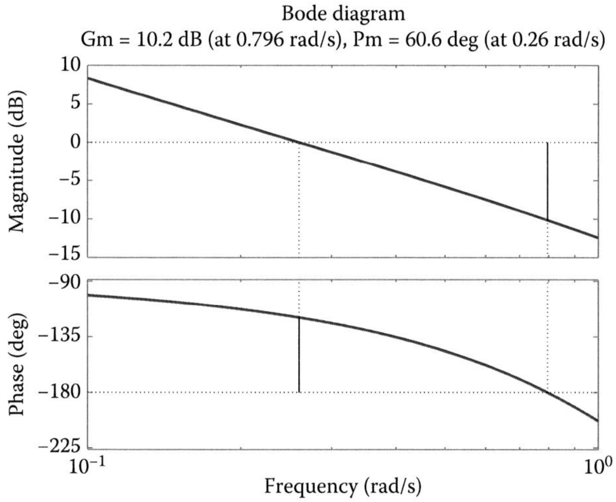

```
% Design the closed-loop step response with the controller
[numCLz,denCLz]=cloop(numLz,denLz,-1);
N=50;
td=0:N-1;
yd=dstep(numCLz,denCLz,N);
stairs(td,yd),grid
```

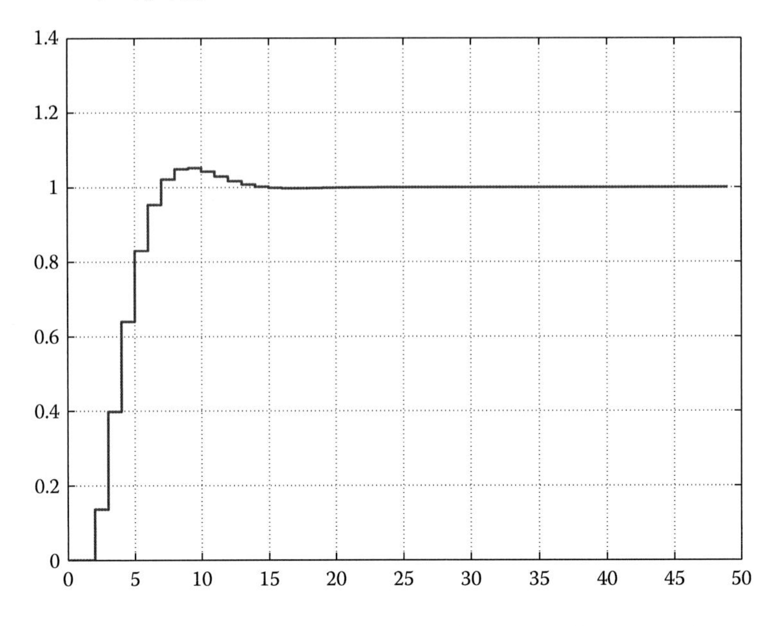

The continuous-time system can be easily simulated via the following Simulink model; parameters: Start Time=0, Stop Time = 20, Min Step Size = 0.1, Max Step Size = 0.1.

GraphU: Time Range=20; y-min=-10, y-max=40; GraphY: Time Range=20; y-min=-0.1, y-max=1.5;

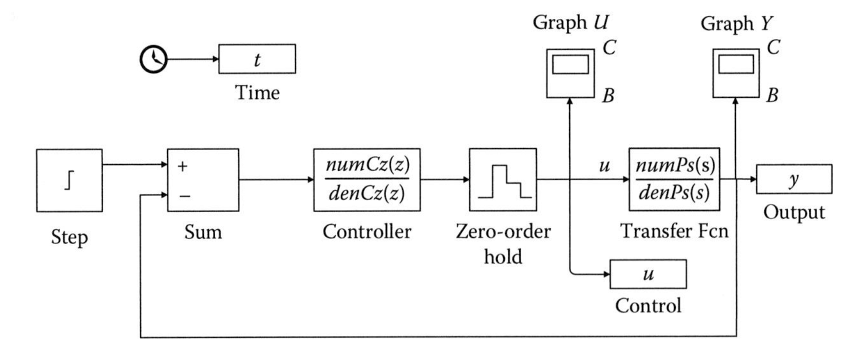

EXERCISE 8.18

A common actuator in automatic control systems is the DC motor. This provides rotational movement and in conjunction with springs or drums and cables, can provide transitional motion. The electric circuit of the armature and the pattern of the rotor are shown in the following scheme.

The values for the physical parameters are

- Motor's moment of inertia.
- Constant damping amplitude of the mechanical system $b = 0.1$ Nms.
- The electro-rotary movement constant $K = Ke = Kt = 0.01$ Nm/Amp.
- Electrical resistance $R = 1\ \Omega$.
- Electric induction $L = 0.5$ H.

Design a digital controller by assuming that the reference voltage is the unit level. Then, the output velocity of the motor should follow the specifications

- The recovery time of less than 2 s.
- The overshoot of less than 5%.
- The steady-state error of less than 1%.

Solution

The motor torque T, relates to the armature current i, by the fixed term K_t. The electromotive force e is related to the rotational velocity via the fixed term K_e.

From the given scheme, and applying the Newton and Kirchhoff laws, the system model is formulated as

$$T(t) = K_t i(t) \tag{8.18.1}$$

$$e(t) = K_e \frac{d\theta(t)}{dt} \tag{8.18.2}$$

$$J\frac{d^2\theta(t)}{dt^2} + b\frac{d\theta(t)}{dt} = K_i i(t) \tag{8.18.3}$$

$$L\frac{di(t)}{dt} + Ri(t) = V(t) - K\frac{d\theta(t)}{dt} \tag{8.18.4}$$

Utilizing the Laplace transform, the system model can be transferred into the s-domain

$$J\frac{d^2\theta(t)}{dt^2} + b\frac{d\theta(t)}{dt} = K_i i(t) \overset{LT}{\Rightarrow} (Js^2 + bs)\Theta(s) = K_i I(s) \tag{8.18.5}$$

$$L\frac{di(t)}{dt} + Ri(t) = V - K\frac{d\theta(t)}{dt} \overset{LT}{\Rightarrow}$$
$$(Ls + R)I(s) = V(s) - Ks\Theta(s) \tag{8.18.6}$$

Cancelling $I(s)$, the following open-loop transfer function is obtained, where the rotational velocity $\Omega(t)$ is the output, while the voltage denotes the input signal:

$$\frac{s\Theta(s)}{V(s)} = \frac{\Omega(s)}{V(s)} = \frac{K}{(Js+b)(Ls+R)+K^2} \tag{8.18.7}$$

This exercise is numerically solved using MATLAB. Define the numerator and denominator of the system transfer function as

```
J=0.01;
b=0.1;
K=0.01;
R=1;
L=0.5;
num=K;
den=[(J*L)   ((J*R)+(L*b))   ((b*R)+K^2)];
```

Adding the subsequent commands, the open-loop system step response is illustrated

```
step(num,den,0:0.1:3)
title('Step Response for the Open Loop System')
```

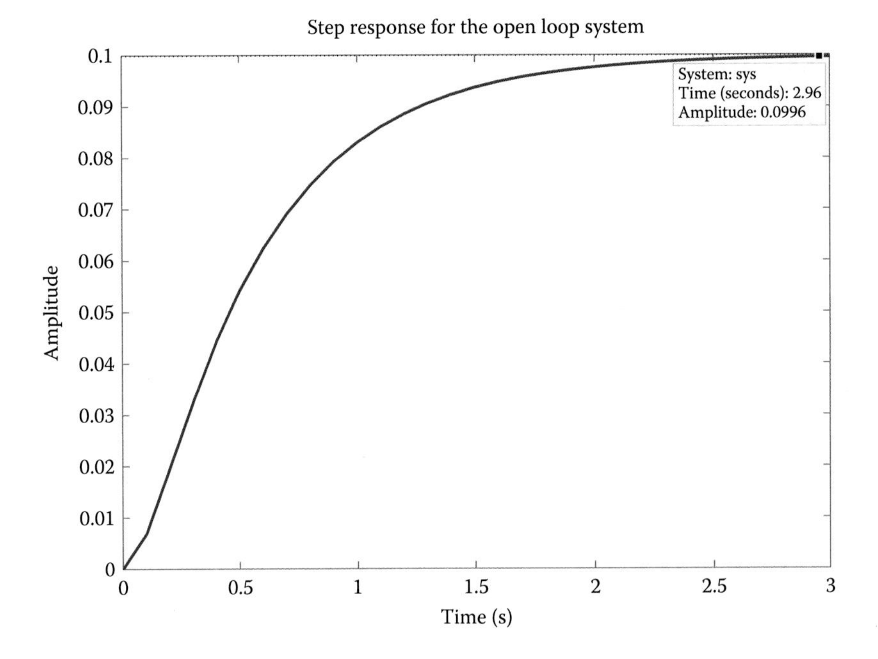

Observe that the motor can achieve a maximum velocity **0.1 rad/s**, 10 times lower than the desired one, when 1 V is applied.

Additionally, **3 s** are required to reach the steady-state velocity, so the 2 s requirement is not satisfied.

Design a PID controller and add it onto the system. The resultant block diagram is

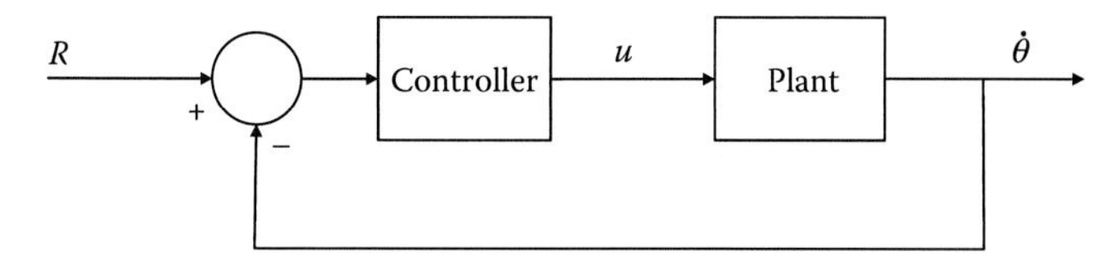

The transfer function of a PID controller is

$$K_p + \frac{K_i}{s} + K_d s = \frac{K_d s^2 + K_p s + K_i}{s} \tag{8.18.8}$$

First, let's try using an *analog* controller with gain 100. The following code is added

```
Kp = 100;
numa = Kp * num;
dena = den;
```

To derive the closed-loop transfer function, use the command **cloop**

```
[numac, denac] = cloop(numa, dena);
where numac and denac denote the numerator and denominator
of the closed-loop transfer function.
```

To design the step response, add the following commands to run the program via the command window:

```
t = 0:0.01:5;
step(numac, denac, t)
title('Step response with Proportional Control')
```

From the resultant plot, which illustrates the closed-loop step response for analog control, observe that the steady-state error and overshoot are both high. Therefore, let's try with another *PID* controller having small Ki and Kd parameters

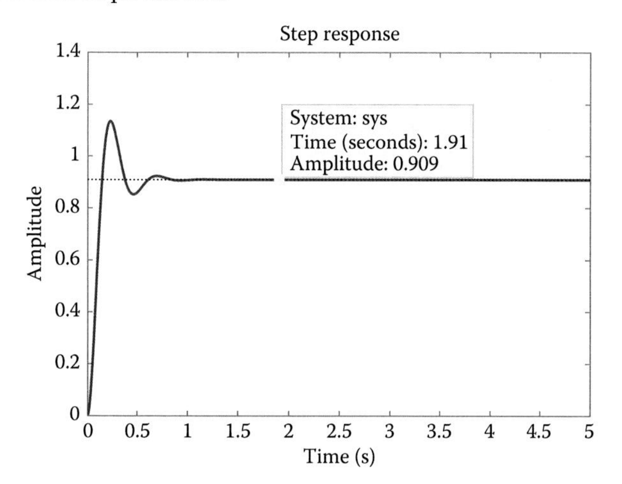

Write the commands

```
J=0.01;
b=0.1;
K=0.01;
R=1;
L=0.5;
num=K;
den=[(J*L) ((J*R)+(L*b)) ((b*R)+K^2)];
Kp=100;
Ki=1;
Kd=1;
numc=[Kd, Kp, Ki];
denc=[1 0];
numa=conv(num,numc);
dena=conv(den,denc);
[numac,denac]=cloop(numa,dena);
step(numac,denac)
title('PID Control with small Ki and Kd')
```

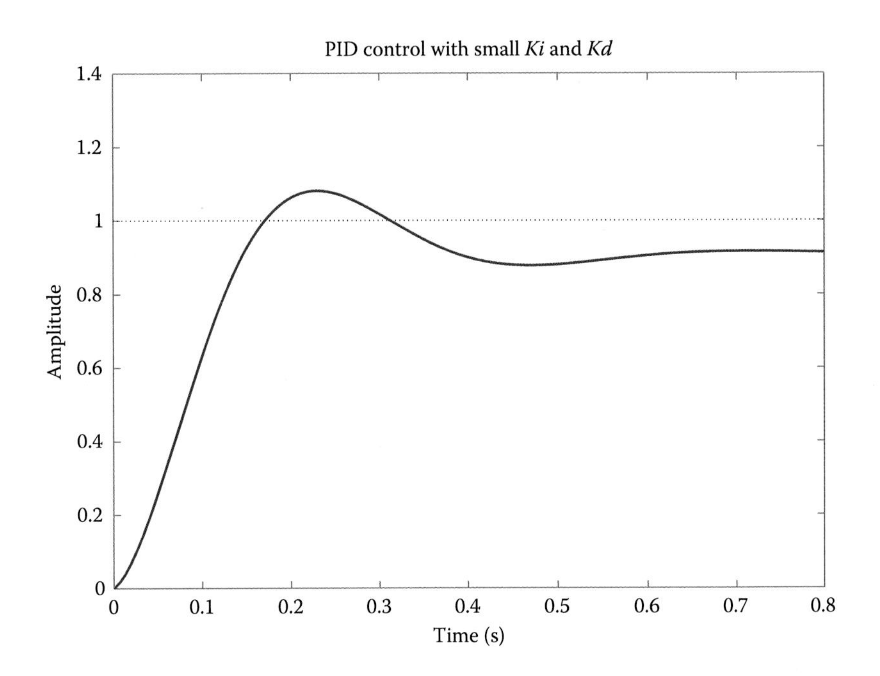

From the resultant plot, it is clear that the response time is extremely high with small *Ki* and *Kd*.

Hence, increase *Ki* to **200** and rerun the program. Observe that the response is emphatically faster than the former one, but the higher *Ki* value worsens the transient response (i.e., high overshoot). Thereby, increase *Kd* to reduce the overshoot effect.

Set $Kd = \mathbf{10}$. Rerun the new program and observe from the resultant plot that the system satisfies all the given specifications.

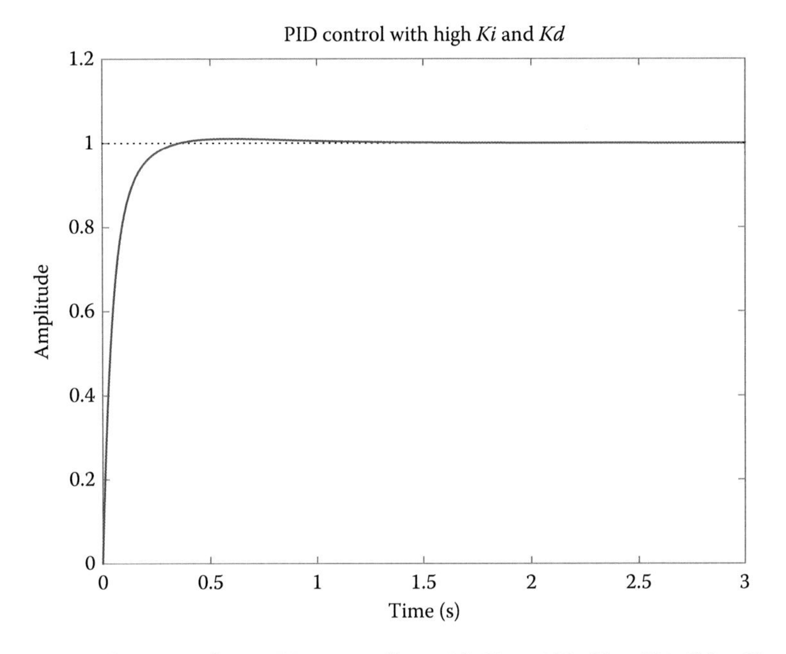

So, in the case when a PID controller with $Kp = 100$, $Ki = 200$, $Kd = 10$ is utilized, the given specifications are satisfied.

The *digital model* of the DC motor can be captured by the conversion of the analog model, as follows. The controller for this example is designed using the *PID* method.

The first step is the continuous-to-*discrete* transfer function conversion via the MATLAB command *c2dm*.

This command requires the following four parameters:

The polynomial numerator (num), the polynomial denominator (den), the sampling time (Ts), and hold circuit type. In this example, a *ZOH* circuit is used.

From the given specifications, set sampling time $Ts = 0.12$ s, which is the 1/10 of the system time constant with recovery time 2 s.

Write the following commands to formulate the discretized transfer function with ZOH:

```
R=1;
L=0.5;
K=0.01;
J=0.01;
b=0.1;
num = K;
den = [(J*L) (J*R)+(L*b) (R*b)+K^2];
Ts = 0.12;
[numz,denz] = c2dm(num,den,Ts,'zoh')
printsys(numz,denz,'z')
```

$$\text{We get}: \frac{\Omega(z)}{V(z)} = \frac{0.0092z + 0.0057}{z^2 - 1.0877z - 0.2369} \tag{8.18.9}$$

First, the closed-loop response is illustrated *without controller.* Add the following code

```
[numz_cl,denz_cl] = cloop(numz,denz);
```

Design the closed-loop step response using the commands dstep and stairs. The command **dstep** computes the discrete output signal vector and the command **stairs** connects these samples:

```
[x1] = dstep(numz_cl,denz_cl,101);
t=0:0.12:12;
stairs(t,x1)
xlabel('Time (seconds)')
ylabel('Velocity (rad/s)')
title('Stairstep Response:Original')
```

Observe that the steady-state response has a very large deviation from the desired one (error of about 90%).

There are several methods to design in the *Z*-domain from the *S*-domain. The most efficient one is by setting $z = e^{Ts}$. However, the transfer function of PID cannot be obtained using this method due to the fact that the discrete-time transfer function would have more zeros than poles; an infeasible condition.

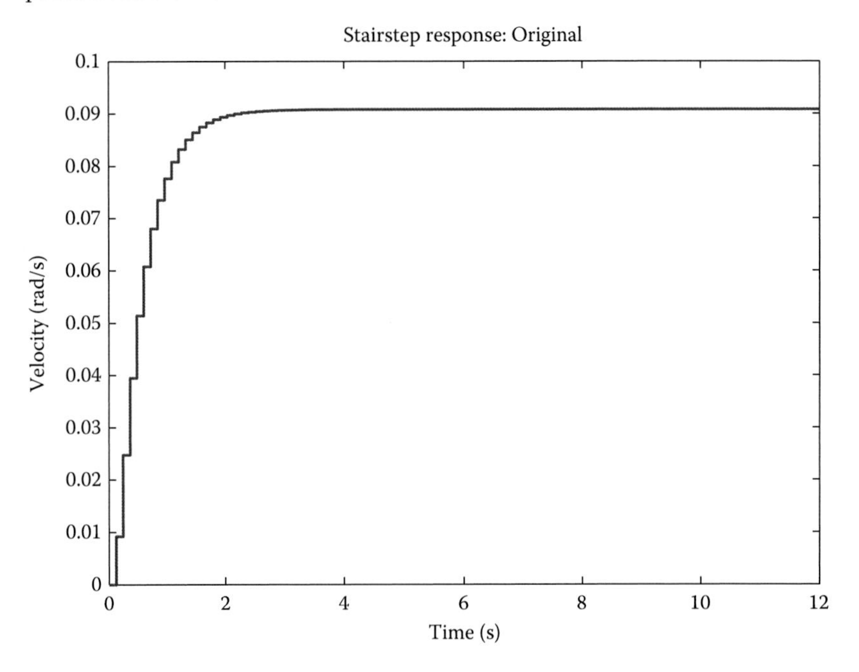

Instead, we use the bilinear transform as follows:

$$s = \frac{2}{T_s}\frac{z-1}{z+1} \tag{8.18.10}$$

Equivalently, the command **c2dm** facilitates the discretization of the PID controller using the *Tustin* method

According to the above analysis, we satisfied that the system parameters $Kp = 100$, $Ki = 200$, and $Kd = 10$ satisfy the given specifications.

We add the following MATLAB commands

```
%Discrete PID controller with bilinear approximation
Kp = 100 ;
Ki = 200 ;
Kd = 10 ;
Ts=0.12;
[dencz, numcz] = c2dm ([1 0], [Kd Kp Ki], Ts, 'tustin') ;
```

Let's see if the performance of the closed loop response with PID controller meets the specifications. Adding the following code, the closed-loop step response is obtained:

```
numaz = conv(numz, numcz) ;
denaz = conv(denz, dencz) ;
[numaz_cl, denaz_cl] = cloop(numaz, denaz) ;
[x2] = dstep (numaz_cl, denaz_cl, 101) ;
t = 0:0.12:12 ;
stairs(t, x2)
xlabel(' Time (seconds) ')
ylabel(' Velocity (rad/sec) ')
title(' Stairstep Response : with PID controller ')
```

From the resultant plot, it is clear that the closed-loop response is *unstable*. Hence, a system error could occur with regard to the compensated system. Design the root locus of the system characteristic equation to extract some useful outcomes

```
rlocus(numaz, denaz)
title(' Root Locus of Compensated System ')
```

From the root locus diagram, it is obvious that the denominator of PID has a pole at –1 in *z*-domain. We know that if the pole is outside the unit circle, then the system is unstable. This compensated system is always unstable for any positive gain since its pole is outside the unit circle.

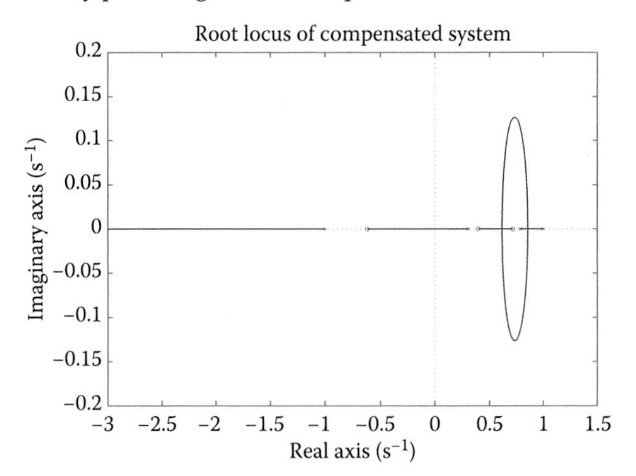

We choose to cancel the zero at –0.62. This will make the system stable for any gain. The appropriate gain can be selected from the root locus diagram that satisfies the given specifications using the command **rlocfind**. Write the following code:

```
dencz = conv([ 1 -1], [1.6 1]);
numaz = conv(numz, numcz);
denaz = conv(denz, dencz);
rlocus(numaz, denaz)
title(' Root Locus of Compensated System ');
[K, poles] = rlocfind (numaz, denaz)
[numaz_cl, denaz_cl] = cloop(K*numaz, denaz);
[x3]= dstep(numaz_cl, denaz_cl, 101);
t = 0:0.12:12;
stairs(t, x3)
xlabel(' Time (seconds) ')
ylabel(' Velocity (rad/sec) ')
title(' Stairstep Response : with PID controller ')
```

The new dencz has a pole at –0.625 instead of –1, which practically cancels the zero of the compensated system. In the MATLAB window, we choose the point based on the root locus diagram. We choose the (x) point, as shown below:

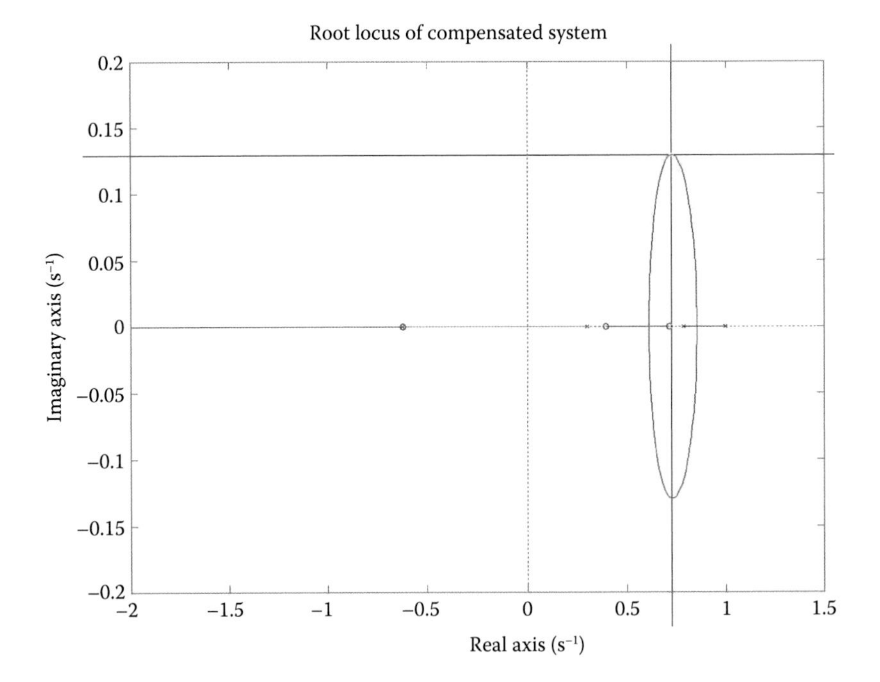

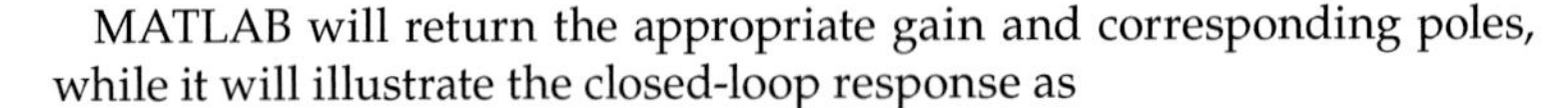
MATLAB will return the appropriate gain and corresponding poles, while it will illustrate the closed-loop response as

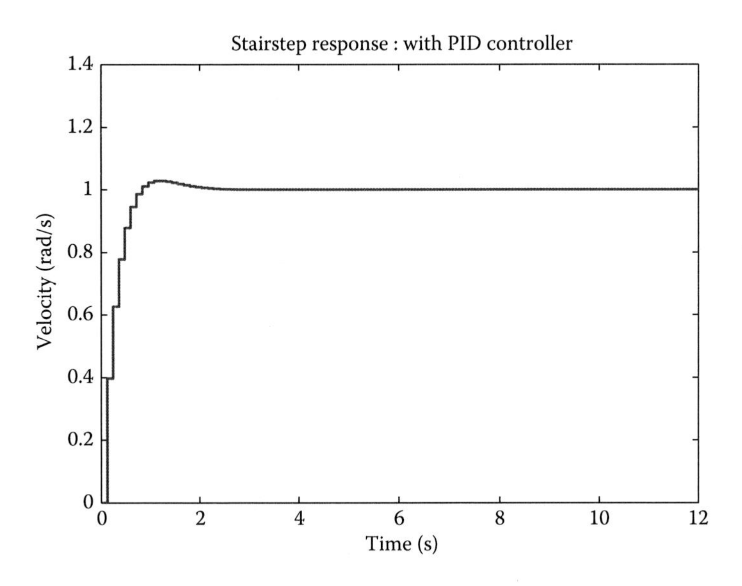

This plot indicates that the rise time is less than 0.2 s, the overshoot percent is approximately 2.9% and the steady-state error is 0. Also, from the root locus diagram, the gain K is approximately 73, which is reasonable. Hence, such a response satisfies the given specifications.

Subsequently, the digital control of the velocity of DC motor follows using *Simulink*. We design the block diagram of the system model.

To simulate the system, select *Parameters* from the *Simulation menu* and insert "3" in the *Stop Time* field.

Insert the following code:

```
J=0.01;
b=0.1;
K=0.01;
R=1;
L=0.5;
```

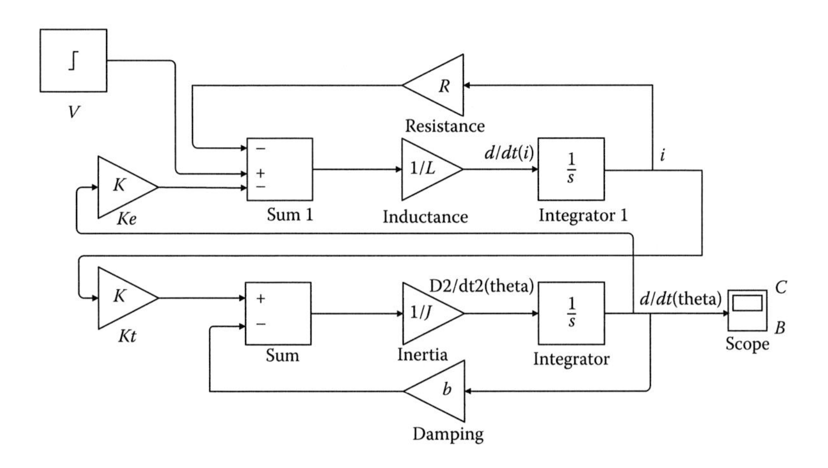

Run the simulation (**Ctrl-t** or **Start** in *Simulation menu*). When the simulation ends, we double-click onto *scope* and *autoscale button*. The waveform of the open-loop step response arises, where we observe that when 1 V is applied on the system, the motor achieves a maximum velocity **0.1 rad/s**, 10 times lower than the desired one. Also, **3 s** are required to reach to the steady-state velocity; hence, the 2 s. condition is not satisfied.

Export of the linear model from Simulink to MATLAB

The linear system model (in state space or in the form of a transfer function) can be exported from Simulink to MATLAB using the commands **In** and **Out** *Connection blocks* and **linmod**.

Substituting the *Step block* and *Scope block* with an *In Connection block* and *Out Connection block*, (these blocks are located in the *Connections block library*). Thus, the system input and output are defined for the exporting process.

The following block diagram arises:

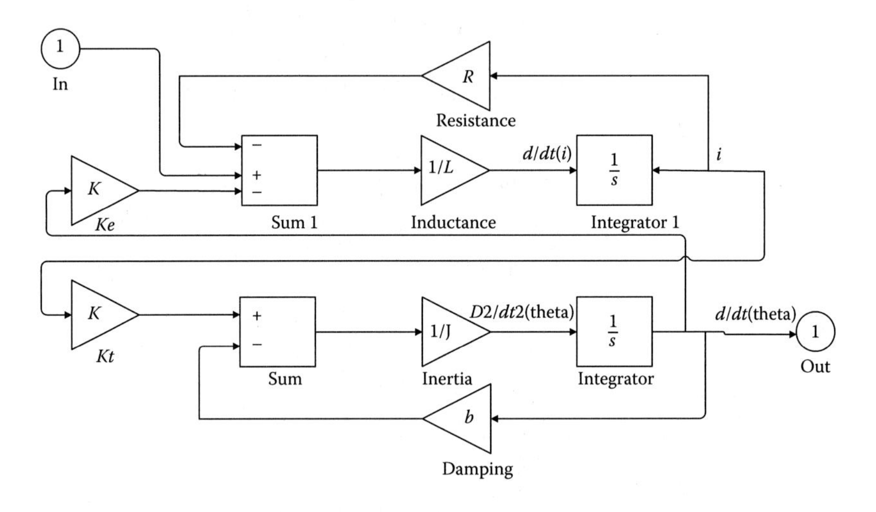

Insert the following code, where the system is modeled in the state space, derive the system transfer function, and design the open-loop system step response. Obviously, the result is identical to the corresponding one derived from Simulink.

```
[A,B,C,D]=linmod('motormod')
[num,den]=ss2tf(A,B,C,D);
step(num,den);
```

Export of the digital model from Simulink to MATLAB

The analog to digital conversion is implemented via the *Zero-Order Hold block* of Simulink. First, we will group all elements of the system (except the Step input and Scope) into a *Subsystem block.*

Select all elements except the Step and Scope blocks. We choose *Create Subsystem* in *Edit menu* (or type **Ctrl-G**). Therefore, all the selected blocks are grouped in a single one. The following scheme arises:

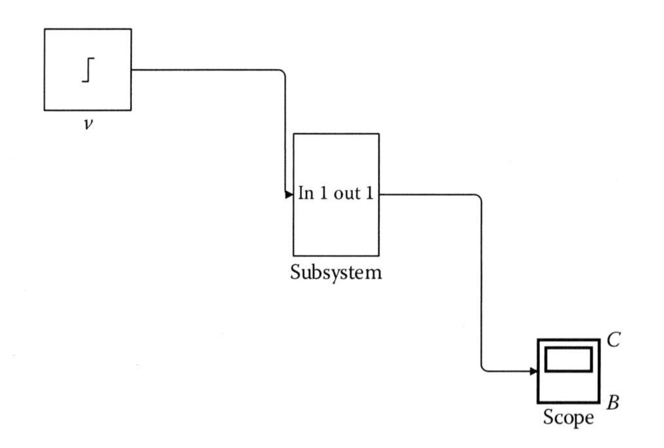

Set the label of *Subsystem block* as "*Continuous Plant.*" Substitute the *Step block* and *Scope block* with *Zero-Order Hold blocks* (derived from *Discrete block library*). The one Zero-Order Hold block is used for the conversion of an analog signal to a quantized signal. The other Zero-Order Hold block is used to collect the discrete output samples of the system.

Set the *Sample Time fields* of *Zero-Order Hold blocks* in 0.001. Connect an *In Connection Block* into the input of the first *Zero-Order Hold block* and an *Out Connection block* into the output of the *Zero-Order Hold block* (these blocks are located at the *Connections block library*).

The following system arises:

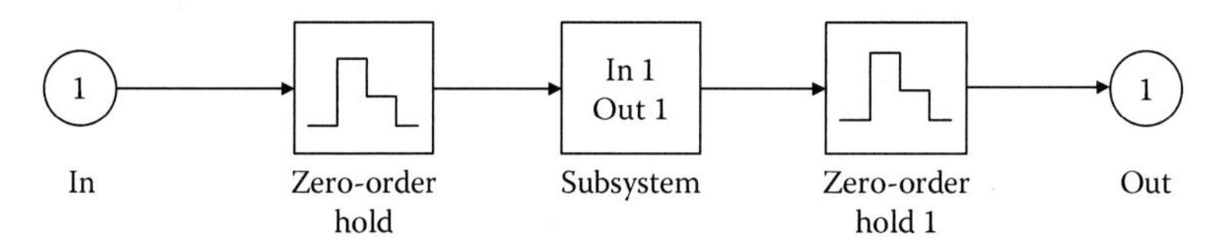

Insert the following code to design the open-loop step response. Observe that the system is unstable.

```
[A,B,C,D]=dlinmod('motorpos',.001)
[num,den]=ss2tf(A,B,C,D);
dstep(num,den);
```

Closed-loop step response

Design the following block diagram of the digital system using Simulink. In this system, a digital controller has been added with the transfer function $450((z-0.85)(z-0.85)/(z+0.98)(z-0.7))$.

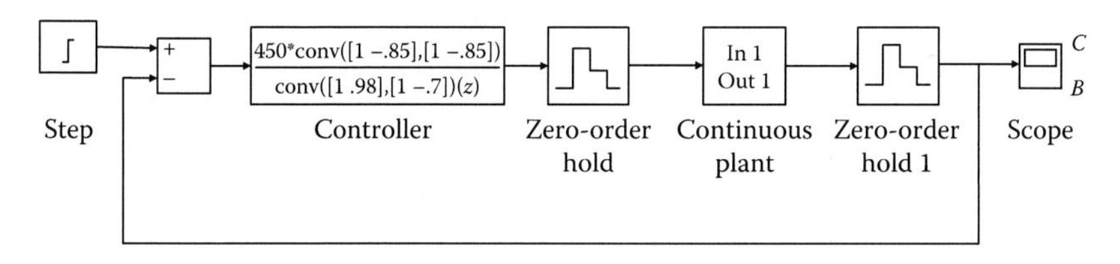

Insert the following code, where the parameter values have been slightly changed so as to provide a better closed-loop response

```
J=3.2284E-6;
b=3.5077E-6;
K=0.0274;
R=4;
L=2.75E-6;
```

Run the simulation (**Ctrl-t** or **Start** in the *Simulation menu*). When the simulation ends, double-click onto *Scope* and *Autoscale button*. From the resultant waveform, observe that all the given specifications are satisfied.

9

Simulation Tools: MATLAB, Simulink, LabVIEW, Comprehensive Control

9.1 Introduction

The **simulation** process offers the control system analyst the ability to design and test the operation, performance, and systems' characteristics without setting them into operation or being intervened in the real system. The simulation substantially contributes to the design, maintenance, and upgrading of control systems. For example, when a controller is being developed to compensate for a system, the examination of its behavior using simulations is a common practice prior to implementation. This allows quick changes and bug fixes before the actual implementation of the system. In the case when the controller has already been implemented, the simulation is critical for the optimal setting of its parameters.

In this chapter, the simulation tools used throughout this textbook will be presented and briefly described; they are **MATLAB®, SIMULINK®, COMPREHENSIVE CONTROL, and LABVIEW.**

9.2 Control Systems Simulation Using MATLAB

MATLAB is a program that uses numerical calculations based on linear algebra. It started as a "Matrix Laboratory" platform, and has been developed enough to become a powerful tool in the implementation of simulations and modeling systems, planning, research, engineering, and communications/telecommunications.

Its technical language is compact and descriptive, which allows the modeling of various systems using an easy-to-learn code. These features make MATLAB an attractive and powerful tool that can be used by the tutor to effectively teach many courses such as the automatic control systems.

The main objective in the design of an automatic control system is to achieve stability and good performance. The performance is essentially ensured by the fulfilment or not of certain basic standards. These specifications are divided in two categories; namely, specifications in the time domain and in frequency domain. The time specifications tend to emerge from the time response of a step system input or other input types. The specifications in the frequency domain are derived from diagrams in the frequency domain. But, to preserve a better performance, we need to adjust or compensate the corresponding system more often, under certain criteria.

MATLAB has a rich collection of functions useful in the field of Automatic Control via the **Control System Toolbox**. The modeling of control systems is implemented by using either transfer functions or state space representation in order to apply classical or optimal control, respectively. Furthermore, it is possible to analyze both continuous and discrete-time systems.

In what follows, a summary of MATLAB capabilities is presented regarding the control systems' standpoint. The analysis is focused on linear **discrete-time invariant systems**.

9.2.1 Analysis and System Modeling

Various MATLAB commands for analysis and modeling of LTI systems are described in Table 9.1.

9.2.2 Control Systems Design

To design SISO systems, we may use commands and the graphical user interface (GUI) from the Control System Toolbox. Some of these commands are presented in Table 9.2.

9.2.3 Simulation of Digital Control Systems Using MATLAB

Both the continuous and discrete systems are described by either the transfer function or state space. MATLAB contains functions tf and ss as well as the zpk for these descriptions (where Ts is the sampling period).

sys=tf(num,den,Ts)	Transfer function
sys=ss(A,B,C,D,Ts)	State space
sys=zpk(Z,P,K,Ts)	System poles–zeros–gain

Also, MATLAB allows the transition from the state space to transfer function and vice versa. Moreover, if the zeros, poles, and gain of transfer function is known, then the transfer function and the state space can be directly calculated and vice versa.

[num,den]=ss2tf(A,B,C,D)	State space to transfer function
[A,B,C,D]=tf2ss(num,den)	Transfer function to state space
[Z,P,K]=ss2zp(A,B,C,D)	State space to poles–zeros–gain
[Z,P,K]=tf2zp(num,den)	Transfer function to poles–zeros–gain
[num,den]=zp2tf(Z,P,K)	Poles–zeros–gain to transfer function
[A,B,C,D]=zp2ss(Z,P,K)	Poles–zeros–gain to state space

TABLE 9.1

Commands for Analysis and Modeling of Control Systems

Command	Description
LTI Model Development	
Tf	Creates a transfer function
ss	Creates a state-space model
zpk	Creates a poles–zeros–gain model
Data Export	
tfdata	Export of numerator and denominator
ssdata	Export of state-space matrix
zpkdata	Export of data for poles, zeros, and gain
Conversions	
Tf	Conversion into a transfer function
ss	Conversion in the state space
zpk	Conversion in the poles–zeros–gain formation
c2d	Analog to discrete conversion
d2c	Discrete to analog conversion
Dynamic Model Characteristics	
pole, eig	System poles
pzmap	Poles and zeros illustration
dcgain	dc gain
damp	Physical frequency and oscillation attenuation factor
pade	Pade approximation for time delays
Time Response	
step	Step response
Impulse	Impulse response
lsim	Response of any type of input signal
initial	System response in the state space when the initial conditions are given
Frequency Response	
bode	Bode diagram
nyquist	Nyquist diagram
nichols	Nichols diagram
margin	Gain and phase margins

TABLE 9.2

Commands to Design Control Systems

Command	Description
Rlocus	Root locus design
Rlocfind	Interactive determination of the gain from the root locus
Zgrid	Creation of grid lines in z-domain for the root locus design or the poles–zeros diagram
Acker	Poles placement for SISO systems
Place	Poles placement for MIMO systems

In the above commands, num and den are the polynomials defining the numerator and denominator, respectively, of the system transfer function. The matrices A, B, C, D are the system state matrices describing the state space and Z, P are the vectors containing the zeros and poles of the system, while K is the corresponding gain.

For calculating the step and impulse response of a discrete system, MATLAB provides the dstep and dimpulse commands, respectively. For any other arbitrary input, the command dlsim is used. Also, for a response to the initial conditions, the command dinitial is available (in this case the system must be defined in state space).

dstep(num,den) dstep(A,B,C,D)	Step response
dimpulse(num,den) dstep(A,B,C,D)	Impulse response
dlsim(num,den,u) dlsim(A,B,C,D,u)	Response to any input
dinitial(A,B,C,D,x0)	Response to initial conditions (only for state space)

where u denotes the input sequence.

In digital control, sometimes, it is necessary to convert a continuous system to a discrete one (discretization). For this procedure, MATLAB provides the following commands:

sysd = c2d(sys,Ts,'method')	Continuous to discrete with method selection
sysd = c2d(sys,Ts)	Continuous to discrete with ZOH
[Ad,Bd,Cd,Dd] = c2dm(A,B,C,D,Ts,'method')	Continuous to discrete (state space)
[numd,dend] = c2dm(num,den,Ts,'method')	Continuous to discrete (transfer function)

where **method** is the discretization method ("**zoh**," "**foh**," "**imp**", "**tustin**," "**prewarp**," "**matched**") and **Ts** is the sampling period. When the **method** parameter is omitted, then it is considered that the discretization is implemented via ZOH. The matched method is used only for

MIMO systems. Also, when prewarp method is used, the critical frequency Wc (rad/s) should be included as an additional argument, that is: sysd=c2d(sys,Ts,"prewarp," Wc).

The system response is determined by its poles and zeros. MATLAB provides the following commands for their calculation:

pole(sys)	System poles
zero(sys)	System zeros
pzmap(sys)	Poles–zeros diagram
minreal(sys)	Common poles–zeros cancellation

Typically, a system is part of a broader set of systems which are linked altogether. MATLAB contains functions that provide the transfer function of two systems, which are connected to each other in series, in parallel, or with a positive (or negative) feedback. These functions/commands are presented as follows:

sys=series(sys1,sys2)	Cascade systems connection
sys=parallel(sys1,sys2)	Parallel systems connection
sys=feedback(sys1,sys2,-1)	Systems connection with negative feedback
sys=feedback(sys1,sys2,+1)	Systems connection with positive feedback

For the study of linear systems in the frequency domain, MATLAB provides the following commands:

bode(sys)	Bode diagrams
[Gm,Pm,Wcg,Wcp] = margin(sys)	Gain margin–phase margin
H=freqresp(sys,w)	Frequency response
nyquist (sys)	Nyquist diagrams
nichols (sys)	Nichols diagrams

where **sys** is the system and **w** is the frequency in rad/s. Moreover, **Gm** and **Pm** denote the gain and phase margins, respectively, while **Wcg** and **Wcp** are the corresponding frequencies.

One of the tools that used in the analysis and design of automatic control systems is the root locus method. To this end, MATLAB provides the following commands (Table 9.3).

rlocus(sys)	Root locus diagram
[K, poles]=rlocfind(sys,P)	Gain and poles calculation

TABLE 9.3

MATLAB Commands for General Use

Command	Description
Who	Displays the variables
Whos	Similar to "who" but with more information
Clc	Clears the MATLAB command window
abs(a)	Returns the absolute value or magnitude of a (in the case when a is real or complex number, respectively)
sign(a)	Returns −1 for a negative a or 1 for a positive a or 0
sqrt(a)	Finds the root of a
sin(a)	Returns the sine of angle a in rads
cos(a)	Returns the cosine of angle a in rads
tan(a)	Returns the tangent of angle a in rads
atan(a)	Returns the arctangent of a
log10(a)	Returns the base-10 logarithm of a
log2(a)	Returns the base-2 logarithm of a
log(a)	Returns the natural logarithm of a
exp(n)	Returns e of n
linspace(a, b, c)	Returns a vector from a to b including c elements
ones(m,n)	Returns an mxn matrix with all elements equal to one
zeros(m,n)	Returns an mxn matrix with all elements equal to zero
eye(m,n)	Returns an mxn identity matrix
size(a)	Returns the dimensions of matrix a
length(a)	Returns the higher dimension of matrix a
flipud(a)	Substitutes the upper with the lower elements of matrix a
fliplr(a)	Substitutes the left with the right elements of matrix a
[x,y] = max(a)	Returns the maximum number of matrix a at x and points its location at y
[x,y] = min(a)	Returns the minimum number of matrix a at x and points its location at y
sum(a)	Sums all elements of matrix a
triu(a)	Makes a matrix a, an upper tridiagonal matrix
tril(a)	Makes a matrix a, a lower tridiagonal matrix
diag(a)	Returns the elements of the diagonal of matrix a
trace(a)	Sums the elements of the diagonal of matrix a
det(a)	Returns the determinant of a square matrix a
inv(a)	Returns the inverse of square matrix a
format short e	Modifies the representation of results
format	Determines the representation of results
plot(x,y,format)	Creates a Cartesian figure where: x denotes the values on the x-axis, y denotes the values on the y-axis, and format specifies the line form
plot(x1,y1,xn,yn,format)	Identical to the above but illustrates many functions in the same figure

(*Continued*)

TABLE 9.3 (*Continued*)

MATLAB Commands for General Use

Command	Description
figure(n)	Creates an n-size empty figure
xlabel('')	Places a text message in the horizontal axis
ylabel('')	Places a text message in the vertical axis
title('')	Places a figure title
text(x,y,'')	Places a certain text in coordinates (x, y)
grid [on/off]	Enables/disables a grid within the figure
axis([xmin xmax ymin ymax])	Illustrates a part of the figure with certain coordinates
legend ('', '', '', ...)	Creates a certain note into the figure
clf, clg	Activates the current figure, returns its number
close(n), close, close all	Closes the figures (nth figure)
polar(θ,ρ,format)	Creates a polar figure with coordinates θ, ρ and specific format
subplot (m,n,k)	Creates multiple subplots into a single figure, each is considered as an mxn matrix and is considered as the kth subplot
real(a)	Returns the real part of complex number a
imag(a)	Returns the imaginary part of complex number a
conj(a)	Returns the conjugate of complex number a
polyval(a,n)	Returns the value of polynomial a for n
angle(a)	Returns the angle formed by complex number a
compass(a)	Creates a polar figure with complex number a as a vector
roots(a)	Returns the roots of polynomial a
poly(a)	Returns the polynomial with the roots given in vector a
conv(a,b)	Multiplication of polynomials a and b
[x,y]=deconv(a,b)	Division of polynomials (a/b), returns the quotient x and the remainder y
[x,y]=polyder(a,b)	Polynomial derivative a/b with numerator a and denominator y after the differentiation
polyint(a)	Integration of polynomial a
eig(a)	Returns a column vector, where its elements are the eigenvalues of square matrix a
[V,D]=eig(a)	Returns matrix V with columns denoting the eigenvectors of matrix a and diagonal matrix D denoting the eigenvalues of matrix a (V and D should be square matrices)
rref(E)	Gauss–Jordan cancellation method, converts matrix E into an upper tridiagonal
rrefmovie(E)	Step-by-step operations to manipulate matrix E
x=sym('a')	Creates a symbolic constant x having the result of a as a value
subs(a,old,new)	Calculates the symbolic entity a, substitutes old with new (it may be a number or expression)
findsym(a)	Returns symbolic variables of symbolic expression a in alphabetical order

(Continued)

TABLE 9.3 (*Continued*)

MATLAB Commands for General Use

Command	Description
expand(a)	Executes the operations of the symbolic expression a
factor(a)	Factorizes the symbolic a
simplify(a)	Simplifies the symbolic a
[x,y]=numden(a)	Executes the operations in fraction a and returns the numerator and denominator
solve(f1)	Solves equations or equation systems
diff(a)	Differentiates a
diff(a,x)	Returns the partial derivative of a with respect to x
diff(a,n)	Returns the nth derivative of a
diff(a,x,n)	Returns the nth derivative of a with respect to x
int(e)	Returns the indefinite integral of e
int(e,x)	Returns the indefinite integral of e with respect to x
int(e,a,b)	Returns the definite integral of e in the range [a, b]
int(e,x,a,b)	Returns the definite integral of e in the range [a, b] with respect to x
syms s t	Creates the symbolic variables s, t
laplace(f)	Returns the Laplace transform of f
ilaplace(F)	Returns the inverse Laplace transform of F

9.3 Simulink

9.3.1 Introduction

Simulink is an extremely useful tool, embedded in MATLAB, which enables the modeling, simulation, and analysis of dynamic systems. Although its use does not require knowledge of MATLAB , this knowledge represents a significant advantage since it enables a more efficient use. One of the major strengths of Simulink is its simplicity. In addition, there is a great variety of features provided for the user.

For modeling, Simulink provides a GUI that allows modeling via block diagrams, using the click, drag, and drop computer mouse options. It includes a great variety of block components (blocks); the most important ones are the sources, sinks, linear continuous components, nonlinear components, and system/signal components. It is also possible to modify and create new components by the user.

The Simulink models are hierarchical (i.e., each model may include some blocks, which in turn may consist of other blocks) and appropriately interconnected to one another in different layers. A system having a hierarchical structure can be seen initially at a high level as a set of interconnected subsystems, each of which is modeled as a block. Then, by double-clicking on the individual blocks, the user has access to lower layers to see an increasing degree of detail. After a model has been defined, the user is able to see the results of simulation as it runs. Also, the user may change the parameters and see the simulation results of particular interest.

After the model creation, its simulation is possible, by using one of various operation methods available in the Simulink environment. Utilizing scopes and other illustration blocks, the monitoring of simulation results is enabled during the process. Moreover, it is possible to export simulation results in the MATLAB workspace for further processing. It is even possible to use Simulink for simulation and real-time systems' control through a certain toolbox, namely, Real Time Workshop.

The key features of Simulink are

- Extensive and expandable collection of libraries with predefined blocks
- Hierarchical modeling
- Open architecture to integrate models from other tools (Application Program Interface)
- Simulation of hybrid systems (continuous–discrete time)
- Support of various forms for simulation boosting
- Full range of diagnostic and debugging tools
- Full interoperability with MATLAB

9.3.2 Model Creation

To enter the Simulink environment, the following command should be written into the MATLAB window:

```
≫simulink
```

where the **Simulink Library Browser** window appears. The user can open an existing Simulink file with the extension **.mdl** or to create a new one. The latter can be realized by using the following three ways:

1. Choosing **file** →**New**→ **Model**
2. Clicking on the upper left icon with the empty page
3. Inserting **Control** + **O**

Various block categories can be used, which can be selected from the following categories:

Commonly Used Blocks	Math Operations	Signal Attributes
Continuous	Model verification	Signal routing
Discontinuities	Model-wide utilities	Sinks
Discrete	Ports & subsystems	Sources

Sources: Contains the following blocks

- Various input signals, such as
    - Step (Step variations)
    - Sine wave (Sine variations)
    - Ramp (Linear variations)
    - Random number (Gaussian-distributed random numbers)
    - Uniform random number (Uniformly-distributed random numbers)
- Various data files as follows:
    - From file
- Various variables from MATLAB workspace as follows:
    - From workspace
    - Clock

Sinks: Contains the following blocks:

- Various capture modules of output signals, such as
    - Display
    - Floating scope
    - Scope
    - XY graph
- Data storage files
    - To file
- Various data variables from the MATLAB workspace
    - To workspace

Coninuous: Contains the following blocks:

- State-space models
    - State space
- Transfer functions
    - Transfer Fcn
    - Zero–pole
    - Integrator
- Downtime-related functions
    - Transport delay
    - Variable time delay
    - Variable transport delay

- Input signal differentiation
    - Derivative

Discrete: These blocks are mainly useful for creating discrete-time models in the form of transfer function or state space, and for implementing discrete PID controller, system discretization with ZOH and FOH, etc. Below, the most commonly used components for the simulation of digital control are presented as Figure 9.1.

Math Operations: Contains a variety of illustration blocks. These are various mathematical functions such as: Abs, Add, Divide, Dot Product, Product, Rounding Function, Sign, Sine Wave Function, Subtract, Trigonometric Function, etc.

Creation of a new block diagram

- In the Simulink Library Browser window click on the icon that shows a white sheet of paper (first icon from upper left of the window), entitled as Create a new model. It will open an empty window entitled as Untitled.

Insertion of a block into the diagram

- From the Simulink Library Browser window, select the item you want and, holding the left button of the mouse, transfer it to the desired location into the block diagram.
- If the desired block is not included in the standard Simulink library, the library name that contains this block is written in the main MATLAB window. Then, the created window is imported as mentioned above.

Block parameters connection, copying, and configuration

- To connect two blocks that we have introduced in the block diagram, we left click in the node of one of the two blocks that we want to connect and holding the mouse button, we move the pointer to the appropriate node in the second block.
- To create a connection branch, click with the right mouse button to the desired point. Then, holding down the button, move the pointer and the branch is created.
- To copy an existing block, use the standard copy–paste procedure.
- By double clicking on a block, it opens the block parameters window. In this window we can change the block parameters, as well as get information about them by clicking Help.

Simulation parameters and displaying results

- In the block diagram window, select the Simulation/Parameters menu. This will open a window where we can regulate, among others, the initial and final time of the simulation and the type of diagram (discrete or continuous).

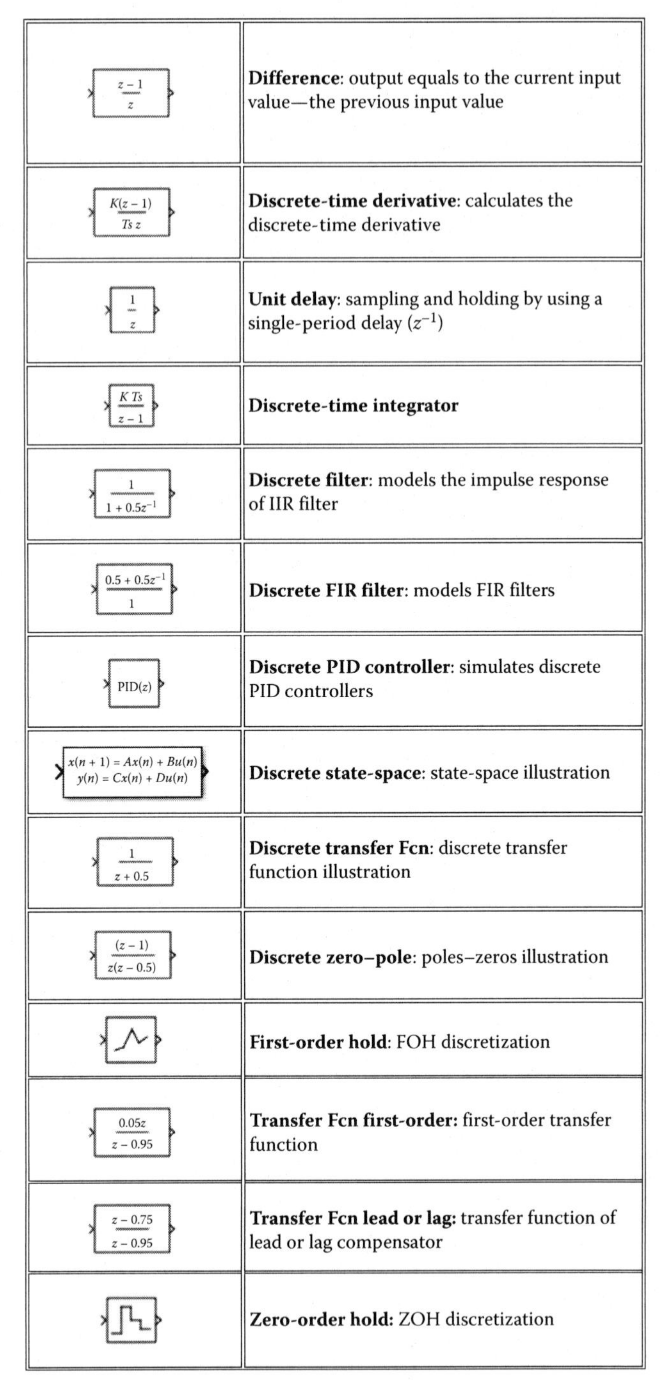

$\frac{z-1}{z}$	**Difference**: output equals to the current input value—the previous input value
$\frac{K(z-1)}{Ts\,z}$	**Discrete-time derivative**: calculates the discrete-time derivative
$\frac{1}{z}$	**Unit delay**: sampling and holding by using a single-period delay (z^{-1})
$\frac{K\,Ts}{z-1}$	**Discrete-time integrator**
$\frac{1}{1+0.5z^{-1}}$	**Discrete filter**: models the impulse response of IIR filter
$\frac{0.5+0.5z^{-1}}{1}$	**Discrete FIR filter**: models FIR filters
PID(z)	**Discrete PID controller**: simulates discrete PID controllers
$x(n+1)=Ax(n)+Bu(n)$ $y(n)=Cx(n)+Du(n)$	**Discrete state-space**: state-space illustration
$\frac{1}{z+0.5}$	**Discrete transfer Fcn**: discrete transfer function illustration
$\frac{(z-1)}{z(z-0.5)}$	**Discrete zero–pole**: poles–zeros illustration
	First-order hold: FOH discretization
$\frac{0.05z}{z-0.95}$	**Transfer Fcn first-order:** first-order transfer function
$\frac{z-0.75}{z-0.95}$	**Transfer Fcn lead or lag:** transfer function of lead or lag compensator
	Zero-order hold: ZOH discretization

FIGURE 9.1
Library: Simulink/discrete.

- To see the simulation results, the output of the block of interest should be connected in a suitable block from the Sinks category. The block type depends on the type of output data (real or complex valued, time domain, or frequency domain).

Open and save a block diagram

- To open an existing block diagram, select the File/Open menu, and then select the name of the block diagram you want to open.
- To save a block diagram, select the File/Save menu.

9.4 LabVIEW

LabVIEW (Laboratory Virtual Instrument Engineering Workbench, http://www.ni.com/labview/) is a powerful and flexible software platform for the design and analysis of various applications developed by National Instruments. Its graphical interface makes it ideal for applications' measurement, automation, instrument control, and data analysis. LabVIEW provides an extensive VI's library functions and libraries for specific applications, appropriate for even someone with little experience in programming. It also includes conventional error correction instruments, in which we can put breakpoint symbols, perform step-by-step program execution, and monitor the data flow.

LabVIEW is an application development program, where even though it is similar to other commercial development platforms (e.g., C or BASIC), it has a significant difference: instead of text-based language, it uses graphical programming language, that is, G, to create programs in a block diagram form. Also, it allows instruments' control, data acquisition, and processing during and after reception.

As a graphical programming language, it is based on graphical symbols rather than text (such as classical programming languages) to describe various operations of the program. LabVIEW goes beyond the classical programming languages that make use of commands, by entering the user in a graphical environment including tools for collecting measurements, control of autonomous instruments, analysis and presentation of measurements.

9.4.1 LabVIEW Environment

1. **VIRTUAL INSTRUMENTS**: the conventional programming languages use functions and subroutines for the development of their applications. In the LabVIEW environment, the corresponding entities are called **virtual instruments** (VI), which provide two programming windows, that is, the front panel and block diagram. There are

also pallets containing options for creating and changing the VI. The front panel is used to display data types "controls and indicators" for the user, while the block diagram contains the code for the virtual instrument that is a representation of the underlying code.

The LabVIEW programs are called virtual instruments because the appearance and function is similar to those of real instruments. However, they operate in accordance to the functions and subroutines belonging to popular programming languages, such as C, FORTRAN, PASCAL, etc. LabVIEW provides mechanisms that allow data to pass easily between the display field (front panel) and the logical flow diagram (block diagram).

2. **FRONT PANEL**: The interacting with the user part of a VI is called front panel because it simulates the appearance and functionality of a physical instrument. It may contain buttons, switches, graphs, and other control buttons and indicators. The user enters data information using the keyboard or mouse, and is able to monitor the results on the computer screen.
3. **BLOCK DIAGRAM**: VI is commanded by a block diagram, which is constructed via the graphic language G. The block diagram of this solution is a graphical illustration of an integrated programming problem. It stands also for the source code of VI.

   With these features, LabVIEW promotes and capitalizes on the idea of modular programming. We divide our application in a number of individual functions, which in turn are divided in smaller units, until a complicated application is converted into a series of simple subapplications. We build a VI to each subapplication and then join all those VIs in another block diagram to achieve the desired primary goal. Finally, we have a high level VI, which contains a collection of sub-VIs representing the functions of the application.
4. **VI EXECUTION**: VIs have hierarchical structure. The sub-VIs can be visualized as VIs used by other Vis, which are placed only in a block diagram and behave like other objects. They belong to the "nodes" category, that is, they have terminal points for inputs and outputs and operate in a special/certain way within the VI. We may visualize them as subroutines belonging to other programming languages. We use the image/connector so that we can convert the VI into a sub-VI, which can call the diagram of any other program.

   VIs are usually executed from the Front Panel. To execute a LabVIEW program, we click on the **Run** button, which is placed on the top of the pallet window. During execution, the Run Button changes its color to black. Also, to execute or terminate the program, there are also the options "Run Continuously", Abort Execution, and Pause.

In the case when the Run arrow is presented as a broken arrow , then some error(s) are present in the program preventing its execution. If we click on this arrow, a debugging list is manifested in order to point out these errors. In the case when all errors have been corrected then the aforementioned arrow returns to the Run status and the program is ready to be executed.

Moreover, we can execute a program from the Block Diagram, where all the aforementioned buttons are placed in the pallet on top of the window , namely: **Highlight Execution** or slow-run button, where we can track the data flow between the involved nodes, which is useful for data error diagnosis, the **Start Single Stepping** and **Step Out** . This usually occurs during the design phase of VI, when we want to check the correct functionality.

5. **FILING OF VI**: The programs in LabVIEW are saved with the suffix *.vi. Many VIs can be saved to a file with the library format ending in *.llb. If we want to create a library, choose **File/Save as/New VI Library**.
6. **LABVIEW HELP**: LabVIEW provides an online assistance guide for any operation we want to achieve. This function is activated by choosing **Help Menu** and then **Show Help**. Activating this function displays a help window showing an icon associated with the selected object and shows the cables that are connected to each terminal.
7. **LABVIEW PALETTES**: The LabVIEW contains graphical palettes in a menu form to further assist the creation and execution of virtual instruments. The three available palettes are **Tools**, **Controls**, and **Functions**.
   - *Tools Palette*: This palette contains tools that are appropriate in the creation and execution of virtual instruments. Each option in the palette contains a suboption with extra control and display buttons, associated with the original choice. If the Tools palette is not visible, then, to reveal it, we choose **Tools palette** from the **View menu**. Touching (with the computer mouse) any tool from the palette of tools over sub-VIs or functions of the block diagram, and the online help for that object is available. The options of this palette are
     - **Operating**: It allows us to change values in the controllers and indicators. Also, we can put into operation buttons, switches, and other objects.
     - **Wiring**: Connects objects in block diagram.
     - **Break point**: Terminates VIs, objects, links, and loops. We use it when we want to close a particular object for a short duration.

- **Positioning**: Select, move, and resize objects.
- **Labeling**: Allows writing labels.
- **Menu** : Opens the shortcut menu of the object.
- **Scrolling**: Moves the window without the sliding rails needing to be used.
- **Probe**: With this tool, we can control the data values transferred in this connection. Usually, it is used when the VI does not give the expected results and we want to check to what point the values are incorrect.
- **Color copy**: Copies the color that will be pasted with color tool.
- **Color**: Colors an object.

- *Controls Palette*: Both the Controls palette and Functions palette consist of subpalettes, each one encloses various programmable objects-pictures used for the development of a VI. Access to these objects is done by activating the image of each subpalette. We can import controllers and indicators in the display field from the Controls palette. Note that if the Controls menu is not visible, you can display it by clicking the right mouse button in a free area of the display field, or choose the Controls palette from the **View** menu. It should also be noted that the palette is active only in the presentation field. There is a "confusion" between controllers and indicators. Examples of indicators are graphs, thermometers, and meters. When they are placed in the front panel, LabVIEW creates the corresponding elements in the block diagram, namely, the terminal elements. An obvious difference between controls and indicators is that the former terminal elements have a thick contour, while the latter have a thin contour.

  The options of this palette are the following:

- **Numeric**: Contains controls and indicators for numerical data.
- **Boolean**: Contains controls and indicators for logical/digital data.
- **String & Path**: Contains controls and indicators for ASCII text format.
- **Array, Matrix,..:** Contains controls and indicators for data aggregation.
- **List & Table**: Contains controls and indicators to create alternative menu options.
- **Graph**: Contains indicators for graphical data representation.

  - **Ring & Enum**: Contains controls and indicators to create alternative menu options.
  - **Refnum**: Contains controls and indicators for file processing.
  - **Decorations**: Contains graphics for contour design in the front panel.
- *Functions Palette*: The Functions palette is active only when we are in the Block Diagram, which contains the graphical source code of a LabVIEW VI (the actual executable code). Each selection of this pallet comprises a number of subtools, some of which are shown below.
  - **Structures**: Contains programming structures, such as While & For Loops.
  - **Array**: Used for matrix processing.
  - **Cluster & Variables**: Contains functions for editing structures of heterogeneous elements. These structures are called clusters in LabVIEW.
  - **Numeric**: Contains arithmetic, logarithmic, and trigonometric operations.
  - **Boolean**: Contains VIs for logic operations.
  - **String**: Contains VIs to edit text in ASCII format.
  - **Comparison**: Contains functions for comparing data, which can be numerical, logical, or characters.
  - **Timing**: Used for timing.
  - **Dialog 7 User...**: Used for interactive windows.
  - **File I/O**: Used for data entry and editing files.
  - **Waveform**: Contains commands for using graphs.

9.4.2 Control Systems in LabVIEW Using the Control Design and Simulation Module

LabVIEW provides some additional modules and toolkits for testing and simulation, such as

- LabVIEW Control Design and Simulation Module
- LabVIEW PID and Fuzzy Logic Toolkit
- LabVIEW System Identification Toolkit
- LabVIEW Simulation Interface Toolkit

Using the LabVIEW Control Design and Simulation Module, we can create and simulate linear, nonlinear, and discrete control systems. From the block diagram of a VI, in the functions palete, we can find the Control Design and Simulation module.

- *Control Design Module*: When the **Control Design and Simulation Module** is installed, the **Control Design** palette is available from the **Functions** palette.

The **Model Construction** palette includes the following functions and/or subpalettes:

- Construct State-Space Model
- Construct Transfer Function Model
- Construct Zero–Pole–Gain Model
- Construct Random Model
- Construct Special Model
  - First-order with (or without) time delay
  - Second-order with (or without) time delay
  - Delay Pade Approximation
  - PID Parallel
  - PID Academic (parallel form)
  - PID Serial
- Draw Transfer Function Equation (to return the transfer function in a suitable form for appearance)
- Draw Zero–Pole–Gain Equation
- Read Model from File
- Write Model from File
- Model Information palette (contains functions to set or get information and properties of our model)

The **Model Conversion** palette has the following functions:

- Convert to State-Space Model
- Convert to Transfer Function Model
- Convert to Zero–Pole–Gain Model
- Convert Delay with Pade Approximation
- Convert Delay to Poles at Origin
- Convert Control Design to Simulation (converts models used in the Control Design Toolkit for use in Simulation Module)
- Convert Simulation to Control Design (the opposite of the latter)

The **Model Interconnection** palette has the following functions and/or subpalettes:

- Serial
- Parallel
- Feedback
- Append
- Rational Polynomial palette

The **Model Reduction** palette has the following functions:

- Minimal Realization
- Model Order Reduction
- Minimal State Realization
- Remove IO (input or output) from Model
- Select IO (input or output) from Model

The **Time Response** palette has the following functions and/or subpalettes:

- Step Response
- Impulse Response
- Initial Response
- Linear Simulation
- Get Time Response Data

The **Frequency Response** palette has the following functions:

- Bode (illustrates a Bode diagram)
- Nyquist
- Nichols
- Singular Values
- All Margins
- Gain and Phase Margin
- Evaluate at Frequency
- Bandwidth
- Get Frequency Response Data

The **Dynamic Characteristics** palette has the following functions:

- Root Locus
- Pole–Zero Map

- Damping Ratio and Natural Frequency
- DC Gain
- Stability
- Norm
- Covariance Response
- Total Delay
- Distribute Delay
- Parametric Time Response

The **State Space Model Analysis** palette has the following functions:

- Controllability Matrix
- Observability Matrix
- Grammians
- Canonical State-Space Realization
- Balance State-Space Model (Diagonal)
- Balance State-Space Model (Grammians)
- Controllability Staircase
- Observability Staircase
- State Similarity Transform

The State Feedback Design palette has the following functions:

- Ackermann
- Pole Placement
- Linear Quadratic Regulator
- Kalman Gain
- State Estimator
- State-Space Controller
- Augment Output with States

- *Simulation Module*: The main features of the Simulation palette are
    - **Control and Simulation loop**: All simulation functions must be placed inside a Control and Simulation loop or in a simulation subsystem.
    - **Continuous Linear Systems Functions**: They are used to represent linear continuous-time systems in the simulation diagram.
    - **Discrete Linear Systems Functions**: They are used to represent linear discrete-time systems in the simulation diagram.

- **Signal Arithmetic Functions**: Used for basic arithmetic functions in a simulation system.
- **Controllers Functions**: Used to implement various types of controllers such as PID, two-degree-of-freedom PID controllers (2 DoF PID), and SIM SISO Controllers.

9.4.3 Simulink—LabVIEW Interconnection

We will give an example of creating a LabVIEW User Interface for a Simulink Model via the LabVIEW Simulation Interface Toolkit. Before LabVIEW can communicate with the Simulink model, we must first formulate our model in Simulink.

Configuring the simulink model:

- Save the files: sinewave.mdl and Sine Wave. VI.
- Open MATLAB and observe the presence of the following message at the command window:

Starting the SIT Server on port 6011
SIT Server started
The Simulation Interface Toolkit automatically installs the SIT server.

- Write "simulink" into the MATLAB command window. Choose File»Open and then the sinewave.mdl.

The resultant Simulation model is illustrated in Figure 9.2.
This Simulink model illustrates a sinusoidal waveform.

- Observe the SignalProbe block in the diagram. We must place a SignalProbe block in the top layer of our diagram. The SignalProbe block is located in the NI SIT Blocks library.

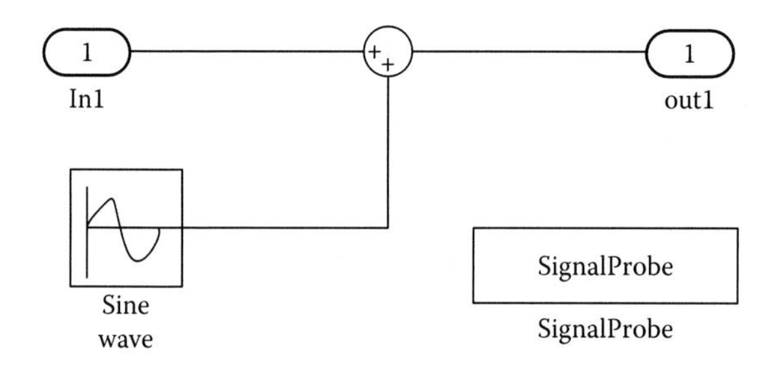

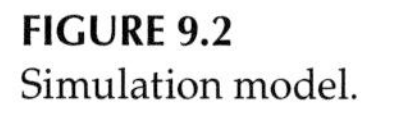

FIGURE 9.2
Simulation model.

- Choosing Simulation≫Simulation parameters, we can see the simulation configuration parameters.
- Save the model in Simulink.

Creating a LabVIEW user interface

- Open LabVIEW and a new VI.
- Place two knob controls on the front panel. We name the one Frequency and the other Amplitude.
- Place a waveform chart on the front panel. We call the y-axis Amplitude and the x-axis Time.
- We can now create the relations between the VI and the Simulink model. From VI, select Tools ≫ SIT Connection Manager.
- In the Model and Host category, select Simulation Environment below the Host Execution. The execution host is the machine where the SIT server runs. If we want to choose Real-Time Target or Driver VI in Local host, we must transform the Simulink model first to a DLL file before the matching.
- From Current Model, select Browse and then sinewave.mdl file.
- From the Project Directory, Browse and choose the folder where we want to save our project.
- From list Category, choose Mappings to display the corresponding Mappings page.
- Highlight the Frequency control to the Current Mappings table and click on the Change Mappings button. So Specify Parameters for Control dialog box will be created.
- From this dialog box, select the Frequency sinewave ≫ Sine Wave tree and click OK.
- We match the Amplitude control in the sinewave ≫ Sine Wave tree.
- We match the Sine Wave indicator to Port 1 from sinewave ≫ Sum tree.
- Click the OK button to close the SIT Connection Manager dialog box and create the code of the block diagram of the VI. Note that there are now Model Controls on the front panel. We can use these to control the model in VI.
- The block diagram now contains code.

We can run the VI and check the simulation with Model Controls. The Frequency and Amplitude knob controls are used to change the output of the sine wave.

9.5 Program CC

Program CC is a special program focused on the analysis and design of linear control systems, which is a product of Systems Technology, Inc. (http://www.programcc.com/).

CC is used to analyze systems in state space and/or in the form of transfer functions. A physical system is described by linear differential equations or difference equations. These equations are transformed into transfer functions or equations in state space and inserted properly into the program. CC has a rich collection of various commands that provide the user the ability to understand the fundamental concepts of automatic control, study, analyze, and evaluate any system of particular interest.

9.5.1 Simulation of Digital Control Systems Using CC

We will analyze some of the basic commands of CC program for the analysis and design of discrete-time control systems. The program is based on the introduction of a series of commands in the active line of the command window.

1. **Import a transfer function**

   All transfer functions need a name, for example, g or g_1, or g_2, etc., and are introduced with a number of ways, such as

   - Algebraic expressions, for example,
     `CC>g=z/((z-1)*(z-.78))`
   - Using the menu and typing **enter** in the command window, for example,
     `CC>enter`

   From this window, we can enter each of the polynomials of the numerator and denominator.

   Notes

   - A fixed term is introduced as a zero-order polynomial.
   - A polynomial (not a transfer function) is inserted by selecting the denominator equal to 1.
   - Using the command **g=enter(coeff1,coeff2,coeff3,...)**, where **coeff** are the transfer function coefficients, namely, the number of polynomials of the numerator, the polynomials which are given by first writing their order and then the term coefficients starting from the highest power of s and continue importing in the same way the polynomials of denominator, such that
     `CC>g=enter(1,0,1,2,1,1,1,1,1,0.15)`

2. **Transfer function forms**

   The program has an option with which we can manipulate the transfer function and convert it into various forms, useful for our study. Below the basic commands are shown, using the transfer function $G(z)=\dfrac{10z+10}{z(z^2+2z+100)}$:

Command	Result
CC>g=(10*z+10)/(z*(z^2+2*z+100)) Inserts discrete transfer function	$g(z)=\dfrac{10z+10}{z(z^2+2z+100)}$
CC>display(g) Returns the transfer function in a polynomial product form (initial form)	$g(z)=\dfrac{10z+10}{z(z^2+2z+100)}$
CC>single(g) Executes the operations in numerator and denominator	$g(z)=\dfrac{10z+10}{z^3+2z^2+100z}$
CC>pzf(g) Returns the transfer function in a poles–zeros product form	$g(z)=\dfrac{10(z+1)}{z[(z+1)^2+9.95^2]}$
CC>shorthand(g) Returns the transfer function in simplified form, that is, the polynomial a*(s+b)*[z^2 + 2*zeta*omega*z + omega^2] Is presented as: a(b)[zeta, omega]	$g(z)=\dfrac{10(1)}{(-0)[0,1,10]}$
CC>unitary(g) The dominant term of z (with highest order) equals to 1	$g(z)=\dfrac{10(z+1)}{z(z^2+2z+100)}$
CC>tcf(g) The constant terms of all polynomials become 1	$g(z)=\dfrac{0.1(z+1)}{z(0.01z^2+0.02z+1)}$
CC>poles(g) Calculates the poles of transfer function	0+0j −1+9,9498744j −1−9,9498744j

3. **Partial fraction expansion of transfer function**

   The command **pfe(g)** converts the transfer function into a partial fraction expansion. Type:

```
CC>pfe(g)
```

   which returns in the computer screen

$$g(z)=0.098+\frac{0.1}{z}-\frac{0.098z^2+0.296z}{[(z+1)^2+9.95^2]}$$

4. **Inverse z-transform**

   The command **izt(g)** calculates the inverse z-transform. Type

```
CC>izt(g)
```

which returns in the computer screen

$$g(n) = 0,098 * delta(n) + 0,1 * delta(n-1) - 0,1 * \\ cos(1,671n - 0,2003) * (10)^n \text{ and } n >= 0$$

5. **Systems interconnection—Transfer function simplification**

   Consider the transfer functions $g_1(z)$ and $g_2(z)$.

   In the case when they are connected in series, the total transfer function is obtained by typing

   ```
   CC>g=g1*g2
   ```

   In the case when they are connected in parallel, the total transfer function is obtained by typing

   ```
   CC>g=g1+g2
   ```

   In the case when they are connected in a negative feedback form, the total transfer function is obtained by typing

   ```
   CC>g=g1/(1+g1*g2)
   ```

   In the case when they are connected in a positive feedback form, the total transfer function is obtained by typing

   ```
   CC>g=g1/(1-g1*g2)
   ```

   Also, the program includes the **feedback operator (|)** through which the closed-loop transfer function can be easily calculated. Namely, by typing **g|h** returns: $g/(1 + h^*g)$, while for a unit feedback (i.e., $h = 1$) the corresponding command is **g|1**. The variables g and h can be in any compatible format (transfer function or state space) with the same dimensions.

6. **Time response**

   The command **time(g)** designs the system step response. The same occurs with the command **(yκ,y)=sim(g)**.

   The commands **time** and **sim** have other syntax forms as well, which can be viewed by typing **help time**.

   ```
   CC>time
   ```

   There is a dialog window, where more scope parameter selections are provided, for example to design more than one curve in the same diagram, giving their name by separating them with commas, in the region that states **Tf**. Pressing the **More** button, the dialog box expands more, enabling us to put labels, choose the color of the scheme, etc.

7. **Bode diagrams—Gain and phase margins**

   The CC program provides the potential design of different types of diagrams (Bode, Nyquist, Nichols, etc).

   Type the command

   ```
   CC>bode
   ```

   where we can choose the type of graph we want.

Subsequently, there are the different ways of writing command **bode(g)** which designs the Bode diagram (amplitude and phase) of the transfer function and the command **margin(g)**, where the gain and phase margins are calculated.

8. **Root locus design**

Command	Description
CC>bode(g)	Illustrates the amplitude (log10(abs(s))) in dB and the phase of g(s), where s is a frequency range that is automatically developed in jw axis. Function g may be a transfer function, a matrix, or a state space system
CC>bode(s,g)	Utilizes the frequencies stored in vector s
CC>bode(s,y)	Illustrates the amplitude abs(log10(w)) with respect to dB(abs(y)) in the case when s, y are vectors
CC>bode(g,'s') CC>bode(s,g,'s') CC>bode(s,y,'s')	Use of certain color, line style, symbol
CC>margin(g)	Calculates the phase, gain, and delay margins, and the local maximum, where g is a transfer function or state-space matrix. It scans all the frequencies from 1e-3 to 1000r/s returning back each available margin type
CC>margin(g,w)	Scans for any margin type in the vicinity of the frequency w rad/s.
margin(g,[wlow,whigh,npts])	Scans from wlow to whigh r/s with npts number of points

phase margin = 180° + the angle where amplitude equals to 1.
delay margin = the delay where phase lag = phase margin.
gain margin = 1/gain where the phase is −180°.
mp margin = the local maximum of $g/(1+g)$.

The command **rootlocus(g)** illustrates the root locus of a characteristic equation for a control system.

The command **rootlocus(k,g)** illustrates the root locus for the gains whose values are given in the vector k.

The command **rootlocus((0,0),g)** illustrates only the poles and zeros of g.

Using the commands **rootlocus(g,'s')** or **rootlocus(k,g,'s')**, we can choose line style, color, and symbols.

9. **State space**

To import a system in state space (quadruple) of the form of: $\dot{x} = Ax + Bu$; $y = Cx + Du$, we use the command **p=pack(a,b,c,d)**, after each matrix has already been inserted. To see the matrix dimensions, we type the command **what(p)**.

A state-space system can be transformed into the form of a transfer function with the commands **fadeeva(p)** or **gep(p)**.

The reverse procedure, that is, a system in the form of a transfer function to be converted into a system in the state space, can be realized by using the commands **ccf(g)** that implements the system in controllable canonical form; **ocf(g)** that implements the system in observable canonical form; and **dcf(g)** that implements the system in a diagonal canonical form.

10. **Time delay systems**

 The command **g=pade(tau,order)** sets $G(s)$ equal to a **Pade** approximation of the exponential term exp(-s*tau), where: tau is the time delay, while order denotes the order of the Pade approximation.

9.5.2 CC Commands

In Table 9.4, all the CC commands are grouped together in alphabetical order. For the operation of each one, one may be consulted by typing Help followed by the command name.

TABLE 9.4

CC Commands in Alphabetical Order

Abs	Acos	acosh	acot	acoth	acsc
acsch	Airplanebw	all	angle	any	asec
asech	Asin	asinh	atan	atan2	atanh
Axis	Balance	balreal	bandwidth	bessel	bilinear
binavg	Blanks	blasche	bode	bodegain	
butter	Cc	ccf	cd	cdim	ceil
Char	Chebyshev	chirp	chol	chpzf	
chsingle	Chst	chtcf	chunitary	clear	close
cls	Cnum	cond	conj	conmat	conv
convert	Copy	copyoption	corrcoef	cos	cosh
cot	Coth	cov	csc	csch	cumprod
cumsum	Data	dB	dcf	deblank	delaymargin
demos	Der	det	dfreqvec	dft	diag
diff	Diophantine	dir	dis	disp	
display	dkbf	dlqr	dlyap	dricc	edit
effdelay	eig	enter	enterbox	eps	error
eval	Exit	exp	expand	expm	eye
fadeeva	Fft	fft2	fftr	fftshift	
fftshift2	fftwin	figure	findstr	fix	floor
format	fprintf	freq	freqvec	fsfb	fsoi
ftext	functionbox	gainmargin	gcd	gcf	geig

(Continued)

TABLE 9.4 (*Continued*)

CC Commands in Alphabetical Order

gep	gepper	ghess	global	grid	gschur
h2	Help	hess	hfa	hold	iden
idim	Ifft	ifft2	ifftr	ilt	ilt2vec
imag	Imc	inner	input	int	integrator
inv	is3d	iscomplex	isempty	isglobal	isint
ismatrix	isnull	isp3d	ispoly	ispolym	isquad
isreal	isscalar	isstr	istf	istfm	isvector
itae	Izt	izt2vec	kbf	laplace	lcm
leadlag	length	lfa	lft	line	
list	Lls	load	log	log10	
logbin	loglog	lower	lpdisp		
lpdisplay	lpeject	lpilt	lpizt	lpoption	
lppfe	lpprintf	lppzf	lpsho	lpshorthand	lpsingle
lptcf	lptext	lpunitary	lqr	lu	lyap
mag	margin	max	maxtc	mean	meansq
median	messagebox	min	mpmargin	name	near
new	nichols	norm	nosmall	notch	null
nullnodisplay	numerator	nyquist	obsmat	ocf	odim
okcancelbox	onepole	ones	onezero	order	outer
pack	Pade	partial	path	pause	pfe
pfe2vec	phase	phasemargin	pid	pinv	plot
plotoption	plotpopup	point	poleplace	poles	primefactors
print	printf	prod	pwd	pzf	qr
quit	Rand	randn	range	rank	
rcond	Rdim	real	reig	rem	
reshape	Resid	ricc	ridf	rl	
rlgain	rootlocus	roots	round	save	scale
schur	Sdim	sec	sech	semilogx	semilogy
senter	Shift	sho	shorthand	sign	sim
similarity	Sin	single	sinh	sisoltf	size
skip	Sort	spectral	sprintf	sqrt	
state	Std	stdisplay	strcat	strcmp	strcmpi
stripchart	strjust	strmatch	strncmp	strncmpi	strrep
strtok	strvcat	subplot	substitute	sum	sumsines
svd	svfreq	tan	tanh	tcf	
text	Tfest	tile	time	timevec	title
tovec	Trace	transpose	tril	triu	twopoles
twozeros	Type	unitary	unpack	upper	var
warning	What	who	whos	winhelp	wplane
xlabel	y2label	yesnobox	ylabel	ylabels	zeroreduce
Zeros	zetaomega	zoh	ztransform		

Bibliography

R. Bateson, *Introduction to Control System Technology*, 7th Edition, Pearson Education, London, England, 2001.

R. Bishop, *Modern Control Systems Analysis and Design Using Matlab*, Addison-Wesley, Boston, MA, 1993.

R. Bishop, *Modern Control Systems Analysis and Design Using Matlab and Simulink*, Addison-Wesley, Boston, MA, 1997.

A. Cavello, R. Setola, F. Vasca, *Using Matlab, Simulink and Control System Tool Box: A Practical Approach*, Prentice Hall, Upper Saddle River, NJ, 1996.

P. Chau, *Process Control: A First Course with Matlab* (Cambridge Series in Chemical Engineering), Cambridge, England, 2002.

F. Colunius, W. Kliemann, L. Grüne, *The Dynamics of Control* (Systems & Control: Foundations & Applications), Birkhäuser, Switzerland, 2000.

R. Dorf, R. Bishop, *Modern Control Systems*, 10th Edition, Prentice Hall, Upper Saddle River, NJ, 2004.

R. Dorf, R. Bishop, *Modern Control Systems*, Addison-Wesley, Boston, MA, 1994.

R. Dukkipati, *Analysis & Design of Control Systems Using MATLAB*, New Age Science, New Delhi, India, 2009.

L. Fenical, *Control Systems Technology*, Delmar Cengage Learning, Boston, MA, 2006.

M. Fogiel, *Automatic Control Systems/Robotics Problem Solver*, 1982.

G. Franklin, J. Powell, A. Emami-Naeini, *Feedback Control of Dynamic Systems*, 6th Edition, Prentice Hall, Upper Saddle River, NJ, 2009.

D. Frederick, J. Chow, *Feedback Control Problems Using MATLAB and the Control System Toolbox* (Bookware Companion Series), Cengage Learning, Boston, MA, 1999.

B. Friedland, *Advanced Control System Design*, Prentice Hall, Upper Saddle River, NJ, 1996.

B. Friedland, *Control System Design: An Introduction to State-Space Methods* (Dover Books on Engineering), Dover Publications, Mineola, NY, 2005.

F. Golnaraghi, B. Kuo, *Automatic Control Systems*, Wiley, Hoboken, NJ, 2009.

D. Hanselman, B. Kuo, *Matlab Tools for Control System Analysis and Design/Book and Disk* (The Matlab Curriculum), Prentice Hall, Upper Saddle River, NJ, 1995.

C. Houpis, J. Dazzo, S. Sheldon, *Linear Control System Analysis and Design with Matlab*, 5th Edition, T&F India, New Delhi, India, 2009.

R. Jacquot, *Modern Digital Control Systems* (Electrical and Computer Engineering), CRC Press, Boca Raton, FL, 1994.

C. Johnson, H. Malki, *Control Systems Technology*, Pearson Education, London, England, 2001.

R. E. Kalman, Contributions to the theory of optimal control. *Bol. Soc. Mat. Mexicana*, 102–119, 2010.

B. Kisacanin, G. Agarwal, *Linear Control Systems: With Solved Problems and MATLAB Examples* (University Series in Mathematics), Springer, NY, 2001.

A. Langill, *Automatic Control Systems Engineering*, Volume I, Prentice Hall, Upper Saddle River, NJ, 1965.

N. Leonard, W. Levine, *Using MATLAB to Analyze and Design Control Systems*, 2nd Edition, Addison Wesley Longman, Boston, MA, 1995.

W. Levine, *The Control Systems Handbook: Control System Advanced Methods*, 2nd Edition (Electrical Engineering Handbook), CRC Press, Boca Raton, FL, 2010.

P. Lewis, C. Yang, *Basic Control Systems Engineering*, Prentice Hall, Upper Saddle River, NJ, 1997.

B. Lurie, *Classical Feedback Control: With Matlab* (Automation and Control Engineering), CRC Press, Boca Raton, FL, 2000.

B. Lurie, P. Enright, *Classical Feedback Control: With MATLAB and Simulink*, 2nd Edition (Automation and Control Engineering), CRC Press, Boca Raton, FL, 2011.

T. Mcavinew, R. Mulley, *Control System Documentation: Applying Symbols and Identification*, ISA: The Instrumentation, Systems, and Automation Society, Pennsylvania, 2004.

N. Nise, *Control Systems Engineering with MATLAB*, Tutorial Version, Wiley, Hoboken, NJ, 1995.

N. Nise, *Control Systems Engineering*, 5th Edition, Wiley, Hoboken, NJ, 2007.

N. Nise, *Matlab Tutorial Update to Version 6 to Accompany Control Systems Engineering*, Wiley, Hoboken, NJ, 2002.

W. Palm, *Control Systems Engineering*, Wiley, Hoboken, NJ, 1986.

E. Rozenwasser, R. Yusupov, *Sensitivity of Automatic Control Systems* (Control Series), CRC Press, Boca Raton, FL, 1999.

D. Sante, *Automatic Control System Technology*, Prentice Hall, Upper Saddle River, NJ, 1980.

B. Shahian, M. Hassul, *Control System Design Using Matlab*, Prentice Hall, Upper Saddle River, NJ, 1993.

S. Shinners, *Modern Control System Theory and Design*, 2nd Edition, Wiley, Hoboken, NJ, 1998.

K. Singh, G. Agnihotri, *System Design through MATLAB, Control Toolbox and SIMULINK*, Springer, NY, 2000.

C. Smith, A. Corripio, *Principles and Practices of Automatic Process Control*, Wiley, Hoboken, NJ, 2005.

R. Stefani, B. Shahian, C. Savant, G. Hostetter, *Design of Feedback Control Systems* (Oxford Series in Electrical and Computer Engineering), OUP, Oxford, England, 2001.

S. Tripathi, *Modern Control Systems: An Introduction* (Engineering Series), Jones & Bartlett Learning, Burlington, MA, 2008.

A. Veloni, A. Palamides, *Control System Problems: Formulas, Matlab (r) and Solutions*, CRC Press, USA, 2011.

L. Wanhammar, *Analog Filters Using Matlab*, Springer, NY, 2009.

R. Williams, D. Lawrence, *Linear State-Space Control Systems*, Wiley, Hoboken, NJ, 2007.

W. Wolovich, *Automatic Control Systems: Basic Analysis and Design* (Oxford Series in Electrical and Computer Engineering), Saunders College Publishing, Philadelphia, 1993.

V. Zakian, *Control Systems Design: A New Framework*, Springer-Verlag, London, UK, 2005.

Index